QUARK–GLUON PLASMA

From Big Bang to Little Bang

This book introduces quark–gluon plasma (QGP) as a primordial matter composed of two types of elementary particles, quarks and gluons, created at the time of the Big Bang. During the evolution of the Universe, QGP undergoes a transition to hadronic matter governed by the law of strong interactions, quantum chromodynamics. After an introduction to gauge theories, various aspects of quantum chromodynamic phase transitions are illustrated in a self-contained manner. The field theoretical approach and renormalization group are discussed, as well as the cosmological and astrophysical implications of QGP, on the basis of Einstein's equations. Recent developments towards the formation of QGP in ultra-relativistic heavy ion collisions are also presented in detail.

This text is suitable as an introduction for graduate students, as well as providing a valuable reference for researchers already working in this and related fields. It includes eight appendices and over 100 exercises.

KOHSUKE YAGI is a professor at the Department of Liberal Arts, Urawa University. He has held positions at the Institute for Nuclear Study at the University of Tokyo, Osaka University and the University of Tsukuba. He has also held several chairs, including in the Japan Nuclear Physics Committee, the Japan–Brookhaven National Laboratory RHIC-PHENIX Project and the International Conference on Quark Matter. He has published 210 articles and written or edited seven books on subatomic physics and general physics, as well as teaching the subject at undergraduate and graduate levels.

TETSUO HATSUDA is a professor in the Department of Physics at the University of Tokyo. He has held positions at the University of Washington, the University of Tsukuba and Kyoto University. He has taught subatomic physics and quantum many-body problems at undergraduate and graduate levels. He has published over 120 scientific articles.

YASUO MIAKE is Professor of Physics at the Institute of Physics, University of Tsukuba. He has conducted research and taught at the University of Tokyo and Brookhaven National Laboratory and the University of Tsukuba. He has experience of teaching electromagnetism and special relativity to undergraduates and subatomic physics to graduates. He has published over 120 scientific articles.

CAMBRIDGE MONOGRAPHS ON PARTICLE PHYSICS NUCLEAR PHYSICS AND COSMOLOGY 23

1. K. Winter (ed.): *Neutrino Physics*
2. J. F. Donoghue, E. Golowich and B. R. Holstein: *Dynamics of the Standard Model*
3. E. Leader and E. Predazzi: *An Introduction to Gauge Theories and Modern Particle Physics, Volume 1: Electroweak Interactions, the 'New Particles' and the Parton Model*
4. E. Leader and E. Predazzi: *An Introduction to Gauge Theories and Modern Particle Physics, Volume 2: CP-Violation, QCD and Hard Processes*
5. C. Grupen: *Particle Detectors*
6. H. Grosse and A. Martin: *Particle Physics and the Schrödinger Equation*
7. B. Anderson: *The Lund Model*
8. R. K. Ellis, W. J. Stirling and B. R. Webber: *QCD and Collider Physics*
9. I. I. Bigi and A. I. Sanda: *CP Violation*
10. A. V. Manohar and M. B. Wise: *Heavy Quark Physics*
11. R. K. Bock, H. Grote, R. Frühwirth and M. Regler: *Data Analysis Techniques for High-Energy Physics. Second edition*
12. D. Green: *The Physics of Particle Detectors*
13. V. N. Gribov and J. Nyiri: *Quantum Electrodynamics*
14. K. Winter (ed.): *Neutrino Physics, Second edition*
15. E. Leader: *Spin in Particle Physics*
16. J. D. Walecka: *Electron Scattering for Nuclear and Nucleon Structure*
17. S. Narison: *QCD as a Theory of Hadrons*
18. J. F. Letessier and J. Rafelski: *Hadrons and Quark–Gluon Plasma*
19. A. Donnachie, H. G. Dosch, P. V. Landshoff and O. Nachtmann: *Pomeron Physics and QCD*
20. A. Hofmann: *The Physics of Synchrotron Radiation*
21. J. B. Kogut and M. A. Stephanov: *The Phases of Quantum Chromodynamics*
22. D. Green: *High P_T Physics at Hadron Colliders*
23. K. Yagi, T. Hatsuda and Y. Miake: *Quark–Gluon Plasma*

QUARK–GLUON PLASMA

From Big Bang to Little Bang

KOHSUKE YAGI
Urawa University

TETSUO HATSUDA
University of Tokyo

YASUO MIAKE
University of Tsukuba

CAMBRIDGE UNIVERSITY PRESS
Cambridge, New York, Melbourne, Madrid, Cape Town, Singapore, São Paulo, Delhi

Cambridge University Press
The Edinburgh Building, Cambridge CB2 8RU, UK

Published in the United States of America by Cambridge University Press, New York

www.cambridge.org
Information on this title: www.cambridge.org/9780521561082

First published 2005
This digitally printed version 2008

A catalogue record for this publication is available from the British Library

ISBN 978-0-521-56108-2 hardback
ISBN 978-0-521-08924-1 paperback

Contents

Preface

Modern physical science provides us with two key concepts: one is the standard model of elementary particles on the basis of the principle of local gauge invariance, and the other is the standard Big Bang cosmology on the basis of the principle of general relativity. These concepts provide us with a clue which may help us to answer the following two questions: (i) what are the building blocks of matter? and (ii) when was the matter created? The main topic of this book is quark–gluon plasma (QGP), which is deeply connected to these questions. In fact, QGP is a primordial form of matter, which existed for only a few microseconds after the birth of the Universe, and it is the root of various elements in the present Universe.

The fundamental theory governing the dynamics of strongly interacting elementary particles (quarks and gluons) is known to be quantum chromodynamics (QCD). QCD suggests that ordinary matter made of protons and neutrons undergoes phase transitions: to a hot plasma of quarks and gluons for temperatures larger than 10^{12} K, and to a cold plasma of quarks for densities larger than 10^{12} kg cm^{-3}. The early Universe, and/or the central core of superdense stars, are the natural places where we expect such phase transitions. It has now become possible to carry out laboratory experiments to produce hot/dense fireballs ("Little Bang") through high-energy nucleus–nucleus collisions using heavy ion accelerators. We expect individual nucleons in the colliding nuclei to dissolve into their constituents to form QGP.

The intention of this book is to introduce the reader who has a limited background in elementary particle physics, nuclear physics, condensed matter physics and astrophysics to the physics of QGP, a fundamental and primordial state of matter. In particular, the authors have in mind advanced undergraduates and beginning graduate students in physics, those not only studying the above-mentioned fields, but also those studying accelerator science and computer science. In addition, the authors hope that the book will serve as a reference text for researchers already working in the fields mentioned above.

Chapter 1 is an introductory chapter, which illustrates the essentials of the physics of QGP and provides a perspective on the discovery of QGP. Methodology quite common to studies of the early structure of the Universe (the Big Bang) and of the structure of QGP (the Little Bang) is emphasized. The text is then divided into three parts.

Part I provides a theoretical background in the physics of QGP and in the QCD phase transitions. Part I may be read independently from the other Parts in order to understand modern gauge field theories, with applications such as color confinement, asymptotic freedom and chiral symmetry breaking in QCD, the basics of thermal field theory and lattice gauge theory, and the physics of phase transitions and critical phenomena.

Part II is devoted to the implications of QGP on cosmology and stellar structures. The physics of an expanding hot Universe and of superdense stars (neutron and quark stars) are discussed with relation to Einstein's theory of general relativity. Appendix D is included for readers who have little knowledge about Riemann space, Einstein's equation, Schwarzschild's solution, etc.

In Part III, the reader will find an overview of the physics of relativistic and ultra-relativistic nucleus–nucleus collisions. This type of collision is the only way of creating and investigating QGP and QCD phase transitions by means of *laboratory experiments*. The relativistic hydrodynamics and the relativistic kinetic theory are introduced in some detail as guiding principles with which to investigate the dynamics of hot/dense matter produced in the collisions. After discussing the various experimental signatures of QGP, the fixed-target experiments are summarized. Then we present the outstanding results achieved with the world's first Relativistic Heavy Ion Collider (RHIC at Brookhaven National Laboratory), for which special emphasis is put on the evidence for a QGP phase. In addition, the special features of detectors used in high-energy heavy ion experiments are discussed.

We have tried to cover topics ranging from fundamentals to frontiers, from theories to experiments, and from the Big Bang and compact stars in the Universe to the Little Bang experiments on Earth. The authors assume that the reader has some familiarity with intermediate level quantum mechanics, the basic methods of quantum field theory, statistical thermodynamics and the special theory of relativity, including the Dirac equation. However, the authors have recapitulated necessary and sufficient introductory elements from these fields. As far as possible, the presentation is self-contained. To accomplish this, the authors have placed key proofs and derivations in eight Appendices and also in about 160 exercises, which may be found at the ends of each chapter.

It was not the authors' intention to provide a complete reference list for the subject of QGP; only references which are particularly useful to students are listed. The reader can find general and up-to-date surveys of the subject in the recent proceedings of the "Quark Matter" Conference series: Heidelberg (1996),

Tsukuba (1997), Torino (1999), Brookhaven – Stony Brook (2001), Nantes (2002) and Berkeley (2004).[1]

Some parts of the original manuscript were used for a series of lectures given to graduate students at the University of Tsukuba and the University of Tokyo; the authors wish to thank the students who attended these lectures. The authors also thank Homer E. Conzett, who carefully read parts of the manuscript and made many grammatical and style suggestions. They also wish to express their gratitude to our editors at Cambridge University Press, Simon Capelin, Tamsin van Essen, Vince Higgs and Irene Pizzie, for a pleasant working relationship. Thanks are due to many friends and colleagues, especially to Masayuki Asakawa, Gordon Baym, Hirotsugu Fujii, Machiko Hatsuda, Tetsufumi Hirano, Kazunori Itakura, Teiji Kunihiro, Tetsuo Matsui, Berndt Müller, Shoji Nagamiya, Atsushi Nakamura, Yasushi Nara, Satoshi Ozaki, Shoichi Sasaki and Hideo Suganuma, who have either provided us with data or were involved in helpful discussions.

QGP forms one of the main areas of research in the physics of QCD which is developing rapidly. In spite of this, the authors hope this book will serve for a long time as a good introduction to the basic concepts of the subject, so that readers can enter the forefront of research without much difficulty.

Although this book is primarily written as a textbook for the physics of QGP, several other teaching options in undergraduate/graduate courses are also recommended.

(a) For a course on an introduction to gauge field theories, we suggest the following sequence: Chapter 2 → Chapter 4 → Chapter 5 → Chapter 6.
(b) For an advanced statistical mechanics and phase transition course, we suggest Chapter 3 → Chapter 4 → Chapter 5 → Chapter 6 → Chapter 7 → Chapter 12.
(c) For a course on an introduction to the applications of general relativity to cosmology and stellar structure, Appendix D → Chapter 8 → Chapter 9.
(d) For an advanced nuclear (hadron) physics course, Chapter 1 → Appendix E → Chapter 9 → Chapter 10 → Chapter 11 → Chapter 13 → Chapter 14 → Chapter 15 → Chapter 16 → Chapter 17.

We would like to thank the American Astronomical Society, publishers of *The Astrophysical Journal*, for permission to reproduce Figs. 8.2, 9.2 and 9.3; the American Physical Society, publishers of *Physical Reviews*, *Physical Review Letters* and *Reviews of Modern Physics*, for permission to reproduce Figs. 3.4, 3.5, 4.10, 5.8, 5.9, 7.6, 8.3, 8.4, 8.10, 13.6, 14.4(b), 15.2, 15.3, 15.12, 16.4, 16.6(a), 16.7, 16.8, 16.9, 16.12, 16.14, 16.15, 16.16, 16.18(a), 16.19, 16.20 and 17.4(b); Springer-Verlag, publishers of *The European Physical Journal*, for

[1] See Braun-Munzinger *et al.* (1996), Hatsuda *et al.* (1998), Riccati *et al.* (1999), Hallman *et al.* (2002), Gutbrod *et al.* (2003) and Ritter and Wang (2004).

permission to reproduce Figs. 15.4, 15.5, 15.6 and 15.11(a); Elsevier Science Publishers B.V., publishers of *Nuclear Physics*, *Physics Letters*, *Physics Reports* and *Nuclear Instruments and Methods in Physics Research*, for permission to reproduce Figs. 3.3, 5.7, 7.3, 7.5, 10.1, 14.9, 15.7, 15.9, 15.10, 15.11(b), 16.3, 16.4, 16.5, 16.10, 16.11, 16.16b, 16.17, 17.2, 17.4(a) and 17.7; Springer-Verlag, publishers of *Lecture Notes in Physics* and *Astronomy and Astrophysics Library*, for permission to reproduce Figs. 3.6 and 9.6; the Institute of Physics, publishers of the *Journal of High Energy Physics* and the *Journal of Physics*, for permission to reproduce Fig. 13.2; and World Scientific, publishers of the *Advanced Series on Directions in High Energy Physics*, for permission to reproduce Fig. 15.8. The source of each figure is given in the caption, and we are grateful to the authors for permission to reproduce or adapt their figures.

Finally, although the authors have tried to eradicate conceptual and typographical errors, they are afraid that some of them may have slipped through. A list of typos and corrections will be posted on the World Wide Web at the following URL: http://utkhii.px.tsukuba.ac.jp/cupbook/. The authors would be grateful if the readers would report/send other errors/comments to this address.

The authors are proud to publish the book in 2005, World Year of Physics (WYP2005), the centennial anniversary of Einstein's three great works on the particle nature of light, the molecular theory of Brownian motion, and the special theory of relativity.

1
What is the quark–gluon plasma?

In this chapter, we present a pedagogical introduction to quantum chromodynamics (QCD), the quark–gluon plasma (QGP), color deconfinement and chiral symmetry restoration phase transitions in QCD, the early Universe and the Big Bang cosmology, the structure of compact stars and the QGP signatures in ultra-relativistic heavy ion collisions. Schematic figures not exploiting any mathematical formulas are utilized. Perspectives on discovering QGP on Earth (Little Bang) are also provided, with an emphasis on the common methodology used to study the early Universe (Big Bang). The issues introduced in this chapter will be elucidated thoroughly in later chapters; the appropriate references to chapters are given in boldface.

1.1 Asymptotic freedom and confinement in QCD

The hydrogen atom is composed of an electron and a proton (Fig. 1.1). Whereas the electron is a point particle within the current experimental resolution, the proton is known to be a composite particle consisting of three quarks. The quarks are fermions having not only the flavor degrees of freedom (up (u), down (d), strange (s), charm (c), bottom (b), top(t)), but also color degrees of

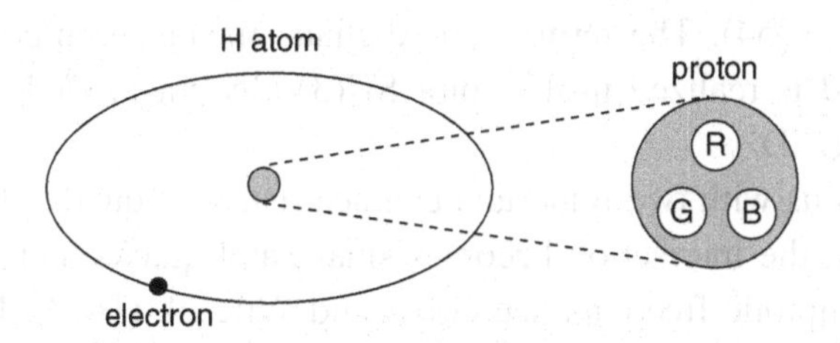

Fig. 1.1. An illustration of a hydrogen atom (H). The proton (p) is a composite object composed of three quarks with primary colors R, B and G glued together by the color gauge field (the gluon). Characteristic sizes of H and p are $\sim 10^{-10}$ m and $\sim 10^{-15}$ m, respectively.

Table 1.1. *Comparison between QED and QCD.*

	QED	QCD
Matter fermions	charged leptons,	quarks
	e.g. e^-, e^+	q^β, $\bar{q}^\beta$ (β = R, B, G)
Gauge bosons	photon (γ)	gluons (g)
	A_μ	A_μ^a $(a = 1,2,\ldots,8)$
Gauge group	U(1)	SU(3)
Charge	electric charge (e)	color charge (g)
Coupling strength	$\alpha = \frac{e^2}{4\pi\hbar c} = O(10^{-2})$	$\alpha_s = \frac{g^2}{4\pi\hbar c} = O(1)$

freedom (red (R), blue (B) and green (G)). Isolated color has never been observed experimentally, which indicates that quarks are always bound together to form color-white composite objects (the hadrons). Baryons (proton, neutron, Λ, Σ, ..., given in **Appendix H**) comprise three valence quarks, and mesons (π, ρ, K, J/ψ, ..., Table H.2) comprise a quark–anti-quark pair. They are the simplest possible constructions of hadrons, but possible multi-quark systems may exist.

The concept of color, as well as the quantum dynamics of color, was first proposed by Nambu (1966), and the theory is now called quantum chromodynamics (QCD); see **Chapter 2**. This is a generalization of quantum electrodynamics (QED) (Brown, 1995), which is a quantum theory of charged particles and the electromagnetic field. QCD (respectively QED) has gluons (the photon) as spin 1 gauge bosons that mediate the force between quarks (charged particles), as shown in Table 1.1. Although QCD and QED look similar, there is a crucial difference: whereas the photon is electrically neutral and therefore transfers no charge, the gluons are not neutral in color. The fact that gluons themselves carry color is related to the fundamental concept of non-Abelian or Yang–Mills gauge theory (Yang and Mills, 1954). The term "non-Abelian" means "non-commutative" as in $AB \neq BA$, and is realized in the color SU(3) algebra in QCD but not in the U(1) algebra in QED.

QCD provides us with two important characteristics of quark–gluon dynamics. At high energies, the interaction becomes small, and quarks and gluons interact weakly (the asymptotic freedom; see Gross and Wilczek (1973), Politzer (1973) and 't Hooft (1985)), while at low energy the interaction becomes strong and leads to the confinement of color (Wilson, 1974); see **Chapter 2**. The asymptotic freedom, which is a unique aspect of non-Abelian gauge theory, is related to the anti-screening of color charge. Because the gauge fields themselves have color, a bare color charge centered at the origin is diluted away in space by the gluons. Therefore, as one tries to find the bare charge by going through the cloud of gluons, one finds a smaller and smaller portion of the charge. This is in sharp contrast to the case of QED, where the screening of a bare charge takes place due to a cloud of, for example, electron–positron pairs surrounding the charge. Shown in Fig. 1.2 is an illustration of the effective (or running) coupling constant in QCD (QED) with the anti-screening (screening) feature. As the typical length scale decreases, or the energy scale increases, the coupling strength decreases in QCD. This is why we can expect QGP at high temperatures, for which the typical thermal energies of the quarks and gluons are large and thus the interactions become weak; see **Chapters 3** and **4**.

Figure 1.2 also indicates that the interaction in QCD becomes stronger at long distances or at low energies. This is a signature of the confinement of color. Indeed, the phenomenological potential between a quark and an anti-quark at large separation increases linearly, as illustrated in Fig. 1.3(a). Consequently, even if we try to separate the quark and the anti-quark, they cannot be forced apart. In reality, beyond some critical distance the potential energy becomes large enough such that a new quark–anti-quark pair pops up from the vacuum. Then, the original quark–anti-quark pair becomes two pairs. In this way, quarks are always confined inside hadrons and can never be isolated in QCD. This feature is shown schematically in Figs. 1.3(b) and (c).

Because the QCD coupling strength, α_s, becomes large at long distances, we encounter a technical difficulty, i.e. we cannot adopt a perturbative method. Wilson's lattice gauge theory (Wilson, 1974) may be used to circumvent this problem. It treats four-dimensional space-time not as a continuum, but as a lattice, just as in crystals, in which quarks occupy lattice points while the gauge field occupies lattice links (Fig. 1.4); see **Chapter 5**. By this lattice discretization, one

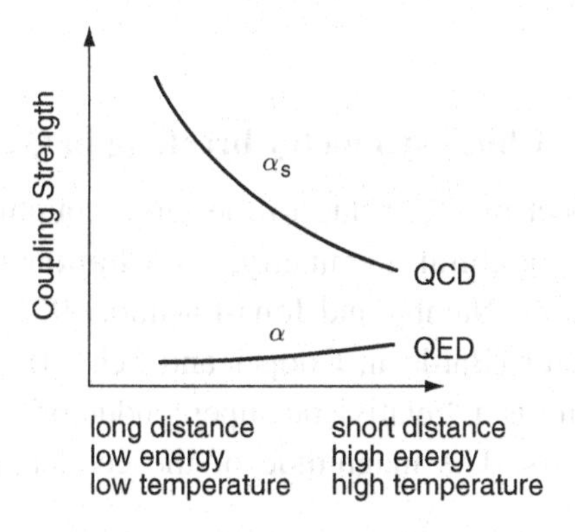

Fig. 1.2. The response of the QCD (QED) coupling strength α_s (α) with variation of the distance scale, the energy scale and the temperature.

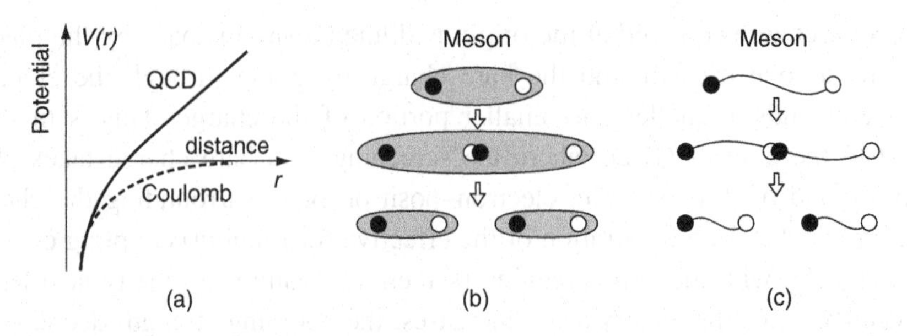

Fig. 1.3. (a) Potential between a quark and an anti-quark as a function of their separation in QCD (solid line) compared with the Coulomb potential (dashed line). (b) The quark confinement mechanism in QCD. The shaded regions represent clouds of the gluon field. The gluon configuration may be approximated as a string with a constant string tension, as shown in (c). For large enough separation, string breaking takes place.

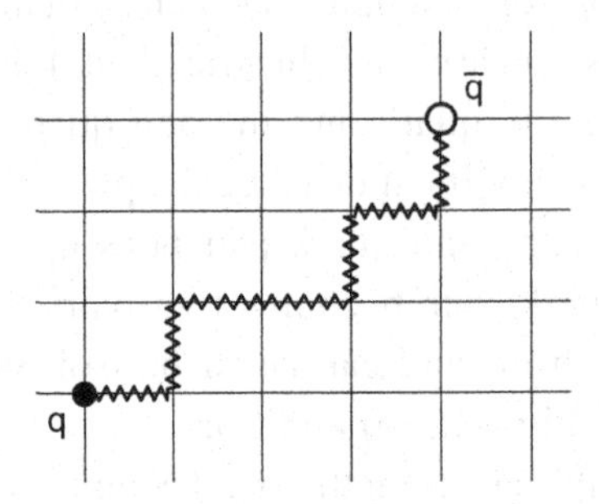

Fig. 1.4. Quarks and gluons on the space-time lattice. Quarks are defined on the lattice sites, and the gluons are defined on the lattice links.

may solve QCD utilizing Monte Carlo numerical simulations. Results confirm that the potential energy is indeed proportional to the length of the string. This agrees with that of the string model, making the idea of confinement feasible.

1.2 Chiral symmetry breaking in QCD

Another important aspect of QCD due to the large coupling at low energies is the dynamical breaking of chiral[1] symmetry; see **Chapter 6**. As first recognized by Nambu (Nambu, 1960; Nambu and Jona-Lasinio, 1961a, b), in analogy with a metallic superconductor (Bardeen, Cooper and Schrieffer, 1957), the strongly interacting QCD vacuum is a "relativistic superconductor" with the condensation of quark–anti-quark pairs. The magnitude of the condensation is measured by

[1] Chiral, originating from "chiro" (= hand in Greek), is a term which refers to the distinction between right-handedness and left-handedness.

the vacuum expectation value, $\langle \bar{q}q \rangle$, which serves as an order parameter. The masses of light hadrons are intimately related to the non-vanishing value of this order parameter.[2] Thus, one may naturally expect, in analogy with metallic superconductors, that there is a phase transition at finite T to the "normal" phase, and that the condensation and the particle spectra will experience a drastic change associated with the phase transition (**Chapter 7**).

1.3 Recipes for quark–gluon plasma

The asymptotic freedom illustrated in Fig. 1.2 immediately suggests two methods for the creation of the quark–gluon plasma (QGP).

(i) Recipe for QGP at high T (Fig. 1.5(a)). We assume that the QCD vacuum is heated in a box. At low temperature, hadrons, such as pions, kaons, etc., are thermally excited from the vacuum. Note that only the color-white particles can be excited by the confinement at low energies. Because the hadrons are all roughly the same size (about 1 fm), they start to overlap with each other at a certain critical temperature, T_c. Above this temperature, the hadronic system dissolves into a system of quarks and gluons (QGP). Note that in the QGP thus produced the number of quarks, n_q, is equal to that of anti-quarks, $n_{\bar{q}}$. The various model calculations and the Monte Carlo lattice QCD simulations yield $T_c = 150 \sim 200$ MeV (**Chapters 3** and **5**). Although this is extremely high in comparison with (for example) the temperature at the center of the Sun, 1.5×10^7 K $= 1.3$ keV, it is

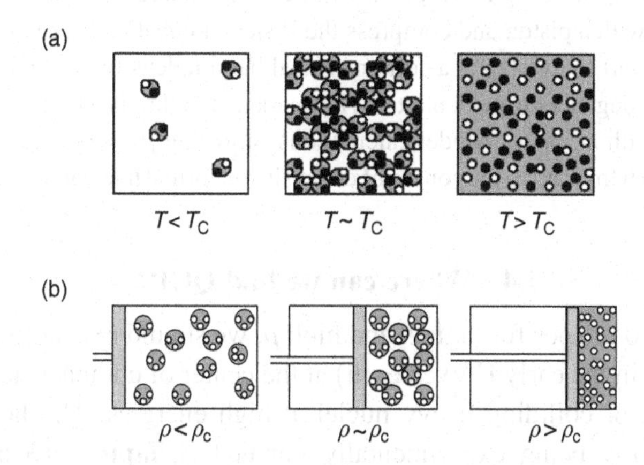

Fig. 1.5. Formation of QGP (a) at high temperature (T) and (b) at high baryon density (ρ).

[2] The concept of the order parameter was first introduced by Landau in 1937 to establish the general theory of phase transitions (see Chap. 14 in Landau and Lifshitz, 1980). Later it was applied to describe various phenomena, such as superconductivity and superfluidity.

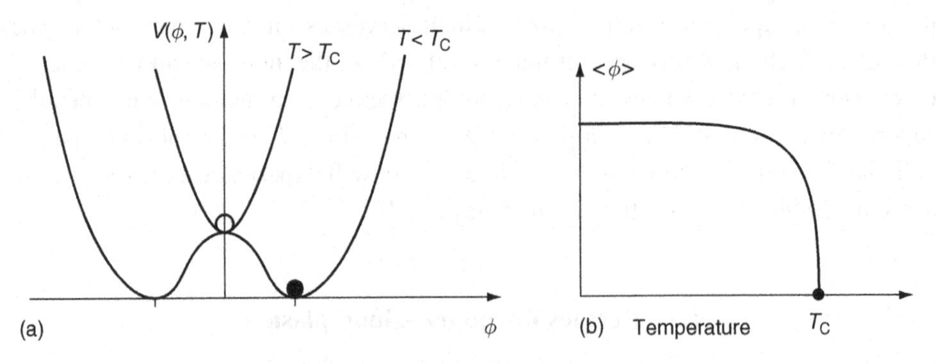

Fig. 1.6. (a) The Ginzburg–Landau potential, $V(\phi, T)$, showing the second order phase transition at $T = T_c$. (b) Behavior of the order parameter, $< \phi >$, as a function of T.

a typical energy scale of hadronic interactions and can be attained in laboratories (**Chapter 10**).

The finite T QCD phase transition may also be described by the order parameter and the corresponding Ginzburg–Landau (GL) potential (or the Landau function) if dynamical symmetry breaking and its restoration are involved. We will see such examples for center symmetry of the color SU(3) gauge group (**Chapter 5**) and global chiral symmetry of quarks (**Chapter 6**). Figure 1.6 is an illustration of the GL potential, V, as a function of an order parameter, ϕ; it also shows the behavior of the minimum of V as a function of T for the second order phase transition.

(ii) Recipe for QGP at high ρ (Fig. 1.5(b)). Let us put a large number of baryons into a cylinder with a piston and compress the system adiabatically, keeping $T \sim 0$. The baryons start to overlap at a certain critical baryon density, ρ_c, and dissolve into a system of degenerate quark matter. The quark matter thus produced is of high baryon density with $n_q \gg n_{\bar{q}}$. Model calculations show that $\rho_c = $ (several) $\times \rho_{nm}$, where $\rho_{nm} = 0.16\,\text{fm}^{-3}$ is the baryon number density of normal nuclear matter (**Chapter 9**).

1.4 Where can we find QGP?

Based on the two recipes for high T and high ρ, we should expect to find QGPs in three places: (i) in the early Universe, (ii) at the center of compact stars and (iii) in the initial stage of colliding heavy nuclei at high energies. The last possibility, which is currently being experimentally pursued (**Chapters 15** and **16**), was already espoused during the mid 1970s (Lee, 1975).

(i) In the early Universe, about 10^{-5} s after the cosmic Big Bang (**Chapter 8**). According to Friedmann's solution (Friedmann, 1922) of Einstein's gravitational equation (**Appendix D**), the Universe experienced an expansion from a singularity at time zero. This scenario has been confirmed by the formulation of Hubble's law for the red shift of distant galaxies (Hubble, 1929). If we extrapolate our expanding Universe backward in time toward the Big Bang, the matter and radiation become hotter and hotter, resulting in the "primordial fireball," as named by Gamow. The discovery of $T \simeq 2.73\,\mathrm{K} \sim 3 \times 10^{-4}\,\mathrm{eV}$ cosmic microwave background (CMB) radiation by Penzias and Wilson (1965) confirmed the remnant light of this hot era of the Universe. In addition, the hot Big Bang theory explains the abundance of light elements (d, He, Li) in the Universe as a result of the primordial nucleosynthesis. This idea was initiated in a paper entitled "The origin of chemical elements" by Alpher, Bethe and Gamow (1948),[3] which reminds us of the book *On the Origin of Species* by Charles Darwin published in 1859. If we go back further in time to $10^{-5} \sim 10^{-4}\,\mathrm{s}$ after its inception, the Universe is likely to have experienced the QCD phase transition at $T = 150 \sim 200\,\mathrm{MeV}$ and an electro-weak phase transition at $T \sim 200\,\mathrm{GeV}$, as shown in Fig. 1.7.

In addition, the discovery of tiny fluctuations of the cosmic temperature by COBE (COsmic Background Explorer) and by WMAP (Wilkinson Microwave Anisotropy Probe; see Bennett *et al.* (2003)) suggests the existence of a preceding inflationary period during which the Universe underwent an exponential expansion (Kazanas, 1980; Guth, 1981, Sato; 1981a, b; Peebles, 1993).

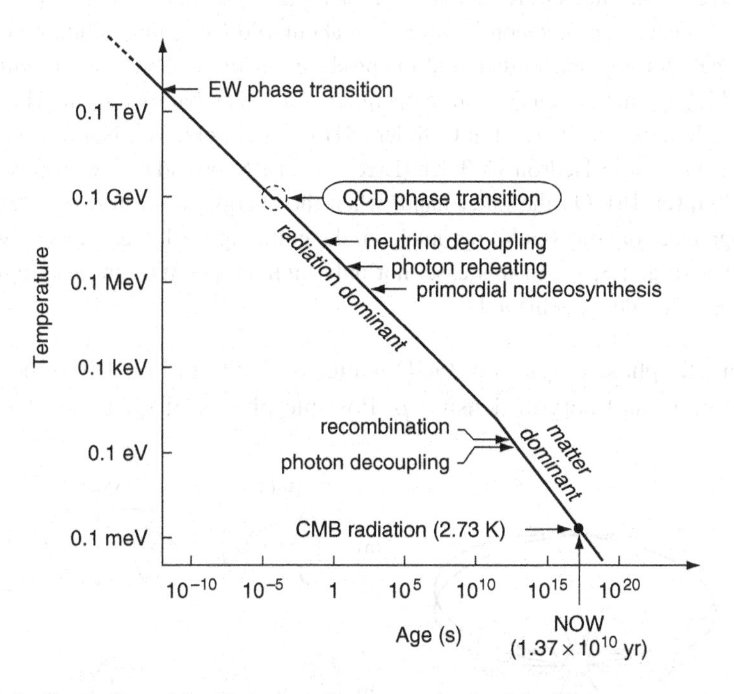

Fig. 1.7. Evolution of the Universe; the temperature of the Universe as a function of time from the moment of creation. ($0.1\,\mathrm{GeV} \simeq 1.1605 \times 10^{12}\,\mathrm{K}$.)

[3] The theory is known as the "alpha-beta-gamma theory" because the initials of the authors' names allude to ABG = ABΓ.

(ii) At the core of superdense stars such as neutron stars and quark stars (**Chapter 9**). There are three possible stable branches of compact stars: white dwarfs, neutron stars and quark stars. The white dwarfs are made entirely of electrons and nuclei, while the major component of neutron stars is liquid neutrons, with some protons and electrons. The first neutron star was discovered as a radio pulsar in 1967 (Hewish *et al.*, 1968). If the central density of the neutron stars reaches 5–10 $\rho_{\rm nm}$, there is a fair possibility that the neutrons will melt into the cold quark matter, as shown in Fig. 1.5(b). There is also a possibility that the quark matter, with an almost equal number of u, d and s quarks (the strange matter), may be a stable ground state of matter; this is called the strange matter hypothesis. If this is true, quark stars made entirely of strange matter become a possibility. In order to elucidate the structure of these compact stars, we have to solve the Oppenheimer–Volkoff (OV) equation (Oppenheimer and Volkoff, 1939), obtained from the Einstein equation, together with the equation of states for the superdense matter (**Appendix D**).

(iii) In the initial stage of the "Little Bang" by means of relativistic nucleus–nucleus collisions with heavy ion accelerators (**Chapter 10**). Suppose we accelerate two heavy nuclei such as Au nuclei ($A = 197$) up to relativistic/ultra-relativistic energies and cause a head-on collision, as shown in Fig. 1.8. In such relativistic energies, the nuclei are Lorentz-contracted as "pancakes." When the center-of-mass energy per nucleon is more than about 100 GeV, the colliding nuclei tend to pass through each other, and the produced matter between the receding nuclei is high in energy density and temperature but low in baryon density (Fig. 1.8(a)). The Relativistic Heavy Ion Collider (RHIC) at Brookhaven National Laboratory and the Large Hadron Collider (LHC) at CERN provide us with this situation (**Chapter 16**). On the other hand, when the energy is at a few to a few tens of giga-electronvolts (GeV) per nucleon, the colliding nuclei tend to stay with each other (Fig. 1.8(b)). In this case, not only high temperature but also high baryon density could be achieved.

A schematic phase diagram of QCD matter is shown in Fig. 1.9 in the plane of temperature, T, and baryon density, ρ. Possible phases of QCD and the precise

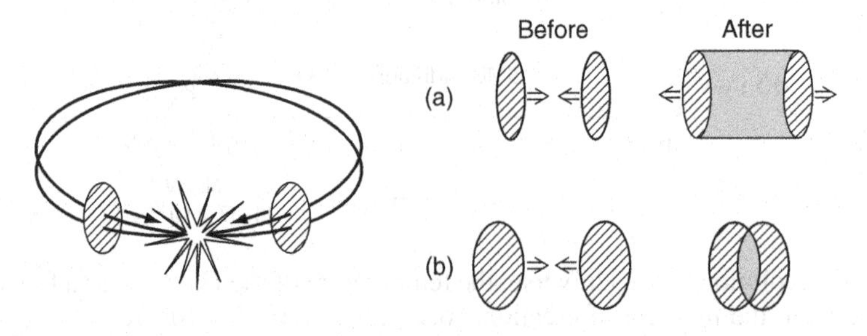

Fig. 1.8. (a) Formation of QGP at high temperature by means of relativistic nucleus–nucleus collisions with a collider-type accelerator. (b) Formation of QGP at high baryon density by means of less energetic collisions than in (a).

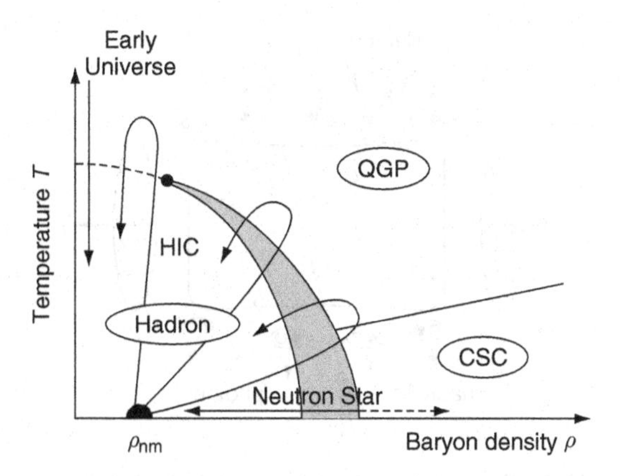

Fig. 1.9. Schematic phase diagram of QCD. "Hadron," "QGP" and "CSC" denote the hadronic phase, the quark–gluon plasma and the color superconducting phase, respectively; ρ_{nm} denotes the baryon density of the normal nuclear matter. Possible locations at which we may find the various phases of QCD include the hot plasma in the early Universe, dense plasma in the interior of neutron stars and the hot/dense matter created in heavy ion collisions (HICs).

locations of critical lines and critical points are currently being actively studied. In fact, unraveling the QCD phase structure is one of the central aims of ongoing and future theoretical and experimental research in the field of hot and/or dense QCD.

1.5 Signatures of QGP in relativistic heavy ion collisions

The relativistic heavy ion collisions are dynamic processes with typical length and time scales of order 10 fm and 10 fm/c respectively. QGP, even if it is created in the initial stage of the collision, cools rapidly, by expanding and emitting various radiation, to a hadron gas through a QCD phase transition (Fig. 1.10). Then the system eventually falls apart into many color-white hadrons (**Chapters 11, 12** and **13**). Therefore, in order to probe the QGP, we need to observe as many as possible particles/radiations emitted during the time history and then to retrace the initial formation of QGP by using the observed data. This is an analogous procedure to probing the early Universe by measuring its remnants, such as the cosmic microwave background, the abundance of atomic elements, etc.

Various QGP signatures and their relation to the real data will be discussed in detail in **Chapters 14, 15** and **16**. In order to capture the essential ideas of the signatures of QCD phase transitions, we illustrate possible candidates of signatures in Fig. 1.11 as a function of the energy density, ε, of a "fireball" initially produced in relativistic nucleus–nucleus collisions. Experimentally, the transverse energy,

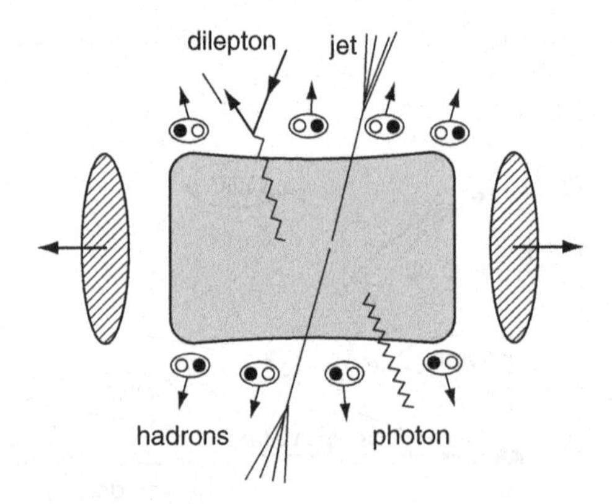

Fig. 1.10. Particles/radiations from the hot quark–hadron matter formed in a central collision between ultra-relativistic heavy nuclei.

dE_T/dy, to be measured in an electromagnetic calorimeter (**Chapter 17**) is closely connected to the energy density, ε.

In the following, (1)–(10) refer to Fig. 1.11.

(1) A second rise in the average transverse momentum of hadrons due to a jump in entropy density at the phase transition.
(2) Measurement of the size of the fireball by particle interferometry with identical hadrons (Hanbury-Brown and Twiss effect).
(3) Enhanced production of strangeness and charm from QGP.
(4) Enhanced production of anti-particles in QGP.
(5) An increase of an elliptic flow (v_2) of hadrons from early thermalization of an anisotropic initial configuration.
(6) Suppression of the event-by-event fluctuations of conserved charges.
(7) Suppression of high-p_T hadrons due to the energy loss of a parton in QGP.
(8) Modification of the properties of heavy mesons (J/ψ, ψ', Υ, Υ') due to the color Debye screening in QGP.
(9) Modifications of the mass and width of the light vector mesons due to chiral symmetry restoration.
(10) Enhancement of thermal photons and dileptons due to the emission from deconfined QCD plasma.

Obviously, the real situation is not as straightforward as that illustrated in Fig. 1.11 due to various backgrounds which tend to hide the possible signals. Also, there are theoretical indications that QGP is not a simple gas of free quarks and gluons, but rather is a strongly interacting system which may modify some of the basic ideas behind the signatures in Fig. 1.11. Nevertheless, as discussed in **Chapters 15** and **16**, the vast number of data from SPS and RHIC already provide quite promising clues which may help in pinning down the nature of QGP.

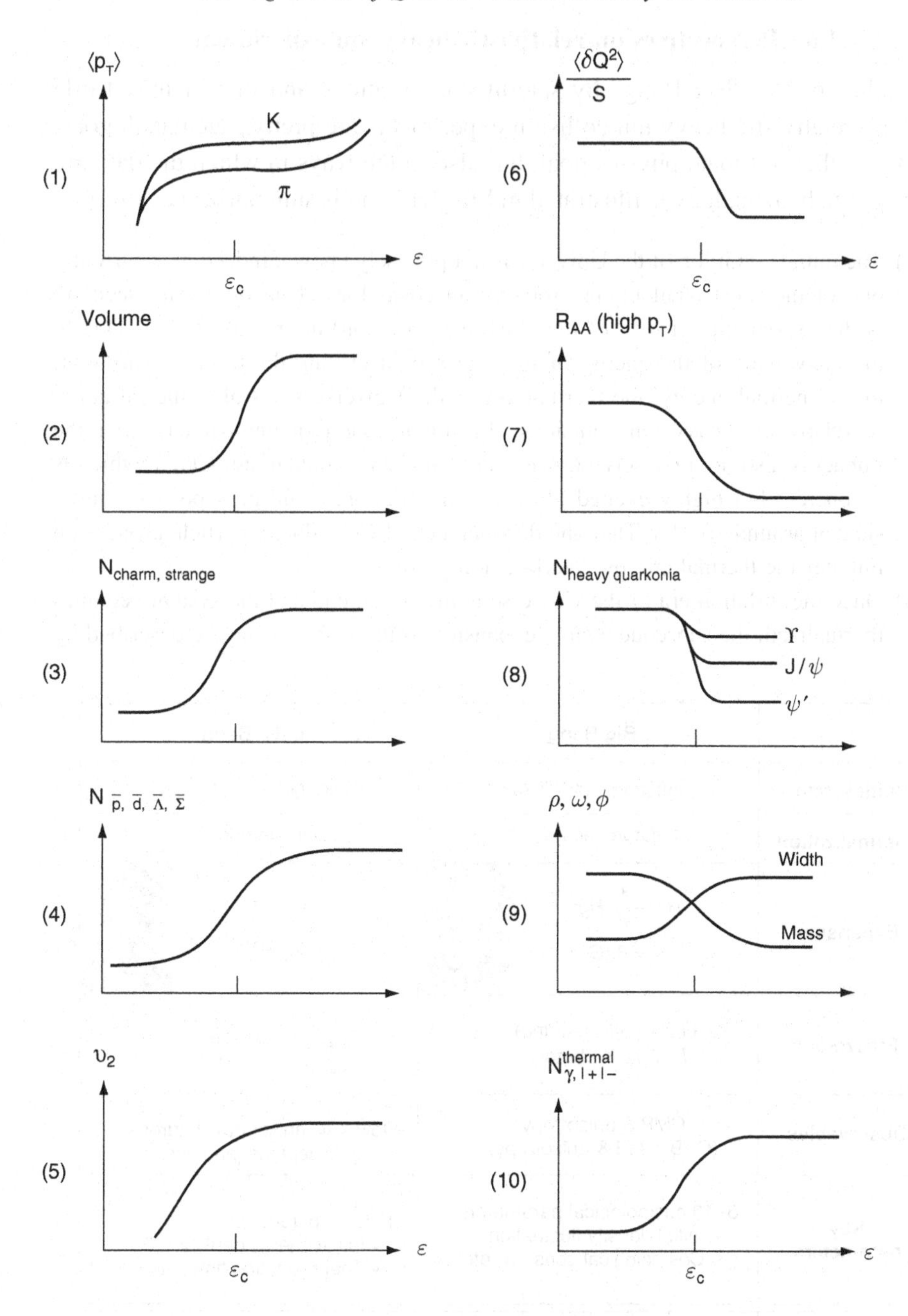

Fig. 1.11. Selected observables as a function of the central energy density ε in the relativistic heavy ion collisions. They are the possible signatures of QCD phase transitions; ε is related to the transverse energy per unit rapidity, dE_{T}/dy; ε_{c} is the critical energy above which the QGP is expected. This figure has been adapted from an original by S. Nagamiya. For an explanation of parts (1)–(10), please see the text.

1.6 Perspectives on relativistic heavy ion experiments

The studies of the "Big Bang" by satellite observations and that of the "Little Bang" by relativistic heavy ion collision experiments are pretty much analogous, not only in their ultimate physics goal, but also in the ways in which the data are analyzed. Such an analogy is illustrated in Fig. 1.12 and is summarized as follows.

(i) The initial condition of the Universe is not precisely known; indeed, it is currently one of the most challenging problems in cosmology. One promising scenario is the exponential growth of the Universe (inflation) at around 10^{-35} s. Due to the conversion of the energy of the scalar field driving the inflation (inflaton) to the thermal energy, the thermal era of the Universe starts after the inflation. In relativistic heavy ion collisions, the initial condition immediately after the impact is also not precisely known. The color glass condensate (CGC), which is a coherent but highly excited gluonic configuration, could be a possible initial state at around 10^{-24} s. Then the decoherence of CGC due to particle production initiates the thermal era, the quark–gluon plasma.

(ii) Once the inflation era of the Universe comes to an end and the system becomes thermalized, the subsequent slow expansion of the Universe can be described by

	Big Bang	**Little Bang**
Initial state	Inflation? (10^{-35} s)	Color glass? (10^{-24} s)
Thermalization	Inflaton decay?	Decoherence?
Expansion	$R^{\mu\nu} - \frac{1}{2} R g^{\mu\nu} = 8\pi G T^{\mu\nu}$ 	$\partial_\mu T^{\mu\nu} = 0$ $\partial_\mu j^\mu_B = 0$ 
Freeze-out	($T = 1.95$ K neutrino) $T = 2.73$ K photon	$T_{chem} \simeq 170$ MeV $T_{therm} \simeq 120$ MeV
Observables	CMB & anisotropy (CνB, CGB & anisotropy)	Flow & its anisotropy, hadrons, jets, leptons, photons
Key parameters	8~10 cosmological parameters • Initial density fluctuation • Cosmological const. Λ, etc.	Plasma parameters • Initial energy density • Thermalization time, etc.
Evolution code	CMBFAST	3D-hydrodynamics

Fig. 1.12. Comparison of the underlying physics and analysis of the Big Bang and the Little Bang. The pictures of Einstein and Landau are taken from http://www-groups.dcs.st-and.ac.uk/history/index.html.

the Friedmann equation with an appropriate equation of state of matter and radiation. In the case of the Little Bang, the expansion of the locally thermalized plasma is governed by the laws of relativistic hydrodynamics originally introduced by Landau (1953). If the constituent particles of the plasma interact strongly enough, one may assume a perfect fluid, which simplifies the hydrodynamic equations.

(iii) The Universe expands, cools down and undergoes several phase transitions such as the electro-weak and QCD phase transitions. Eventually, the neutrinos and photons decouple (freeze-out) from the matter and become sources of the cosmic neutrino background (CνB) and cosmic microwave background (CMB). Even the cosmic gravitational background (CGB) could be produced. These backgrounds carry not only information about the thermal era of the Universe, but also information about the initial conditions before the thermal era. In the case of the Little Bang, the system also expands, cools down and experiences the QCD phase transition. The plasma eventually undergoes a chemical freeze-out and a thermal freeze-out, and then falls apart into many hadrons. Not only hadrons, but also photons, dileptons and jets emerge from the various stages of the expansion. These carry information about the thermal era and the initial conditions.

(iv) What we want to know is the state of matter in the early epochs of the Big Bang and the Little Bang. The CMB data and its anisotropy from the Big Bang is analyzed in the following way. First we define certain key cosmological parameters (usually eight to ten parameters), such as the initial density fluctuations, the cosmological constant, the Hubble constant, etc. Then we make a detailed comparison of the data with the theoretical CMB obtained by solving the Boltzmann equation for the photons (an example of the fast numerical code for this purpose is CMBFAST).[4] By doing this, we can bridge the gap between what happened in the past to what is observed now. WMAP data provide an impressively precise determination of many of the cosmological parameters using this method; see Table 8.1 in **Chapter 8**. The strategy used in the Little Bang is similar. We first define a few key plasma parameters, such as the initial energy density and its profile, the initial thermalization time, the freeze-out temperatures, etc. Then a full three-dimensional relativistic hydrodynamics code is solved to relate these parameters to the plentiful data obtained from laboratory experiments. Such precision studies have now begun to be possible under the assumption of the perfect fluid (Hirano and Nara, 2004).

Figure 1.13 shows a comparison between the study of the Big Bang and that of the Little Bang from the point of view of past, present and future facilities. COBE, launched by NASA in 1989 (SPS at CERN started in 1987), exposed tantalizing evidence of the initial state of the Universe (heavy ion collisions). WMAP, launched by NASA in 2001 (RHIC at BNL started in 2000), provides better images of the newly born state and has initiated precision cosmology

[4] See the CMBFAST website at http://www.cmbfast.org/.

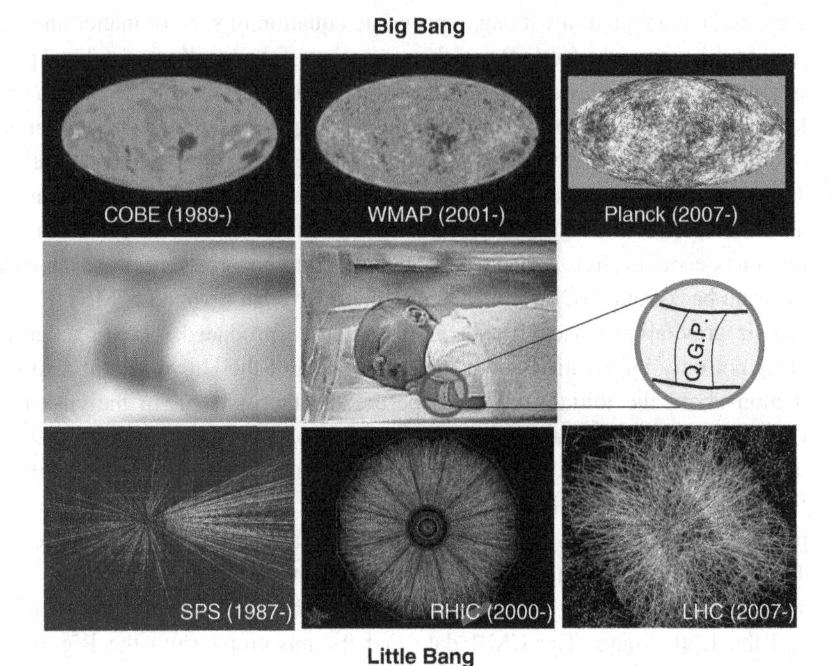

Fig. 1.13. Past, present and future facilities used to study the Big Bang and the Little Bang. COBE, WMAP and infant images are taken from http://map.gsfc.nasa.gov/index.html (courtesy of NASA/WMAP Science Team). The simulation picture at Planck is taken from http://www.rssd.esa.int (courtesy of NASA/WMAP Science Team). The SPS and LHC images are CERN photos taken from http://cdsweb.cern.ch (courtesy of CERN).

(precision QGP physics). The Planck satellite is due to be launched by ESA in 2007 (LHC at CERN is to be started in 2007); it is expected that these facilities will shed further light on the initial conditions and the origin of dark energy (initial conditions and the dynamics of the hot quark–gluon system).

1.7 Natural units and particle data

Throughout this book we use the natural units $\hbar = c = 1$. In addition we set the Boltzmann constant to unity, $k_B = 1$. This unit system is described in **Appendix A**. Tables of particles and their properties are given in **Appendix H**.

I

Basic Concept of Quark–Gluon Plasma

2
Introduction to QCD

In this chapter, we outline the basic concepts of quantum chromodynamics (QCD), its classical and quantum aspects and its symmetry and vacuum structure; various non-perturbative approaches are summarized for later purposes.

2.1 Classical QCD action

The classical Lagrangian density of QCD contains quark and gluon fields as fundamental degrees of freedom; also, it is designed to have a local color $SU_c(3)$ symmetry (Nambu, 1966). For a quark with mass m, the Lagrangian density is given by

$$\mathcal{L}_{\rm cl} = \bar{q}^{\alpha}(i \not{D}_{\alpha\beta} - m\delta_{\alpha\beta})q^{\beta} - \frac{1}{4}F^{a}_{\mu\nu}F^{\mu\nu}_{a}. \tag{2.1}$$

The quark (gluon) field q^{α} (A^{a}_{μ}) belongs to the $SU_c(3)$ triplet (octet). Therefore, α runs from 1 to 3, while a runs from 1 to 8. Note that summation over repeated indices is assumed unless otherwise stated.

We define $\not{D} \equiv \gamma^{\mu}D_{\mu}$, where D_{μ} is a covariant derivative acting on the color-triplet quark field:

$$D_{\mu} \equiv \partial_{\mu} + igt^{a}A^{a}_{\mu}. \tag{2.2}$$

Here g is the dimensionless coupling constant in QCD; the t^{a} denote the fundamental representation of $SU_c(3)$ Lie algebra (See Appendix B.3). They are traceless 3×3 hermitian matrices satisfying the following commutation relation and normalization:

$$[t^{a}, t^{b}] = if_{abc}t^{c}, \ \mathrm{tr}(t^{a}t^{b}) = \frac{1}{2}\delta^{ab}. \tag{2.3}$$

Explicit forms of t^a and f_{abc} (the structure constants) are given in Appendix B.3 together with some basic relations. For later convenience, we also define the covariant derivative acting on the color-octet field:

$$\mathcal{D}_\mu \equiv \partial_\mu + igT^a A^a_\mu. \tag{2.4}$$

Here T^a are the adjoint representations of the $SU_c(3)$ Lie algebra. They are traceless 8×8 hermitian matrices given by $(T^a)_{bc} = -if_{abc}$. Some basic relations of T^a are also summarized in Appendix B.3.

The field strength tensor of the gluon $F^a_{\mu\nu}$ is defined as

$$F^a_{\mu\nu} = \partial_\mu A^a_\nu - \partial_\nu A^a_\mu - gf_{abc} A^b_\mu A^c_\nu. \tag{2.5}$$

By introducing $A_\mu \equiv t^a A^a_\mu$ and $F_{\mu\nu} \equiv t^a F^a_{\mu\nu}$, we may simplify Eq. (2.5) as follows:

$$F_{\mu\nu} = \partial_\mu A_\nu - \partial_\nu A_\mu + ig[A_\mu, A_\nu] = \frac{-i}{g}[D_\mu, D_\nu]. \tag{2.6}$$

Color electric and magnetic fields may be defined from $F_{\mu\nu}$ in analogy with the standard electromagnetic field,

$$E^i = F^{i0}, \quad B^i = -\frac{1}{2}\epsilon_{ijk} F^{jk}, \tag{2.7}$$

where ϵ_{ijk} is a complete antisymmetric tensor with $\epsilon_{123} = 1$. The classical equations of motion are obtained immediately from Eq. (2.1) as follows:

$$(i\not{D} - m)q = 0, \tag{2.8}$$

$$[D_\nu, F^{\nu\mu}] = gj^\mu \quad \text{or} \quad \mathcal{D}^{ab}_\nu F^{\nu\mu}_b = gj^\mu_a, \tag{2.9}$$

where $j^\mu = t^a j^\mu_a$ and $j^\mu_a = \bar{q}\gamma^\mu t^a q$. These are simply the Dirac equation and the Yang–Mills equation for quarks and gluons (See Exercise 2.1).

The Lagrangian density, Eq. (2.1), is invariant under the $SU_c(3)$ gauge transformation (See Exercise 2.2(1)):

$$q(x) \to V(x)q(x), \quad gA_\mu(x) \to V(x)(gA_\mu(x) - i\partial_\mu)V^\dagger(x), \tag{2.10}$$

where $V(x) \equiv \exp(-i\theta^a(x)t^a)$. To show this gauge invariance, it is useful to remember that $F_{\mu\nu}$ and D_μ transform covariantly; i.e.

$$F_{\mu\nu}(x) \to V(x)F_{\mu\nu}(x)V^\dagger(x), \quad D_\mu(x) \to V(x)D_\mu(x)V^\dagger(x). \tag{2.11}$$

A small increment under the infinitesimal gauge transformation is given by

$$\delta q(x) = -i\theta(x)q(x), \qquad \delta(gA_\mu(x)) = [D_\mu, \theta(x)], \tag{2.12}$$

where $\theta(x) = t^a\theta^a(x)$. The latter can be also written as $\delta(gA^a_\mu(x)) = \mathcal{D}^{ab}_\mu \theta^b(x)$.

In principle, Eq. (2.1) may contain a gauge-invariant term, $\epsilon_{\mu\nu\lambda\rho}F^{\mu\nu}_a F^{\lambda\rho}_a \propto E^a_i B^a_i$. The existence of such a term violates the time-reversal invariance or, equivalently, the CP (charge conjugation + parity) invariance. Although the measurement of the electric dipole moment of a neutron shows no sign of such a CP-violating term in the strong interaction, the fundamental reason for its absence is still not clear, and this is called the strong CP problem.

Because of the gauge invariance, terms such as $A^a_\mu A^\mu_a$ are prohibited. As a consequence, the gluons are massless. On the other hand, quark masses are not constrained by the gauge symmetry and they are indeed finite, as shown in Table H.1 in Appendix H. Moreover, quarks with different flavors (up (u), down (d), strange (s), charm (c), bottom (b) and top (t)) carry different masses. For N_f flavors, we treat the quark field q (the quark mass m) in Eq. (2.1) as a vector with N_f components (an $N_f \times N_f$ matrix). In the standard model, the origin of the quark masses is ascribed to the Yukawa coupling of quarks to the Higgs field. However, the reason why there exist so many varieties of quark masses, ranging from a few mega-electronvolts (MeV) to 175 GeV, is not understood.

2.2 Quantizing QCD

The classical Lagrangian density, $\mathcal{L}_{\rm cl}$, in Eq. (2.1) does not tell us much about the real dynamics of QCD at low energies. This is in contrast to quantum electrodynamics (QED), where the Maxwell equations (as a classical limit of QED) have wide applications in our daily lives. The basic difference between QCD and QED is that the quantum effects become more (less) important at low energies in QCD (QED). To see this difference explicitly, we need to quantize QCD and to study the effects of vacuum polarizations. For a more complete discussion on quantization and renormalization of QCD, see, for example, Ynduráin (1993) and Muta (1998).

There are several different ways of quantizing gauge theories. In this book, we will follow the quantization based on Feynman's functional integral (see Appendix C), which is best suited for making a connection to the classical limit and also for carrying out numerical simulations to study the full quantum aspects of QCD. Another useful quantization procedure is the covariant canonical operator formalism (Kugo and Ojima, 1979).

In the functional integral method, one starts with the partition function under an external source, J:

$$Z[J] = \langle 0+ \mid 0- \rangle_J = \int [dA\ d\bar{q}\ dq]\ \mathrm{e}^{i\int d^4x(\mathcal{L}_{\mathrm{cl}}+J\Phi)}. \qquad (2.13)$$

The physical meaning of $Z[J]$ is the transition amplitude from the vacuum at $t \to -\infty$ to that at $t \to \infty$. The functional integration is carried out over the c-number field, $A^a_\mu(x)$, and the Grassmann fields, $q^\alpha(x)$ and $\bar{q}^\alpha(x)$. The external source is defined as $J\Phi = \bar{\eta}q + \bar{q}\eta + j^\mu_a A^a_\mu$, where η and $\bar{\eta}$ are two independent Grassmann external fields and j^μ_a is a c-number external field.

Note that $Z[0]$ is manifestly gauge-invariant since the integration measure, $[dA\ d\bar{q}\,dq]$, and the action, $S = \int d^4x\,\mathcal{L}_{\mathrm{cl}}$, are both gauge-invariant (See Exercise 2.2(2)). Therefore, each gauge field has an infinite number of relatives related by gauge transformations. Also, they all contribute to $Z[0]$ with the same weight, which leads to a divergence of the functional integral. To avoid such multiple counting, we need to fix the gauge so that we can pick up one representative among the relatives. This situation is illustrated in Fig. 2.1. Note, however, that gauge-fixing is not necessary if one carries out the functional integral numerically over the discretized space-time lattice (Wilson, 1974; Creutz, 1985). This is called lattice QCD simulation, which will be discussed further in Chapter 5.

A neat way to fix the gauge is to insert a decomposition of unity into Eq. (2.13) (Faddeev and Popov, 1967):

$$1 = \int [dV]\Delta_{\mathrm{FP}}[A]\delta(G(A^V)), \qquad (2.14)$$

where $A^V = VAV^\dagger$, $G(A)$ is a gauge-fixing function and $\Delta_{\mathrm{FP}}[A]$ is a Faddeev–Popov (FP) Jacobian (determinant) which makes the integral unity; dV is an invariant measure (the Haar measure) in the group space which satisfies $d(VV') = d(V'V) = dV$ (Creutz, 1985; Gilmore, 1994). (See Exercise 5.2.) By substituting

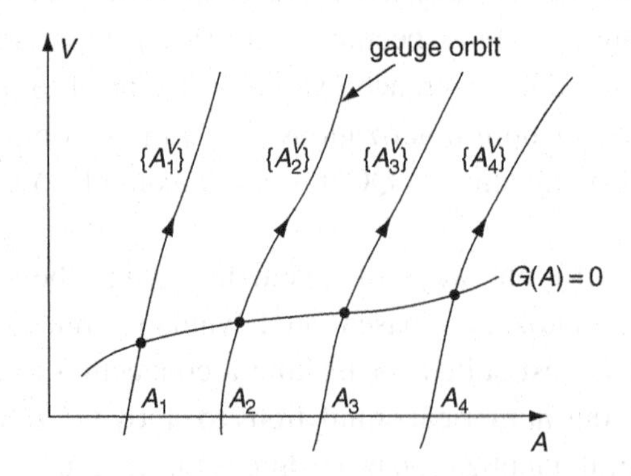

Fig. 2.1. Gauge orbits and the gauge-fixing condition $G(A) = 0$.

the identity Eq. (2.14) into Eq. (2.13) and making an inverse gauge rotation, we arrive at

$$Z[0] = \left(\int [dV]\right) \times \int [dA\, d\bar{q}\, dq]\, \Delta_{\rm FP}[A]\, \delta(G(A))\, {\rm e}^{i\int d^4x\, \mathcal{L}_{\rm cl}}. \tag{2.15}$$

Note that $\delta(G(A))$ simply picks up a representative from the given gauge orbit in Fig. 2.1. The explicit form of $\Delta_{\rm FP}[A]$ is obtained from

$$\Delta_{\rm FP}[A] = \det \left. \frac{\delta G(A^V)}{\delta V} \right|_{V=1}, \tag{2.16}$$

where the determinant is taken for both color and space-time indices.

The first factor on the right-hand side of Eq. (2.15) is the finite gauge volume multiplied by the space-time volume, which is infinite. Since this factor has isolated nicely and is just a multiplicative constant in $Z[J]$, we may simply drop it without affecting the vacuum expectation value of the field products (the Green's functions).

Choice of G is arbitrary as long as it can properly pick up representatives. Commonly used ones are the axial gauge ($G = n^\mu A_\mu$ with $n^2 < 0$), the light-cone gauge ($G = n^\mu A_\mu$ with $n^2 = 0$), the Fock–Schwinger gauge ($G = x^\mu A_\mu$), the Coulomb gauge ($G = \partial^i A_i$), the temporal gauge ($G = A_0$) and the covariant gauge

$$G(A) = \partial^\mu A_\mu - f(x), \tag{2.17}$$

where $f(x)$ is an arbitrary function of space-time (Exercise 2.2(3)).

In the case of Eq. (2.17), we may further multiply the identity, $1 = \int [df] {\rm e}^{-if^2/2\xi}$, to $Z[0]$ so that $f(x)$ is eliminated. Here ξ is called the gauge parameter. The FP determinant may be exponentiated by introducing two independent Grassmann fields, $c^a(x)$ and $\bar{c}^a(x)$, called the ghost and the anti-ghost, respectively. Then we arrive at the final form:

$$Z[J] = {\rm e}^{iW[J]} = \int [dA\, d\bar{q}\, dq][d\bar{c}\, dc]\, {\rm e}^{i\int d^4x (\mathcal{L} + J\Phi)}, \tag{2.18}$$

$$\begin{aligned} \mathcal{L} = {} & \bar{q}^\alpha (i \not{D}_{\alpha\beta} - m\delta_{\alpha\beta}) q^\beta - \frac{1}{4} F^a_{\mu\nu} F^{\mu\nu}_a \\ & - \bar{c}_a \partial^\mu \mathcal{D}^{ab}_\mu c_b - \frac{1}{2\xi} (\partial^\mu A^a_\mu)^2. \end{aligned} \tag{2.19}$$

Although the gauge-fixed Lagrangian density, $\mathcal{L}$, is no longer invariant under the *classical* gauge transformation, Eq. (2.10), it has *quantum* gauge invariance under the Becchi–Rouet–Stora–Tyutin (BRST) transformation, $\delta_{\rm BRST}$

(Becchi *et al.*, 1976; Iofa and Tyutin, 1976); See Exercise 2.2(4). The transformations are given by

$$\delta_{\text{BRST}} q = -ig\lambda c q, \qquad \delta_{\text{BRST}} A_\mu = \lambda[D_\mu, c], \tag{2.20}$$

$$\delta_{\text{BRST}} \bar{c} = -\lambda \xi \partial^\mu A_\mu, \qquad \delta_{\text{BRST}} c = -\frac{i}{2} g\lambda[c, c], \tag{2.21}$$

where $c = c^a t^a$ and λ is a space-time-independent Grassmann number. The transformations in Eq. (2.20) are simply obtained from Eq. (2.12) by the replacement $\theta \to g\lambda c$. The BRST transformation is nilpotent, i.e. $\delta^2_{\text{BRST}} = 0$. Canonical quantization of QCD can be beautifully carried out on the basis of the BRST symmetry (Kugo and Ojima, 1979).

The standard perturbation theory employed to evaluate $Z[J]$ or $W[J]$ can now be developed, starting from Eq. (2.19), by decomposing it into a free part and an interaction part:

$$\mathcal{L} = \mathcal{L}_0 + \mathcal{L}_{\text{int}}, \tag{2.22}$$

where $\mathcal{L}_0$ is obtained by setting $g = 0$ in $\mathcal{L}$. The Feynman rules in the Euclidean space-time with this decomposition will be given in Chapter 4. Note that $Z[J]$ is a generating functional of the full Green's function, while $W[J]$ is a generating functional for the one-particle irreducible (1PI) Green's functions. We may also introduce an effective action, $\Gamma[\varphi]$, by a Legendre transform, $\Gamma[\varphi] = W[J] - J\varphi$, where $\varphi \equiv \delta W/\delta J$; $\Gamma[\varphi]$ is the generating functional for the 1PI proper vertices. These will be covered in more detail in Chapter 6 when we discuss the critical phenomena associated with the second order phase transition.

2.3 Renormalizing QCD

In field theories, the quantum corrections (loops) calculated in perturbation theory have ultraviolet divergences originating from the intermediate states with high momenta. In renormalizable field theories, such as QCD, these divergences can always be combined with the *bare* parameters of the Lagrangian and are absorbed in the *renormalized* parameters.[1]

The energy scale at which the divergences are renormalized is called the renormalization point, and is denoted by κ throughout this book; κ is an arbitrary parameter. Any observables, such as the proton mass, the pion decay constant,

[1] Renormalizability of the QCD Lagrangian is not an accident but rather is a consequence of a large gap between the typical QCD scale and the scale beyond QCD. Non-renormalizable terms, although they exist in principle, become irrelevant at low energies. Note also that general non-renormalizable theories have predictive powers as long as one has a systematic low-energy expansion: a well known example is the chiral perturbation theory for low-energy QCD. For more details on the modern concepts of renormalization and effective field theories, consult Lepage (1990), Weinberg (1979), Polchinski (1992) and Kaplan (1995).

etc., do not depend on κ, whereas the renormalized coupling, g, the quark mass, m, and the gauge parameter, ξ, do depend on κ.

The renormalization in QCD is summarized as a general statement that the partition function defined in Eq. (2.18) can be made finite by the redefinition of the parameters. Namely,

$$Z[J_{\mathrm{B}}; g_{\mathrm{B}}, m_{\mathrm{B}}, \xi_{\mathrm{B}}] = Z[J(\kappa); g(\kappa), m(\kappa), \xi(\kappa); \kappa], \tag{2.23}$$

where the quantities with suffix B are bare parameters. The external field, J_{B}, is also regarded as one of the bare parameters. The precise relation between the bare and renormalized quantities will be discussed in detail in Chapter 6. Since the left-hand side of Eq. (2.23) is κ-independent, Z satisfies

$$\kappa \frac{d}{d\kappa} Z = 0. \tag{2.24}$$

The renormalization group equations for Green's functions are simply obtained from this master equation by taking certain derivatives with respect to J.

To grasp the meaning of κ in perturbation theory, let us consider the $\mathrm{e}^+\mathrm{e}^-$ annihilation into hadrons at a high center-of-mass energy, $\sqrt{s}$. The perturbative calculation of $\mathrm{e}^+\mathrm{e}^- \to q\bar{q}$ for a massless quark yields the following cross-section:

$$\sigma(s) = \sigma_0(s) \left[1 + c_1(\sqrt{s}/\kappa) \frac{\alpha_{\mathrm{s}}(\kappa)}{\pi} + c_2(\sqrt{s}/\kappa) \left(\frac{\alpha_{\mathrm{s}}(\kappa)}{\pi} \right)^2 + \cdots \right]. \tag{2.25}$$

In this formula, $\sigma_0(s) = (4\pi\alpha^2/3s) \cdot 3 \cdot Q_q^2$ is the lowest order cross-section for producing a pair of massless quarks with electric charge Q_q; See Section 14.3.1. The factor $\alpha_{\mathrm{s}} \equiv g^2/4\pi$ ($\alpha \equiv e^2/4\pi$) is the fine structure constant in the strong (electromagnetic) interaction; c_i are dimensionless constants. It happens that $c_1 = 1$, but in general $c_{i \geq 2}$ depend on $\ln(\sqrt{s}/\kappa)$.

The cross-section is an observable and thus cannot depend on κ. However, in the perturbation theory this is guaranteed only order by order. To make the higher order corrections behave well in Eq. (2.25), we may choose, for example, $\kappa = \sqrt{s}$, so that large logs at $\sqrt{s} \gg \kappa$ inside c_i disappear. At the same time, $\alpha_{\mathrm{s}}(\kappa)$ becomes the running coupling $\alpha_{\mathrm{s}}(\sqrt{s})$, which decreases as s increases in QCD, as will be shown in a moment. This makes the perturbation series well behaved at high s.

As we will see in Chapter 4, the free-energy density of massless quarks and gluons in the quark–gluon plasma at high temperature (T) has the following form:

$$\begin{aligned} f(T) = -\frac{8\pi^2}{45} T^4 \Bigg[& f_0 + f_2 \, \frac{\alpha_{\mathrm{s}}(\kappa)}{\pi} + f_3 \left(\frac{\alpha_{\mathrm{s}}(\kappa)}{\pi} \right)^{3/2} \\ & + f_4 \left(\ln \frac{\pi T}{\kappa}, \ln \alpha_{\mathrm{s}}(\kappa) \right) \left(\frac{\alpha_{\mathrm{s}}(\kappa)}{\pi} \right)^2 + \cdots \Bigg]. \end{aligned} \tag{2.26}$$

The factor $f(T)|_{\alpha_s \to 0}$ is called the Stefan–Boltzmann limit, and it contains contributions from a non-interacting gas of quarks and gluons. The coefficients f_0, f_2 and f_3 are κ-independent constants, while $f_{i\geq 4}$ depend on κ through either $\ln(\kappa/\pi T)$ or $\ln \alpha_s(\kappa)$. By choosing, for example, $\kappa \sim \pi T$, we can suppress large logs when $T \gg \kappa$ and consequently make the series well behaved. The explicit form of f_i is given in Eqs. (4.111)–(4.115).

2.3.1 Running coupling constants

What is the behavior of the running coupling g as a function of κ? This can be answered by looking at the solution of the *flow* equation,

$$\kappa \frac{\partial g}{\partial \kappa} = \beta. \tag{2.27}$$

The right-hand side is called the β-function and may be calculated in perturbation theory if g is small enough. All the manipulations become particularly simple if we adopt the modified minimal subtraction scheme ($\overline{\text{MS}}$) with the dimensional regularization (Muta, 1998). In this scheme, β depends only on g and has a series expansion of the following form (Gross, 1981; Muta, 1998)

$$\beta(g) = -\beta_0 g^3 - \beta_1 g^5 + \cdots \tag{2.28}$$

$$\beta_0 = \frac{1}{(4\pi)^2}\left(11 - \frac{2}{3}N_f\right), \quad \beta_1 = \frac{1}{(4\pi)^4}\left(102 - \frac{38}{3}N_f\right). \tag{2.29}$$

Factors β_0 and β_1 are subtraction-scheme-independent and are both positive for $N_f \leq 8$. A negative β, together with Eq. (2.27), implies that $g(\kappa)$ decreases as κ increases; this is called the ultraviolet asymptotic freedom (Gross and Wilczek, 1973; Politzer, 1973; 't Hooft, 1985). Note that the β-function in QED is given by $\beta(e) = e^3/(12\pi^2) + e^5/(64\pi^4) + \cdots$, which shows that $e(\kappa)$ increases as κ increases. Among the renormalizable quantum field theories in four space-time dimensions, only the non-Abelian gauge theories are asymptotically free (Coleman and Gross, 1973).

It is a straightforward matter to extract the explicit form of $g(\kappa)$ in the lowest order by taking β_0 and β_1. The result is as follows:

$$\alpha_s(\kappa) = \frac{1}{4\pi\beta_0 \ln(\kappa^2/\Lambda_{\text{QCD}}^2)}\left[1 - \frac{\beta_1}{\beta_0^2}\frac{\ln(\ln(\kappa^2/\Lambda_{\text{QCD}}^2))}{\ln(\kappa^2/\Lambda_{\text{QCD}}^2)} + \cdots\right]. \tag{2.30}$$

Here Λ_{QCD} is called the QCD *scale parameter*, to be determined from experiments. Since it is related to the integration constant of the differential equation, Eq. (2.27), it is κ-independent (Exercise 2.3).

It is important to specify the subtraction scheme and the number of active flavors when quoting the actual value of $\Lambda_{\rm QCD}$; for example, $\Lambda_{\overline{\rm MS}}^{(N_f=5)} = (217 \pm 24)$ MeV (Eidelman *et al.* 2004). Equation (2.30) tells us that the running coupling decreases logarithmically as κ increases. Therefore, our use of perturbation theory is justified as long as κ is large enough. This does not mean, however, that the expansion in terms of g is convergent for large κ. Instead, the expansion is at most asymptotic. The running coupling determined by data from several experiments is shown in Fig. 2.2. Values of the fine structure constant, α_s, for typical values of κ are as follows:

$$\alpha_s(100\ \text{GeV}) = 0.1156 \pm 0.0020, \qquad \alpha_s(10\ \text{GeV}) = 0.1762 \pm 0.0048,$$
$$\alpha_s(2\ \text{GeV}) = 0.300 \pm 0.015, \qquad \alpha_s(1\ \text{GeV}) = 0.50 \pm 0.06\ . \tag{2.31}$$

The quark mass, m, is a parameter similar to the coupling constant, g; it obeys a flow equation,

$$\kappa \frac{d}{d\kappa} m = -\gamma_m(g) m, \tag{2.32}$$

$$\gamma_m(g) = \gamma_{m0} g^2 + \gamma_{m1} g^4 + \cdots, \ \gamma_{m0} = \frac{8}{(4\pi)^2}. \tag{2.33}$$

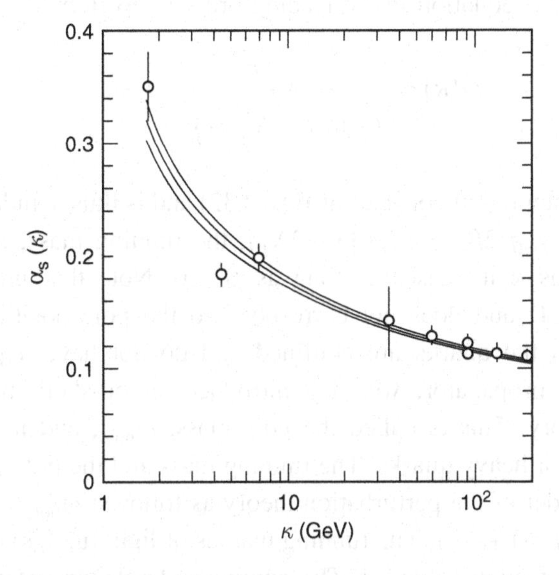

Fig. 2.2. The running coupling determined from τ decay, Υ decay, deep inelastic scattering, e^+e^- annihilation and Z-boson resonance shape and width. The figure is adapted from Eidelman *et al.* (2004).

Table 2.1. *Running masses in the $\overline{\text{MS}}$ scheme and the scale-independent ratios.*

$m_{\text{u,d,s}}\ (\kappa = 1\,\text{GeV}) \simeq 1.35\ m_{\text{u,d,s}}\ (\kappa = 2\,\text{GeV})$.

	Running mass[a]	Mass ratio[b]
Up: m_{u}	1.5 to 4.5 MeV ($\kappa = 2\,\text{GeV}$)	$m_{\text{u}}/m_{\text{d}} = 0.553 \pm 0.043$
Down: m_{d}	5.0 to 8.5 MeV ($\kappa = 2\,\text{GeV}$)	
Strange: m_{s}	80 to 155 MeV ($\kappa = 2\,\text{GeV}$)	$m_{\text{s}}/m_{\text{d}} = 18.9 \pm 0.8$

[a] From Eidelman *et al.* (2004).
[b] From Leutwyler (2001a).

Table 2.2. *Running masses in the $\overline{\text{MS}}$ scheme and the pole masses for the charm, bottom and top quarks.*

	Running mass[a]	Pole mass[a]
Charm: m_{c}	1.0 to 1.4 GeV ($\kappa = m_{\text{c}}$)	1.5 to 1.8 GeV
Bottom: m_{b}	4.0 to 4.5 GeV ($\kappa = m_{\text{b}}$)	4.6 to 5.1 GeV
Top: m_{t}	$\sim$175 GeV ($\kappa = m_{\text{t}}$)	$\sim$175 GeV

[a] From Eidelman *et al.* (2004).

The solution of this equation in the leading order is given by

$$m(\kappa) = \frac{\bar{m}}{\left(\frac{1}{2}\ln(\kappa^2/\Lambda^2_{\text{QCD}})\right)^{\gamma_{m0}/2\beta_0}}, \tag{2.34}$$

where $\bar{m}$ is an integration constant of Eq. (2.32) and is thus κ-independent (Exercise 2.4). Since $\gamma_{m0}/2\beta_0 = 12/(33-2N_f)$, the running mass, $m(\kappa)$, decreases logarithmically as κ increases as long as $\beta < 0$. Note that $m(\kappa)$ is simply a parameter of QCD and does not correspond to the pole position of the propagator. Although the quarks are confined and do not have a pole in the full non-perturbative propagator, we may introduce an on-shell mass, at least in perturbation theory. This is called the pole mass, m_{pole}, and it is useful in the phenomenology of heavy quarks. The running mass and the pole mass are related in the leading order of the perturbation theory as follows: $m_{\text{pole}} = m(\kappa = m)(1 + (4/3)(\alpha_{\text{s}}(\kappa = m)/\pi) + \cdots)$. The running masses of light (u, d, s) quarks and their ratios are summarized in Table 2.1. The running and pole masses of heavy (c, b, t) quarks are summarized in Table 2.2. A more extensive and updated listing may be found in Eidelman *et al.* (2004).

2.3.2 More on asymptotic freedom

The polarization of the vacuum screens (anti-screens) a test charge in QED (QCD). The screening (anti-screening) may be interpreted as the *effective charge*, which is a sum of the test charge and the induced charge, increases (decreases) as one approaches the test charge. In terms of the dielectric constant, ϵ, of the vacuum, $\epsilon > 1 (< 1)$ corresponds to the screening (anti-screening) in analogy with dielectric substances (Jackson, 1999).

This intuitive picture is justified either by studying the dielectric constant, ϵ, of the QCD vacuum under an external color electric field (Hughes, 1981) or by studying the magnetic permeability, μ, under a color magnetic field (Nielsen, 1981; Huang, 1992). Since Lorentz invariance imposes $\epsilon\mu = 1$, these are two equivalent ways of seeing the same physics; namely, screening, $\epsilon > 1$ (anti-screening, $\epsilon < 1$) corresponds to diamagnetism, $\mu < 1$ (paramagnetism, $\mu > 1$).

Let us consider the energy density of the vacuum, ε_{vac}, under a uniform magnetic field, B, coupled with massless particles with a bare charge, e_0, and spin S $(= 0, 1/2, 1)$. Finding the Landau levels under the uniform magnetic field and summing over the zero-point fluctuations lead us to the following formula:

$$\varepsilon_{\text{vac}}(B) \simeq \left(\frac{1}{2} - \frac{b(S)}{2} e_0^2 \ln \frac{\Lambda^2}{e_0 B}\right) B^2 \equiv \frac{1}{2\mu(B)} B^2. \tag{2.35}$$

Here Λ is an ultraviolet cutoff of the vacuum fluctuation and $\mu(B)$ is interpreted as the magnetic permeability of the vacuum. (One may also introduce the magnetic susceptibility, $\chi(B) = 1 - \mu(B)$.) The coefficient of the zero-point fluctuation is given by

$$b(S) = \frac{(-1)^{2S}}{8\pi^2} \left[(2S)^2 - \frac{1}{3}\right], \tag{2.36}$$

which has a simple interpretation. The first term on the right-hand side of Eq. (2.36) is from the Pauli paramagnetism related to the alignment of the particle spin along the external magnetic field. The second term is from the Landau diamagnetism related to the orbital motion of the charged particles under the magnetic field (Exercise 2.5). The sign factor, $(-1)^{2S}$, reflects the particle statistics (zero-point energy is positive for bosons and negative for fermions). This formula indicates that the $S = 1$ vector boson leads to paramagnetism, $\mu > 1$, while the $S = 0$ boson and the $S = 1/2$ fermion lead to diamagnetism, $\mu < 1$.

The relation between $b(S)$ and the β-function becomes clear if we define an effective coupling constant, $e^2(\sqrt{B}) = e_0^2/\epsilon = \mu e_0^2$, in analogy with standard electromagnetism. Then it is straightforward to see that $\kappa \partial e/\partial\kappa = -b(S)e^3$, where $\kappa = \sqrt{B}$, which implies that b is simply the first coefficient of the β-function, β_0, defined in Eq. (2.28).

To make a quantitative connection with QCD, B may be identified with the external color magnetic field. Its couplings to the $S=1$ and $S=1/2$ fields are generated by the gluon self-coupling and the gluon–quark coupling, respectively. Then, after taking into account appropriate color and flavor factors (Nielsen, 1981; Huang, 1992), we arrive at the familiar formula for asymptotic freedom,

$$\beta_0(\text{QCD}) = \frac{3}{2}b(S=1) + \frac{N_f}{2}b\left(S=\frac{1}{2}\right) = \frac{1}{(4\pi)^2}\left(11 - \frac{2N_f}{3}\right). \quad (2.37)$$

Thus we find that the spin of the gluon and the gluon self-coupling make the QCD vacuum behave like a paramagnetic substance. This result, combined with Lorentz invariance, implies that the QCD vacuum exhibits anti-screening of color charges. For a system at finite temperature and density, the situation changes due to the existence of matter. In fact, quark–gluon plasma exhibits the screening of color charges, rather than anti-screening, as will be discussed in Chapter 4.

In summary, we have found that the running coupling, g, becomes small at high energies, and this is known as the ultraviolet (UV) asymptotic freedom. In this case, the perturbation theory becomes more and more reliable at high energies once an appropriate choice of κ is made. On the other hand, at low energies, g grows, and the perturbation theory is no longer reliable. This feature is the origin of color confinement and is called the *infrared (IR) slavery*. UV asymptotic freedom and IR slavery are just opposite sides of the same coin.

2.4 Global symmetries in QCD

Besides $SU_c(3)$ local gauge symmetry, classical Lagrangian density, Eq. (2.1), has other global symmetries, such as the chiral symmetry and dilatational symmetry.

2.4.1 *Chiral symmetry*

Let us first introduce the left-handed and right-handed quarks as two eigenstates of the chirality operator, γ_5, with the eigenvalues ± 1:

$$q_{\text{L}} = \frac{1}{2}(1-\gamma_5)q, \quad q_{\text{R}} = \frac{1}{2}(1+\gamma_5)q. \quad (2.38)$$

For free massless quarks, the chirality is equivalent to the helicity, $\boldsymbol{\sigma}\cdot\hat{\boldsymbol{p}}$. Now, consider QCD with N_f flavors and write the quark field as a vector with N_f components as follows:

$$^t q = (\text{u}, \text{d}, \text{s}, \ldots). \quad (2.39)$$

Then the quark mass, m, becomes an $N_f \times N_f$ matrix and $\mathcal{L}_{\rm cl}$ may be decomposed as follows:

$$\mathcal{L}_{\rm cl} = \mathcal{L}_{\rm cl}(q_{\rm L}, A) + \mathcal{L}_{\rm cl}(q_{\rm R}, A) - (\bar{q}_{\rm L} m q_{\rm R} + \bar{q}_{\rm R} m q_{\rm L}). \tag{2.40}$$

It is clear from this expression that $\mathcal{L}_{\rm cl}$ in Eq. (2.1) and also $\mathcal{L}$ in Eq. (2.19) for $m = 0$ are invariant under the $\mathrm{U}_{\rm L}(N_f) \times \mathrm{U}_{\rm R}(N_f)$ global transformation

$$q_{\rm L} \to \mathrm{e}^{-i\lambda^j \theta^j_{\rm L}} q_{\rm L}, \quad q_{\rm R} \to \mathrm{e}^{-i\lambda^j \theta^j_{\rm R}} q_{\rm R}, \tag{2.41}$$

where the $\theta^j_{\rm L,R}$ $(j = 0, 1, \ldots, N_f^2 - 1)$ are space-time-independent parameters and $\lambda^0 = \sqrt{2/N_f}$, $\lambda^j = 2t^j$ $(j = 1, \ldots, N_f^2 - 1)$. This is called chiral symmetry.

It is also convenient to define vector and axial-vector transformations as

$$q \to \mathrm{e}^{-i\lambda^j \theta^j_{\rm V}} q, \quad q \to \mathrm{e}^{-i\lambda^j \theta^j_{\rm A} \gamma_5} q, \tag{2.42}$$

where $\theta_{\rm V} = \theta_{\rm L} = \theta_{\rm R}$ and $\theta_{\rm A} = -\theta_{\rm L} = \theta_{\rm R}$. There are two $\mathrm{U}(1)$ transformations in the above: $\mathrm{U}_{\rm B}(1)$, which is related to the baryon number ($\theta^j_{\rm V} \propto \delta^{j0}$, $\theta^j_{\rm A} = 0$), and $\mathrm{U}_{\rm A}(1)$, which corresponds to the flavor-singlet axial rotation ($\theta^j_{\rm V} = 0$, $\theta^j_{\rm A} \propto \delta^{j0}$).

Because of Noether's theorem, which relates symmetries and conservation laws, the vector and axial currents, $J^j_\mu = \bar{q}\gamma_\mu \lambda^j q$ and $J^j_{\mu 5} = \bar{q}\gamma_\mu \gamma_5 \lambda^j q$, have the following relations (Exercise 2.6):

$$\partial^\mu J^j_\mu = i\bar{q}[m, \lambda^j]q, \quad (j = 0, \ldots, N_f^2 - 1), \tag{2.43}$$

$$\partial^\mu J^j_{\mu 5} = i\bar{q}\{m, \lambda^j\}\gamma_5 q, \quad (j = 1, \ldots, N_f^2 - 1), \tag{2.44}$$

$$\partial^\mu J^0_{\mu 5} = \sqrt{2/N_f}\left(2i\bar{q}m\gamma_5 q - 2N_f \frac{g^2}{32\pi^2} F^{\mu\nu}_a \tilde{F}^a_{\mu\nu}\right), \tag{2.45}$$

where $\tilde{F}^a_{\mu\nu} \equiv \frac{1}{2}\epsilon_{\mu\nu\lambda\rho} F^{\lambda\rho}_a$ $(\epsilon_{0123} = 1)$ is the dual field strength tensor. As can be seen in Eq. (2.45), the conservation of the flavor-singlet axial current, $J^0_{\mu 5}$, is broken not only by the quark mass matrix, m, but also by the quantum effect called the *axial anomaly* (Bertlemann, 1996). From the functional integral point of view, it originates from the non-invariance of the path integral measure $[d\bar{q}\, dq]$ under the $\mathrm{U}_{\rm A}(1)$ transformation (Fujikawa, 1980a, b). See Exercise 2.7.

2.4.2 Dilatational symmetry

Consider the scale transformation

$$q(x) \to \sigma^{3/2} q(\sigma x), \quad A^a_\mu(x) \to \sigma A^a_\mu(\sigma x), \tag{2.46}$$

where σ is a space-time-independent parameter. Note that $\mathcal{L}_{\rm cl}$ is invariant under Eq. (2.46) in the chiral limit ($m = 0$). This symmetry is, however, explicitly broken by the quantum effect, and we arrive at

$$\partial_\mu \Delta^\mu = T^\mu_\mu = \frac{\beta}{2g} F^a_{\mu\nu} F^{\mu\nu}_a + (1+\gamma_m) \sum_q \bar{q} m_q q, \tag{2.47}$$

where $T_{\mu\nu}$ is the energy-momentum tensor and Δ_μ is the dilatational current defined by $x_\nu T^{\nu\mu}$. The second term on the right-hand side of Eq. (2.47) shows that the dilatational current is not conserved if $g \neq 0$. Since it is related to the trace of the energy-momentum tensor, this is called the *trace anomaly* (Collins *et al.*, 1977; Nielsen, 1977). (For a heuristic derivation, see Exercise 2.8.)

2.5 QCD vacuum structure

Because the QCD coupling, g, becomes strong in the IR region, quarks and gluons interact non-perturbatively at low energies. As a result, the QCD vacuum acquires a non-trivial structure, such as the quark and gluon condensates.

Analyses of the mass spectrum of charmoniums using QCD spectral sum rules (Shifman *et al.*, 1979; Colangelo and Khodjamirian, 2001) indicate that gluons have non-perturbative condensation:

$$\left\langle \frac{\alpha_s}{\pi} F^a_{\mu\nu} F^{\mu\nu}_a \right\rangle_{\rm vac} \sim (300\ \text{MeV})^4. \tag{2.48}$$

This has an interesting physical consequence if it is combined with the trace anomaly in Eq. (2.47). First, the vacuum expectation value of the energy-momentum tensor can be written as $\langle T^{\mu\nu} \rangle_{\rm vac} = -\varepsilon_{\rm vac} g^{\mu\nu}$ using Lorentz invariance, where $\varepsilon_{\rm vac}$ is the energy density of the QCD vacuum. Taking the chiral limit, $m = 0$, for simplicity and expanding $\beta(g)$ up to the leading order yields

$$\begin{aligned} \varepsilon_{\rm vac} &\simeq -\frac{11 - 2N_f/3}{32} \left\langle \frac{\alpha_s}{\pi} F^a_{\mu\nu} F^{\mu\nu}_a \right\rangle_{\rm vac} \\ &\simeq -0.3\ \text{GeV fm}^{-3}. \end{aligned} \tag{2.49}$$

Namely, the QCD vacuum has a smaller energy density than that of the perturbative vacuum. Very often, one introduces the notation $\varepsilon_{\rm vac} = -B$; B is called the *bag constant* for historical reasons, as explained in Section 2.6. The color confinement must be related to the gluonic structure of the non-perturbative vacuum. Although many attempts to bridge this link have been made, a satisfactory and clear explanation remains elusive. The anti-screening nature we have encountered in perturbation theory ($\epsilon < 1$ and $\mu > 1$) is already a signature of the color confinement. In fact, the QCD vacuum may well be a perfect anti-dielectric or equivalently perfect paramagnetic substance; i.e.

$$\epsilon_{\rm vac} = 0, \quad \mu_{\rm vac} = \infty. \tag{2.50}$$

What about the quark structure of the QCD vacuum? The Gell-Mann–Oakes–Renner (GOR) relation, which is obtained solely by chiral symmetry, yields (Exercise 2.9(2, 3))

$$f_\pi^2 m_{\pi^\pm}^2 = -\hat{m}\langle \bar{\rm u}{\rm u} + \bar{\rm d}{\rm d}\rangle_{\rm vac} + O(\hat{m}^2), \tag{2.51}$$

$$f_\pi^2 m_{\pi^0}^2 = -\langle m_{\rm u}\bar{\rm u}{\rm u} + m_{\rm d}\bar{\rm d}{\rm d}\rangle_{\rm vac} + O(\hat{m}^2). \tag{2.52}$$

Here $\hat{m} = (m_{\rm u} + m_{\rm d})/2$ is the averaged mass of u and d quarks, $f_\pi(= 93\,{\rm MeV})$ is the pion decay constant and $m_{\pi^\pm} \simeq 140\,{\rm MeV}$ ($m_{\pi^0} \simeq 135\,{\rm MeV}$) is the charged (neutral) pion mass. Using the quark mass from Table 2.1, we obtain $\hat{m}(\kappa = 1\,{\rm GeV}) \sim 5.6\,{\rm MeV}$, and we have

$$\langle(\bar{\rm u}{\rm u} + \bar{\rm d}{\rm d})/2\rangle_{\rm vac} \sim -(250\,{\rm MeV})^3 \quad \text{at} \quad \kappa = 1\,{\rm GeV}. \tag{2.53}$$

This implies that the QCD vacuum involves condensation of quark–anti-quark pairs. It was Nambu who developed the theory of dynamical breaking of chiral symmetry by noticing its close analogy with Cooper pairs and superconductivity in metals (Nambu and Jona-Lasinio, 1961a, b).

Since $\bar{q}q(= \bar{q}_{\rm L}q_{\rm R} + \bar{q}_{\rm R}q_{\rm L})$ is not invariant under the ${\rm SU_L}(N_f) \times {\rm SU_R}(N_f)$ transformation given in Eq. (2.41), it serves as an *order parameter* of the dynamical breaking of chiral symmetry. Indeed, the pattern of the dynamical chiral symmetry breaking in a QCD vacuum is

$${\rm SU_L}(N_f) \times {\rm SU_R}(N_f) \to {\rm SU_V}(N_f), \tag{2.54}$$

where ${\rm SU_V}(N_f)$ is a vector symmetry remaining in the vacuum. At extremely high temperature, the restoration of chiral symmetry takes place, as will be discussed in Chapter 6. At extremely high baryon density, the system exhibits a different type of symmetry breaking pattern (the color superconductivity), which will be discussed briefly in Chapter 9.

The energy density of the QCD vacuum as a function of the order parameter looks like that given in Fig. 14.4. It has a minimum at the bottom of the "wine bottle" and thus breaks symmetry. The amplitude fluctuation of $\langle\bar{q}q\rangle$ costs energy and thus corresponds to a massive excitation, called the σ-meson (Hatsuda and Kunihiro, 2001). On the other hand, the phase fluctuation of $\langle\bar{q}q\rangle$ costs no energy in the chiral limit and corresponds to the massless pions. The latter statement can be made more rigorous, and is known as the Nambu–Goldstone theorem (Exercise 2.9(1)).

2.6 Various approaches to non-perturbative QCD

Various models of QCD have been proposed, such as the bag model, the potential model and the Nambu–Jona-Lasinio model. Although one must be cautious not to extrapolate too far beyond the applicability of the models, they are quite useful in elucidating some of the essential features of QCD in a non-perturbative regime. Methods more closely related to QCD are the QCD sum rules and the chiral perturbation theory. In addition, direct numerical simulations of QCD defined on the space-time lattice have become an extremely useful tool in recent years. We will briefly summarize these approaches in the following.

The bag model

Hadrons may be considered as bags embedded in a non-perturbative QCD vacuum (Chodos *et al.*, 1974). A special boundary condition is introduced for quarks and gluons to incorporate the confinement so that color charge does not leak out from the bag. Inside the bag, no condensates are assumed and quarks and gluons are treated perturbatively. An intuitive picture of the model is shown in Fig. 2.3.

The mass of the proton in this model is given by

$$M_{\mathrm{p}} = \frac{3x}{R_{\mathrm{bag}}} + \frac{4\pi}{3} B R_{\mathrm{bag}}^3 + \cdots , \tag{2.55}$$

where R_{bag} is the radius of a spherical bag; x/R_{bag} is the kinetic energy of each quark confined inside the bag ($x \sim 2.04$ for the lowest-energy mode of a massless quark). The term proportional to B (called the bag constant) is the volume energy of the bag, which represents the energy necessary to create a perturbative region in the non-perturbative vacuum. Thus, $-B$ can be identified with Eq. (2.49). The "$\cdots$" in Eq. (2.55) contains interactions among quarks inside the bag, the Casimir energy of quarks and gluons in the spherical cavity, the effect of the

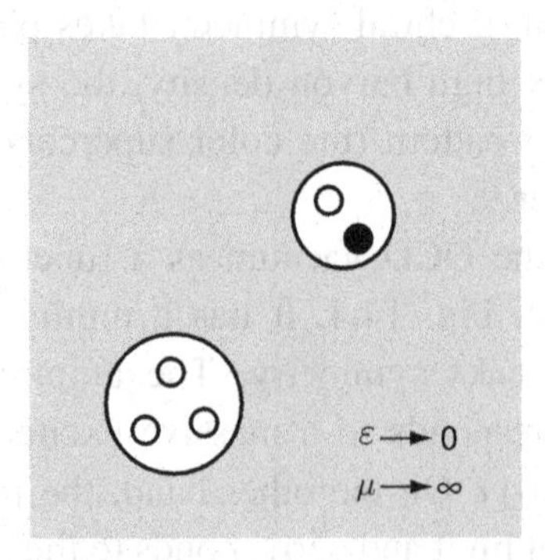

Fig. 2.3. Baryon and meson bags in a non-perturbative QCD vacuum which has a perfect anti-dielectric (perfect paramagnetic) property.

meson cloud surrounding the bag, and so on. The model can describe the mass spectra of light hadrons composed of u, d and s quarks reasonably well with a suitable choice of the bag constant. An estimate using Eqs. (2.48) and (2.49), together with the definition $B = -\varepsilon_{\text{vac}}$, yields

$$B \sim (220 \text{ MeV})^4. \tag{2.56}$$

This will be used in Chapter 3 to construct a schematic model of the phase transition between the hadronic matter and the quark–gluon plasma.

The potential model

In order to describe hadrons composed of heavy quarks, such as charm and bottom quarks, the non-relativistic Schrödinger equation with static inter-quark potential may be a good starting point. The potential picture is justified when the motion of the valence quarks is much slower than the typical frequency of gluons exchanged between quarks. A typical form of the phenomenological $q\bar{q}$ potential is the Coulomb + linear type:

$$V_{q\bar{q}}(R) = -\frac{a}{R} + KR + \cdots, \tag{2.57}$$

where R is the distance between the quark and the anti-quark, and K is the string tension representing the strength of the quark confinement force. Its empirical value obtained by fitting the charmonium and bottomnium spectra (Bali, 2001) is given by

$$K \simeq (0.42 \text{ GeV})^2 = 0.9 \text{ GeV fm}^{-1}. \tag{2.58}$$

This value is consistent with that obtained from the Regge trajectories of light mesons. In perturbation theory, the coefficient of the Coulomb term, $-a/R$, becomes $a = 4\alpha_s/3$. The "$\cdots$" in Eq. (2.57) represents the relativistic corrections such as the hyper-fine term, the spin-orbit term, and so on. Application of the model to the heavy quark bound states at finite T will be discussed in Chapter 7.

There is no a priori reason to believe that the potential picture also works for hadrons composed of light quarks (u, d, s). Nevertheless, it has a remarkable phenomenological success in describing mass spectra and electro-magnetic properties (De Rujula *et al.*, 1975) if we assume that the u, d and s quarks have the *constituent masses*

$$M_{\text{u,d}} \sim 340 \text{ MeV and } M_{\text{s}} \sim 540 \text{ MeV}. \tag{2.59}$$

These are much larger than the current masses, m_q, listed in Table 2.1; M_q may be determined, for example, by fitting the magnetic moments of light baryons, and is now believed to be an effective mass generated by the dynamical breaking of chiral symmetry.

The Nambu–Jona-Lasinio (NJL) model

This model is based on a Lagrangian which captures essential aspects of chiral symmetry in QCD (Vogl and Weise, 1991; Klevansky, 1992; Hatsuda and Kunihiro, 1994). The typical form of the Lagrangian density for N_f flavors is given by (Nambu and Jona-Lasinio, 1961a,b)

$$\mathcal{L}_{\rm NJL} = \bar{q}(i\gamma_\mu \partial^\mu - m)q + \frac{g_{\rm NJL}}{2} \sum_{j=0}^{N_f^2-1} \left[(\bar{q}\lambda^j q)^2 + (\bar{q} i\gamma_5 \lambda^j q)^2 \right]. \tag{2.60}$$

Equation (2.60) includes the quark kinetic term and a chiral symmetric four-fermi interaction with a dimensionful coupling constant, $g_{\rm NJL}$. The dynamical breaking of chiral symmetry ($\langle \bar{q}q \rangle_{\rm vac} \neq 0$) occurs when $g_{\rm NJL}$ exceeds some critical value. In addition, the constituent quark masses, M_q, in Eq. (2.59) are generated dynamically:

$$M_q = m_q - 2g_{\rm NJL} \langle \bar{q}q \rangle_{\rm vac}. \tag{2.61}$$

The model can be generalized to incorporate the $\mathrm{U_A}(1)$ axial anomaly by introducing an extra term, $\det_{k,l} \bar{q}_k (1+\gamma_5) q_l + \mathrm{h.c.}$, which is called the Kobayashi–Maskawa–'t Hooft vertex ('t Hooft, 1986; Hatsuda and Kunihiro, 1994).

It is a straightforward matter to apply this model to study the restoration of broken chiral symmetry at high temperature and/or baryon density. It is amazing to see that this simple model captures the essential feature of the QCD phase diagram in the real world, a point that will be discussed further in Chapter 6.

Chiral perturbation theory

Because of the dynamical breaking of chiral symmetry, the pion as a Nambu–Goldstone boson becomes the lightest hadron. Therefore, at low energies, we may introduce an effective description of the QCD partition function, $Z[0]$, solely in terms of the pions (Weinberg, 1979; Gasser and Leutwyler, 1984, 1985):

$$Z[0] = \int [dU]\, \mathrm{e}^{i \int d^4x (\mathcal{L}^{(2)}(U) + \mathcal{L}^{(4)}(U) + \mathcal{L}^{(6)}(U) + \cdots)}, \tag{2.62}$$

where $U(x)$ is a field in the $\mathrm{SU}(N_f)$ space and is related to the pion field as $U = \exp(i\lambda^j \pi^j)$ with $j = 1 \sim N_f^2 - 1$.

Each term in the effective Lagrangian is ordered in terms of the number of derivatives and the pion mass divided by the typical scale of chiral symmetry breaking, $\Lambda_\chi = 4\pi f_\pi / \sqrt{N_f}$; namely, $\mathcal{L}^{(l)}(U) \sim O((m_\pi/\Lambda_\chi)^{l_1} (\partial/\Lambda_\chi)^{l_2})$ with $l_1 + l_2 = l$. Lorentz invariance, locality and chiral symmetry are the general constraints with which we can form an explicit construction of $\mathcal{L}^{(l)}(U)$. For example, $\mathcal{L}^{(2)}$ is simply given by

$$\mathcal{L}^{(2)} = \frac{f^2}{4} \mathrm{tr}\left(\partial_\mu U \partial^\mu U^\dagger + h^\dagger U + U^\dagger h \right), \tag{2.63}$$

where tr is taken over the flavor indices. The free parameters associated with $\mathcal{L}^{(l)}(U)$, such as f and h in Eq. (2.63), are determined by experimental inputs order by order in the chiral expansion. Once the parameters are fixed, we can make a systematic prediction of other physical quantities (Leutwyler, 2001b). As will be shown in Chapter 7, this method is also useful in analyzing the thermodynamic properties of hadrons and their interactions, although applicability is limited to low temperatures (Gerber and Leutwyler, 1989).

The QCD sum rules

This approach relates the hadronic properties to the quark and gluon condensates in the QCD vacuum. The method is a generalization of the Thomas–Reiche–Kuhn sum rule in quantum mechanics, in which various spectral sum rules are derived on the basis of the completeness of state vectors (Ring and Schuck, 2000). In QCD, the dispersion relation and the Wilson's operator product expansion are the basic ingredients necessary to derive sum rules (Shifman *et al.*, 1979). For example, the time-ordered correlation function of the electromagnetic current in the momentum-space satisfies the subtracted dispersion relation:

$$\Pi(q^2) = \frac{1}{\pi} \int_0^\infty \frac{\rho(s)}{s - q^2 - i\delta}\, ds - \text{subtraction}, \tag{2.64}$$

where the last factor on the right-hand side is the subtraction term required to make the integral finite; $\rho(s) = \mathrm{Im}\Pi(s)$ carries all the information about the hadronic spectrum coupled to the electromagnetic current and is called the *spectral function* (see Section 4.4). By taking an asymptotic expansion of both sides by $1/q^2$ in the deep-Euclidean region ($q^2 \to -\infty$), we may relate certain integrals of $\rho(s)$ to the vacuum condensates.

Assuming that the spectral function has a sharp resonance at low s and a smooth continuum at high s, and taking appropriate values of the vacuum condensates, one can extract the masses of hadrons, couplings and electromagnetic properties, etc., with reasonable accuracy (Colangelo and Khodjamirian, 2001). Generalizing the method to finite temperature and/or density is also possible, and is useful to make a connection between the spectral integrals and the condensates in the medium (Hatsuda and Lee, 1992; Hatsuda *et al.*, 1993).

The lattice QCD

A key concept of the lattice QCD approach is to discretize the Euclidean space-time, x_i $(i = 1, 2, 3, 4)$, into a lattice, $x_i = an_i$, where a is the lattice constant. Then the quark field, $q(n)$, is defined on a lattice site labeled by n and the gluon field, $U_\mu(n) = \exp(iagA_\mu(n))$, is defined on a lattice link labeled by n and μ. The lattice constant, a, serves as a natural ultraviolet cutoff required to regularize the theory in a gauge-invariant manner (Wilson, 1974).

The QCD partition function on the lattice may be schematically written as a discretized form:

$$Z[0] = \int \prod_{n,\mu} [dU_\mu(n) d\bar{q}(n) dq(n)] \mathrm{e}^{-[S_{\rm g}(U)+S_q(q,\bar{q},U)]}. \tag{2.65}$$

The explicit form of the gluon action, $S_{\rm g}$, and that of the quark action, S_q, will be given in Chapter 5. A systematic expansion, such as the weak coupling and strong coupling expansions on the lattice, may be developed, and one can carry out the direct numerical integration of Eq. (2.65) using the method of important sampling (Creutz, 1985). Because of the recent increase in computer power, as well as theoretical developments, lattice QCD simulations have become the most powerful technique utilized in solving QCD from first principles.

A considerable amount of work on QCD phase transition at a finite temperature has been done on the lattice (Karsch, 2002). For example, the critical temperature, $T_{\rm c}$, and critical energy density, $\varepsilon_{\rm crit}$, for a system with massless u and d quarks ($N_f = 2$) are extracted as (see Section 3.8)

$$T_{\rm c} \sim 175 \text{ MeV}, \quad \varepsilon_{\rm crit} \sim 1 \text{ GeV fm}^{-3}. \tag{2.66}$$

We will discuss the basic notion and applications of lattice QCD in more detail in Chapter 5.

Exercises

2.1 Classical Yang–Mills equation. Derive the classical equations of motion, Eqs. (2.8) and (2.9) from the classical Lagrangian density, $\mathcal{L}_{\rm cl}$, in Eq. (2.1).

2.2 Gauge invariance.

(1) Show that $\mathcal{L}_{\rm cl}$ is invariant under the gauge transformation in Eq. (2.10).

(2) Show that the measure $[dA][d\bar{q}\, dq]$ is invariant under the gauge transformation Eq. (2.10).

(3) Derive the explicit form of the Faddeev–Popov determinant, $\Delta_{\rm FP}[A]$, for the covariant gauge Eq. (2.17). What about the case for the axial gauge?

(4) Show the invariance of $\mathcal{L}$ in Eq. (2.19) under the BRST transformation Eqs. (2.20) and (2.21).

2.3 Running coupling and scale parameter.

(1) Solve the flow equation, Eq. (2.27), in the leading order of the β-function, $\beta(g) = -\beta_0 g^3$, and show that Eq. (2.30), without the ln ln term, is indeed the solution.

(2) Derive the following relation between the renormalization scale, κ, and the QCD scale parameter in the leading order:

$$\Lambda_{\text{QCD}} = \kappa \exp(-1/2\beta_0 g^2).$$

(3) Repeat the above derivation with the next-to-leading order β-function, $\beta(g) = -\beta_0 g^3 - \beta_1 g^5$.

2.4 **Running mass.** Solve the flow equation, Eq. (2.32), in the leading order, $\gamma_m(g) = \gamma_{m0} g^2$, and show that Eq. (2.34) is indeed the solution.

2.5 **Pauli and Landau magnetism.** The Pauli paramagnetism arises from spin polarization of electrons in a magnetic field, while the Landau diamagnetism arises from the cyclotron motion of electrons in a magnetic field. Refer to a textbook on solid state physics and derive the following relation between the two magnetic susceptibility: $\chi_{\text{Landau}} = -\chi_{\text{Pauli}}/3 < 0$.

2.6 **Noether's theorem.** The partition function, $Z[J=0]$, in Eq. (2.18) must be invariant under the change of variable $q(x) \to q'(x)$. Making the change of variables, such as in Eq. (2.42), with space-time-dependent parameters, $\theta^j_{\text{V,A}}(x)$, and demanding that $Z[J=0]$ is invariant under this change, derive the partial conservation laws of vector and axial currents in Eqs. (2.43) and (2.44). This constitutes the functional integral derivation of Noether's theorem.

2.7 **Axial anomaly.** By carefully defining the Jacobian (called the Fujikawa Jacobian) of the functional measure $[d\bar{q}\, dq]$ under the change of variables, and by expanding the quark field in terms of the eigenfunctions of the Dirac operator, $i\not{D}$, show that there appears the axial anomaly term for the divergence of the flavor-singlet axial current, Eq. (2.45) (Fujikawa, 1980a, b).

2.8 **Trace anomaly.** Let us derive the trace anomaly equation, Eq. (2.47), in a heuristic manner. (For a rigorous derivation, see Collins *et al.* (1977) and Nielsen (1977)). For simplicity, we consider the vacuum expectation value of Eq. (2.47) in the chiral limit $m = 0$.

(1) The partition function of the gauge field without gauge-fixing is given by

$$Z = e^{iW} = \int [d\bar{A}]\, e^{-\frac{i}{4g^2}\bar{F}^2},$$

where the QCD coupling constant, g, is absorbed in the field as $\bar{A}^a_\mu \equiv gA^a_\mu$ and $\bar{F}^a_{\mu\nu} \equiv gF^a_{\mu\nu}$. Show that W is related to the vacuum energy density, ε_{vac}, as $W = -\varepsilon_{\text{vac}} V_4$, with V_4 being the space-time four-volume. By differentiating Z with respect to g, show that $\partial\varepsilon_{\text{vac}}/\partial g = \langle \bar{F}^2 \rangle/(-2g^3)$.

(2) Renormalization group (RG) invariance of the partition function, Eq. (2.24), or equivalently the RG invariance of ε_{vac}, implies

$\kappa d\varepsilon_{\text{vac}}/d\kappa = (\kappa\partial/\partial\kappa + \beta\partial/\partial g)\varepsilon_{\text{vac}} = 0$. Show that this relation, together with the fact that ε_{vac} has mass-dimension four, leads to $\partial\varepsilon_{\text{vac}}/\partial g = -(4/\beta)\langle\varepsilon_{\text{vac}}\rangle$.

(3) Due to the Lorentz invariance of the vacuum, the trace of the energy-momentum tensor is related to ε_{vac} as $\langle T^{\mu}_{\mu}\rangle = 4\varepsilon_{\text{vac}}$ (Section 2.5). Combining this with the results in (1) and (2), derive the desired relation, $\langle T^{\mu}_{\mu}\rangle = (\beta/2g^3)\langle\bar{F}^2\rangle = (\beta/2g)\langle F^2\rangle$.

2.9 Nambu–Goldstone theorem and soft pion theorem.

(1) Consider the matrix element of the axial current, $J^a_{5\mu}$, in Eq. (2.44):

$$\langle 0|J^a_{5\mu}|\pi^b(k)\rangle \equiv if_\pi k_\mu \, \mathrm{e}^{-ik\cdot x}\delta^{ab}.$$

Here a and b are flavor indices running from 1 to $N_f^2 - 1$; $|0\rangle$ is the vacuum state normalized as $\langle 0|0\rangle = 1$, while the one-pion state obeys the covariant normalization, $\langle\pi^a(k)|\pi^b(k')\rangle = 2E(k)\delta^{ab}(2\pi)^3\delta^3(k-k')$ with $E(k) = \sqrt{\boldsymbol{k}^2 + m_\pi^2}$. The factor $f_\pi(\simeq 93\,\text{MeV})$ is called the pion decay constant since the above matrix element appears in calculating the weak decay of the pion, such as $\pi^- \to \mu^- + \bar{\nu}_\mu$. Consider the chiral limit where the current quark masses (m) are all zero. Show that $f_\pi m_\pi = 0$ holds by using the conservation of the axial current. If the pion exists and couples to the axial current with non-vanishing magnitude, it implies that the pion mass is zero when $m = 0$. This is the simplest proof of the Nambu–Goldstone theorem. For a general proof applicable to the system at finite temperature, see Section 7.2.2.

(2) Starting from the matrix element $\langle 0|\mathrm{T}J^a_{5\mu}(x)\hat{O}(y)|0\rangle$, where $\hat{O}$ is an arbitrary operator at y and T is the time-ordered product (Eq. (4.47)), deduce the following soft pion theorem in the chiral limit ($m = 0$):

$$\lim_{k\to 0}\langle\pi^a(k)|\hat{O}|0\rangle = \frac{-i}{f_\pi}\langle 0|[Q^a_5, \hat{O}]|0\rangle,$$

where $Q^a_5(= \int d^3x\, J^a_{5,0})$ is the axial charge operator. Generalize the idea to the arbitrary matrix element, $\langle\alpha|\mathrm{T}J^a_{5\mu}(x)\hat{O}(y)|\beta\rangle$, and study the subtleties when baryons are involved in the states. For further details, consult, for example, Chap. 19 in Weinberg (1996).

(3) Consider the matrix element $\langle\pi^a(k=0)|\hat{H}_{\text{QCD}}|\pi^b(k=0)\rangle$, where $\hat{H}_{\text{QCD}}$ is the QCD Hamiltonian. Using the soft pion theorem twice, derive the Gell-Mann–Oakes–Renner relation given in Eqs. (2.51) and (2.52). Note that $\hat{H}_{\text{QCD}}$ commutes with Q^a_5, except for the quark mass term, $\int d^3x\, \bar{q}(x)mq(x)$. Note also that we are using the covariant normalization of the state vectors.

3

Physics of the quark–hadron phase transition

At low temperature (T) and low baryon density (ρ), QCD exhibits dynamical breaking of chiral symmetry and confinement, which are both intimately related to the non-perturbative structure of the QCD vacuum, as we have seen in Chapter 2. On the other hand, if T and/or $\rho^{1/3}$ are much larger than the QCD scale parameter $\Lambda_{\text{QCD}} \sim 200\,\text{MeV}$, the QCD running coupling constant given in Eq. (2.30), $\alpha_s(\kappa \sim T,\rho^{1/3})$, becomes small. Furthermore, the long-range color electric force receives plasma screening and becomes short-ranged. This is why the system at T and/or $\rho^{1/3}$ is expected to behave as a weakly interacting gas of quarks and gluons: the quark–gluon plasma (QGP) (Collins and Perry, 1975).

The above considerations suggest that the QCD vacuum undergoes a phase change at some values of T and ρ. In fact, various model approaches and numerical simulations of QCD strongly indicate the existence of a transition from the hadronic phase to the quark–gluon phase, which is relevant to the early Universe at the age of $\sim 10^{-5}$ s where $T \sim 170\,\text{MeV} \sim 10^{12}$ K, and the deep interior of neutron stars where $\rho \sim$ several $\times \rho_{\text{nm}} \sim 10^{12}\,\text{kg cm}^{-3}$. It may also be realized at and after the impact of the relativistic nucleus–nucleus collisions. Such experiments are conducted by using the Super Proton Synchrotron (SPS) at the European Organization for Nuclear Research (CERN), the Relativistic Heavy Ion Collider (RHIC) at Brookhaven National Laboratory (BNL) and the planned Large Hadron Collider (LHC) at CERN.

In this chapter, we discuss the basic concepts of the quark–hadron phase transition in a hot environment ($T \gg \rho^{1/3}$, Λ_{QCD}). The phase transition in a dense environment ($\rho^{1/3} \gg T$, Λ_{QCD}) will be discussed in Chapter 9 with respect to the physics of compact stars.

3.1 Basic thermodynamics

In this section, we briefly summarize the basic statistical and thermodynamic relations, following standard textbooks such as the one by Landau and Lifshitz (1980).

For a statistical system in equilibrium with volume V, temperature T and chemical potential μ, we introduce the grand-canonical density operator, $\hat{\rho}$, the grand-canonical partition function, $Z(T, V, \mu)$, and the grand potential, $\Omega(T, V, \mu)$, in the natural units $k_B = \hbar = c = 1$:

$$\hat{\rho} = \frac{1}{Z}\, e^{-(\hat{H}-\mu\hat{N})/T}, \tag{3.1}$$

$$\begin{aligned} Z(T, V, \mu) &= \mathrm{Tr}\; e^{-(\hat{H}-\mu\hat{N})/T} \\ &= \sum_n \langle n| e^{-(\hat{H}-\mu\hat{N})/T} |n\rangle \equiv e^{-\Omega(T,V,\mu)/T}. \end{aligned} \tag{3.2}$$

Here $\hat{H}$ and $\hat{N}$ are the Hamiltonian operator and the number operator, respectively. The trace, Tr, is taken over a complete set of quantum states labeled by n. Note that $\mathrm{Tr}\ \hat{\rho} = 1$ holds by definition. We may also introduce an entropy operator,

$$\hat{S} = -\ln\, \hat{\rho}. \tag{3.3}$$

Since the thermal average of an arbitrary operator, $\hat{O}$, is given by $\langle \hat{O} \rangle = \mathrm{Tr}[\hat{\rho}\hat{O}]$, the averaged energy, E, the averaged particle number, N, and the averaged entropy, S, are given by

$$E = \langle \hat{H} \rangle, \quad N = \langle \hat{N} \rangle, \quad S = \langle \hat{S} \rangle = -\mathrm{Tr}\,[\hat{\rho}\ \ln\hat{\rho}]\,. \tag{3.4}$$

Equation (3.2), combined with Eq. (3.4), yields the following thermodynamic relations:

$$\Omega(T, V, \mu) = E - TS - \mu N, \tag{3.5}$$

$$d\Omega = -S\ dT - P\ dV - N\ d\mu, \tag{3.6}$$

$$dE = T\ dS - P\ dV + \mu\ dN. \tag{3.7}$$

Equations (3.5) and (3.6) are obtained by the definition of S together with the definition of the pressure, $P = -d\Omega/dV|_{T,\mu}$. Equation (3.7) which is called the *first law of thermodynamics*, is an immediate consequence of Eqs. (3.5) and (3.6). (See Exercise 3.1(1).)

To consider a system with a fixed number of particles and/or pressure, it is useful to introduce the Helmholtz free energy, $F(T, V, N)$, and the Gibbs free energy, $G(T, P, N)$. They are different from the grand potential, $\Omega(T, V, \mu)$, only in their independent parameters, and are therefore obtained as Legendre transforms of $\Omega(T, V, \mu)$:

$$F(T, V, N) = \Omega + \mu N = E - TS, \tag{3.8}$$

$$dF(T, V, N) = -S\, dT - P\, dV + \mu\, dN, \tag{3.9}$$

$$G(T, P, N) = F + PV, \tag{3.10}$$

$$dG(T, P, N) = -S\, dT + V\, dP + \mu\, dN. \tag{3.11}$$

Since $\Omega(T, V, \mu)$ $(G(T, P, N))$ has only one extensive variable V (N), it is linearly proportional to V (N). Then the coefficient, which is an intensive variable, is extracted from Eq. (3.6) (Eq. (3.11)) to give

$$\Omega = -PV, \quad G = \mu N. \tag{3.12}$$

The Gibbs–Duhem relation, $S\, dT - V\, dP + N\, d\mu = 0$, is a direct consequence of Eqs. (3.6) and (3.12).

For a spatially uniform system, it is convenient to introduce the energy density, $\varepsilon = E/V$, the number density, $n = N/V$, and the entropy density, $s = S/V$. Then, after using Eq. (3.12), Eqs. (3.5), (3.6) and (3.7) are rewritten as follows:

$$-P = \varepsilon - Ts - \mu n, \tag{3.13}$$

$$dP = s\, dT + n\, d\mu, \tag{3.14}$$

$$d\varepsilon = T\, ds + \mu\, dn. \tag{3.15}$$

The following relations are also useful (Exercise 3.1(2)):

$$\varepsilon = \frac{T}{V}\left(\left.\frac{\partial \ln Z}{\partial \ln T}\right|_{V,\mu} + \left.\frac{\partial \ln Z}{\partial \ln \mu}\right|_{V,T} \right), \quad P = T \left.\frac{\partial \ln Z}{\partial V}\right|_{T,\mu},$$
$$s = \frac{1}{V}\left(1 + \frac{\partial}{\partial \ln T}\right) \ln Z \Big|_{V,\mu}. \tag{3.16}$$

The *second law of thermodynamics* (the principle of increase of entropy) implies that $F(T, V, N)$, if its arguments are fixed, takes a minimum at equilibrium. Similar statements also hold for $G(T, P, N)$ and $\Omega(T, V, \mu)$. For example, consider a system (labeled by "sys") coupled to a heat reservoir (labeled by "res") at temperature T. Suppose that "sys + res" departs from equilibrium by a small variation $E_{\rm sys}$, with $V_{\rm sys}$ and $N_{\rm sys}$ fixed. Because of the energy conservation, $\delta E_{\rm sys} = -\delta E_{\rm res}$, and the second law of thermodynamics, $\delta S_{\rm sys} + \delta S_{\rm res} \leq 0$, we find that

$$\delta F_{\rm sys} = \delta(E_{\rm sys} - TS_{\rm sys}) = -T\delta(S_{\rm sys} + S_{\rm res}) \geq 0. \tag{3.17}$$

Namely, the Helmholtz free energy has its minimum at equilibrium. A similar proof holds for G (for fixed T, P, N) and for Ω (for fixed T, V, μ) (Exercise 3.2(1)).

One can also study the stability of a system on the basis of the second law of thermodynamics. The resultant relations are called *thermodynamic inequalities*. Among others, some useful relations are

$$C_{\rm P} \geq C_{\rm V} \geq 0, \quad \kappa_{\rm T} \geq 0, \tag{3.18}$$

where $C_{\rm V} = T\partial S/\partial T|_{V,N}$ ($C_{\rm P} = T\partial S/\partial T|_{P,N}$) is the specific heat with constant volume (constant pressure) and $\kappa_{\rm T} = -(1/V)\partial V/\partial P|_{T,N}$ is the isothermal compressibility (Exercise 3.2(2)).

If the temperature of the system approaches zero, only a few quantum states become available in the summation in Z, which leads to the *third law of thermodynamics*, namely $S = -\partial\Omega/\partial T$ approaches a finite constant as $T \to 0$.

If two thermodynamic systems I and II have thermal and chemical contact with each other, the equilibrium conditions are written as follows:

$$P_{\rm I} = P_{\rm II}, \quad T_{\rm I} = T_{\rm II}, \quad \mu_{\rm I} = \mu_{\rm II}. \tag{3.19}$$

They are obtained by maximizing the total entropy, $dS_{\rm I} + dS_{\rm II} = 0$, under the conditions that the total volume, energy and particle numbers are fixed: $dV_{\rm I} + dV_{\rm II} = 0$, $dE_{\rm I} + dE_{\rm II} = 0$, $dN_{\rm I} + dN_{\rm II} = 0$ (Exercise 3.3(1)).

If chemical reactions such as $\rm A + B \leftrightarrow C$ (or equivalently $\rm A + B - C = 0$) take place, the last condition in Eq. (3.19) should be modified. The generalized reaction may be given by (Exercise 3.3(2))

$$\nu_1 {\rm A}_1 + \nu_2 {\rm A}_2 + \cdots = \sum_j \nu_j {\rm A}_j = 0, \tag{3.20}$$

where ${\rm A}_j$ ($j = 1, 2, \ldots$) distinguish the different particles and ν_j counts the number of particles (with appropriate sign) participating in the single reaction. For example, $\rm 2p + 2n \leftrightarrow {}^4He$ implies $\rm A_1 = p$, $\rm A_2 = n$, $\rm A_3 = {}^4He$ and $\nu_1 = \nu_2 = 2$, $\nu_3 = -1$. The condition for chemical equilibrium then becomes

$$\sum_j \nu_j \mu_j = 0, \tag{3.21}$$

where μ_j is a chemical potential for the jth particle. This is called the *generalized law of mass action*. In Chapter 9, we will consider the phase equilibrium of cold quark matter with cold neutron matter. In this case, the reaction $3q \leftrightarrow \mathrm{N(nucleon)}$ is relevant, and chemical equilibrium is realized when $3\mu_q = \mu_{\mathrm{N}}$.

3.2 System with non-interacting particles

The grand-canonical partition function for non-interacting massive bosons with d internal degrees of freedom is given by

$$Z_{\mathrm{B}} = \prod_k \left(\sum_{l=0}^{\infty} \mathrm{e}^{-l(E(k)-\mu)/T} \right)^d = \prod_k \left(1 - \mathrm{e}^{-(E(k)-\mu)/T}\right)^{-d}. \tag{3.22}$$

Here the infinite product is taken for all possible momentum states and l is the occupation number for each quantum state with energy $E(k) = \sqrt{\boldsymbol{k}^2 + m^2}$. For anti-particles, μ should be replaced by $-\mu$. The factor Z_{B} in Eq. (3.22) is finite as long as $\mu \le m$. The case $\mu = m$ is related to the physics of the Bose–Einstein condensation (BEC), which are actively studied both experimentally and theoretically in atomic systems (Pethick and Smith, 2001).

For $\mu = 0$, the grand potential is given by

$$\frac{\Omega_{\mathrm{B}}(T, V, 0)}{V} = d \int \frac{d^3k}{(2\pi)^3} T \ln\left(1 - \mathrm{e}^{-E(k)/T}\right) \tag{3.23}$$

$$= -d \int \frac{d^3k}{(2\pi)^3} \frac{1}{3} \boldsymbol{v}\cdot\boldsymbol{k} \frac{1}{e^{E(k)/T} - 1} \tag{3.24}$$

$$\underset{m=0}{\longrightarrow} -d \frac{\pi^2}{90} T^4, \tag{3.25}$$

where $\boldsymbol{v} = \partial E / \partial \boldsymbol{k} = \boldsymbol{k}/E$ is the velocity of the particle. Equation (3.24) yields an intuitive interpretation of the pressure, $P(= -\Omega/V)$, to be the averaged momentum transfer perpendicular to the unit area of a wall of the container per unit time. Equation (3.25) is obtained by using the integral formula in Exercise 3.4(1). The black body radiation of the photon (*the Stefan–Boltzmann (SB) law*) corresponds to $d = 2$ (two polarization directions) in Eq. (3.25). It is straightforward to derive the pressure, P, the energy density, ε, and the entropy density, s, from Eq. (3.25).

The grand-canonical partition function for non-interacting massive fermions with d internal degrees of freedom is given by

$$Z_{\mathrm{F}} = \prod_k \left(\sum_{l=0,1} \mathrm{e}^{-l(E(k)-\mu)/T} \right)^d = \prod_k \left(1 + \mathrm{e}^{-(E(k)-\mu)/T}\right)^d, \tag{3.26}$$

where d counts the spin and other internal degrees of freedom. For anti-particles, μ should be replaced by $-\mu$. When $\mu = 0$, the grand potential is written as

$$\frac{\Omega_{\rm F}(T,V,0)}{V} = -d\int \frac{d^3k}{(2\pi)^3} T\ln\left(1+{\rm e}^{-E(k)/T}\right) \tag{3.27}$$

$$= -d\int \frac{d^3k}{(2\pi)^3}\frac{1}{3}\boldsymbol{v}\cdot\boldsymbol{k}\frac{1}{e^{E(k)/T}+1} \tag{3.28}$$

$$\underset{m=0}{\longrightarrow} -d\frac{7}{8}\frac{\pi^2}{90}T^4. \tag{3.29}$$

This is the Stefan–Boltzmann law for fermions. The factor $7/8$ in Eq. (3.29) reflects the fact that the Bose–Einstein (BE) distribution, $n_{\rm B}$, at $\mu = 0$ is larger than the Fermi–Dirac (FD) distribution, $n_{\rm F}$, at $\mu = 0$ for fixed $E(k)$ (Exercise 3.4(2)):

$$n_{\rm B}(k) = \frac{1}{{\rm e}^{(E(k)-\mu)/T}-1}, \quad n_{\rm F}(k) = \frac{1}{{\rm e}^{(E(k)-\mu)/T}+1}. \tag{3.30}$$

3.3 Hadronic string and deconfinement

The deconfinement transition at finite T may be understood from the property of a hadronic string. To see this, let us first remember the *energy–entropy argument* relating to phase transitions at finite T (Goldenfeld, 1992). Consider the Helmholtz free energy, $F = E - TS$. At low temperature, F is dominated by the internal energy, E, i.e. the system favors an ordered state with small E. On the other hand, at high temperature, F is dominated by the entropy factor, $-TS$, i.e. a disordered state with large entropy is favored. This implies that there should be a phase transition from the ordered state to the disordered state at a certain T.

Let us apply this argument to the string picture of hadrons (Fig. 3.1) and demonstrate the possible existence of a deconfinement phase transition (Patel, 1984). For simplicity, we consider an open string with string tension K. To make the calculation tractable, we discretize the D-dimensional space into a square lattice with lattice spacing a. The partition function of a string defined on the lattice links is given by

$$Z = \sum_L \sum_{\rm config} \exp\left(-\frac{LKa}{T}\right), \tag{3.31}$$

where we have assumed that one end of the string is attached to a certain lattice site and the sum over "config" is taken for all possible shapes of the string with length La.

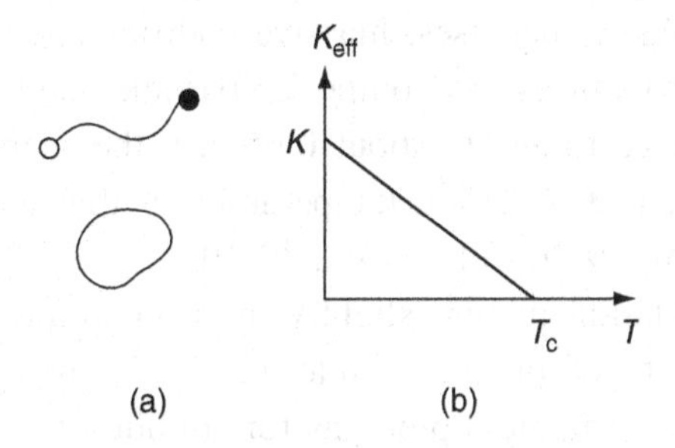

Fig. 3.1. (a) An open string corresponding to the $q\bar{q}$ meson and a closed string corresponding to the glueball. (b) Effective string tension, K_{eff}, as a function of T.

Since the string defined on the lattice may be considered as a non-backtracking random walk, the number of strings with length L is $(2D-1)^L$. Therefore,

$$Z = \sum_L \exp\left[-\frac{1}{T}(LKa - LT\ln(2D-1))\right]. \tag{3.32}$$

The free energy of the open string is given by (Exercise 3.5)

$$F = E - TS = LKa - TL\ln(2D-1). \tag{3.33}$$

As T increases, the entropy factor from the fluctuating string starts to increase, and the partition function Z becomes singular at

$$T_c = \frac{Ka}{\ln(2D-1)}. \tag{3.34}$$

This is a sign of the phase transition. An effective string tension may be defined as

$$K_{\mathrm{eff}} = K - \frac{T}{a}\ln(2D-1), \tag{3.35}$$

which is illustrated in Fig. 3.1. Since K_{eff} decreases as T approaches T_c, the q and $\bar{q}$ attached to the end points of the open string lose correlation towards the critical point. This suggests a phase transition to the quark–gluon plasma at T_c. Taking $D=3$, $K=0.9\,\mathrm{GeV\,fm^{-1}}$ (Eq. (2.58)) and choosing a to be, for example, 0.5 fm, we have $T_c \simeq 280\,\mathrm{MeV}$. We have not considered the string breaking due to the thermal excitations of $q\bar{q}$ pairs. Such an effect reduces T_c.

3.4 Percolation of hadrons

In the discussion in the previous section, the effects of dynamical $q\bar{q}$ pairs, which are excited from the vacuum at finite T, have been neglected. An intuitive picture of the QCD phase transition at finite T with a special emphasis on this point has been illustrated in Fig. 1.5. At low T, only the pion (the lightest hadron) is thermally excited. As T increases, massive hadrons (such as K, η, ρ, ω, ...) are excited too. Since hadrons are composite particles and have finite size (for example the pion charge radius is about 0.65 fm), the thermal hadrons start to overlap with each other at a certain temperature so that the system becomes a color-conductor (Baym, 1979; Celik *et al.*, 1980).

To make this percolation picture slightly more quantitative, let us first evaluate the number density of pions at finite T, $n_\pi(T)$, using the Bose–Einstein distribution. Assuming a massless pion gas for simplicity, we obtain

$$n_\pi(T) = 3\int \frac{d^3k}{(2\pi)^3}\frac{1}{e^{k/T}-1} = \frac{3\zeta(3)}{\pi^2}T^3. \tag{3.36}$$

Here the factor 3 on the right-hand side reflects three possible states (π^+, π^-, π^0); $\zeta(3) \simeq 1.202$ is the Riemann zeta function $\zeta(n=3)$. To obtain the final expression, we have used the integral formula for $I_4^-(0)$ in Exercise 3.4(1).

We assume further that a single pion occupies a volume $V_\pi = (4\pi/3)R_\pi^3$ with a radius R_π ($\simeq$0.65 fm). In the percolation problem of classical inter-penetrating spheres randomly distributed in three spatial dimensions (the continuum percolation) (Isichenko, 1992; Stauffer and Aharony, 1994), the critical density of particles multiplied by the volume of the sphere is about 0.35, which is substantially smaller than the close-packing situation. Applying this to our case of percolation transition from color-insulator to color-conductor, the critical temperature is given by

$$n_\pi(T_c)\cdot V_\pi = 0.35 \quad\rightarrow\quad T_c = \left(\frac{\pi}{4\zeta(3)}\right)^{1/3}\frac{0.35}{R_\pi} \simeq 186 \text{ MeV}. \tag{3.37}$$

Note that the close-packing takes place at $T_{cp} = 1.4T_c \simeq 263$ MeV. Although Eq. (3.37) is a crude estimate based on a simple model, it may be compared with the result of first principle lattice QCD simulations in Eq. (2.66) for $N_f = 2$.

3.5 Bag equation of state

So far, we have seen how two toy models may be used to approach the critical point from the low-T side. To describe low-T and high-T phases simultaneously, we discuss in this section an approach based on the bag model. The basic idea of the bag model was briefly introduced in Section 2.6.

In the following, we take the chiral limit ($m_q = 0$) for simplicity. Then the dominant excitation in the hadronic phase is the massless pion, while that in the quark–gluon plasma is the massless quark and gluon. At extremely low temperature, $T \ll \Lambda_{QCD}$, the typical momenta of pions are small and the interactions among pions are suppressed by the powers of $T/4\pi f_\pi$ (see Eq. (2.62)). At extremely high T, the typical momenta of quarks and gluons are high, and the running coupling $\alpha_s(\kappa \sim T)$ becomes weak due to asymptotic freedom. Therefore, one may assume a free pion gas (free quark–gluon gas) in the low- (high-) T limit as a first approximation.

Pressure, energy density and entropy density of the massless pion gas are easily obtained from the grand potential given in Eq. (3.25):

$$P_{\rm H} = d_\pi \frac{\pi^2}{90} T^4, \tag{3.38}$$

$$\varepsilon_{\rm H} = 3 d_\pi \frac{\pi^2}{90} T^4, \tag{3.39}$$

$$s_{\rm H} = 4 d_\pi \frac{\pi^2}{90} T^3. \tag{3.40}$$

Here d_π is the number of massless Nambu–Goldstone bosons in N_f flavors, as discussed in Chapter 2:

$$d_\pi = N_f^2 - 1. \tag{3.41}$$

Since Eqs. (3.38)–(3.40) determine the thermodynamic properties of the system, they are called the *equations of state* (EOS).

In the quark–gluon plasma (QGP) phase, we have

$$P_{\rm QGP} = d_{\rm QGP} \frac{\pi^2}{90} T^4 - B, \tag{3.42}$$

$$\varepsilon_{\rm QGP} = 3 d_{\rm QGP} \frac{\pi^2}{90} T^4 + B, \tag{3.43}$$

$$s_{\rm QGP} = 4 d_{\rm QGP} \frac{\pi^2}{90} T^3. \tag{3.44}$$

Here $d_{\rm QGP}$ is an effective degeneracy factor of the quarks and gluons in the QGP phase:

$$d_{\rm QGP} = d_{\rm g} + \frac{7}{8} d_q, \tag{3.45}$$

$$d_{\rm g} = 2_{\rm spin} \times (N_c^2 - 1), \tag{3.46}$$

$$d_q = 2_{\rm spin} \times 2_{q\bar{q}} \times N_c \times N_f. \tag{3.47}$$

The factor 7/8 in Eq. (3.45) originates from the difference in statistics: see Eqs. (3.25) and (3.29).

Table 3.1. *Degeneracy factors of the "pion," d_π, the quark, d_q, the gluon, d_g, and $d_{QGP} = d_g + \frac{7}{8} d_q$ for $N_c = 3$ and with massless N_f flavors.*

N_f	0	2	3	4
d_π	0	3	8	15
d_g	16	16	16	16
d_q	0	24	36	48
d_{QGP}	16	37	47.5	58

The parameter B is the bag constant previously introduced in Eq. (2.56) in Chapter 2. It parametrizes the difference between the energy density of the true QCD vacuum and the perturbative QCD vacuum: $B = \varepsilon_{pert} - \varepsilon_{vac} > 0$. Because of the Lorentz invariance of the vacuum at zero temperature and density, the vacuum expectation value of the energy-momentum tensor is given by $\langle T^{\mu\nu} \rangle = \text{diag}(\varepsilon, P, P, P) \propto g^{\mu\nu}$. Therefore, B also represents the difference in pressure: $B = P_{vac} - P_{pert}$. Unless general relativity is involved, it is a matter of convenience to put B into the grand potential of the hadronic phase or into that of the QGP phase . We choose to put it in the QGP phase here, so that the grand potential is normalized to be zero at $T = 0$. In Table 3.1, d_π and d_{QGP} for $N_c = 3$ and $N_f = 0, 2, 3, 4$ are summarized. The degeneracy factor increases by an order of magnitude from the hadronic phase to the QGP phase due to the liberation of color degrees of freedom.

We are now in a position to describe the phase transition in the bag model. The pressures of the two phases are shown as a function of T, in Fig. 3.2(a). The lines with arrows indicate the actual behavior of P. The hadronic pressure is larger than the QGP pressure and is favored at low T because of $B > 0$. On the other hand, the QGP phase is favored at high T because of its large degeneracy factor,

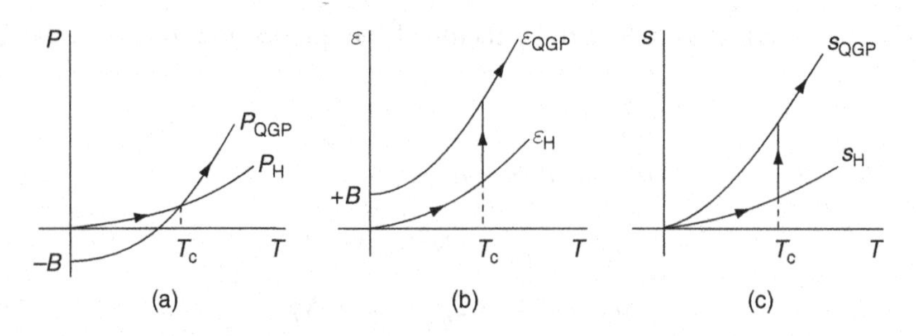

Fig. 3.2. The equations of state in the bag model at finite T with zero chemical potential: (a) the pressure, (b) the energy density, (c) the entropy density. The arrows show how the system evolves as an adiabatic increase of T.

$d_{\rm QGP} \gg d_\pi$. The critical point is obtained from the phase equilibrium condition, Eq. (3.19), for $\mu = 0$:

$$P_{\rm H}(T_{\rm c}) = P_{\rm QGP}(T_{\rm c}), \tag{3.48}$$

which leads to

$$T_{\rm c}^4 = \frac{90}{\pi^2} \frac{B}{d_{\rm QGP} - d_\pi}. \tag{3.49}$$

Using the value $B^{1/4} \sim 220\,\text{MeV}$ in Eq. (2.56) with $N_c = 3$, we obtain $T_{\rm c}(N_f = 2) \sim 160\,\text{MeV}$, which may be compared with the result of first-principle lattice QCD simulations in Eq. (2.66) for $N_f = 2$ (Exercise 3.6).

The energy density and the entropy density as a function of T are also shown in Figs. 3.2(b) and (c), in which sudden jumps at $T_{\rm c}$ are seen. This is the typical behavior of the *first order phase transition*, where the derivative of the grand potential with respect to its arguments (e.g. T) is discontinuous across the phase boundary. In the present case, the entropy density, $s = \partial P/\partial T$, is discontinuous at $T = T_{\rm c}$. (See Chapter 6 for more discussions on the general theory of phase transitions.)

The latent heat, L, released at $T_{\rm c}$ is given by

$$\begin{aligned} L &\equiv T_{\rm c}\left(s_{\rm QGP}(T_{\rm c}) - s_{\rm H}(T_{\rm c})\right) \\ &= \varepsilon_{\rm QGP}(T_{\rm c}) - \varepsilon_{\rm H}(T_{\rm c}) = 4B. \end{aligned} \tag{3.50}$$

Since $L \gg \varepsilon_{\rm H}(T_{\rm c})$, the critical energy density required to realize QGP is estimated to be

$$\varepsilon_{\rm crit} \equiv \varepsilon_{\rm QGP}(T_{\rm c}) \sim 4B \sim 1.2\ \text{GeV fm}^{-3}, \tag{3.51}$$

which is an order of magnitude larger than the energy density of normal nuclear matter, $\varepsilon_{\rm nm} \simeq 0.15\,\text{GeV fm}^{-3}$; see Eq. (11.3).

Although the bag model seems to capture the essential feature of the phase transition, the effects of the interactions among particles are completely neglected. Such interactions can be taken into account in perturbative theory, at least at extremely low T, using chiral perturbation theory, and at extremely high T using perturbative QCD. The physics near the critical point, however, may well be modified by non-perturbative interactions. To consider the phenomena near the critical point more rigorously, the lattice QCD simulations (to be discussed in Chapter 5) and the renormalization group method (to be discussed in Chapter 6) are necessary.

3.6 Hagedorn's limiting temperature

As long as T is substantially smaller than the pion decay constant, $f_\pi = 93\,\mathrm{MeV}$, the chiral perturbation theory discussed in Chapter 2 provides a systematic improvement of the EOS in Eqs. (3.38)–(3.40). This is possible because the typical momentum of the massless pion, $k \sim 3T$, constitutes a small dimensionless parameter, $k/4\pi f_\pi \sim T/4f_\pi$, in the chiral limit. We quote here only the results of the next-to-leading order calculation for $N_f = 2$ in the chiral limit (Gerber and Leutwyler, 1989):

$$P_{\mathrm{H}}/P_{\mathrm{SB}} = 1 + \frac{T^4}{36 f_\pi^4} \ln \frac{\Lambda_{\mathrm{p}}}{T} + O(T^6), \tag{3.52}$$

$$\varepsilon_{\mathrm{H}}/\varepsilon_{\mathrm{SB}} = 1 + \frac{T^4}{108 f_\pi^4} \left(7 \ln \frac{\Lambda_{\mathrm{p}}}{T} - 1 \right) + O(T^6), \tag{3.53}$$

$$s_{\mathrm{H}}/s_{\mathrm{SB}} = 1 + \frac{T^4}{144 f_\pi^4} \left(8 \ln \frac{\Lambda_{\mathrm{p}}}{T} - 1 \right) + O(T^6), \tag{3.54}$$

where P_{SB}, $\varepsilon_{\mathrm{SB}}$ and s_{SB} are the leading order Stefan–Boltzmann EOS given in Eqs. (3.38)–(3.40), and $\Lambda_{\mathrm{p}} (= 275 \pm 65\,\mathrm{MeV})$ is a quantity related to the pion–pion scattering.

As the temperature exceeds 100 MeV, the contributions from massive hadrons other than the pion start to become important. A crude estimate of such contributions may be given by $n(T)/T^3$ (the number density of hadrons divided by T^3), as shown in Fig. 3.3. The interactions among hadrons are neglected in the figure. The "massive resonances" denote contributions from hadrons other than

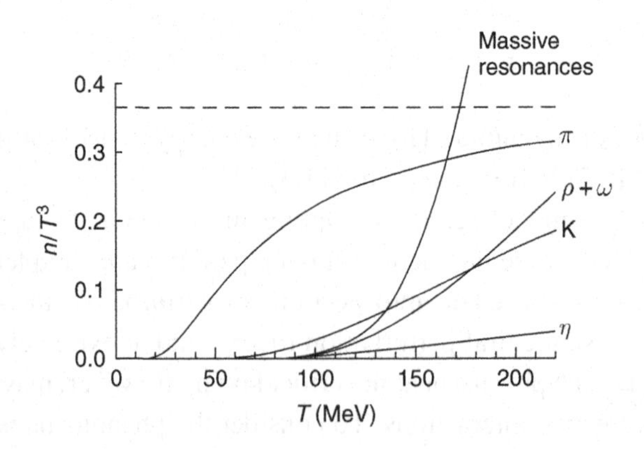

Fig. 3.3. The number density of hadrons, $n(T)$, calculated from the Bose–Einstein and Fermi–Dirac distributions at finite T. The dashed line is a contribution from massless pions. The figure is adapted from Gerber and Leutwyler (1989).

π, K, η, ρ and ω (Appendix H). Although the pion is the only relevant excitation for $T < 100\,\mathrm{MeV}$, other contributions start to dominate for $T > 160\,\mathrm{MeV}$.

A natural question to ask is why so many massive hadrons can be excited at high T. The reason is that the state density, $\tau(M)$, of hadrons with mass M may increase exponentially as M increases, which compensates for the Boltzmann suppression factor, $\exp(-M/T)$ (Hagedorn, 1985):

$$\tau(M \to \infty) \to \frac{c}{M^a}\, \mathrm{e}^{+M/T_0}, \tag{3.55}$$

$$n_{\mathrm{tot}}(T) = \int_0^\infty dM\; n(T; M)\; \tau(M), \tag{3.56}$$

where $n_{\mathrm{tot}}(T)$ is the total number density of hadrons in the system and $n(T; M)$ is the number density of a hadron with mass M. From the fit to the observed hadronic spectrum, $a = 2\text{–}3$ and $T_0 = 150\text{–}200\,\mathrm{MeV}$. We are no longer allowed to neglect the massive hadrons at $T \sim T_0$. Indeed, the integral in Eq. (3.56) diverges when $T > T_0$, namely T_0 is considered to be the ultimate temperature of the hadron gas. Even if we put heat into the system from outside, it is consumed to excite resonances and cannot contribute to an increase in temperature. This is called *Hagedorn's limiting temperature*. In QCD, we can go beyond this limiting temperature since hadrons have a finite size and may overlap with each other to form the quark–gluon plasma. This is simply the percolation we have already discussed in Section 3.4.

3.7 Parametrized equation of state

In this section, we try to construct an EOS beyond the bag model to study the generic features of the QCD phase transition at finite T (Asakawa and Hatsuda 1997). A key observation here is that the pressure and the energy density are simply related to the entropy density $s(T)$ as follows:

$$P(T) = \int_0^T s(t)\; dt, \tag{3.57}$$

$$\varepsilon(T) = Ts(T) - P(T), \tag{3.58}$$

where $P(T)$ is normalized as $P(T=0) = 0$. One may also obtain the sound velocity, c_{s}, from $s(T)$, as shown in Appendix G.1:

$$c_{\mathrm{s}}^2 = \frac{\partial P}{\partial \varepsilon} = \left[\frac{\partial \ln s(T)}{\partial \ln T}\right]^{-1}. \tag{3.59}$$

Therefore, if one can make a reasonable parametrization of $s(T)$, other quantities are obtained automatically. The simplest possible parametrization satisfying the thermodynamic inequality $\partial s(T)/\partial T \geq 0$ (Eq. (3.18)) and the third law of thermodynamics $s(T \to 0) \to$ constant is

$$s(T) = f(T)s_{\rm H}(T) + (1 - f(T))s_{\rm QGP}(T), \tag{3.60}$$

$$f(T) = \frac{1}{2}\left[1 - \tanh\left((T - T_{\rm c})/\Gamma\right)\right]. \tag{3.61}$$

Here $s_{\rm H}(T) = 4d_\pi(\pi^2/90)T^3$ ($s_{\rm QGP}(T) = 4d_{\rm QGP}(\pi^2/90)T^3$) is the entropy density of the non-interacting pion gas (the quark–gluon plasma) in the low- (high-) T limit; $s(T)$ exhibits a jump in the interval $|T - T_{\rm c}| < \Gamma$.

The EOS obtained from Eqs. (3.57) and (3.58) combined with Eq. (3.60) has several advantages over the bag EOS, Eqs. (3.38)–(3.44), in the sense that (i) the phenomenological parameter, such as B, does not have to be introduced, and (ii) not only the first order phase transition, but also a smooth crossover, can be treated in a thermodynamically consistent way.

In Fig. 3.4(a), $\varepsilon(T)/T^4$ and $3P(T)/T^4$ are shown as functions of T for $\Gamma/T_{\rm c} = 0.05$ and $N_f = 2$ (Exercise 3.7). Note that $3P/T^4$ increases rather slowly above $T_{\rm c}$ because $P(T)$ is an integral of $s(T)$ and cannot be discontinuous at $T_{\rm c}$. On the other hand, $\varepsilon(T)/T^4$ increases rapidly around $T_{\rm c}$, reflecting a rapid increase of $s(T)$ at $T_{\rm c}$. In Fig. 3.4(b), $c_{\rm s}^2$, calculated from Eq. (3.59), is shown in the chiral limit (the solid line). Since P (ε) is a slowly (rapidly) varying function of T across the critical point, the sound velocity slows down. This is the reason why $c_{\rm s}^2$ drops suddenly in the narrow region $|T - T_{\rm c}| < \Gamma$. Away from $T_{\rm c}$, $c_{\rm s}$ approaches the relativistic limit, $1/\sqrt{3}$. The dashed line in Fig. 3.4(b) illustrates the case for finite quark mass, where the sound velocity approaches zero at $T = 0$.

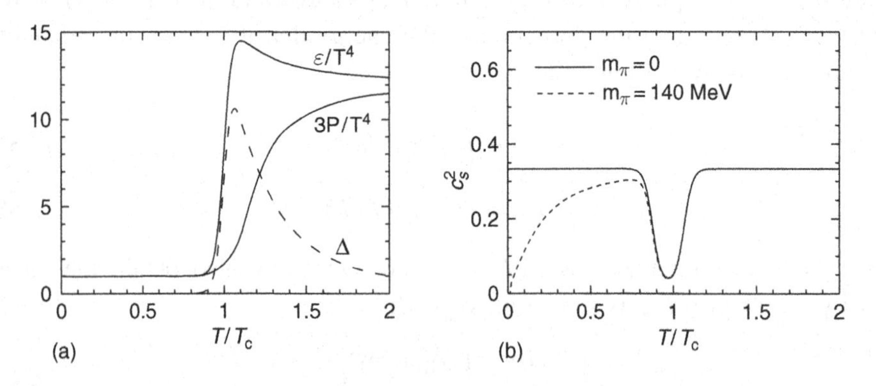

Fig. 3.4. (a) ε/T^4 and $3P/T^4$ obtained from the parametrized entropy density, Eq. (3.60), with $\Gamma/T_{\rm c} = 0.05$ and $N_f = 2$; $\Delta \equiv (\varepsilon - 3P)/T^4$ is shown by the dashed line. (b) Sound velocity squared as a function of T with the same parametrization for the entropy. The figures are adapted from Asakawa and Hatsuda (1997).

The generic features discussed above provide us with a good guide when we examine and interpret the EOS obtained from the lattice QCD simulations in Section 3.8.

3.8 Lattice equation of state

The ultimate way to study the QCD phase transition is the numerical simulation defined on a Euclidean lattice with box size L and lattice constant a. Following the first attempts to find the deconfinement phase transition at finite T in $SU_c(2)$ Yang–Mills theory (Kuti *et al.*, 1981; McLerran and Svetitsky, 1981a), there have been many improvements both in numerical techniques and computer power. Some of the recent results of the EOS at finite T on the lattice are shown in the following. A more detailed account of the basic formulation of lattice QCD will be given in Chapter 5.

Figure 3.5 presents ε/T^4 and $3P/T^4$ for the $SU_c(3)$ pure Yang–Mills theory (gauge theory without fermions or equivalently $N_f = 0$) (Okamoto *et al.*, 1999). The curves behave in a similar fashion to those discussed in Section 3.7; ε/T^4 is suppressed below T_c, and exhibits a large jump in a narrow interval in temperature, while $3P/T^4$ has a smooth change across T_c.

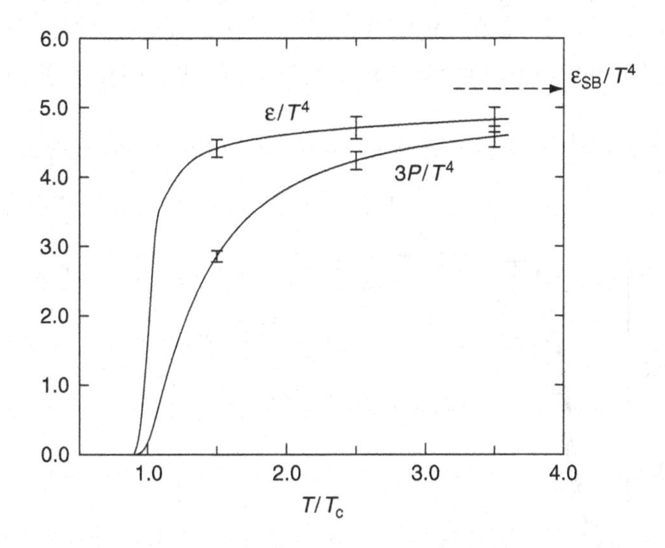

Fig. 3.5. Equations of state for the pure Yang–Mills theory in Monte Carlo simulations. The error bars indicate uncertainties from statistical errors in the Monte Carlo simulations as well as systematic errors due to the extrapolation to the infinite volume limit and the continuum limit. The dashed arrow indicates the Stefan–Boltzmann limit of the energy density. The figure is adapted from Okamoto *et al.* (1999).

In the pure Yang–Mills theory, massive glueballs (having masses more than 1 GeV) can only be excited below T_c. Therefore, ε and P are suppressed much more than those expected for the pion gas. Above T_c, the system is expected to be in a deconfined gluon plasma. The arrow in Fig. 3.5 shows the Stefan–Boltzmann limit, ε_{SB}/T^4, corresponding to the non-interacting gluon gas. The deviation of ε/T^4 from the arrow indicates that the gluons are interacting above T_c. Although the order of the phase transition is not clear from Fig. 3.5, finite size scaling analysis, in which observables as a function of the lattice volume are studied, shows that the phase transition is of first order and thus that there is a discontinuous jump in ε (Fukugita *et al.* 1989, 1990). This has indeed been anticipated from a theoretical argument concerning the center symmetry in pure Yang–Mills theories (Yaffe and Svetitsky, 1982). We will come back to this point in Chapter 5. The critical temperature of the phase transition turns out to be $T_c/\sqrt{K} \simeq 0.650$, with K being the string tension (Okamoto *et al.*, 1999). Taking the value of K given in Eq. (2.58), we obtain

$$T_c(N_f = 0) \simeq 273 \text{ MeV}, \tag{3.62}$$

with a statistical and systematic errors not more than a few percent.

Shown in Fig. 3.6 is the result of lattice QCD simulations with dynamical quarks ($N_f \neq 0$) (Karsch, 2002). Quark masses employed in the figure are $m_{u,d}/T = 0.4$ for $N_f = 2$, $m_{u,d,s}/T = 0.4$ for $N_f = 3$ and $m_{u,d}/T = 0.4$, $m_s/T = 1.0$ for $N_f =$ "2 + 1." Sudden jumps of the energy density at T_c are seen in all three cases.

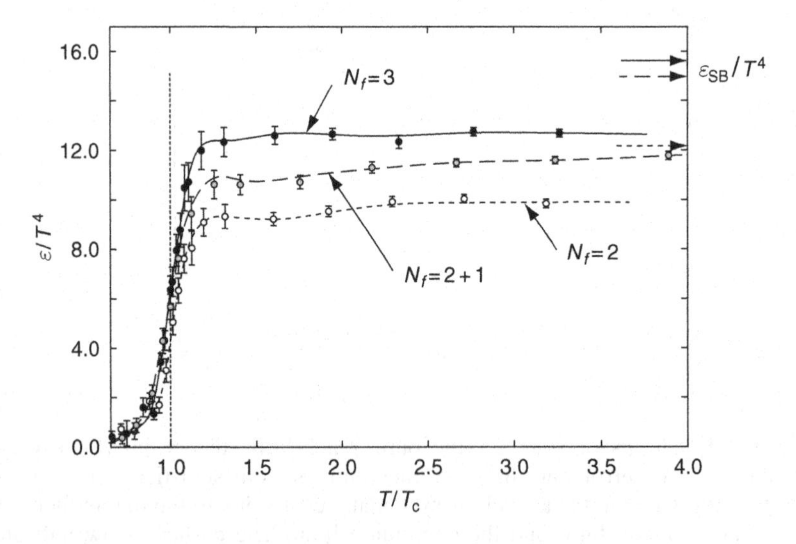

Fig. 3.6. The energy density of QCD with dynamical quarks in lattice Monte Carlo simulations. The figure is adapted from Karsch (2002).

The deviations from the Stefan–Boltzmann limit are also seen at high T. The critical temperatures extrapolated into the chiral limit ($m_q = 0$) are found to be

$$T_c(N_f = 2) \simeq 175 \text{ MeV}, \quad T_c(N_f = 3) \simeq 155 \text{ MeV}, \tag{3.63}$$

with at least ± 10 MeV statistical and systematic errors at the time of writing. The corresponding critical energy density can be read off from Fig. 3.6 as

$$\varepsilon_{\text{crit}} \sim 1 \text{ GeV fm}^{-3}. \tag{3.64}$$

The behavior of the chiral condensate, $\langle \bar{q}q \rangle$, on the lattice shows evidence of a second (first) order transition for $N_f = 2$ ($N_f = 3$). This is consistent with the results of renormalization group analysis (Pisarski and Wilczek, 1984). We will discuss the physics of chiral symmetry restoration in more detail in Chapter 6.

The amazing thing is that bag EOS as well as the parametrized EOS discussed in the previous sections capture the essential features of the lattice QCD results shown in this section. Finding the EOS in hot QCD is an essential step when studying matter in the early Universe and in relativistic heavy ion collisions. The Einstein equation (the relativistic hydrodynamics equation) determines the space-time evolution of the hot plasma for the former (the latter), and will be discussed more in Chapter 8 (Chapter 11).

Exercises

3.1 Thermodynamic relations.

(1) Derive Eqs. (3.5)–(3.7), starting from Eq. (3.2).

(2) Derive the relations in Eq. (3.16).

3.2 Thermodynamic stability.

(1) Generalize the proof of $\delta F_{\text{sys}} \geq 0$ given in Eq. (3.17) to the case for grand potential Ω and the Gibbs free energy G.

(2) Derive the thermodynamic inequalities, Eq. (3.18).

3.3 Phase equilibrium.

(1) Derive the conditions for phase equilibrium, Eq. (3.19), utilizing $\delta(S_{\text{I}} + S_{\text{II}}) = 0$ and the first law of thermodynamics, Eq. (3.7).

(2) Derive the generalized law of mass action, Eq. (3.21), utilizing the Helmholtz free energy, F, with fixed T and V.

3.4 Bose and Fermi integrals. Consider the following integral:

$$I^{\pm}_{n+1}(y) = \frac{1}{\Gamma(n+1)} \int_0^{\infty} dx \, \frac{x^n}{(x^2+y^2)^{1/2}} \frac{1}{e^{(x^2+y^2)^{1/2}} \pm 1},$$

where $\Gamma(n) = (n-1)!$ is the gamma function.

(1) Taking $y = 0$, show that the following formula is valid for integer n (≥ 2):

$$I^{\pm}_{n+1}(0) = \frac{1}{n}\zeta(n)a^{\pm}_n,$$

with $a^+_n = 1 - 2^{1-n}$, $a^-_n = 1$. Note that $\zeta(n)$ is the Riemann zeta function, $\zeta(n) = \sum_{j=1}^{\infty} j^{-n}$, with $\zeta(2) = \pi^2/6$, $\zeta(3) = 1.202$, $\zeta(4) = \pi^4/90$, $\zeta(5) = 1.037, \ldots$ For $n = 1$, an explicit evaluation shows $I^+_2(0) = \ln 2$.

(2) Show that the factor 7/8 in Eq. (3.29) and Eq. (3.45) arises from the ratio a^+_4/a^-_4 in the above formula.

(3) Derive the recursion relation, $dI^{\pm}_{n+1}(y)/dy = -(y/n)I^{\pm}_{n-1}(y)$. Using this, derive the high-temperature expansion of the grand potential in terms of m/T (Kapusta, 1989):

$$\Omega_{\rm B}(T,V,0)/V$$
$$= -d\left[\frac{\pi^2}{90}T^4 - \frac{m^2T^2}{24} + \frac{m^3T}{12\pi} + \frac{m^4}{64\pi^2}\left(\ln\left(\frac{m^2}{(4\pi T)^2}\right) + C\right) + \cdots\right],$$

$$\Omega_{\rm F}(T,V,0)/V$$
$$= -\frac{7}{8}d\left[\frac{\pi^2}{90}T^4 - \frac{m^2T^2}{42} - \frac{m^4}{56\pi^2}\left(\ln\left(\frac{m^2}{(\pi T)^2}\right) + C\right) + \cdots\right],$$

with $C = 2\gamma - 3/2 \simeq -0.346$, and γ, the Euler constant, defined as

$$\gamma = \lim_{n\to\infty}\left(\sum_{k=1}^{n}\frac{1}{k} - \ln\, n\right) = \sum_{n=2}^{\infty}(-1)^n\frac{\zeta(n)}{n} = 0.5772156649\ldots$$

3.5 Fluctuating string at finite T. Evaluate the partition function, Eq. (3.31), for a closed string. Consult a review (Chandrasekhar, 1943) if necessary to obtain the entropy of the closed string.

3.6 $T_{\rm c}$ in the bag model.

(1) Derive the critical temperature, $T_{\rm c}$, in the bag model, Eq. (3.49).

(2) Taking $N_c = 3$ and assuming that the bag constant B is independent of N_f, study the behavior of $T_{\rm c}$ as a function of N_f.

3.7 EOS from entropy density. Reproduce Fig. 3.4 by using the parametrized form of the entropy density, Eqs. (3.60) and (3.61).

4

Field theory at finite temperature

In this chapter, we summarize the basic idea of thermal perturbation theory and its application to QCD at high temperature. We start with the derivation of the Feynman rules at finite temperature in the imaginary-time (Matsubara) formalism. This is necessary in order to study the static and dynamical properties of the quark–gluon plasma.

We have assumed a non-interacting gas of quarks and gluons at high temperature to derive the bag equation of state in Chapter 3. But, in practice, properties of the plasma constituents are modified due to QCD interactions and they develop collective excitations such as the plasmon (gluon-like mode) and the plasmino (quark-like mode). We derive the dispersion relation of these quasi-particles in the plasma on the basis of the hard thermal loop (HTL) approximation and show that they acquire effective masses of $O(gT)$ in the hot environment. The HTL approximation and the HTL resummation are two of the most important ingredients in the development of a consistent scheme of thermal perturbation theory.

Although the thermal perturbation works well at low orders, it is known to break down if one goes to high enough orders. This is due to the infrared property of the magnetic part of the gluon fields. As an example of such phenomena, we discuss the pressure of the quark–gluon plasma at high T.

An alternative approach to the study of the dynamical properties of a plasma is the kinetic theory, in which coupled equations of "hard" plasma constituents with "soft" collective modes are formulated and solved in a self-consistent manner. In particular, they lead to a straightforward and intuitive derivation of the hard thermal loops originally obtained in diagrammatic method.

4.1 Path integral representation of Z

The transition amplitude from a state $|m\rangle$ at time $t_{\rm I}$ to a state $|n\rangle$ at time $t_{\rm F}$ (the Feynman kernel) is given by

$$K_{nm}(t_{\rm F}, t_{\rm I}) = \langle n|{\rm e}^{-i\hat{H}(t_{\rm F}-t_{\rm I})}|m\rangle. \tag{4.1}$$

The path integral representation of this amplitude is obtained by discretizing the time interval, $t_I < t < t_F$, into small pieces and inserting complete sets as shown in Appendix C. On the other hand, the grand-canonical partition function, Z, introduced in Chapter 3 is given by

$$Z = \sum_n \langle n | e^{-(\hat{H}-\mu\hat{N})/T} | n \rangle . \tag{4.2}$$

This is alternatively obtained by the following replacement in Eq. (4.1): $t_F \to t_I - i/T$, $\hat{H} \to \hat{H} - \mu\hat{N}$ and $|m\rangle \to |n\rangle$. For the path integral representation, one needs to choose a path, C, connecting the initial time, t_I, and the final time, t_F, in the complex time plane ($t = \mathrm{Re}\, t + i\mathrm{Im}\, t$). There is a constraint that $\mathrm{Im}\, t$ on C should not be an increasing function (Mills, 1969; Landsman and van Weert, 1987). (See Exercise 4.1.) A typical path with such a property is shown in Fig. 4.1, where t_I is taken to be at the origin.

A particularly useful path is the straight line connecting $t_I = 0$ and $t_F = -i/T$ on the imaginary axis ($t_0 = 0$ in Fig. 4.1). We call this the Matsubara path since it corresponds to the imaginary-time operator formalism for Z developed by Matsubara (1955) and Ezawa *et al.* (1957). Another useful path is $t_0 > 0$ and $t_1 = -i/(2T)$ in Fig. 4.1, which corresponds to the double-field operator formalism (thermo field dynamics) developed by Takahashi and Umezawa (1996). Since the Matsubara path can be characterized by a real parameter τ ($0 \le \tau \le 1/T$), the partition function for a one-component scalar field, ϕ, is given by

$$Z = \int [d\phi]\, e^{-\int_0^{1/T} d\tau \int d^3x\, \mathcal{L}_E(\phi(\tau,\boldsymbol{x}),\partial\phi(\tau,\boldsymbol{x}))}, \tag{4.3}$$

where $\phi(\tau, \boldsymbol{x})$ and $\mathcal{L}_E$ are, respectively, the scalar field and the Lagrangian density defined in the Euclidean space-time. Integration over ϕ should be carried out with a boundary condition, $\phi(1/T, \boldsymbol{x}) = \phi(0, \boldsymbol{x})$, because the initial state and final state must be the same, as is obvious from Eq. (4.2).

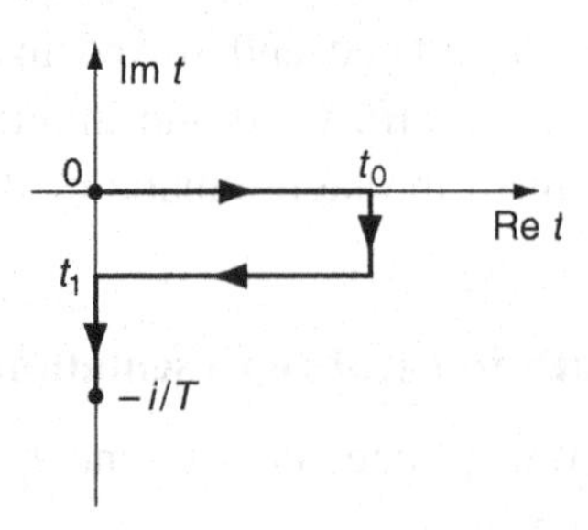

Fig. 4.1. A typical path in the complex time plane.

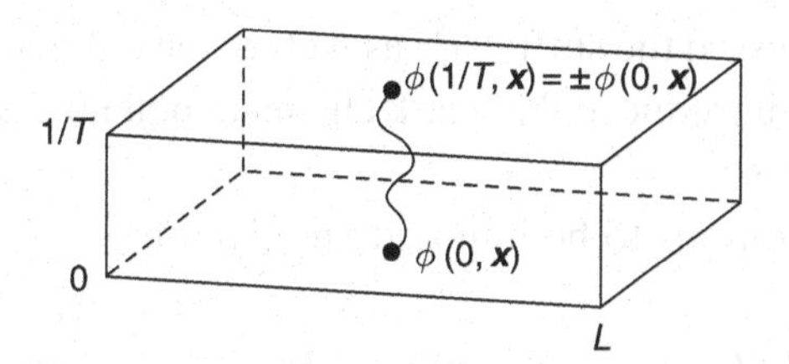

Fig. 4.2. Euclidean space-time volume with spatial size L and temporal size $1/T$. The bosonic (fermionic) field obeys periodic (anti-periodic) boundary conditions in the temporal direction.

Equation (4.3) may be interpreted as a field theory in a four-dimensional slab. Its spatial size is infinite, while the temporal size is $1/T$, as shown in Fig. 4.2. The limit $T \to 0$ corresponds to the Euclidean field theory at zero temperature. Formal correspondence between Eq. (4.3) and the partition function in the Minkowski space with real t and Lagrangian $\mathcal{L}_{\mathrm{M}}$ is clear: one obtains the former from the latter by the following replacements: $t \to -i\tau$, $\phi(t, \boldsymbol{x}) \to \phi(\tau, \boldsymbol{x})$ and $\mathcal{L}_{\mathrm{M}} \to -\mathcal{L}_{\mathrm{E}}$. For fermion fields, $\psi(\tau, \boldsymbol{x})$, the temporal boundary condition should be taken to be anti-periodic, $\psi(1/T, \boldsymbol{x}) = -\psi(0, \boldsymbol{x})$. This originates from an unusual nature of the Trace in Eq. (4.2) written in terms of the Grassmann variables.

Obtaining the grand-canonical partition function, Z, in QCD is a straightforward generalization of Eq. (4.3), if we accept its formal correspondence with the vacuum-to-vacuum amplitude in Eqs. (2.18) and (2.19) in Chapter 2. First, it is convenient to introduce Euclidean vectors having only lower cases:

$$\begin{aligned}
(x_\mu)_{\mathrm{E}} &= (\tau, \boldsymbol{x}), & (\partial_\mu)_{\mathrm{E}} &= (\partial_\tau, \nabla), \\
(\gamma_\mu)_{\mathrm{E}} &= (\gamma_4 = i\gamma^0, \boldsymbol{\gamma}), & (A^a_\mu)_{\mathrm{E}} &= (A_4 = iA^0, \boldsymbol{A}^a), \\
(D_\mu)_{\mathrm{E}} &= (\partial_\mu - igt^a A^a_\mu)_{\mathrm{E}}, & (\mathcal{D}_\mu)_{\mathrm{E}} &= (\partial_\mu - igT^a A^a_\mu)_{\mathrm{E}}, \\
(F_{\mu\nu})_{\mathrm{E}} &= \tfrac{i}{g}([D_\mu, D_\nu])_{\mathrm{E}},
\end{aligned} \tag{4.4}$$

where $\{(\gamma_\mu)_{\mathrm{E}}, (\gamma_\nu)_{\mathrm{E}}\} = -2\delta_{\mu\nu}$. Note that the opposite signs in front of g in the above expressions from those in the Minkowski expressions, Eqs. (2.2), (2.4) and (2.6), stem from the fact that $(D_i)_{\mathrm{E}} \equiv (D^i)_{\mathrm{M}}$.

On making the replacement $t \to -i\tau$ in Eqs. (2.18) and (2.19) and rewriting the result in terms of the variables defined in Eq. (4.4), we obtain

$$Z = \int [dA\, d\bar{q}\, dq\, d\bar{c}\, dc]\, \mathrm{e}^{-\int_0^{1/T} d\tau \int d^3x\, \mathcal{L}}, \tag{4.5}$$

$$\mathcal{L} = \bar{q}(-i\gamma_\mu D_\mu + m + i\gamma_4 \mu)q + \frac{1}{4} F_{\mu\nu} F_{\mu\nu} + \bar{c}\partial_\mu \mathcal{D}_\mu c + \frac{1}{2\xi}(\partial_\mu A_\mu)^2, \tag{4.6}$$

where we have suppressed the suffix "E" as well as the color and spinor indices for simplicity. (We have introduced the quark chemical potential, μ, for completeness in the above formula.)

The boundary conditions to be imposed are given by

$$q(1/T, \boldsymbol{x}) = -q(0, \boldsymbol{x}), \quad \bar{q}(1/T, \boldsymbol{x}) = -\bar{q}(0, \boldsymbol{x}), \tag{4.7}$$

$$A_\mu(1/T, \boldsymbol{x}) = A_\mu(0, \boldsymbol{x}), \tag{4.8}$$

$$c(1/T, \boldsymbol{x}) = c(0, \boldsymbol{x}), \quad \bar{c}(1/T, \boldsymbol{x}) = \bar{c}(0, \boldsymbol{x}). \tag{4.9}$$

Ghost fields obey the periodic boundary condition, although they are Grassmann fields. This is because they are introduced to exponentiate the Faddeev–Popov (FP) determinant, which is a function of A_μ being periodic in τ.

4.2 Black body radiation

Let us study how the black body radiation formula at finite T for non-interacting quarks and gluons is obtained from the QCD partition function. Taking $g = 0$ in $\mathcal{L}$ of Eq. (4.6) and adopting the Feynman gauge ($\xi = 1$), $\mathcal{L}$ becomes

$$\mathcal{L}_0 = \bar{q}(-i\gamma \cdot \partial + m + i\gamma_4 \mu)q - \frac{1}{2} A_\mu \partial^2 A_\mu + \bar{c}\partial^2 c. \tag{4.10}$$

Then, the Gaussian integrals (Appendix C) for the corresponding partition function, Z_0, can be performed as follows:

$$Z_0 \sim \left[\left(\frac{1}{\sqrt{\det \partial^2}} \right)^4 \cdot (\det \partial^2) \right]^{N_c^2 - 1} \cdot \left[\det(-i\gamma \cdot \tilde{\partial} + m) \right]^{N_c N_f} \tag{4.11}$$

$$\sim \left[\left(\frac{1}{\sqrt{\det \partial^2}} \right)^{4-2} \right]^{N_c^2 - 1} \cdot \left[\det(-i\gamma \cdot \tilde{\partial} + m) \right]^{N_c N_f}, \tag{4.12}$$

where "det" implies the determinant in the functional space on which ∂ acts, and $\tilde{\partial}_\nu = \partial_\nu - \mu\delta_{\nu 4}$. The first, second and third "det" in Eq. (4.11) originate from the integrations of the gluon, the ghost and the quark, respectively. Equation (4.12) shows explicitly that the ghost field correctly cancels the unphysical degrees of freedom and leaves two physical polarization states of the gluon: 2 (physical gauge fields) = 4 (full gauge fields) −2 (ghosts).

We now evaluate Z_0 by making a Fourier decomposition of the fields as

$$\phi(\tau, \boldsymbol{x}) = \sqrt{\frac{1}{V/T}} \sum_{n=-\infty}^{+\infty} \sum_{\boldsymbol{k}} \mathrm{e}^{i(k_4\tau + \boldsymbol{k}\cdot\boldsymbol{x})} \, \phi_n(\boldsymbol{k}), \tag{4.13}$$

where ϕ denotes either gluon, ghost or quark fields. The factor V/T is the volume of the Euclidean box in which the fields are defined. Note that k_4 takes discrete values (Matsubara frequencies) because the temporal direction of the box is finite. Depending on the boundary conditions, we obtain

$$-k_4 \equiv \begin{cases} \omega_n = 2n\pi T & \text{for gluon and ghost,} \\ \nu_n = (2n+1)\pi T & \text{for quark,} \end{cases} \tag{4.14}$$

where n takes integer values. (The minus sign in front of k_4 is just a convention.) Note also that $A_{\mu,n}(\boldsymbol{k}) = A_{\mu,-n}(-\boldsymbol{k})$ because $A_\mu(\tau, \boldsymbol{x})$ is real.

The path integral measure is then given by

$$\begin{aligned} [dA \, d\bar{q} \, dq \, d\bar{c} \, dc] &\equiv \mathcal{N} \prod_{\tau, \boldsymbol{x}} dA_\mu \cdot d\bar{c} \cdot dc \cdot d\bar{q} \cdot dq \\ &= \mathcal{N} J \prod_{n, \boldsymbol{k}} dA_{\mu,n}(\boldsymbol{k}) \cdot d\bar{c}_n(\boldsymbol{k}) \cdot dc_n(\boldsymbol{k}) \cdot d\bar{q}_n(\boldsymbol{k}) \cdot dq_n(\boldsymbol{k}), \end{aligned} \tag{4.15}$$

where $\mathcal{N}$ is a constant and J is a Jacobian associated with the change of variables from $\phi(\tau, \boldsymbol{x})$ to $\phi_n(\boldsymbol{k})$. Then we arrive at

$$\begin{aligned} Z_0 = \mathcal{N} J \prod_{n, \mathbf{k}} \Big[\int dA_{\mu,n}(\boldsymbol{k}) \, \mathrm{e}^{-\frac{1}{2} A_{\mu,n}(\boldsymbol{k})(\omega_n^2 + \boldsymbol{k}^2) A_{\mu,n}(\boldsymbol{k})} \\ \times \int d\bar{c}_n(\boldsymbol{k}) dc_n(\boldsymbol{k}) \, \mathrm{e}^{-\bar{c}_n(\boldsymbol{k})(\omega_n^2 + \boldsymbol{k}^2) c_n(\boldsymbol{k})} \\ \times \int d\bar{q}_n(\boldsymbol{k}) dq_n(\boldsymbol{k}) \, \mathrm{e}^{-\bar{q}_n(\boldsymbol{k})(-\gamma_4(\nu_n - i\mu) + \boldsymbol{\gamma}\cdot\boldsymbol{k} + m) q_n(\boldsymbol{k})} \Big] \end{aligned} \tag{4.16}$$

$$\begin{aligned} &= \mathcal{N}' \prod_{n, \boldsymbol{k}} [(\omega_n^2 + \boldsymbol{k}^2)]^{-(N_c^2 - 1)} \\ &\quad \times [((\nu_n - i\mu)^2 + \boldsymbol{k}^2 + m^2)]^{2N_c N_f}, \end{aligned} \tag{4.17}$$

where $\mathcal{N}' = \mathcal{N} \times J \times$ (constant from the Gaussian integration).

To evaluate Eq. (4.17), we differentiate $\ln Z_0$ with respect to $E_g = |\boldsymbol{k}|$ and $E_q = \sqrt{\boldsymbol{k}^2 + m^2}$, and perform the summation over n using the formula

$$\sum_{n=-\infty}^{+\infty} \frac{1}{a^2 + \bar{n}^2} = \frac{\pi}{2a} \times \begin{cases} \coth(\pi a/2) & \text{for } \bar{n} = 2n \\ \tanh(\pi a/2) & \text{for } \bar{n} = 2n+1. \end{cases} \tag{4.18}$$

By integrating the result with respect to $E_{\rm g}$ and E_q, we obtain

$$\begin{aligned}\Omega_0(T, V, \mu) &= -T \ln Z_0 \\ &= V \int \frac{d^3k}{(2\pi)^3} \left[2(N_c^2-1) \left(\frac{E_{\rm g}(\boldsymbol{k})}{2} + T \ln(1-\mathrm{e}^{-E_{\rm g}(\boldsymbol{k})/T}) \right) \right. \\ &\quad + 2N_cN_f \left(-\frac{E_q(\boldsymbol{k})}{2} - T \ln(1+\mathrm{e}^{-(E_q(\boldsymbol{k})-\mu)/T}) \right) \\ &\quad \left. + 2N_cN_f \left(-\frac{E_q(\boldsymbol{k})}{2} - T \ln(1+\mathrm{e}^{-(E_q(\boldsymbol{k})+\mu)/T}) \right) \right]. \end{aligned} \tag{4.19}$$

In the above formula, $2(N_c^2-1) =$ (spin) $\times$ (color) and $2N_cN_f =$ (spin) $\times$ (color) $\times$ (flavor) are the degeneracy factors for the gluon and the quark, respectively; $E_{\rm g}(\boldsymbol{k})/2$ and $-E_q(\boldsymbol{k})/2$ are the zero-point energies. The terms with ln are the entropy of the free quarks and gluons. Careful treatment of integration measure together with $\mathcal{N}$ and J shows that there arises no additive constant in Eq. (4.19). (See Exercise 4.2.)

4.3 Perturbation theory at finite T and μ

To go beyond the leading order, and to take into account thermal and quantum fluctuations, we need to formulate the perturbation theory. First we decompose the Lagrangian density into free and interaction parts:

$$S = \int_0^{1/T} d\tau \int d^3x \, (\mathcal{L}_0 + \mathcal{L}_{\rm I}) = S_0 + S_{\rm I}. \tag{4.20}$$

Then, $\ln Z$ can be expanded as

$$\frac{Z}{Z_0} = \frac{\sum_{n=0}^{\infty} \frac{1}{n!} \int [d\phi] \, (-S_{\rm I})^n \, \mathrm{e}^{-S_0}}{\int [d\phi] \, \mathrm{e}^{-S_0}} \tag{4.21}$$

$$\equiv \sum_{n=0}^{\infty} \frac{1}{n!} \langle (-S_{\rm I})^n \rangle_0 = \exp\left[\sum_{n=1}^{\infty} \frac{1}{n!} \langle (-S_{\rm I})^n \rangle_0^{\rm c} \right], \tag{4.22}$$

where $[d\phi]$ is an abbreviation of the measure $[dA \, d\bar{q} \, dq \, d\bar{c} \, dc]$. Note that $\langle (-S_{\rm I})^n \rangle_0$ implies the thermal average taken with respect to the free action S_0; $\langle (-S_{\rm I})^n \rangle_0^{\rm c}$ denotes the "connected" part of the Feynman diagrams of $\langle (-S_{\rm I})^n \rangle_0$. The linked cluster theorem assures that all the Feynman diagrams with disconnected parts are generated by exponentiating the connected ones (Exercise 4.3).

The perturbative expansion of an expectation value of some operator $\hat{O}$ is formulated in the same manner:

$$\langle \hat{O} \rangle = \frac{\int [d\phi]\, \hat{O}\, \mathrm{e}^{-S}}{\int [d\phi]\, \mathrm{e}^{-S}} \tag{4.23}$$

$$= \sum_{n=0}^{\infty} \frac{1}{n!} \frac{\langle \hat{O} \cdot (-S_{\mathrm{I}})^n \rangle_0}{Z/Z_0} = \sum_{n=0}^{\infty} \frac{1}{n!} \langle \hat{O} \cdot (-S_{\mathrm{I}})^n \rangle_0^{\mathrm{c}}. \tag{4.24}$$

In QCD, $-\mathcal{L}_{\mathrm{I}}$ contains different powers of the coupling constant, $O(g)$ and $O(g^2)$, as shown later in Eqs. (4.34)–(4.36). Therefore, we need to reorganize the expansion Eq. (4.24) further to obtain a correct expansion in terms of g.

4.3.1 Free propagators

To construct Feynman rules, we first need to define free propagators through S_0. For example, the coordinate-space gluon propagator in the Feynman gauge is given by

$$\begin{aligned} D^0_{\mu\nu}(x-y) &= \langle A_\mu(x) A_\nu(y) \rangle_0 \\ &= Z_0^{-1} \int [dA]\, A_\mu(x) A_\nu(y)\, \mathrm{e}^{-\int_0^{1/T} d^4z [-\frac{1}{2} A_\lambda(z) \partial^2 A_\lambda(z)]} \\ &= Z_0^{-1} \left[\frac{\delta}{\delta J_\mu(x)} \frac{\delta}{\delta J_\mu(y)} \int [dA]\, \mathrm{e}^{-\int_0^{1/T} d^4z [-\frac{1}{2} A_\lambda \partial^2 A_\lambda - J_\lambda A_\lambda]} \right]_{J\to 0} \\ &= \left[\frac{\delta}{\delta J_\mu(x)} \frac{\delta}{\delta J_\mu(y)}\, \mathrm{e}^{-\int d^4z [\frac{1}{2} J_\lambda \frac{1}{\partial^2} J_\lambda]} \right]_{J\to 0} \\ &= -\frac{1}{\partial^2} \delta^4(x-y). \end{aligned} \tag{4.25}$$

Here, the J factor introduced in the intermediate steps is an external field taken to be zero at the end. The propagator in the momentum-space is defined through the Fourier transform. If $\phi(x)$ is a periodic (anti-periodic) function in τ, its Fourier transform is given by

$$\phi(\tau, \boldsymbol{x}) = T \sum_n \int \frac{d^3q}{(2\pi)^3}\, \mathrm{e}^{iq_4\tau + i\boldsymbol{q}\cdot\boldsymbol{x}} \phi(q_4, \boldsymbol{q}), \tag{4.26}$$

where $q_4 = -\omega_n$ or $-\nu_n$ according to Eq. (4.14). For simplicity, ϕ in the coordinate-space and that in the momentum-space are distinguished only by their argument.

For later purposes, we introduce the Euclidean 4-vectors

$$Q_\mu = (q_4, \boldsymbol{q}) = (-\omega_n \text{ or } -\nu_n, \boldsymbol{q}), \tag{4.27}$$

$$\tilde{Q}_\mu = (q_4 + i\mu, \boldsymbol{q}) = (-\nu_n + i\mu, \boldsymbol{q}). \tag{4.28}$$

Then the gluon propagator in the momentum-space in the covariant gauge becomes

$$D^0_{\mu\nu}(Q) = \frac{1}{Q^2}\left(\delta_{\mu\nu} - (1-\xi)\frac{Q_\mu Q_\nu}{Q^2}\right), \tag{4.29}$$

with $Q^2 = \omega_n^2 + \boldsymbol{q}^2$.

Carrying out similar calculations for the quark and ghost fields by introducing the Grassmann external fields, we arrive at the following propagators in the coordinate-space:

$$S^0(x-y) = \langle q(x)\bar{q}(y)\rangle_0 = \frac{1}{-i\gamma\cdot\partial + m}\delta^4(x-y), \tag{4.30}$$

$$G^0(x-y) = \langle c(x)\bar{c}(y)\rangle_0 = \frac{1}{\partial^2}\delta^4(x-y); \tag{4.31}$$

and those in the momentum-space:

$$S^0(Q) = \frac{1}{-\gamma_4(\nu_n - i\mu) + \boldsymbol{\gamma}\cdot\boldsymbol{q} + m} = \frac{1}{\gamma\cdot\tilde{Q} + m}, \tag{4.32}$$

$$G^0(Q) = \frac{-1}{\omega_n^2 + \boldsymbol{q}^2} = \frac{-1}{Q^2}. \tag{4.33}$$

4.3.2 Vertices

The explicit form of the quark–gluon, gluon–gluon and ghost–gluon vertices can be easily seen from the definition $\mathcal{L}_\mathrm{I} = \mathcal{L} - \mathcal{L}_0$:

$$-\mathcal{L}_\mathrm{I} = +g\bar{q}\gamma_\mu t^a A^a_\mu q \tag{4.34}$$

$$-g f_{abc}(\partial_\mu A^a_\nu)A^b_\mu A^c_\nu + \frac{g^2}{4} f_{abc} f_{ade} A^b_\mu A^c_\nu A^d_\mu A^e_\nu \tag{4.35}$$

$$+g f_{abc}(\partial_\mu \bar{c})A^b_\mu c^c. \tag{4.36}$$

In the momentum-space, these vertices are translated into the following forms:

$$qqg : +g\gamma_\mu (t^a)_{ji}, \tag{4.37}$$

$$ggg : +ig f_{abc}[\delta_{\mu\nu}(K-P)_\rho + \delta_{\nu\rho}(P-Q)_\mu + \delta_{\rho\mu}(Q-K)_\nu], \tag{4.38}$$

$$gggg : -g^2[f_{abe} f_{cde}(\delta_{\mu\nu}\delta_{\nu\sigma} - \delta_{\mu\sigma}\delta_{\nu\rho}) + ((b,\nu) \leftrightarrow (c,\rho)) + ((b,\nu) \leftrightarrow (d,\sigma))], \tag{4.39}$$

$$ccg : -igP_\mu f_{abc}, \tag{4.40}$$

where the overall minus sign of $-S_\mathrm{I}$ in Eqs. (4.21) and (4.23) is taken into account in the vertices defined above.

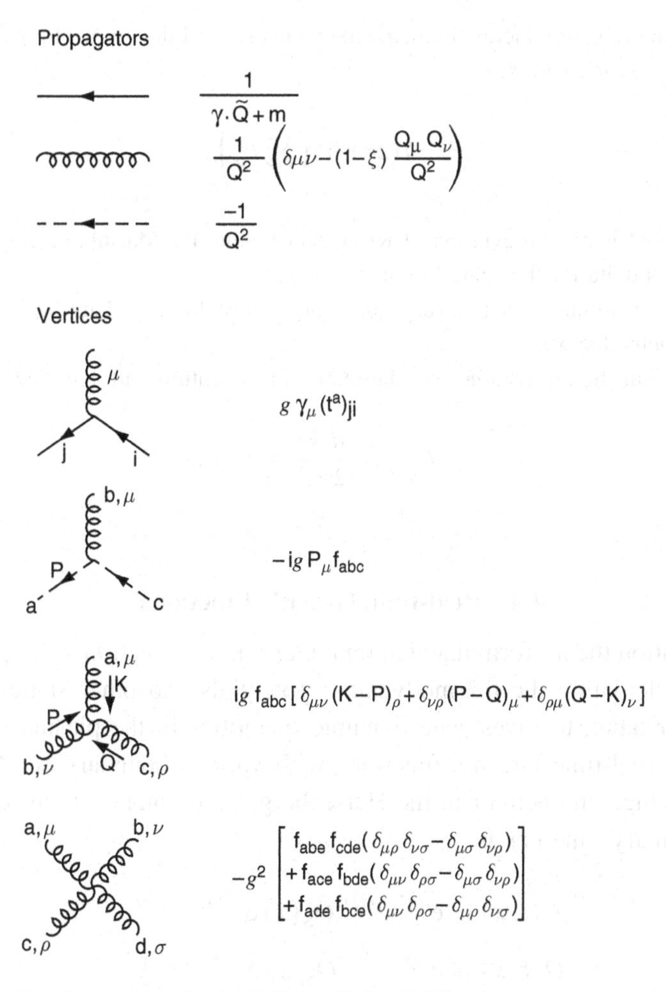

Fig. 4.3. The Feynman rules in QCD at finite T. The four-momentum is defined as $Q_\mu = (q_4, \boldsymbol{q}) = (-\omega_n, \boldsymbol{q})$ for gluons (spiral lines) and ghosts (broken lines), while $Q_\mu = (q_4, \boldsymbol{q}) = (-\nu_n, \boldsymbol{q})$ and $\tilde{Q}_\mu = (q_4 + i\mu, \boldsymbol{q}) = (-\nu_n + i\mu, \boldsymbol{q})$ for quarks (solid lines).

The propagators and vertices, with proper assignment of the momenta, spin and color indices, are summarized in Fig. 4.3.

4.3.3 Feynman rules

With all the above ingredients in place, we are now in a position to write down the Feynman rules.

(a) Draw all the topologically independent connected diagrams for given order in g.
(b) Assign the free propagator summarized in Fig. 4.3 to each internal line.

(c) Assign the vertex factor summarized in Fig. 4.3 and the momentum-conservation factor to each vertex:

$$T(2\pi)^3\delta^4\left(\sum_i P_\mu^i\right), \tag{4.41}$$

where δ^4 is an abbreviation of Kronecker's delta for Matsubara frequencies and Dirac's delta for the spatial momenta.

(d) Assign a minus sign to every quark and ghost loop, and multiply appropriate symmetry factors.

(e) Carry out the integration over the internal momentum with the measure

$$T\sum_n \int \frac{d^3k}{(2\pi)^3} \equiv \int (dK). \tag{4.42}$$

4.4 Real-time Green's functions

The perturbation theory formulated in imaginary time in Section 4.3 is particularly useful for calculating the thermodynamic potentials and other static quantities. On the other hand, to investigate dynamic quantities in the plasma we need to consider the real-time Green's functions with various boundary conditions. Let us first introduce an operator in the Heisenberg representation $\hat{O}$ in real time (t) and in imaginary time (τ) as

$$\hat{O}(t, \boldsymbol{x}) = \mathrm{e}^{i(\hat{H}-\mu\hat{N})t}\hat{O}_{\mathrm{S}}(\boldsymbol{x})\,\mathrm{e}^{-i(\hat{H}-\mu\hat{N})t}, \tag{4.43}$$

$$\hat{O}(\tau, \boldsymbol{x}) = \mathrm{e}^{(\hat{H}-\mu\hat{N})\tau}\hat{O}_{\mathrm{S}}(\boldsymbol{x})\,\mathrm{e}^{-(\hat{H}-\mu\hat{N})\tau}, \tag{4.44}$$

where $\hat{O}_{\mathrm{S}}(\boldsymbol{x})$ is an operator in the Schrödinger representation.

Retarded (R), advanced (A), causal (T) and imaginary time (T_τ) products of operators $\hat{O}_1$ and $\hat{O}_2$ are defined as follows:

$$\mathrm{R}\hat{O}_1(t, \boldsymbol{x})\hat{O}_2(0) = \theta(t)[\hat{O}_1(t, \boldsymbol{x}), \hat{O}_2(0)]_{\mp}, \tag{4.45}$$

$$\mathrm{A}\hat{O}_1(t, \boldsymbol{x})\hat{O}_2(0) = -\theta(-t)[\hat{O}_1(t, \boldsymbol{x}), \hat{O}_2(0)]_{\mp}, \tag{4.46}$$

$$\mathrm{T}\hat{O}_1(t, \boldsymbol{x})\hat{O}_2(0) = \theta(t)\hat{O}_1(t, \boldsymbol{x})\hat{O}_2(0) \pm \theta(-t)\hat{O}_2(0)\hat{O}_1(t, \boldsymbol{x}), \tag{4.47}$$

$$\mathrm{T}_\tau\hat{O}_1(\tau, \boldsymbol{x})\hat{O}_2(0) = \theta(\tau)\hat{O}_1(\tau, \boldsymbol{x})\hat{O}_2(0) \pm \theta(-\tau)\hat{O}_2(0)\hat{O}_1(\tau, \boldsymbol{x}), \tag{4.48}$$

where $[\ ,\ ]_{\mp}$ denotes the commutator (anti-commutator) for the $-$ sign ($+$ sign). If both $\hat{O}_1$ and $\hat{O}_2$ are bosonic (fermionic) operators, the upper (lower) sign is taken in the above definitions.

The thermal averages of these products are called the retarded, advanced, causal (Feynman) and imaginary time (Matsubara) Green's functions, and they are given by $\mathcal{G}^{\mathrm{R}}(t,\boldsymbol{x})$, $\mathcal{G}^{\mathrm{A}}(t,\boldsymbol{x})$, $\mathcal{G}^{\mathrm{F}}(t,\boldsymbol{x})$ and $\mathcal{G}(\tau,\boldsymbol{x})$, respectively. For example,

$$\mathcal{G}^{\mathrm{R}}(t,\boldsymbol{x}) = i\,\mathrm{Tr}\left[\mathrm{e}^{-(\hat{H}-\mu\hat{N})/T}\,\mathrm{R}\hat{O}_1(t,\boldsymbol{x})\hat{O}_2(0)\right] \tag{4.49}$$
$$\equiv \int \frac{d^4k}{(2\pi)^4}\,\mathcal{G}^{\mathrm{R}}(k^0,\boldsymbol{k})\,\mathrm{e}^{-ik_0t+i\boldsymbol{k}\cdot\boldsymbol{x}}\ ,$$

$$\mathcal{G}(\tau,\boldsymbol{x}) = \mathrm{Tr}\left[\mathrm{e}^{-(\hat{H}-\mu\hat{N})/T}\,\mathrm{T}_\tau\hat{O}_1(\tau,\boldsymbol{x})\hat{O}_2(0)\right] \tag{4.50}$$
$$\equiv T\sum_n\int \frac{d^3k}{(2\pi)^3}\,\mathcal{G}(k_4,\boldsymbol{k})\,\mathrm{e}^{ik_4\tau+i\boldsymbol{k}\cdot\boldsymbol{x}}\ ,$$

with $k_4=-\omega_n(-\nu_n)$; $\mathcal{G}^{\mathrm{A}}(k^0,\boldsymbol{k})$ and $\mathcal{G}^{\mathrm{F}}(k^4,\boldsymbol{k})$ are defined in a similar manner.

To see the relation between the retarded and imaginary time Green's functions, we insert a complete set into Eq. (4.49):

$$\mathcal{G}^{\mathrm{R}}(\omega,\boldsymbol{q}) = i\int d^4x\,\mathrm{e}^{-i\omega t+i\boldsymbol{q}\cdot\boldsymbol{x}}\theta(t)\sum_{n,m}\frac{\mathrm{e}^{-(E_n-\mu N_n)/T}}{Z}$$
$$\times\Big[\langle n|\hat{O}_1(0)|m\rangle\langle m|\hat{O}_2(0)|n\rangle\,\mathrm{e}^{-i\omega_{mn}t+i\boldsymbol{P}_{mn}\cdot\boldsymbol{x}}$$
$$\mp\langle n|\hat{O}_2(0)|m\rangle\langle m|\hat{O}_1(0)|n\rangle\,\mathrm{e}^{i\omega_{mn}t-i\boldsymbol{P}_{mn}\cdot\boldsymbol{x}}\Big], \tag{4.51}$$

where $(\omega_{mn},\boldsymbol{P}_{mn})=(E_m-E_n-\mu(N_m-N_n),\boldsymbol{P}_m-\boldsymbol{P}_n)$.

Using the identity for $\delta>0$,

$$\int_0^\infty \mathrm{e}^{i(z+i\delta)t}\,dt = \frac{i}{z+i\delta}, \tag{4.52}$$

and on comparing the real and imaginary parts we immediately obtain the following *spectral representation* or the dispersion relation:

$$\mathcal{G}^{\mathrm{R}}(\omega,\boldsymbol{q}) = \int_{-\infty}^{+\infty}\frac{\rho(\omega',\boldsymbol{q})}{\omega'-\omega-i\delta}\,d\omega', \tag{4.53}$$

with the spectral function

$$\rho(\omega,\boldsymbol{q}) = \frac{1}{\pi}\,\mathrm{Im}\mathcal{G}^{\mathrm{R}}(\omega,\boldsymbol{q}) \tag{4.54}$$
$$=(2\pi)^3\sum_{n,m}\frac{\mathrm{e}^{-(E_n-\mu N_n)/T}}{Z}\langle n|\hat{O}_1(0)|m\rangle\langle m|\hat{O}_2(0)|n\rangle$$
$$\times(1\mp\mathrm{e}^{-\omega_{mn}/T})\delta(\omega-\omega_{mn})\delta^3(\boldsymbol{q}-\boldsymbol{P}_{mn}). \tag{4.55}$$

We have implicitly assumed that the spectral integral on the right-hand side of Eq. (4.53) converges. If this is not the case, we need to take a suitable number of derivatives with respect to ω until the integral converges. The resultant spectral representation is sometimes called the subtracted dispersion relation.

The spectral function, ρ, has the following useful property:

$$\rho(\omega, \boldsymbol{q}) = \rho_{1\leftrightarrow 2}(-\omega, -\boldsymbol{q}), \tag{4.56}$$

where $\rho_{1\leftrightarrow 2}(\omega, \boldsymbol{q})$ is the spectral function obtained by interchanging $\hat{O}_1$ and $\hat{O}_2$.

The spectral representation of the imaginary time Green's function, $\mathcal{G}$, is also obtained by inserting a complete set into Eq. (4.50):

$$\mathcal{G}(q_4, \boldsymbol{q}) = \int_{-\infty}^{\infty} \frac{\rho(\omega', \boldsymbol{q})}{\omega' - i\omega_n(i\nu_n)}\, d\omega'. \tag{4.57}$$

The spectral representation of the advanced Green's function is obtained by a replacement, $\delta \to -\delta$, in Eq. (4.53). Also, it is easy to find the relation

$$\mathcal{G}^{\mathrm{F}}(\omega, \boldsymbol{q}) = \frac{1}{1 \mp \mathrm{e}^{-\omega/T}} [\mathcal{G}^{\mathrm{R}}(\omega, \boldsymbol{q}) \mp \mathrm{e}^{-\omega/T} \mathcal{G}^{\mathrm{A}}(\omega, \boldsymbol{q})]. \tag{4.58}$$

Note that $\mathcal{G}^{\mathrm{R}}$, $\mathcal{G}^{\mathrm{A}}$ and $\mathcal{G}^{\mathrm{F}}$ differ only in their imaginary parts. Equation (4.53) implies that $\mathcal{G}^{\mathrm{R}}$ ($\mathcal{G}^{\mathrm{A}}$) is an analytic function in the upper (lower) half of the complex-ω-planes, while Eq. (4.58) shows that $\mathcal{G}^{\mathrm{F}}$ has poles and cuts in both the upper and lower planes.

A comparison of Eq. (4.53) and Eq. (4.57) tells us that we obtain $\mathcal{G}$ once we know $\mathcal{G}^{\mathrm{R,A}}$:

$$\text{for Im}\ \omega > 0, \qquad \mathcal{G}^{\mathrm{R}}(\omega \to i\omega_n(i\nu_n), \boldsymbol{q}) = \mathcal{G}(q_4, \boldsymbol{q}), \tag{4.59}$$

$$\text{for Im}\ \omega < 0, \qquad \mathcal{G}^{\mathrm{A}}(\omega \to i\omega_n(i\nu_n), \boldsymbol{q}) = \mathcal{G}(q_4, \boldsymbol{q}). \tag{4.60}$$

Suppose that $\mathcal{G}^{\mathrm{R}}(\omega, \boldsymbol{q})$ ($\mathcal{G}^{\mathrm{A}}(\omega, \boldsymbol{q})$) is an analytic function in the upper (lower) half ω-plane that does not grow faster than $\mathrm{e}^{\pi|\omega|}$ for large $|\omega|$. In this case, $\mathcal{G}^{\mathrm{R}}(\omega, \boldsymbol{q})$ ($\mathcal{G}^{\mathrm{A}}(\omega, \boldsymbol{q})$) is obtained as a unique analytic continuation of $\mathcal{G}$ to the upper (lower) half plane owing to Carlson's theorem (Titchmarsh, 1932). (See Exercise 4.4.) Since $\mathcal{G}$ can be calculated in a systematic perturbation theory, as shown in Section 4.3, we can extract the real-time Green's functions from the imaginary-time information (Abrikosov *et al.*, 1959; Fradkin, 1959; Baym and Mermin, 1961). For a generalization of this result to the n-point functions, see Baier and Niégawa (1994).

An alternative approach to obtaining real-time Green's function is a formalism in which perturbation theory can be formulated directly in real time without using analytic continuations. In this case, the complex-time path C in Fig. 4.1 needs to be chosen appropriately. Interested readers should consult Mills (1969), Landsman and van Weert (1987), Takahashi and Umezawa (1996) for details.

The method of the real-time Green's functions and the spectral functions developed in this section will be applied in the following sections in this chapter and in Chapter 7.

4.5 Gluon propagator at high T and zero μ

As an application of the thermal perturbation theory, let us consider the gluon self-energy at high temperature with zero baryon chemical potential, $\mu = 0$. The self-energy is defined as the difference of the full gluon propagator, $D_{\mu\nu}$, and the free one, $D^0_{\mu\nu}$:

$$(D_{\mu\nu}(Q))^{-1} = (D^0_{\mu\nu}(Q))^{-1} + \Pi_{\mu\nu}(Q); \tag{4.61}$$

or, equivalently, $D = D^0 - D^0 \Pi D^0 + D^0 \Pi D^0 \Pi D^0 + \cdots$, as shown in Fig. 4.4. The diagrams corresponding to $\Pi_{\mu\nu}(Q)$ at order g^2 are shown in Fig. 4.5. In this section, we will focus on the hard thermal loop (HTL) approximation, in which

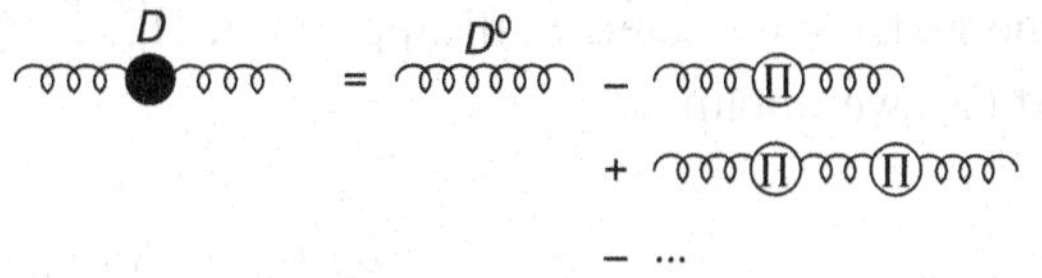

D = D^0 − Π + Π Π − ...

Fig. 4.4. The full gluon propagator, D, in terms of the free gluon propagator, D^0, and the one-particle irreducible self-energy, Π.

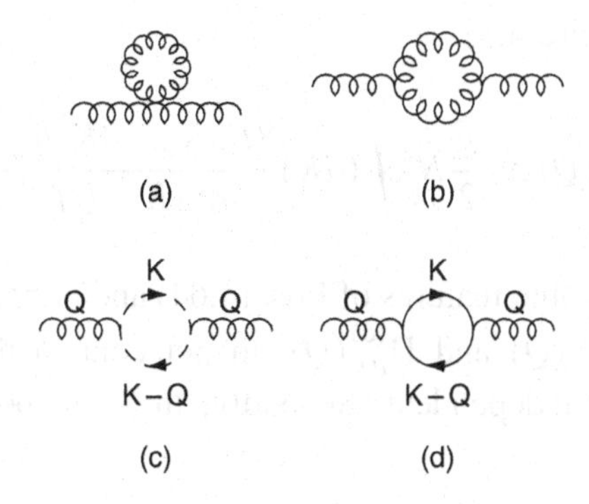

Fig. 4.5. The gluon self-energy in one-loop order: (a) and (b) the gluon loop, (c) the ghost loop and (d) the quark loop.

T is assumed to dominate over all the other scales in the loop. This implies that we need to extract the contributions proportional to g^2T^2 in $\Pi_{\mu\nu}(Q)$.

As an illustration, let us consider the ghost-loop diagram, Fig. 4.5(c);

$$\Pi^{(c)}_{\mu\nu}(Q) \simeq (-)(-)\int (dK)(igK_\mu)(igK_\nu)(-N_c)\frac{-1}{K^2}\frac{-1}{(K-Q)^2} \tag{4.62}$$

$$= g^2 N_c \int (dK)\frac{K_\mu K_\nu}{K^2(K-Q)^2}, \tag{4.63}$$

where $Q_\mu = (q_4, \boldsymbol{q}) = (-\omega_l, \boldsymbol{q})$ and $K_\mu = (k_4, \boldsymbol{k}) = (-\omega_n, \boldsymbol{k})$. On the right-hand side of Eq. (4.62), the first minus sign is from the definition in Eq. (4.61); the second minus sign is from the ghost loop. To pick up the leading high-T contribution in the integral, we have approximated $K_\mu - Q_\mu$ in the ghost–ghost–gluon vertex by K_μ. Here one may wonder why we can make the approximation $K \gg Q$ even for the time component in which Q takes only discrete values. Indeed, the approximation is justified only after making the analytic continuation to the real frequency, $i\omega_l \to \omega$. The factor $-N_c$ is from the color summation, $f_{abc}f_{cba'} = -N_c\delta_{aa'}$, which is obtained from the quadratic Casimir operator in the adjoint representation, $(T^aT^a)_{bc} = C_{\rm A}\delta_{bc}$, with $(T^a)_{bc} = -if_{abc}$ and $C_{\rm A} = N_c$ (see Appendix B.3).

After making a similar calculation for Fig. 4.5(a) and (b) in the HTL approximation, we find that the gauge parameter, ξ, disappears in $\Pi^{(\rm a+b)}_{\mu\nu}(Q)$ (Exercise 4.5). Adding (a), (b) and (c), we obtain

$$\Pi^{(\rm a+b+c)}_{\mu\nu}(Q) \simeq -g^2 N_c \int (dK)\frac{4K_\mu K_\nu - 4K^2\delta_{\mu\nu}}{K^2(K-Q)^2}, \tag{4.64}$$

where $K_\mu = (k_4, \boldsymbol{k}) = (-\omega_n, \boldsymbol{k})$. The contribution from the quark loop is obtained in a similar way (Exercise 4.5):

$$\Pi^{(\rm d)}_{\mu\nu}(Q) \simeq \frac{g^2}{2}N_f \int (dK)\frac{8K_\mu K_\nu - 2K^2\delta_{\mu\nu}}{K^2(K-Q)^2}. \tag{4.65}$$

There are two remarkable features of Eqs. (4.64) and (4.65) in the HTL approximation. First, $\Pi^{(\rm a+b+c)}_{\mu\nu}(Q)$ and $\Pi^{(\rm d)}_{\mu\nu}(Q)$ do not contain the gauge parameter, ξ, and thus are gauge-independent. Secondly, they are both *conserved* in the sense that

$$Q_\mu \Pi^{(\rm a+b+c)}_{\mu\nu}(Q) = 0, \quad Q_\mu \Pi^{(\rm d)}_{\mu\nu}(Q) = 0. \tag{4.66}$$

Equation (4.66) can be easily checked by multiplying Q_μ by Eqs. (4.64) and (4.65), using the relation $K^2-(K-Q)^2 \simeq 2K\cdot Q$ in the HTL approximation, and making the change of variables $K-Q \to K$.

Because of Eq. (4.66), we may introduce two invariant self-energies, $\Pi_{\rm T}$ and $\Pi_{\rm L}$, as follows:

$$\Pi_{\mu\nu} = \Pi_{\rm T}({\rm P_T})_{\mu\nu} + \Pi_{\rm L}({\rm P_L})_{\mu\nu}. \tag{4.67}$$

Here the projection operators, ${\rm P_{T,L}}$, in Euclidean space[1] have the following properties:

$$Q_\mu({\rm P_T})_{\mu\nu} = Q_\mu({\rm P_L})_{\mu\nu} = 0, \tag{4.68}$$

$$({\rm P_T})^2 = ({\rm P_L})^2 = 1, \quad {\rm P_T P_L} = {\rm P_L P_T} = 0, \tag{4.69}$$

$$({\rm P_T}+{\rm P_L})_{\mu\nu} = \delta_{\mu\nu} - \frac{Q_\mu Q_\nu}{Q^2}, \tag{4.70}$$

$$({\rm P_T})_{ij} = \delta_{ij} - \frac{Q_i Q_j}{q^2}, \quad ({\rm P_T})_{44} = ({\rm P_T})_{4j} = ({\rm P_T})_{j4} = 0, \tag{4.71}$$

where $q \equiv |\boldsymbol{q}|$. Then we also have the relations

$$\Pi_{\rm L} = \frac{Q^2}{q^2}\Pi_{44}, \quad \Pi_{\rm T} = \frac{1}{2}(\Pi_{\mu\mu} - \Pi_{\rm L}). \tag{4.72}$$

The gluon propagator in the HTL approximation in the covariant gauge is given by

$${}^*D_{\mu\nu}(Q) = \frac{({\rm P_T})_{\mu\nu}}{Q^2+\Pi_{\rm T}} + \frac{({\rm P_L})_{\mu\nu}}{Q^2+\Pi_{\rm L}} + \xi\frac{Q_\mu Q_\nu}{Q^4}, \tag{4.73}$$

where we have put $*$ preceding $D_{\mu\nu}(Q)$ to emphasize that it is obtained in the HTL approximation.

The evaluation of $\Pi_{\mu\mu}$ and Π_{44} proceeds as follows. For example, from Eq. (4.64), we obtain

$$\Pi_{\mu\mu}^{(\rm a+b+c)} = 4g^2 N_c I_1, \quad \Pi_{44}^{(\rm a+b+c)} = 4g^2 N_c (I_2 - I_1/2), \tag{4.74}$$

with

$$I_1 = \int (dK)\frac{1}{K^2}, \quad I_2 = \int (dK)\frac{k^2}{K^2(K-Q)^2}. \tag{4.75}$$

[1] The projection operators in Minkowski space can be defined in a similar way: $({\rm P_T})_{ij} = -g_{ij} - q_i q_j/|\boldsymbol{q}|^2$; $({\rm P_T})_{0j} = ({\rm P_T})_{i0} = ({\rm P_T})_{00} = 0$; $({\rm P_L})_{\mu\nu} + ({\rm P_T})_{\mu\nu} = -g_{\mu\nu} + q_\mu q_\nu/q_\lambda^2$.

By using the formula shown in Exercise 4.6, we can perform a summation over the Matsubara frequency and obtain an integration with the Bose–Einstein factor, $n_{\rm B}(k) = 1/(e^{k/T} - 1)$. For I_1 it is easily evaluated as

$$I_1 = T\sum_n \int \frac{d^3k}{(2\pi)^3} \frac{1}{\omega_n^2 + k^2} \tag{4.76}$$

$$\to \frac{-2}{2\pi i} \int \frac{d^3k}{(2\pi)^3} \int_{-i\infty+\delta}^{i\infty+\delta} dk_0 \frac{1}{k_0^2 - k^2} \frac{1}{e^{\beta k_0} - 1} \tag{4.77}$$

$$= \frac{1}{2\pi^2} \int_0^\infty dk\, k\, n_{\rm B}(k) = \frac{T^2}{12}, \tag{4.78}$$

where we have picked up only the leading term at high T and used the integral $I_3^-(0)$ in Exercise 3.4. The contribution to the loop integral from the vacuum polarization is not proportional to T^2 and is simply renormalized away. The same result may be obtained by using Eq. (4.18) directly.

For I_2, we obtain

$$I_2 = T\sum_n \int \frac{d^3k}{(2\pi)^3} \frac{k^2}{(\omega_n^2 + k^2)[(\omega_n - \omega_l)^2 + |\boldsymbol{k} - \boldsymbol{q}|^2]} \tag{4.79}$$

$$\to -\int \frac{d^3k}{(2\pi)^3} \frac{k^2}{4E_1E_2} \left[\left(\frac{1}{E_1 - E_2 + i\omega_l} - \frac{1}{E_1 + E_2 + i\omega_l} \right) n_{\rm B}(E_1) \right.$$
$$\left. + (E_2 \leftrightarrow E_1, \omega_l \to -\omega_l)\, n_{\rm B}(E_2) \right], \tag{4.80}$$

where $E_1 \equiv |\boldsymbol{k}|$ and $E_2 \equiv |\boldsymbol{k} - \boldsymbol{q}|$. We have again taken only the term which depends on T. There are two contributions in Eq. (4.80) which produce a $O(T^2)$ term. One is a term proportional to $n_{\rm B}(E_1) + n_{\rm B}(E_2) \sim 2n_{\rm B}(k)$ and the other is proportional to $n_{\rm B}(E_1) - n_{\rm B}(E_2) \sim (E_1 - E_2)\partial n_{\rm B}(k)/\partial k$, with $E_1 - E_2 \sim k - (k - q\cos\theta) = q\cos\theta$. Thus,

$$I_2 = -\frac{1}{8\pi^2} \int_0^\infty dk\, k^2 \int_{-1}^1 dy \left[\frac{y}{y + i\omega_l/q} \frac{\partial n_{\rm B}(k)}{\partial k} - \frac{1}{k} n_{\rm B}(k) \right] \tag{4.81}$$

$$= \frac{I_1}{2} \cdot \left(3 - \frac{i\omega_l}{q} \ln \left| \frac{i\omega_l/q + 1}{i\omega_l/q - 1} \right| \right). \tag{4.82}$$

Combining the equations and making the analytic continuation $i\omega_l \to \omega + i\delta$ with the Minkowski four-momentum, $Q^\mu = (\omega, \boldsymbol{q})$, we obtain

$${}^*D^{\rm R}_{\mu\nu} = \frac{-({\rm P_T})_{\mu\nu}}{Q^2 - \Pi_{\rm T}} + \frac{-({\rm P_L})_{\mu\nu}}{Q^2 - \Pi_{\rm L}} + \xi \frac{Q_\mu Q_\nu}{Q^4}, \tag{4.83}$$

with

$$\Pi_{\rm L}(Q) = -\omega_{\rm D}^2 \frac{Q^2}{q^2}(1 - F(\omega/q)), \tag{4.84}$$

$$\Pi_{\rm T}(Q) = \frac{\omega_{\rm D}^2}{2}\left[1 + \frac{Q^2}{q^2}(1 - F(\omega/q))\right], \tag{4.85}$$

$$F(x) \equiv \frac{x}{2}\left[\ln\left|\frac{x+1}{x-1}\right| - i\pi\theta(1-|x|)\right]. \tag{4.86}$$

Note that $F(x)$ has a cut for $|x| < 1$ and has an imaginary part in this interval; $\omega_{\rm D}$ is the Debye screening mass defined as

$$\omega_{\rm D}^2 = \frac{1}{3}g^2T^2\left(N_c + \frac{1}{2}N_f\right) \qquad \text{(QCD)}, \tag{4.87}$$

$$= \frac{1}{3}e^2T^2 \qquad \text{(QED)}. \tag{4.88}$$

In the uniform medium at rest at finite T, only the rotational invariance O(3) in three-dimensional space remains out of the full Lorentz invariance. Therefore, the non-analytic ratio, such as ω/q, can appear in the self-energies. This causes the phenomenon that the two limits $\omega \to 0$ and $q \to 0$ do not commute.

Let us consider the gluons in the time-like region ($\omega > q$). The long-wavelength limit ($\omega =$ finite, $q = 0$) corresponds to $x \to \infty$. Since $F(x \to \infty) \to 1 + 1/(3x^2) + 1/(5x^4) + \cdots$, we obtain

$$\Pi_{\rm L,T}(\omega, q=0) = \frac{1}{3}\omega_{\rm D}^2 \equiv \omega_{\rm pl}^2, \tag{4.89}$$

$$^*D_{\rm L,T}^{\rm R}(\omega, q=0) = [\omega^2 - \omega_{\rm pl}^2]^{-1}, \tag{4.90}$$

where $^*D_{\rm L,T}^{\rm R}(Q) \equiv [Q^2 - \Pi_{\rm L,T}(Q)]^{-1}$. This implies that the non-static gluon fluctuations (both longitudinal and transverse) in the hot plasma oscillate with a characteristic frequency (the *plasma frequency*), given by

$$\omega_{\rm pl} = \frac{1}{\sqrt{3}}\omega_{\rm D}. \tag{4.91}$$

We now consider the more general case where q is finite. The pole positions of the gluon propagator are obtained from

$$Q^2 = \Pi_{\rm L}(\omega, q), \quad Q^2 = \Pi_{\rm T}(\omega, q). \tag{4.92}$$

Equation (4.92) has real solutions in the time-like region ($\omega > q$) which are called the plasma oscillations, or plasmons. For $q \ll \omega_{\rm D}$,

$$\omega^2 = \omega_{\rm pl}^2 + \frac{3}{5}q^2 + \cdots \quad \text{(longitudinal mode)}, \tag{4.93}$$

$$\omega^2 = \omega_{\rm pl}^2 + \frac{6}{5}q^2 + \cdots \quad \text{(transverse mode)}. \tag{4.94}$$

On the other hand, for large q ($\omega_{\rm D} \ll q \ll T$), both modes approach the light-cone region $\omega \simeq q$. The dispersion relations obtained above in the time-like region ($\omega > q$) are shown by the solid lines in Fig. 4.6(a).

In the space-like region ($\omega < q$), the self-energy has an imaginary part because $F(x)$ in Eq. (4.86) is complex for $|x| < 1$. The physical origin of the imaginary part is simple. Consider Fig. 4.7, in which a gluon with four-momentum, $(\omega, \boldsymbol{q})$, is scattered by the massless constituents inside the plasma. Because of the energy–momentum conservation, this scattering takes place under the following condition:

$$\omega^2 - \boldsymbol{q}^2 = (k_0 - p_0)^2 - (\boldsymbol{k} - \boldsymbol{p})^2 = -2pk\,(1 - \cos\theta) \le 0, \tag{4.95}$$

where θ is the angle between $\boldsymbol{p}$ and $\boldsymbol{k}$. This is a damping associated with the energy transfer from the collective modes to the plasma constituents; it is called the Landau damping. In the HTL approximation, the plasmons reside in the time-like region. Therefore, they do not suffer from the Landau damping and behave as long-lived quasi-particles. However, in higher orders, the damping takes place through the scattering of the plasmon with other plasma constituents; this will be discussed in Section 4.7.

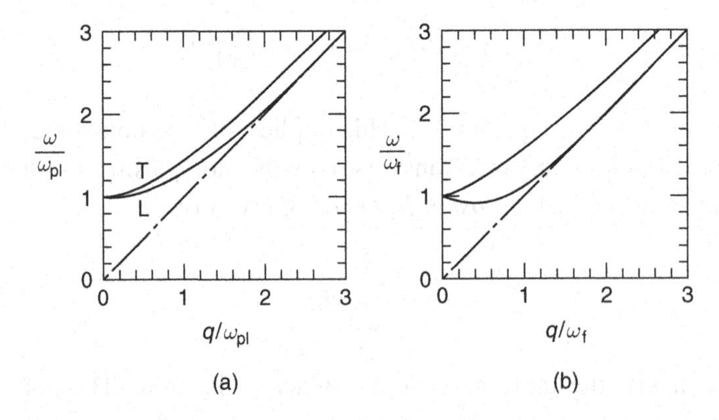

Fig. 4.6. Dispersion relations for (a) gluons and (b) quarks in the hard thermal loop approximation. T and L represent the transverse and longitudinal modes, respectively.

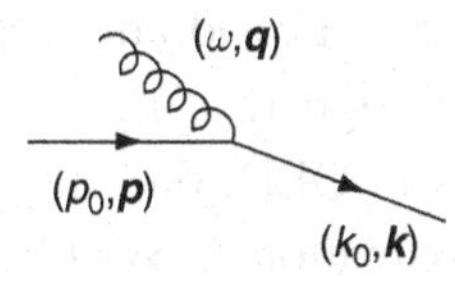

Fig. 4.7. A diagram corresponding to Landau damping, where the space-like gluon is absorbed by the thermal constituents of the plasma.

Because of the Landau damping, the HTL propagators near the static limit $\omega \ll q$ ($x \to 0$) up to the leading order of x are given by

$$^*D_{\mathrm{L}}^{\mathrm{R}}(\omega \ll q, q) = \left(q^2 + \omega_{\mathrm{D}}^2 + i\frac{\pi\omega_{\mathrm{D}}^2}{2}\frac{\omega}{q}\right)^{-1}, \tag{4.96}$$

$$^*D_{\mathrm{T}}^{\mathrm{R}}(\omega \ll q, q) = \left(q^2 - i\frac{\pi\omega_{\mathrm{D}}^2}{4}\frac{\omega}{q}\right)^{-1}. \tag{4.97}$$

Equation (4.96) shows that the longitudinal gluon becomes "massive" in the static limit ($\omega = 0$) because of the Debye screening. In the coordinate-space, this leads to a Yukawa potential between heavy quarks, with a length scale characterized by $1/\omega_{\mathrm{D}}$. This has an important implication for the properties of heavy-quark bound states, such as J/ψ and Υ in the quark–gluon plasma, as will be discussed in Chapter 7. On the other hand, Eq. (4.97) shows that the transverse gluon is "massless" and is long-ranged in the static limit ($\omega = 0$), but receives dynamical screening for finite ω due to Landau damping. This dynamical screening also plays important roles in taming the singularities in the scattering processes with the exchange of transverse gluons.

4.6 Quark propagator at high T and zero μ

A similar calculation in the HTL approximation can be performed for the fermion through the diagram shown in Fig. 4.8. The quark self-energy, with $Q_\mu = (q_4, \boldsymbol{q}) = (-\nu_l, \boldsymbol{q})$ and $K_\mu = (k_4, \boldsymbol{k}) = (-\omega_n, \boldsymbol{k})$, with zero chemical potential, is given by

$$\Sigma(Q) = S(Q)^{-1} - S_0(Q)^{-1} \tag{4.98}$$

$$\simeq -g^2 C_{\mathrm{F}} \int (dK)\ \gamma_\mu (\gamma \cdot K) \gamma_\mu \frac{1}{K^2 (Q+K)^2}, \tag{4.99}$$

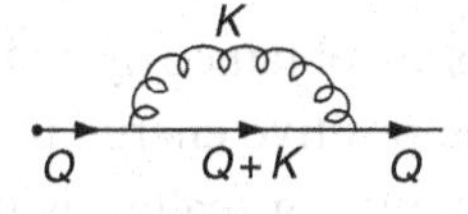

Fig. 4.8. One-loop self-energy of a quark in the plasma.

where $C_F = (N_c^2 - 1)/2N_c = 3/2$ is a quadratic Casimir invariant for the fundamental representation defined through $(t^a t^a)_{ij} = C_F \delta_{ij}$ (Appendix B.3). For the self-energy of massless fermions in QED, one simply replaces g by e and C_F by 1 in Eq. (4.99). The HTL approximation $K \gg Q$ has been used in the numerator in Eq. (4.99). Also, the gauge-dependent part of the gluon propagator, $-(1-\xi)K_\mu K_\nu/K^4$, does not contribute in this approximation. (See Exercise 4.7(1).)

After evaluating the integral in the same way as we did for the gluon self-energy and making the analytic continuation to the Minkowski space, $i\nu_l \to q^0 + i\delta$, with $-i\gamma_4 = \gamma^0$ and $Q^\mu = (q^0, \boldsymbol{q})$, we obtain the retarded quark propagator in the HTL approximation, ${}^*S^R(Q) = -[\gamma \cdot Q - \Sigma(Q)]^{-1}$, with the self-energy (Exercise 4.7(2)) given by

$$\Sigma(\omega, q) = a(\omega, q)\gamma^0 + b(\omega, q)\boldsymbol{\gamma}\cdot\hat{\boldsymbol{q}}, \tag{4.100}$$

$$a(\omega, q) = \frac{\omega_f^2}{\omega} F(\omega/q), \tag{4.101}$$

$$b(\omega, q) = \frac{\omega_f^2}{q}[1 - F(\omega/q)]. \tag{4.102}$$

The factor ω_f is a characteristic fermion frequency in the hot plasma:

$$\omega_f^2 = \frac{g^2}{8} C_F T^2 \qquad \text{(QCD)}, \tag{4.103}$$

$$= \frac{e^2}{8} T^2 \qquad \text{(QED)}. \tag{4.104}$$

The collective mode defined as a pole of ${}^*S^R(Q)$ is called the *plasmino*, which is a fermionic analog of the plasmon (Klimov, 1981; Weldon, 1982). The dispersion relations for the plasmino for small momentum ($q \ll \omega_f$) and large momentum ($\omega_f \ll q \ll T$) are obtained from $F(x)$ for $x \to \infty$ and for $x \to 0$, respectively. The behavior of the positive energy solutions is given by

$$\omega \simeq \omega_f \pm \frac{1}{3} q \qquad \text{(small } q\text{)}, \tag{4.105}$$

$$\simeq q \qquad \text{(large } q\text{)}, \tag{4.106}$$

and the negative energy solutions are obtained by making the substitution $\omega \to -\omega$.

The dispersion relation for the positive energy plasminos for a wide range of q is shown in Fig. 4.6(b). One interesting feature is that the velocity, $v_f = \partial\omega/\partial q$, of the lower branch changes sign as q increases. Note also that the non-vanishing mass gap, $\omega(q=0) = \pm\omega_{\rm f}$, does not imply the breaking of chiral symmetry. Indeed, the self-energy, Eq. (4.100), is proportional to γ^μ. Therefore, it is chiral-invariant, $\{\Sigma, \gamma_5\} = 0$.

4.7 HTL resummation

Hard thermal loops (HTLs) appear not only in self-energies, such as $\Pi_{\rm L,T}$ and Σ, but also in vertices. Indeed, it has been shown that vertices with n external gluons, and those with $(n-2)$ external gluons and two external quarks, have HTLs of $O(T^2)$ (Braaten and Pisarski, 1990a). Furthermore, these vertices can be summarized as the HTL effective Lagrangian in imaginary time (Braaten and Pisarski, 1992b; Frenkel and Taylor, 1992):

$$\mathcal{L}_{\rm HTL} = \omega_{\rm f}^2 \int \frac{d\Omega}{4\pi} \bar{q} \frac{\gamma \cdot v}{-iv \cdot D} q + \frac{1}{2}\omega_{\rm D}^2 \,{\rm tr} \int \frac{d\Omega}{4\pi} F_{\mu\lambda} \frac{v_\lambda v_\rho}{(v \cdot \mathcal{D})^2} F_{\mu\rho}, \tag{4.107}$$

where $v_\mu = (i, \boldsymbol{v})$ with $\boldsymbol{v}^2 = 1$ and $\int d\Omega$ is an integration over the direction of $\boldsymbol{v}$. Also, $F_{\mu\nu} = F^a_{\mu\nu} t^a$, and tr is taken over the color indices. Factors D and $\mathcal{D}$ are the Euclidean covariant derivatives defined in Eq. (4.4); $\omega_{\rm f}$ and $\omega_{\rm D}$ are defined in Eq. (4.103) and Eq. (4.87), respectively. On expanding Eq. (4.107) in terms of g and making the momentum-space representation, $\Pi_{\rm L,T}$ and Σ are obtained in the leading order and HTL vertices are obtained in higher orders (Exercise 4.8). An important aspect of $\mathcal{L}_{\rm HTL}$ is that it is gauge-invariant; this is why the gauge-parameter dependence disappears in the calculation of the gluon and quark propagators in the HTL approximation in the previous sections. Also, $\mathcal{L}_{\rm HTL}$ is a non-local function in the coordinate-space.

For hard particles with momentum $q \sim T$, the HTL contributions in $\mathcal{L}_{\rm HTL}$ are suppressed by powers of g compared to the classical Lagrangian, $\mathcal{L}_{\rm cl}$, given in Eq. (2.1). On the other hand, for soft particles with momentum $q \sim gT$, $\mathcal{L}_{\rm HTL}$ and $\mathcal{L}_{\rm cl}$ are the same order because $\omega_{\rm f}^2/\partial \sim (gT)^2/(gT) \sim \partial$ and $\omega_{\rm D}^2/\partial^2 \sim (gT)^2/(gT)^2 \sim 1$. Therefore, for soft particles, it is necessary to reorganize the perturbation theory by defining effective propagators and vertices with HTLs included; this is called the *HTL resummation* (Braaten and Pisarski, 1990a). It is written schematically as

$$\mathcal{L}_{\rm cl} = (\mathcal{L}_{\rm cl} + \mathcal{L}_{\rm HTL}) - \mathcal{L}_{\rm HTL} = \mathcal{L}_{\rm eff} - \mathcal{L}_{\rm HTL}, \tag{4.108}$$

where $\mathcal{L}_{\rm eff}$ defines the effective propagators and vertices, while $-\mathcal{L}_{\rm HTL}$ avoids the double counting of the same contributions. Historically, the HTL resummation

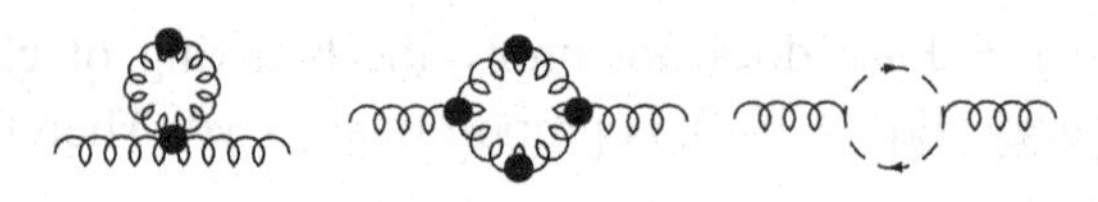

Fig. 4.9. The self-energy in the reorganized perturbation theory for a soft gluon ($q \sim gT$). Filled circles denote the effective propagators and vertices in the HTL resummation.

plays a crucial role is the gluon (quark) damping rate $\gamma(q)$, which is defined by the imaginary part of the pole position of the gluon (quark) propagator: $\omega = \omega(q) - i\gamma(q)$ (Braaten and Pisarski, 1990b, 1992a). For example, the damping rate of a gluon at rest for $N_f = 0$ is calculated from the diagram in Fig. 4.9 to be

$$\gamma_{\rm g}(q=0) \simeq 1.1 N_c \alpha_{\rm s} T, \tag{4.109}$$

whereas the damping rate of a quark at rest for $N_f = 2$ is given by $\gamma_q(q = 0) \simeq 1.4 C_{\rm F} \alpha_{\rm s} T$. For more detailed discussions and applications of the HTLs, see Thoma (1995), Le Bellac (1996) and Kraemmer and Rebhan (2004).

4.8 Perturbative expansion of the pressure up to $O(g^5)$

We turn now to study the evaluation of thermodynamic quantities of the quark–gluon plasma at high T with zero μ. Among others, the pressure, $P(T) = -\Omega(T, V)/V$, is the fundamental quantity which can be calculated by evaluating connected Feynman diagrams with no external legs (Exercise 4.3). Here we recapitulate the result for massless N_f flavors in the $\overline{\rm MS}$ scheme up to $O(g^5)$:

$$\begin{aligned} P(T) \simeq \frac{8\pi^2}{45} T^4 \Big[& c_0 + c_2\, \bar{g}^2(\kappa) + c_3\, \bar{g}^3(\kappa) \\ & + c_4 \left(\ln \frac{\kappa}{T}, \ln \bar{g} \right) \bar{g}^4(\kappa) + c_5 \left(\ln \frac{\kappa}{T} \right) \bar{g}^5(\kappa) \Big], \end{aligned} \tag{4.110}$$

where κ is a renormalization scale and $\bar{g}^2 \equiv g^2/4\pi^2 = \alpha_{\rm s}/\pi$.

We also define the free energy in the Stefan–Boltzmann (SB) limit as $f_{\rm SB} \equiv -c_0(8\pi^2/45)T^4$. The coefficients c_2 and c_3 were calculated by Shuryak (1978a) and Kapusta (1979), respectively. The coefficient c_4 was calculated by Arnold and Zhai (1995), and c_5 was calculated by Zhai and Kastening (1995) and Braaten and Nieto (1996a). (The coefficient of $\bar{g}^6 \ln \bar{g}$ was also calculated by Kajantie *et al.* (2003). These coefficients are given by

$$c_0 = 1 + \frac{21}{32} N_f, \tag{4.111}$$

$$c_2 = -\frac{15}{4} \left(1 + \frac{5}{12} N_f\right), \tag{4.112}$$

$$c_3 = 30 \left(1 + \frac{1}{6} N_f\right)^{3/2}, \tag{4.113}$$

$$\begin{aligned} c_4 = {} & 237.2 + 15.97 N_f - 0.4150 N_f^2 \\ & + \frac{135}{2} \left(1 + \frac{1}{6} N_f\right) \ln \left[\bar{g}^2 \left(1 + \frac{1}{6} N_f\right)\right] \\ & - \frac{165}{8} \left(1 + \frac{5}{12} N_f\right) \left(1 - \frac{2}{33} N_f\right) \ln \left(\frac{\kappa}{2\pi T}\right), \end{aligned} \tag{4.114}$$

$$\begin{aligned} c_5 = {} & \left(1 + \frac{1}{6} N_f\right)^{1/2} \left[-799.2 - 21.96 N_f - 1.926 N_f^2 \right. \\ & \left. + \frac{495}{2} \left(1 + \frac{1}{6} N_f\right) \left(1 - \frac{2}{33} N_f\right) \ln \left(\frac{\kappa}{2\pi T}\right)\right]. \end{aligned} \tag{4.115}$$

Naive perturbation theory using the Feynman rule given in Fig. 4.3 generates an expansion with only even powers of $\bar{g}$ for the pressure. Because of the massless gluon propagator, such a naive expansion becomes infrared-divergent from $O(\bar{g}^4)$. The problem is cured by reorganizing the perturbation series by taking into account the Debye screening of the longitudinal gluon propagator. This corresponds to a resummation of a certain set of diagrams and leads to non-analytic terms in Eq. (4.110), such as $|g|^3$, $g^4 \ln \bar{g}$ and $|g|^5$. Examples of non-analytic terms with similar physical origin are found, for example, in the Debye–Hückel classical theory for strong electrolytes (Debye and Hückel, 1923) and in the Gell-Mann–Brueckner resummation for the correlated electron system (Gell-Mann and Brueckner, 1957).

The existence of non-analytic terms also implies that the expansion has a zero radius of convergence and is, at most, an asymptotic series. (Roughly speaking, if $F_n(x) \equiv \sum_{i=1}^n f_i(x)$ is an asymptotic expansion of $F(x)$ around $x = 0$, $F_n(x)$ becomes a good approximation to $F(x)$ for $x \to 0$ with n fixed. On the other hand, if $F_n(x)$ is a convergent expansion, it is a good approximation to $F(x)$ for $n \to \infty$ with x fixed.) The asymptotic nature is common in high order terms in various problems in quantum mechanics and in quantum field theories (Le Guillou and Zinn-Justin, 1990).

To check the reliability of the expansion, Eq. (4.110), let us define a dimensionless ratio,

$$R \equiv \frac{P(T)}{P_{\rm SB}(T)}. \tag{4.116}$$

Here, $P_{\rm SB}(T)$ is the pressure in the SB limit ($g \to 0$). In Fig. 4.10(a) R is shown as a function of $\alpha_{\rm s} \equiv g^2/4\pi$ for $N_f = 4$; κ is chosen to be $2\pi T$ so that the dangerous large logs in coefficients c_4 and c_5 are suppressed. The solid lines in Fig. 4.10(a) denote R calculated up to $O(g^n)$ ($n = 2, 3, 4, 5$). It is clear that R oscillates around $R = 1$ if $\alpha_{\rm s}$ is relatively large.

Figure 4.10(b) shows α_s as a function of T, obtained by solving the flow equation, Eq. (2.27), with the two-loop β-function, Eq. (2.29), and $N_f = 4$. The boundary condition of the flow equation is set to be $\alpha_{\rm s}(\kappa = 5\,\text{GeV}) = 0.21$. The range $0.2 < \alpha_{\rm s} < 0.4$ corresponds to $1\,\text{GeV} > T > 0.15\,\text{GeV}$, which is relevant for the quark–gluon plasma accessible in relativistic heavy ion collisions at RHIC and LHC. In this range, the perturbative expansion of $P(T)$ oscillates extremely rapidly, as seen in Fig. 4.10(a). Therefore, one cannot trust the expansion literally. In other words, the asymptotic expansion, Eq. (4.110), is reliable only for $\alpha_{\rm s} < 0.1$, which corresponds to $T > O(10^2)$ GeV.

Despite the above fact being true for perturbation theory, non-perturbative lattice QCD simulations (Figs. 3.5 and 3.6) indicate that the pressure does not deviate from the SB limit by more than 20% for $T > (2-3) \times T_{\rm c}$. This gives us

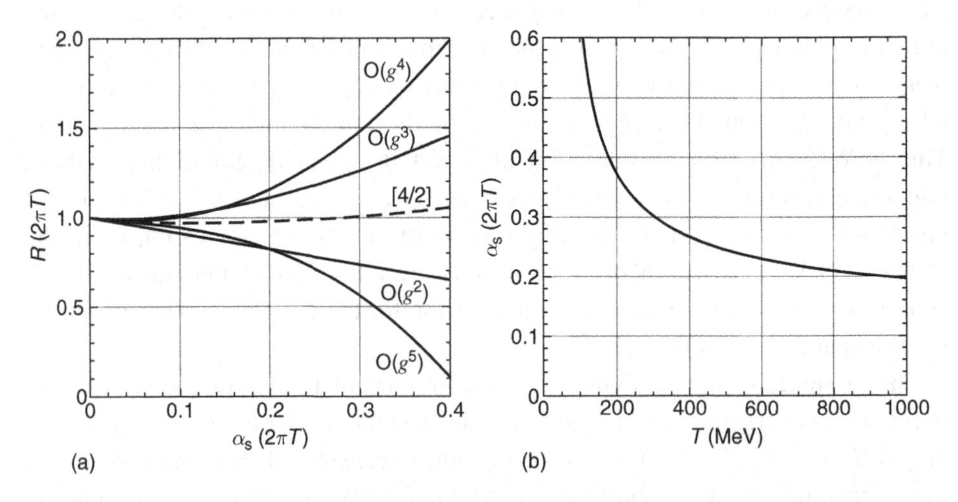

Fig. 4.10. (a) Solid lines show the ratio, R (perturbative pressure, $P(T)$, divided by its Stefan–Boltzmann limit, $P_{\rm SB}(T)$) calculated up to $O(g^n)$ ($n = 2, 3, 4, 5$). The dashed line is R from the [4/2] Padé approximant constructed from the perturbative series. (b) Running coupling constant as a function of T with the two-loop β-function and $N_f = 4$. Adapted from Hatsuda (1997).

hope that further resummation of the perturbation series may lead to a reliable result, even for $T \sim T_c$. Various methods, such as the Padé approximants, the optimized perturbation, the HTL resummed perturbation, etc., have been proposed for this purpose (Exercise 4.9). Further details of these approaches are given in Kraemmer and Rebhan (2004).

4.9 Infrared problem of $O(g^6)$ and beyond

So far we have discussed the perturbative expansion of $P(T)$ calculated up to $O(g^5)$. It is known that the perturbative expansion in terms of g breaks down at $O(g^6)$, where one encounters an infrared (IR) problem, even after taking the Debye screening of the longitudinal gluons into account.

To see this explicitly, let us examine the diagram shown in Fig. 4.11, which was originally discussed by Linde (1980). The diagram has $(l+1)$ loops, $2l$ vertices and $3l$ propagators. Since a possible infrared singularity arises from the propagators with zero Matsubara frequency, $\omega_n = 0$, the dominant contribution to the pressure in Fig. 4.11 may be schematically given by

$$P_{(2l)} \sim g^{2l} \left(T \int_m^T d^3k \right)^{l+1} \left(\frac{1}{k^2} \right)^{3l} k^{2l}, \tag{4.117}$$

where the IR regulator, m, and the UV cutoff, T, are introduced to make the integral well defined. The last factor, k^{2l}, comes from the ggg vertex, which has a derivative (see Eqs. (4.38)).

Depending on the number of loops, the above expression behaves as follows:

$$l < 3: \quad g^{2l} T^4, \tag{4.118}$$

$$l = 3: \quad g^6 T^4 \ln^4(T/m), \tag{4.119}$$

$$l > 3: \quad g^6 T^4 (g^2 T/m)^{l-3}. \tag{4.120}$$

If the gluon propagators in Fig. 4.11(a) are all longitudinal (electric), m can be chosen to be the Debye mass, $m = m_{\rm el} = \omega_{\rm D} \propto gT$. In this case, no IR divergence

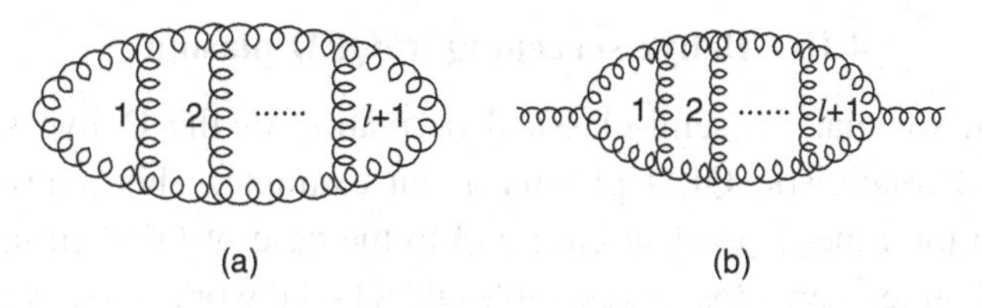

Fig. 4.11. An $(l+1)$-loop diagram for (a) the pressure and (b) the gluon self-energy.

occurs, and higher loops give higher order contributions in terms of g. If the gluon propagators in Fig. 4.11(a) are all transverse (magnetic), one needs to assign a magnetic screening mass, m_{mag}, to m. However, in perturbation theory, m_{mag} does not appear in $O(gT)$, as we have seen in Eq. (4.97). Therefore, the higher order terms corresponding to $O(g^{n\geq 6})$ diverge. The situation is slightly improved if m_{mag} appears in $O(g^2T)$. In this case, the pressure is finite, but all the terms for $l > 3$ give the same contribution to the pressure, which implies that the perturbation theory breaks down.

The same IR problem arises in the calculation of m^2_{mag} in perturbation theory. Consider the diagram shown in Fig. 4.11(b), which is similar to Fig. 4.11(a) except for the external lines. IR counting, similar to Eq. (4.117), shows that the perturbation theory starts to break down at $O(g^4)$:

$$l = 1 : \quad g^4T^2 \ln^2(T/m), \tag{4.121}$$

$$l > 1 : \quad g^4T^2(g^2T/m)^{l-1}. \tag{4.122}$$

The IR problem of the QCD perturbation, which is closely connected to the transverse (magnetic) sector of the gluons, has not yet been solved. It is not clear if one can develop a systematic resummation procedure to cure this problem. An alternative approach to the problem (Braaten and Nieto, 1996b) is to decompose the pressure into three parts by introducing intermediate cutoffs, $\Lambda_{\text{el}} \sim gT$ and $\Lambda_{\text{mag}} \sim g^2T$:

$$P(T)/T^4 = p_{\text{el}}(g; \Lambda_{\text{el}}) + p_{\text{mag}}(g; \Lambda_{\text{el}}, \Lambda_{\text{mag}})g^3 + p_{\text{g}}(g; \Lambda_{\text{mag}})g^6. \tag{4.123}$$

Note that p_{el} contains contributions from the hard momentum ($k > \Lambda_{\text{el}}$), whereas p_{mag} contains those from the soft momentum ($\Lambda_{\text{el}} > k > \Lambda_{\text{mag}}$). Due to the presence of the cutoffs, p_{el} and p_{mag} are free from the IR problem and are calculable in any high orders of g in perturbation theory. Up to $O(g^5)$, $p_{\text{el}}(g) + p_{\text{mag}}(g)g^3$ reproduces the known results, Eqs. (4.110), (4.111)–(4.115). The last term, p_{g}, is calculable only by non-perturbative methods, such as the lattice QCD simulations. It contains the contributions from the super-soft momentum ($\Lambda_{\text{mag}} > k$).

4.10 Debye screening in QED plasma

In this section we make a semi-classical derivation of the Debye screening by taking hot and relativistic QED plasma as an example. This approach will be generalized to the time-dependent case and to the case of QCD in later sections. The derivation given here closely resembles the classic work of Debye and Hückel (1923); see also Fetter and Walecka (1971).

We start with the Maxwell equation,

$$\partial_\nu F^{\nu\mu}(x) = j^\mu(x). \tag{4.124}$$

For the time-independent case, the $\mu = 0$ component of Eq. (4.124) reduces to the Gauss law,

$$\nabla \cdot \boldsymbol{E}(\boldsymbol{x}) = -\nabla^2 \phi(\boldsymbol{x}) = j^0(\boldsymbol{x}) = \rho_{\rm ind}(\boldsymbol{x}) + \rho_{\rm ext}(\boldsymbol{x}), \tag{4.125}$$

where $\rho_{\rm ext}$ is the charge density of an external source and $\rho_{\rm ind}$ is the charge density induced by the source.

We assume that the system is composed of equal numbers of electrons and positrons in thermal equilibrium and that the temperature is much larger than the electron mass (relativistic plasma). Then, by neglecting $m_{\rm e}$, the induced charge density can be written as the difference of "distorted" Fermi–Dirac distributions, $n_\pm(\boldsymbol{p}, \boldsymbol{x})$, under the influence of the scalar potential $\phi(\boldsymbol{x})$:

$$\rho_{\rm ind}(\boldsymbol{x}) \simeq 2e \int \frac{d^3k}{(2\pi)^3} [n_+(\boldsymbol{p}, \boldsymbol{x}) - n_-(\boldsymbol{p}, \boldsymbol{x})], \tag{4.126}$$

$$n_\pm(\boldsymbol{p}, \boldsymbol{x}) \equiv \frac{1}{{\rm e}^{(|\boldsymbol{p}| \pm e\phi(\boldsymbol{x}))/T} + 1}. \tag{4.127}$$

The factor 2 on the right-hand side of Eq. (4.126) corresponds to the two spin states. Use of Eqs. (4.126) and (4.127) is justified under the assumption that the system is static and collisionless, which will be discussed in Section 4.11.

The field $\phi(\boldsymbol{x})$ should be determined self-consistently by solving Eq. (4.126) and Eq. (4.125) simultaneously. Assuming $T \gg e\phi$ and linearizing Eq. (4.127), we obtain

$$\rho_{\rm ind}(\boldsymbol{x}) \simeq 4e^2 \phi(\boldsymbol{x}) \int \frac{d^3p}{(2\pi)^3} \frac{dn_{\rm F}(p)}{dp} \tag{4.128}$$

$$= -\frac{e^2 T^2}{3} \phi(\boldsymbol{x}), \tag{4.129}$$

where $n_{\rm F} \equiv n_\pm|_{\phi=0}$ is the standard Fermi–Dirac distribution and the following integral has been used (Exercise 3.4(1)):

$$\int \frac{d^3p}{(2\pi)^3} \frac{dn_{\rm F}(p)}{dp} = -\frac{T^2}{\pi^2} \int_0^\infty dx \frac{x}{{\rm e}^x + 1} = -\frac{T^2}{12}. \tag{4.130}$$

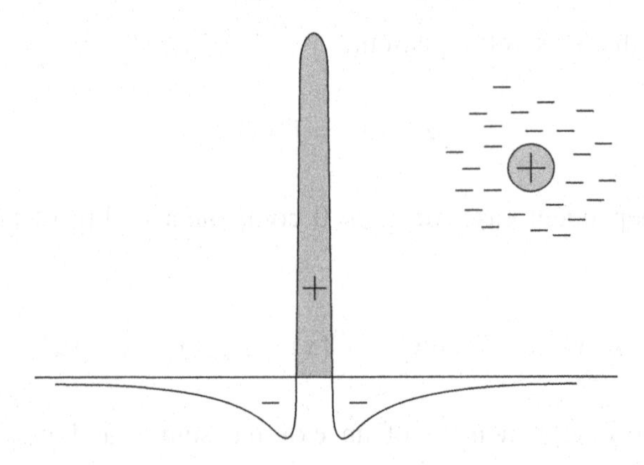

Fig. 4.12. Debye screening of an external positive charge by plasma particles.

Then the Gauss law in Eq. (4.125) is given by

$$(-\nabla^2 + \omega_{\rm D}^2)\phi(\boldsymbol{x}) = \rho_{\rm ext}(\boldsymbol{x}) \quad \text{with} \quad \omega_{\rm D} = \frac{1}{\sqrt{3}} eT, \tag{4.131}$$

where $\omega_{\rm D}$ is simply the Debye screening mass we encountered in Eq. (4.88).

To see the physical meaning of $\omega_{\rm D}$, let us consider a heavy external source located at the origin with charge $Q_{\rm ext} > 0$: $\rho_{\rm ext} = Q_{\rm ext}\delta^3(\boldsymbol{x})$. Then, the resultant scalar potential and the induced charge are given by

$$\phi(\boldsymbol{x}) = \frac{Q_{\rm ext}}{4\pi R}\, {\rm e}^{-\omega_{\rm D} R}, \qquad \rho_{\rm ind}(\boldsymbol{x}) = -\frac{\omega_{\rm D}^2 Q_{\rm ext}}{4\pi R}\, {\rm e}^{-\omega_{\rm D} R}. \tag{4.132}$$

This implies that a negative charge distribution is induced around the positive source in the medium, which is shown schematically in Fig. 4.12 for an external source with finite spatial size. The potential, $V(r)$, between two external sources with charge $Q_{\rm ext}$ and $-Q_{\rm ext}$, respectively, is screened at a distance characterized by $\lambda_{\rm D} = 1/\omega_{\rm D}$:

$$V(R) = -\frac{Q_{\rm ext}^2}{4\pi R}\, {\rm e}^{-\omega_{\rm D} R}. \tag{4.133}$$

A similar classical argument can be made for time-dependent plasma oscillations (Tonks and Langmuir, 1929; Fetter and Walecka, 1971). Instead of returning to it here, we will consider the discussion on Vlasov equations, which allow us to analyze more general cases.

4.11 Vlasov equations for QED plasma

Consider again a relativistic QED plasma and a space-time-dependent distribution function, $n_{\pm}(\boldsymbol{p}, x)$, with $x^\mu = (t, \boldsymbol{x})$. The Vlasov equations are a set of equations for the distribution function of plasma particles under the slowly varying (soft) electromagnetic field:

$$\partial_\nu F^{\nu\mu}(x) = j^\mu_{\rm ind}(x) + j^\mu_{\rm ext}(x), \tag{4.134}$$

$$j^\mu_{\rm ind}(x) = 2e \int \frac{d^3p}{(2\pi)^3}\, v^\mu \left[n_+(\boldsymbol{p},x) - n_-(\boldsymbol{p},x)\right], \tag{4.135}$$

$$\left[v_\mu \partial^\mu_x \pm e(\boldsymbol{E}(x) + \boldsymbol{v}\times\boldsymbol{B}(x))\cdot\nabla_{\boldsymbol{p}}\right] n_\pm(\boldsymbol{p},x) = 0, \tag{4.136}$$

where

$$v^\mu = (1, \boldsymbol{v}), \tag{4.137}$$

and $\boldsymbol{v}$ is the group velocity of the relativistic plasma particles, $\boldsymbol{v} = dE/d\boldsymbol{p} = \boldsymbol{p}/|\boldsymbol{p}| \equiv \hat{\boldsymbol{p}}$. Note also that $v_\mu \partial^\mu_x \equiv v\cdot\partial_x = \partial_t + \boldsymbol{v}\cdot\nabla$.

Equations (4.134) and (4.136) are straightforward generalizations of Eqs. (4.125) and (4.126), respectively. Equation (4.136) describes the motion of the particles in the phase-space under the influence of the electromagnetic mean-field. It corresponds to the collisionless limit of the Boltzmann equation, which will be discussed in Chapter 12.

Equations (4.134)–(4.136) constitute self-consistent equations. Let us first solve Eq. (4.136) by making the following ansatz similar to Eq. (4.127):

$$n_\pm(\boldsymbol{p},x) = n_{\rm F}(|\boldsymbol{p}| \pm e\Phi(x,\boldsymbol{v})), \tag{4.138}$$

where $\Phi(x,\boldsymbol{v})$ is a "potential function" analogous to the scalar potential, $\phi(\boldsymbol{x})$, in the static case. Substituting Eq. (4.138) into Eq. (4.136), we obtain

$$v\cdot\partial_x\Phi(x,\boldsymbol{v}) = -\boldsymbol{v}\cdot\boldsymbol{E}(x), \tag{4.139}$$

where the magnetic field does not contribute because $\boldsymbol{v}\cdot(\boldsymbol{v}\times\boldsymbol{B}) = 0$.

If the fluctuation of the distribution function, $\delta n_\pm(\boldsymbol{p},x) = n_\pm(\boldsymbol{p},x) - n_{\rm F}(p)$, is small, we may write

$$\delta n_\pm(\boldsymbol{p},x) \simeq \pm e\Phi(x,\boldsymbol{v})\frac{\partial n_{\rm F}(p)}{\partial p}. \tag{4.140}$$

Then the induced current in Eq. (4.135) in the same approximation is given by

$$j^\mu_{\rm ind}(x) \simeq -\omega_{\rm D}^2 \int \frac{d\Omega}{4\pi} v^\mu \Phi(x,\boldsymbol{v}). \tag{4.141}$$

The potential function, Φ, in Eq. (4.139) can be solved explicitly under the boundary condition that the mean-field vanishes in the far past:

$$n_\pm(\boldsymbol{p},x)|_{t\to-\infty} \to n_{\rm F}(p). \tag{4.142}$$

Then the solution is given by

$$\Phi(x, v) = -\int_{-\infty}^{t} dt' \; \boldsymbol{v} \cdot \boldsymbol{E}(t', \boldsymbol{x} - \boldsymbol{v}(t - t')) \tag{4.143}$$

$$= -\int_{0}^{\infty} d\tau \; \mathrm{e}^{-\delta\tau} \boldsymbol{v} \cdot \boldsymbol{E}(x - v\tau), \tag{4.144}$$

where δ is a small positive number, which has an effect equivalent to the condition $\boldsymbol{E}(t \to -\infty, \boldsymbol{x}) \to 0$. The integral in Eq. (4.143) is simply the work done on a particle moving with a constant velocity, $\boldsymbol{v}$, during the time interval $-\infty \to t$.

Using $\boldsymbol{E} = -\nabla A_0 - \partial_t \boldsymbol{A}$, the induced current, j_μ^{ind}, on the right-hand side of Eq. (4.141) may be rewritten in coordinate-space and momentum-space as follows:

$$j^\mu_{\mathrm{ind}}(x) = \int d^4y \; \Pi^{\mu\nu}(x-y) A_\nu(y), \quad j^\mu_{\mathrm{ind}}(Q) = \Pi^{\mu\nu}(Q) A_\nu(Q), \tag{4.145}$$

with $Q^\mu = (\omega, \boldsymbol{q})$.

With a useful formula,

$$\int d^4x \; \mathrm{e}^{iQx} \int_0^\infty d\tau \; \mathrm{e}^{-\delta\tau} f(x - v\tau) = \frac{i f(Q)}{v \cdot Q + i\delta}, \tag{4.146}$$

we find

$$\Pi_{\mu\nu}(Q) = -\omega_{\mathrm{D}}^2 \left[\delta_{\mu 0}\delta_{\nu 0} - \omega \int \frac{d\Omega}{4\pi} \frac{v_\mu v_\nu}{v \cdot Q + i\delta} \right], \tag{4.147}$$

where $\int d\Omega$ is an integration over the spatial angle of $\boldsymbol{v}$. After this angular integration, it is easy to see that the $\Pi^{\mu\nu}(Q)$ derived here is equivalent to the retarded self-energy of the photon in the HTL approximation: Eq. (4.67) with Eqs. (4.84), (4.85) and (4.88).

As we have seen before, the two limits $\omega \to 0$ and $\boldsymbol{q} \to 0$ do not commute:

$$\Pi_{\mu\nu}(0, \boldsymbol{q}) = -\omega_{\mathrm{D}}^2 \delta_{\mu 0}\delta_{\nu 0}, \quad \Pi_{\mu\nu}(\omega, \boldsymbol{0}) = \omega_{\mathrm{pl}}^2 \delta_{\mu i}\delta_{\nu j}, \tag{4.148}$$

which are obtained directly from Eq. (4.147) by using an angular average, $\int v_i v_j \; d\Omega/(4\pi) = \delta_{ij}/3$. Therefore, $j_\mu^{\mathrm{ind}}(\boldsymbol{x}) = -\omega_{\mathrm{D}}^2 \delta_{\mu 0} A_0(\boldsymbol{x})$ for a static field $A_\mu(\boldsymbol{x})$, while $j_\mu^{\mathrm{ind}}(\boldsymbol{x}) = \omega_{\mathrm{pl}}^2 \delta_{\mu i} A_i(t)$ for a homogeneous field $A_\mu(t)$.

Now, substituting the above induced current into the Maxwell equation, Eq. (4.134), with $j_\mu^{\mathrm{ext}} = 0$, we obtain

$$(-\nabla^2 + \omega_{\mathrm{D}}^2)\boldsymbol{E}(\boldsymbol{x}) = 0, \quad -\nabla^2 \boldsymbol{B}(\boldsymbol{x}) = 0, \tag{4.149}$$

for a static case. This implies that the electric (magnetic) field is screened (unscreened) in the plasma. For a homogeneous but non-static case, we obtain

$$(\partial_t^2 - \omega_{\mathrm{pl}}^2)\boldsymbol{E}(t) = 0, \quad (\partial_t^2 - \omega_{\mathrm{pl}}^2)\boldsymbol{B}(t) = 0. \tag{4.150}$$

Namely, both electric and magnetic oscillations in the plasma have the plasma frequency, $\omega_{\rm pl}$. For general propagation of the electromagnetic waves with finite ω and $\boldsymbol{q}$, we need to solve the coupled Vlasov equations (Exercise 4.10).

4.12 Vlasov equations for QCD plasma

The previous discussions on QED can be generalized to QCD. A detailed derivation of the QCD Vlasov equations with respect to the Kadanoff–Baym formalism (Kadanoff and Baym, 1962) can be seen in Blaizot and Iancu (2002). We repeat only the final results here, with a special emphasis on the similarity to the QED case.

Considering the soft gluonic mean-field interacting with the hard plasma particles (quarks and gluons), the covariant and linearized forms of the QCD Vlasov equations are given by

$$[D_\nu, F^{\nu\mu}(x)]^a = j_{\rm ind}^{\mu,a}, \tag{4.151}$$

$$j_{\rm ind}^{\mu,a} = -\omega_{\rm D}^2 \int \frac{d\Omega}{4\pi} v^\mu \Phi^a(x, \boldsymbol{v}), \tag{4.152}$$

$$[v \cdot D_x, \Phi(x, \boldsymbol{v})]^a = -\boldsymbol{v} \cdot \boldsymbol{E}^a(x), \tag{4.153}$$

$$\delta n_\pm^a(\boldsymbol{p}, x) = \pm g \Phi^a(x, \boldsymbol{v}) \frac{dn_{\rm F}}{dp}, \tag{4.154}$$

$$\delta N^a(\boldsymbol{p}, x) = g \Phi^a(x, \boldsymbol{v}) \frac{dn_{\rm B}}{dp}. \tag{4.155}$$

Here $n_\pm^a$ and N^a are the distribution functions for quarks and gluons, respectively; $D_\mu = \partial_\mu + igt^a A_\mu^a$ is the covariant derivative acting on $F_{\mu\nu} \equiv F_{\mu\nu}^a t^a$; and $\Phi = \Phi^a t^a$.

Equation (4.151) is the Yang–Mills equation with a source current, which is analogous to the Maxwell equation, Eq. (4.134), in the QED plasma. The induced current, $j_{\rm ind}^{\mu,a}$, on the right-hand side, however, receives contributions from quarks and gluons in the QCD plasma since they are both colored. Equations (4.152) and (4.153) are analogous to Eqs. (4.141) and (4.139), respectively. The small fluctuation around the equilibrium distributions of quarks and gluons, Eqs. (4.154) and (4.155), corresponds to the similar expression in Eq. (4.140).

The solution of Eq. (4.153) is inferred from Eq. (4.143), and is easily obtained as

$$\Phi^a(x, \boldsymbol{v}) = -\int_0^\infty d\tau \, {\rm e}^{-\delta\tau} \, U^{ab}(x, x - v\tau) \boldsymbol{v} \cdot \boldsymbol{E}^b(x - v\tau), \tag{4.156}$$

where U is the Wilson line defined as a path-ordered product of the gauge field,

$$U(x, x - v\tau) = \mathrm{P}\exp\left(-ig\int_0^\tau dt\ v\cdot A(x - v(\tau - t))\right). \qquad (4.157)$$

Since $A_\mu(x) = A_\mu^a(x)t^a$ is a matrix and $[A_\mu(x),\ A_\mu(y)] \neq 0$ for $x \neq y$, we need the path ordering, P. Under the local gauge transformation, we have $U(x, y) \to V(x)U(x, y)V^\dagger(y)$. Therefore, Φ^a and $\boldsymbol{E}^a$ transform in a covariant way.

A crucial difference between Φ^a in QCD and Φ in QED is that Φ^a is a non-linear function of the gauge field because of the Wilson line, U^{ab}. By expanding U in terms of A_μ, we have

$$j_{\rm ind}^{\mu,a} = \Pi_{ab}^{\mu\nu} A_\nu^b + \frac{1}{2}\Gamma_{abc}^{\mu\nu\lambda} A_\nu^b A_\lambda^c + \cdots. \qquad (4.158)$$

Coefficients Π, Γ and so on are equivalent to the HTL self-energies and vertices which are generated from Eq. (4.107) (see also Exercise 4.8). Thus we have two different but equivalent ways of deriving HTLs: the diagrammatic approach and the kinetic theory approach. For further discussions and applications of the latter, see the extensive review by Blaizot and Iancu (2002).

Exercises

4.1 Complex time path. Consider a real scalar field defined on a complex time, t, by $\hat{\phi}(t, \boldsymbol{x}) = \mathrm{e}^{i\hat{H}t}\hat{\phi}(0, \boldsymbol{x})\ \mathrm{e}^{-i\hat{H}t}$. Define an ordered product on a complex-time path, C, as

$$\mathrm{T}_C\ \hat{\phi}(x_1)\ \hat{\phi}(x_2) = \theta(s_1 - s_2)\ \hat{\phi}(x_1)\ \hat{\phi}(x_2) + \theta(s_2 - s_1)\ \hat{\phi}(x_2)\ \hat{\phi}(x_1),$$

where $x_1 = (t_1, \boldsymbol{x}_1)$ and $x_2 = (t_2, \boldsymbol{x}_2)$, with $t_{1,2} \in C$. Every point t on C is characterized by a real parameter, s, which increases uniformly as one moves from the initial point, $t_{\rm I}$, to the final point, $t_{\rm F}$. Show that Im t ($t \in C$) should not be an increasing function of s for the thermal average of $\mathrm{T}_C\ \hat{\phi}(x_1)\ \hat{\phi}(x_2)$ to be well defined.

4.2 Grand potential in quantum mechanics. Taking a harmonic oscillator in quantum mechanics at finite T and treating the path-integral measure carefully, show that there arises no additive constant to the grand potential, Ω_0, in Eq. (4.19).

4.3 Linked cluster theorem. Let us prove the linked cluster theorem for the grand potential $\Omega = -T\ln Z$ by using the replica trick (Negele and Orland, 1998).
(1) Derive the identity

$$\ln\left(\frac{Z}{Z_0}\right) = \lim_{x\to 0}\frac{d(Z/Z_0)^x}{dx}.$$

(2) Show that the path integral representation of $(Z/Z_0)^l$ is obtained by introducing an l replica of the field, ϕ, namely ϕ_σ ($\sigma = 1, 2, \ldots, l$). (Different replica fields do not interact with each other.)

(3) Show that the connected Feynman diagrams for $(Z/Z_0)^l$ contain a factor l, while the disconnected diagrams for $(Z/Z_0)^l$ contain a factor l^2 or higher powers of l. Then prove that $\ln(Z/Z_0)$ has connected diagrams only.

4.4 Carlson's theorem and retarded Green's function. The theorem goes as follows (see Sect. 5.8 of Titchmarsh (1932)). Consider a regular function, $f(z)$, for Re $z \geq 0$. Suppose the function is bounded as $|f(z)| < M e^{k|z|}$ for Re $z \geq 0$ with some constant M and $k < \pi$. Suppose also that $f(z) = 0$ for $z = 0, 1, 2, \ldots$ Then, $f(z) = 0$ identically. By using this theorem, prove that the analytic continuation of the imaginary-time Green's function, $\mathcal{G}(q_4, \boldsymbol{q})$, to the retarded Green's function, $\mathcal{G}^{\mathrm{R}}(\omega, \boldsymbol{q})$ (advanced correlation $\mathcal{G}^{\mathrm{A}}(\omega, \boldsymbol{q})$), in the upper half (lower half) ω-plane is unique.

4.5 The gluon self-energy at finite T. Evaluate the diagrams in Figs. 4.5(a), (b) and (d) in the hard thermal loop approximation and derive Eqs. (4.64) and (4.65).

4.6 Matsubara summation.

(1) Prove the following summation formula for bosons ($\omega_n = 2n\pi T$):

$$
\begin{aligned}
T \sum_{n=-\infty}^{+\infty} f(k_0 = i\omega_n) &= \frac{T}{2\pi i} \oint_C dk_0\, f(k_0) \frac{1}{2T} \coth(k_0/2T) \\
&= \frac{1}{2\pi i} \int_{-i\infty}^{+i\infty} dk_0 \left[f(k_0) + f(-k_0) \right] / 2 \\
&\quad + \frac{1}{2\pi i} \int_{-i\infty+\delta}^{+i\infty+\delta} dk_0 \left[f(k_0) + f(-k_0) \right] \frac{1}{e^{k_0/T} - 1},
\end{aligned}
$$

where the contour C encircles the imaginary axis in the complex k_0-plane, and $f(k_0)$ is assumed to have no singularities along the imaginary axis. The summation is separated into the vacuum part and the T-dependent part in the last expression.

(2) Prove the following summation formula for fermions ($\nu_n = (2n+1)\pi T$):

$$
\begin{aligned}
T \sum_{n=-\infty}^{+\infty} f(k_0 = i\nu_n + \mu) &= -\frac{1}{2\pi i} \int_{-i\infty+\mu+\delta}^{+i\infty+\mu+\delta} dk_0\, f(k_0) \frac{1}{e^{(k_0-\mu)/T} + 1} \\
&\quad - \frac{1}{2\pi i} \int_{-i\infty+\mu-\delta}^{+i\infty+\mu-\delta} dk_0\, f(k_0) \frac{1}{e^{-(k_0-\mu)/T} + 1} \\
&\quad + \frac{1}{2\pi i} \oint_\Gamma dk_0\, f(k_0) + \frac{1}{2\pi i} \int_{-i\infty}^{+i\infty} dk_0\, f(k_0),
\end{aligned}
$$

where Γ is a rectangular contour in the k_0-plane: $0-i\infty \to \mu - i\infty \to \mu + i\infty \to 0+i\infty \to 0-i\infty$. The last term on the right-hand side is the vacuum contribution, which is independent of T and μ.

4.7 Quark self-energy at finite *T*.

(1) Evaluate the diagrams in Fig. 4.8 in the hard thermal loop approximation and derive Eq. (4.99); show the gauge-parameter independence of $\Sigma(Q)$.

(2) Evaluate the integral Eq. (4.99) using the formula in Exercise 4.6 and derive Eqs. (4.100)–(4.102).

4.8 Hard thermal loop vertices.

(1) Keeping only the leading terms in the coupling constant and making an analytic continuation to the Minkowski space, derive the following HTL self-energies from Eq. (4.107):

$$\Pi_{\mu\nu}(Q) = -\omega_{\mathrm{D}}^2 \left[\delta_{\mu 0}\delta_{\nu 0} - \omega \int \frac{d\Omega}{4\pi} \frac{v_\mu v_\nu}{v\cdot Q + i\delta} \right],$$

$$\Sigma(Q) = \omega_{\mathrm{f}}^2 \int \frac{d\Omega}{4\pi} \frac{\gamma\cdot v}{v\cdot Q + i\delta},$$

where $Q^\mu = (\omega, \boldsymbol{q})$ and $v^\mu = (1, \boldsymbol{v})$ with $\boldsymbol{v}^2 = 1$. Show further that the above expressions are equivalent to Eqs. (4.84), (4.85) and (4.100).

(2) Expand Eq. (4.107) in terms of g and derive the general form of the HTL vertices in momentum-space.

4.9 Padé appoximant for Euler series. The $[L/M]$ Padé approximant of a formal power series, $f(z) = \sum_{N=0}^{\infty} c_N z^N$, is defined as $f(z) = [L/M] + O(z^{L+M+1})$, with

$$[L/M] \equiv \frac{a_0 + a_1 z + \cdots + a_L z^L}{b_0 + b_1 z + \cdots + b_M z^M}.$$

The $[L/M]$ Padé approximant is a unique rational fraction defined up to a common multiple such that its Taylor series up to z^{L+M} agrees with the original series, $f(z)$. Consider the following integral and its asymptotic expansion for $\arg(z) \neq \pm\pi$ (the Euler series):

$$E(z) = \int_0^\infty \frac{\mathrm{e}^{-t}}{1+zt}\, dt$$
$$\simeq 1 - z + 2!\, z^2 - 3!\, z^3 + 4!\, z^4 - 5!\, z^5 + \cdots.$$

$E(z)$ has a cut along the negative real z-axis. Therefore, the Euler series has a zero radius of convergence, but it is known to be a unique asymptotic expansion.

(1) Plot the Euler series as a function of real positive z for various values of N. Then compare the results with $E(z)$ obtained by numerical integration to see the asymptotic nature of the series.

(2) Construct the $[M/M]$ Padé approximant from the Euler series for various M and compare the results with $E(z)$ obtained by numerical integration. The $[L/M]$ Padé approximants with $L-M \geq -1$ are known to converge to $E(z)$ for $-\pi < \arg(z) < \pi$. See Chap. 5 of Baker and Graves-Morris (1996) for more details.

4.10 Dielectric constants. Define the electric flux density, $\boldsymbol{D}$, by $\boldsymbol{D} = \boldsymbol{E} + \boldsymbol{P}$, where $\boldsymbol{P}$ is the polarization given by $\partial \boldsymbol{P}/\partial t = \boldsymbol{j}_{\rm ind}$ and $\nabla \cdot \boldsymbol{P} = -j^0_{\rm ind}$. Then the dielectric tensor, ϵ_{kl}, in momentum-space is defined through $D_k(\omega, \boldsymbol{q}) = \epsilon_{kl}(\omega, \boldsymbol{q}) E_l(\omega, \boldsymbol{q})$. One may also introduce the longitudinal and transverse components as $\epsilon_{kl} = (\delta_{kl} - \hat{q}_k \hat{q}_l)\epsilon_{\rm T} + \hat{q}_k \hat{q}_l \epsilon_{\rm L}$.

(1) Derive the relations between the dielectric constants, $\epsilon_{\rm T,L,}$ and the self-energies given in Eqs. (4.84) and (4.85),

$$\epsilon_{\rm L} = 1 - \frac{\Pi_{\rm L}}{\omega^2 - \boldsymbol{q}^2}, \quad \epsilon_{\rm T} = 1 - \frac{\Pi_{\rm T}}{\omega^2}.$$

(2) Plot the real and imaginary parts of the above dielectric constants and discuss the dielectric properties of the plasma in various regions of frequency and momentum.

5

Lattice gauge approach to QCD phase transitions

In Chapter 4, we studied the properties of QCD at high T using perturbation theory in terms of the QCD coupling, g. Although such an approach provides many insights into the qualitative features of the quark–gluon plasma, there are also some drawbacks. First, perturbation theory never reveals the physics of a phase transition, which is intrinsically non-perturbative. Also, perturbation theory has its own problems, even at extremely high T, as we saw in Section 4.9.

A powerful method that overcomes these difficulties is the lattice QCD approach originally proposed by Wilson (1974). The key concept is to define QCD on a space-time lattice. This is a neat method which leads to a gauge-invariant regularization of the ultraviolet (UV) divergences and simultaneously allows non-perturbative numerical simulations. In this chapter we will introduce the basic concepts of lattice QCD. Then we apply the method to the physics of QCD phase transition.

5.1 Basics of lattice QCD

5.1.1 The Wilson line

To define the gauge field on the lattice, it is not enough to discretize the standard QCD action. To maintain gauge invariance on the discretized space-time, we need to use a special (yet natural) basic variable for the gauge field. It is called the link variable.

Let us start with the gauge theory in continuous space-time and consider a path, P, connecting Euclidean space-time points, y_μ and x_μ, as shown in Fig. 5.1(a). The path may be characterized by a coordinate, $z_\mu(s)$, parametrized as $z_\mu(s=0)=y_\mu$ and $z_\mu(s=1)=x_\mu$. Now define the path-ordered product of the gauge field,

$$U_P(x, y; A) = \mathrm{P}\exp\left(ig\int_P dz_\mu\, A_\mu\right) = \mathrm{P}\exp\left(ig\int_0^1 ds\, \lambda_\mu A_\mu\right)$$

$$= \sum_{n=0}^{\infty}\frac{(ig)^n}{n!}\int_0^1 ds_1\, ds_2 \cdots ds_n\, \mathrm{P}[\lambda\cdot A(s_1)\cdots\lambda\cdot A(s_n)], \quad (5.1)$$

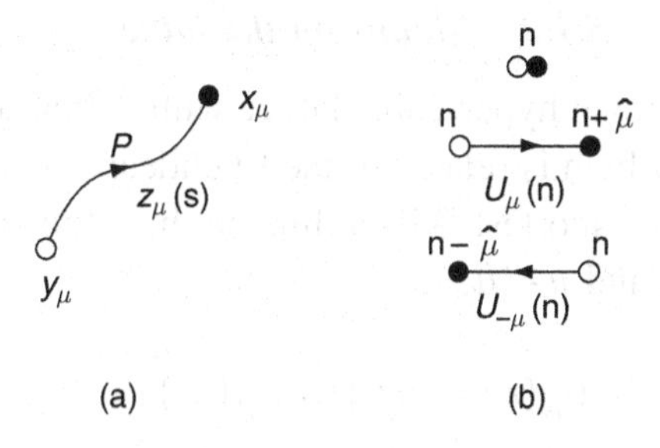

Fig. 5.1. (a) The Wilson line in Euclidean space-time. (b) Basic quark bilinears with gauge invariance.

where $\lambda_\mu = dz_\mu/ds$. The path-ordered symbol, P, is a generalization of the time-ordered symbol, T, and is necessary because $A_\mu = A^a_\mu t^a$ is a matrix in color-space; $U_P(x, y; A)$ is called the *Wilson line* or the Schwinger line. Note also that $U_P(x, y; A)$ is a mathematical analog of the time-evolution operator in quantum mechanics in the interaction picture:

$$U(t, t') = \mathrm{T}\exp\left(-i\int_{t'}^{t} H_{\mathrm{I}}(s)\, ds\right). \tag{5.2}$$

The Wilson line has the following properties, which can be proved from the definition of U_P (Exercise 5.1).

(i) It can be broken into parts at any arbitrary points on the path:

$$U_P(x, y; A) = U_{P_2}(x, z(s); A)U_{P_1}(z(s), y; A). \tag{5.3}$$

(ii) It satisfies a differential equation

$$\frac{d}{ds}U_P(z(s), y; A) = \left[ig\lambda_\mu(s)A_\mu(z(s))\right]\ U_P(z(s), y; A). \tag{5.4}$$

(iii) Under the local gauge transformation, it transforms covariantly:

$$U_P(x, y; A) \to U_P(x, y; A^V) = V(x)U_P(x, y; A)V^\dagger(y), \tag{5.5}$$

where the gauge transformation in the Euclidean space-time is $A^V_\mu(x) = V(x)$ $[A_\mu(x) + (i/g)\partial_\mu]V^\dagger(x)$.

The Wilson line is a useful tool in defining the non-local gauge-invariant objects. In particular, the gauge-invariant quark bilinear, $\bar{q}(x)U_P(x, y; A)q(y)$, and the gauge-invariant Wilson loop, tr $U_P(x, x; A)$, turn out to be important. (Here "tr" implies trace over color indices.) They are indeed building blocks which lead to the definition of QCD action on the lattice, as we will see very soon.

5.1.2 Gluons on the lattice

Consider a four-dimensional hyper-cubic lattice with lattice spacing a. We specify the lattice sites by n_μ, which is related to the Euclidean coordinates by $x_\mu = an_\mu$, as shown in Fig. 5.2. The shortest Wilson line on the lattice is the one connecting the neighboring sites n and $n+\hat{\mu}$,

$$U_\mu(n) = \exp\left(igaA_\mu(n)\right), \tag{5.6}$$

which is called the link variable. Here, $\hat{\mu}$ implies a vector pointing in the direction of μ, with length a; $U_\mu(n)$ is shown by an arrow on the link in Fig. 5.2. Since it is the minimal Wilson line, we do not need the path-ordering symbol P. Also, any non-minimal Wilson line on the lattice is represented by a product of link variables. Because $U_\mu(n)$ is a unitary matrix, the link pointing along the opposite direction can be written as $U_{-\mu}(n+\hat{\mu}) = [U_\mu(n)]^\dagger$.

Let us define a smallest closed loop,

$$U_{\mu\nu}(n) = U_\nu^\dagger(n)U_\mu^\dagger(n+\hat{\nu})U_\nu(n+\hat{\mu})U_\mu(n), \tag{5.7}$$

which transforms covariantly under a local gauge transformation as $U_{\mu\nu}(n) \to U_{\mu\nu}^V(n) = V(n)U_{\mu\nu}(n)V^\dagger(n)$. Also, it approaches the field strength tensor in the continuum limit ($a \to 0$):

$$U_{\mu\nu}(n) - 1 \underset{a\to 0}{\longrightarrow} ia^2 gF_{\mu\nu}(n). \tag{5.8}$$

This is obtained by using the Baker–Campbell Hausdorff formula, $\exp A \exp B = \exp(A+B+[A,B]/2+\cdots)$.

The trace tr $U_{\mu\nu}(n)$ is a minimal gauge-invariant object (the smallest Wilson loop), which is called the plaquette. We can construct a gauge-invariant gluon

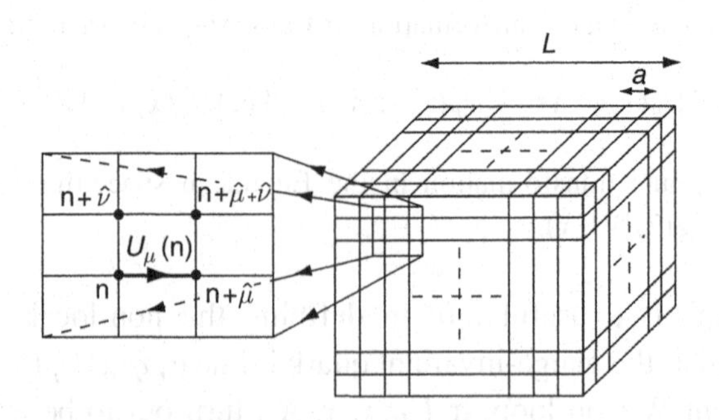

Fig. 5.2. A hyper-cubic lattice in Euclidean space-time with lattice constant a and lattice size L. Quarks, $q(n)$ (gluons, $U_\mu(n)$) are defined on the sites (links).

action from the plaquette as follows:

$$S_{\rm g} = \frac{2N_c}{g^2}\sum_{\rm p}\left(1-\frac{1}{N_c}\ {\rm Re\ tr}\ U_{\mu\nu}(n)\right) \tag{5.9}$$

$$\underset{a\to 0}{\longrightarrow} \frac{1}{4}\int d^4x F^b_{\mu\nu}(x)^2. \tag{5.10}$$

Here, $\sum_{\rm p}$ denotes the summation over all plaquettes with a definite orientation, namely

$$\sum_{\rm p} = \sum_n \sum_{1\le\mu<\nu\le 4} = \frac{1}{2}\sum_n \sum_{1\le\mu\ne\nu\le 4}. \tag{5.11}$$

Note also that $a^4\sum_n \simeq \int d^4x$ for small a.

The lattice gluon action, $S_{\rm g}$, is not unique in the sense that one may add arbitrary non-minimal terms which vanish in the continuum limit, $a\to 0$. One may even utilize this fact to accelerate the approach to the continuum limit. The lattice action with non-minimal terms is called the improved action, and is useful in obtaining precise results on a relatively coarse lattice in numerical simulations. For more details and references, see Kronfeld (2002).

5.1.3 Fermions on the lattice

Analogous to the gluon case, we first look for the small-size gauge-invariant objects. Up to the nearest neighbor coupling, one has three such objects (Fig. 5.1(b)):

$$\bar{q}(n)q(n),\quad \bar{q}(n+\hat{\mu})U_\mu(n)q(n),\quad \bar{q}(n-\hat{\mu})U_{-\mu}(n)q(n). \tag{5.12}$$

Here one may put any γ-matrices between $\bar{q}$ and q without spoiling the color gauge invariance. A special combination of the above terms is called Wilson's fermion action,

$$\begin{aligned} S_{\rm W} &= a^4\sum_n\Bigg[m\bar{q}(n)q(n) - \frac{1}{2a}\sum_\mu \bar{q}(n+\hat{\mu})\Gamma_\mu U_\mu(n)q(n) \\ &\quad -\frac{r}{2a}\sum_\mu\left(\bar{q}(n+\hat{\mu})U_\mu(n)q(n)-\bar{q}(n)q(n)\right)\Bigg] \\ &\equiv a^4\sum_{n',n}\bar{q}(n')\left(m\delta_{n',n}+D_{\rm W}(n',n;r)\right)q(n), \end{aligned} \tag{5.13}$$

where Wilson's Dirac operator is given by

$$D_{\rm W}(n',n;r) = -\frac{1}{2a}\sum_\mu\left[\delta_{n',n+\hat{\mu}}(r+\Gamma_\mu)U_\mu(n) - r\delta_{n',n}\right]. \tag{5.14}$$

The summation over μ is taken for both positive and negative directions; $\sum_{\mu} \equiv \sum_{\mu=\pm1,\pm2,\pm3,\pm4}$. The Γ_{μ} s are the hermitian γ-matrices satisfying $\Gamma_{\mu}^{\dagger} = \Gamma_{\mu}$, $\Gamma_{-\mu} = -\Gamma_{\mu}$ and $\{\Gamma_{\mu}, \Gamma_{\nu}\} = 2\delta_{\mu\nu}$. They are related to the Euclidean γ-matrices in Eq. (4.4) as $\Gamma_{\mu} = -i(\gamma_{\mu})_{\mathrm{E}}$; see Appendix B.1.

Taking the continuum limit of Eq. (5.13) with $(f(x+a)-f(x-a))/2a = f'(x) + O(a^2)$ and $(f(x+a)+f(x-a)-2f(x))/a^2 = f''(x)+O(a^2)$, we obtain

$$S_{\mathrm{W}} \xrightarrow[a \to 0]{} \int d^4x\, \bar{q}(x) \left(m - i\gamma \cdot D - \frac{ar}{2} D^2 \right) q(x). \tag{5.15}$$

Therefore, m corresponds to the quark mass, while the parameter r controls the magnitude of the higher derivative term, which vanishes in the continuum limit.

The parameter r is introduced to S_{W} to avoid the fermion-doubling problem on the lattice. To see this, consider the free fermion ($U_{\mu}(n) = 1$) and make a Fourier transform (F.T.): $S(p) = [\mathrm{F.T.}(m\delta_{n',n} + D_{\mathrm{W}}(n', n; r))]^{-1}$ in Eq. (5.13). Assuming that the space-time volume is infinite, we obtain

$$S(p)^{-1} = m(p) + i \sum_{\mu>0} \bar{p}_{\mu} \Gamma_{\mu}, \tag{5.16}$$

$$m(p) = m + \frac{r}{a} \sum_{\mu>0} \left(1 - \cos(p_{\mu}a)\right), \tag{5.17}$$

where p_{μ} is a continuous momentum restricted to the first Brillouin zone, $-\pi/a \le p_{\mu} \le \pi/a$, and $\bar{p}_{\mu} = a^{-1} \sin(p_{\mu}a)$. In fact $S(p)$ is the Euclidean fermion propagator on the lattice, and $m(p)$ acts as a momentum-dependent mass.

Since $\sin(p_{\mu}a)$ becomes zero for $p_{\mu}a = (0,0,0,0)$, $(\pi,0,0,0)$, $(0,\pi,\pi,\pi)$, (π,π,π,π), there arise $2^4 = 16$ degenerate fermions for $r=0$. This is called the fermion-doubling problem on the lattice. In fact, there is a no-go theorem by Nielsen and Ninomiya (1981a, b) which is as follows.

The fermion doubling always exists, if the free fermion action on the lattice has (i) bilinearity in the quark field, (ii) translational invariance, (iii) hermiticity (in the Minkowski space-time), (iv) locality in space-time, and (v) exact chiral symmetry.

For a simplified proof of the theorem using the Poincaré–Hopf theorem, see Karsten (1981).

If $r \neq 0$ in D_{W}, the second term on the right-hand side of Eq. (5.17) leads to a mass splitting of 16 fermions:

$$m(p) \simeq \begin{cases} m & ({}^{\forall}p_{\mu} \to 0) \\ m + \frac{2r}{a} N_{\pi} & ({}^{\exists}p_{\mu} \to \pi/a), \end{cases} \tag{5.18}$$

where $N_{\pi} (= 1,2,3,4)$ is the number of πs in $p_{\mu}a$. This implies that we can select only one light fermion by choosing $m \simeq 0$ and that all the other 15 fermions have masses of $O(1/a)$ for positive r. The price we have to pay is that the non-zero r breaks chiral symmetry explicitly for finite a, $\{\gamma_5, D_{\rm W}\} \neq 0$, and only $\mathrm{SU_V}(N_f) \times \mathrm{U_B}(1)$ symmetry is preserved in $S_{\rm W}$ (Exercise 5.3(1)). Namely, the Nielsen–Ninomiya theorem is evaded by breaking condition (v).

Another method that allows us to avoid fermion doubling is the staggered fermion formulation, in which 16 doublers are reinterpreted as four-component Dirac spinor × four-flavors (Susskind, 1977). Again, this breaks chiral symmetry explicitly, although some remnant symmetry, $\mathrm{U_V}(N_f/4) \times \mathrm{U_A}(N_f/4)$, remains.

Wilson's fermion action, Eq. (5.13), is conveniently given by

$$S_{\rm W} = \sum_{n',n} \bar{\psi}(n') F_{\rm W}(n', n) \psi(n), \tag{5.19}$$

$$F_{\rm W}(n', n) = \delta_{n'n} - \kappa \sum_{\mu} \delta_{n', n+\hat{\mu}} (r + \Gamma_\mu) U_\mu(n), \tag{5.20}$$

where we have redefined the quark field as $\psi = a^{3/2} q / \sqrt{2\kappa}$, where $\kappa = [2(ma + 4r)]^{-1}$ is the hopping parameter. If the quark mass, m, is large, κ is small, and the "hopping" to neighboring lattice site is suppressed.

Since the Wilson fermion breaks chiral symmetry explicitly by r for finite a, care must be taken when we study the chiral properties of the system, such as the dynamical breaking of chiral symmetry, the properties of Nambu–Goldstone bosons and the chiral phase transition at finite temperature. It would be better to define a generalized chiral transformation which has an exact symmetry, even for finite a, and reduces to the standard chiral symmetry for $a \to 0$.

Let us consider a transformation

$$q \to \mathrm{e}^{-i\theta_{\rm A}\hat{\gamma}_5} q, \qquad \bar{q} \to \bar{q} \mathrm{e}^{-i\theta_{\rm A}\gamma_5}, \tag{5.21}$$

$$\hat{\gamma}_5 = \gamma_5 (1 - 2aD_{\rm GW}). \tag{5.22}$$

These reduce to the standard axial rotation for $a \to 0$. Note that $D_{\rm GW}$ is a generalized Dirac operator, which is constructed such that $\bar{q} D_{\rm GW} q$ is invariant under Eq. (5.21) for finite a:

$$\gamma_5 D_{\rm GW} + D_{\rm GW} \hat{\gamma}_5 = 0. \tag{5.23}$$

This can also be written as $\{\gamma_5, D_{\rm GW}\} = 2aD_{\rm GW}\gamma_5 D_{\rm GW}$, which is called the Ginsparg–Wilson relation (Ginsparg and Wilson, 1982).

An explicit form of $D_{\rm GW}$ may be constructed as follows (Exercise 5.3(2)):

$$D_{\rm GW} = \frac{1}{2a}\left(1 + \frac{X}{\sqrt{X^\dagger X}}\right), \tag{5.24}$$

$$X \equiv D_{\rm W}^{(r=1)} - m_0, \tag{5.25}$$

where $m_0 a$ is a dimensionless parameter of $O(1)$. Unlike the case of m in the Wilson fermion in Eq. (5.13), m_0 is not directly related to the physical fermion mass, as is evident from the minus sign on the right-hand side of Eq. (5.25). Nevertheless, if we choose the region $0 < m_0 a < 2$, there exists an *exact* massless mode for $N_\pi = 0$ for finite a, and another 15 modes have a large mass, $(2/a)(2N_\pi - m_0 a) > 0$ (Exercise 5.3(3)).

It should be noted that $D_{\rm GW}$ breaks condition (v) of the no-go theorem in such a way that the definition of chiral symmetry is modified. This new lattice fermion, together with its parent fermion defined in a five-dimensional space-time (called the domain-wall fermion), is actively studied theoretically and numerically (Neuberger, 2001; Lüscher, 2002).

5.1.4 Partition function on the lattice

Once we have set up the actions on the lattice based on the gauge-invariant variables, the next step is to quantize the theory. This can be achieved in a straightforward way by the functional integral over quarks and gluons. Since the gauge field is represented by the group element U, the appropriate measure must be the Haar measure, dU (Exercise 5.2). Thus, the complete form of the partition function is give by

$$
\begin{aligned}
Z &= \int [dU][d\bar{\psi}\, d\psi]\, \mathrm{e}^{-S_{\rm g}(U)-S_q(\bar{\psi},\psi,U)} \\
&= \int [dU]\, \mathrm{Det}\, F[U]\, \mathrm{e}^{-S_{\rm g}(U)},
\end{aligned}
\tag{5.26}
$$

where the gluon action, $S_{\rm g}$, is given in Eq. (5.9), while the fermion action, S_q, has a general form, $S_q = \sum_{n',n} \bar{\psi}(n')F(n', n)\psi(n)$. (For the Wilson fermion, $F = F_{\rm W}$.) The "Det" in Eq. (5.26) should be taken for all indices of F, namely color, flavor, spin and space-time coordinates. Since F is a functional of U, Det F represents the effect of a quark loop under an arbitrary gauge-field background, as illustrated in Fig. 5.3.

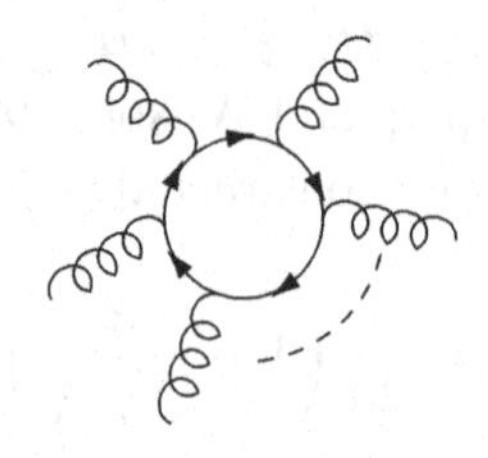

Fig. 5.3. The fermion determinant, Det F, under the gluon background.

Setting Det $F[U]$ = constant is called the quenched approximation, in which all the virtual quark–anti-quark excitations are ignored. A useful approximation for the Wilson fermion is a hopping parameter expansion, in which Det $F_{\rm W}$ and other quantities (such as the fermion propagator, $F_{\rm W}^{-1}$) are expanded in a power series of κ. This is an expansion valid for heavy quarks since $\kappa = [2(ma+4r)]^{-1}$.

The strong coupling expansion in terms of $1/g^2$ and the weak coupling expansion in terms of g are also methods useful for analyzing Z. We will discuss these expansions in relation to the quark confinement and the asymptotic scaling.

5.2 The Wilson loop

Whether the QCD vacuum leads to quark confinement or not can be tested by studying the expectation value of a non-local operator called the Wilson loop (Wilson, 1974). Consider a situation in which a heavy quark and an anti-quark are embedded in the non-perturbative QCD vacuum in Minkowski space. A gauge-invariant operator for such a pair may be written as $\mathcal{M}(x,y) = \bar{Q}(x)\gamma U_P(x,y;A)Q(y)$. Here, $Q(x)$ is a heavy quark with mass $m_{\rm Q}$, γ is an arbitrary γ-matrix, and $U_P(x,y;A)$ is the Wilson line in the Minkowski space. The mass of this pair is extracted from the correlation function,

$$I = \langle \mathcal{M}(y',x')\mathcal{M}^\dagger(x,y)\rangle, \tag{5.27}$$

with $x_0 = y_0 = 0$, $y'_0 = x'_0 = \mathcal{T} > 0$ and $|\boldsymbol{x}-\boldsymbol{y}| = R$; $\langle\cdots\rangle$ denotes the average with respect to the QCD partition function in Minkowski space without virtual quark loops (the quenched approximation).

In the leading order of the heavy quark mass $m_{\rm Q}$, the spatial momentum of the quarks can be neglected, and the quark propagator, $S(x',x) = \langle Q(x')\bar{Q}(x)\rangle$, satisfies

$$\left[\gamma_0\left(i\frac{\partial}{\partial x'_0} - gA_0(x')\right) - m_{\rm Q}\right]S(x',x) = \delta^4(x'-x). \tag{5.28}$$

This first order differential equation is easily solved as follows (Exercise 5.4):

$$\begin{aligned} iS(x',x) &= U_P(x',x;A_0)\delta^3(\boldsymbol{x}'-\boldsymbol{x}) \\ &\quad\times\left[\theta(x'_0-x_0)\,\mathrm{e}^{-im_{\rm Q}(x'_0-x_0)}\Lambda_+ + \theta(x_0-x'_0)\,\mathrm{e}^{im_{\rm Q}(x'_0-x_0)}\Lambda_-\right], \end{aligned} \tag{5.29}$$

where $\Lambda_\pm = (1\pm\gamma_0)/2$ are the projection operators for the positive and negative energy states. We adopt the boundary condition that the positive (negative) energy quark propagates forward (backward) in time. Since $m_{\rm Q}$ is large, the quark and the anti-quark do not move in space and only oscillate in time under the influence of A_0.

Using Eq. (5.29), we may evaluate Eq. (5.27) as follows:

$$I \propto \mathrm{tr_s}\left(\Lambda_- \gamma \Lambda_+ \bar{\gamma}\right) \mathrm{e}^{-2im_Q \mathcal{T}} \langle \mathrm{tr\ P\ e}^{-ig \oint_C dz_\mu A^\mu(z)} \rangle \tag{5.30}$$

$$\propto \exp\left[-i(2m_Q + V(R))\mathcal{T}\right], \tag{5.31}$$

where C is a Minkowski contour, as shown in Fig. 5.4, $\bar{\gamma} = \gamma_0 \gamma^\dagger \gamma_0$ and $\mathrm{tr_s}$ (tr) is a trace over spin (color). We have taken only the leading contribution for large $\mathcal{T}$ in the exponent in Eq. (5.31). The exponent $2m_Q + V(R)$ is simply the total energy of the quark and anti-quark separated by a distance R; thus, $V(R)$ is the potential energy (Brown and Weisberger, 1979; Eichten and Feinberg, 1981).

Making a continuation of the last factor in Eq. (5.30) to Euclidean space, we arrive at the Wilson loop,

$$\langle W(C) \rangle \equiv \langle \mathrm{tr\ P\ e}^{ig \oint_C dz_\mu A_\mu} \rangle \tag{5.32}$$

$$\propto \mathrm{e}^{-V(R)\mathcal{T}} \simeq \exp\left[-\left(KR + b + \frac{c}{R} + \cdots\right)\mathcal{T}\right], \tag{5.33}$$

where we have taken a rectangular path, as shown in Fig. 5.4, and have taken a limit $\mathcal{T} \gg R \to \infty$ in Eq. (5.33).

Since $V(R)$ is a potential between Q and $\bar{\mathrm{Q}}$, $K \neq 0$ implies the existence of a string-like linear confining potential. It also implies the area law of the Wilson loop, $\langle W(C) \rangle \sim \exp(-KA)$, where $A = R \times \mathcal{T}$ is simply the area inside the path C. In full QCD, where virtual $q\bar{q}$ pairs are allowed, the linear rising potential becomes flat at long distances because of the breaking of the string, $\mathrm{Q\bar{Q}} \to (\mathrm{Q}\bar{q})(q\bar{\mathrm{Q}})$.

On the lattice, the Wilson loop is written as a product of link variables defined on the path C:

$$\langle W(C) \rangle = \langle \mathrm{tr} \prod_{\mathrm{link} \in C} U_\mu(n) \rangle. \tag{5.34}$$

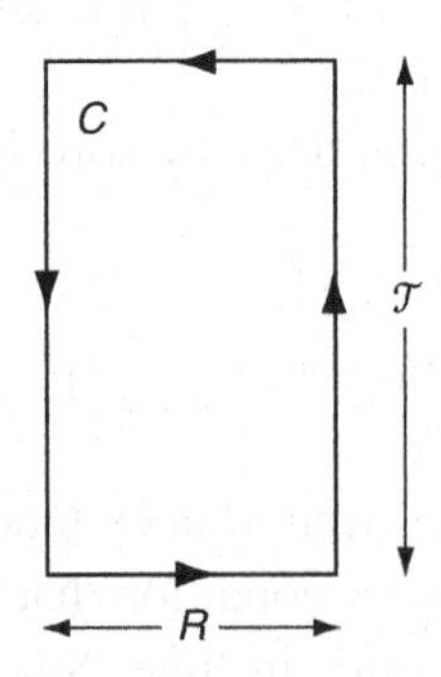

Fig. 5.4. A rectangular Wilson loop with the temporal (spatial) size $\mathcal{T}$ (R). The contour, C, can be either in Minkowski space or in Euclidean space.

In general, the $Q\bar{Q}$ potential as a function of R may be defined as

$$V(R) = -\lim_{\mathcal{T}\to\infty}\left[\frac{1}{\mathcal{T}}\ln\langle W(C)\rangle\right]_{\mathcal{T}\gg R}. \tag{5.35}$$

It is proved that $V(R)$ cannot increase faster than R at large distances (Seiler, 1978). The weak coupling perturbative expansion of $\langle W(C)\rangle$ in terms of g does not show the area law in any finite orders, while the strong coupling expansion in terms of $1/g$ shows the area law in the leading order (Section 5.3). Non-perturbative numerical simulations on the lattice provide clear evidence of the linear rising, $V(R)$, not only in the strong coupling, but also in the weak coupling regime in the quenched approximation, as will be shown in Fig. 5.7.

5.3 Strong coupling expansion and confinement

Let us evaluate the Wilson loop in the strong coupling limit, $g\to\infty$. Since $S_{\rm g}$ is proportional to $1/g^2$, we can expand $\exp(-S_{\rm g}) = 1 - S_{\rm g} + S_{\rm g}^2/2 + \cdots$ to obtain

$$\langle W(C)\rangle = \frac{1}{Z}\int [dU]\ \mathrm{tr}\prod_{\mathrm{link}\in C} U_\mu(n)\ \sum_{l=0}^{\infty}\frac{1}{l!}(-S_{\rm g})^l. \tag{5.36}$$

Only the three following integrals are necessary to extract a leading contribution to $\langle W(C)\rangle$ in the strong coupling:

$$\int dU = 1, \qquad \int dU\ U_{ij} = 0, \qquad \int dU\ U_{ij}U^\dagger_{kl} = \frac{1}{N_c}\delta_{il}\delta_{jk}, \tag{5.37}$$

where U_{ij} $(i, j = 1, 2, \ldots, N_c)$ is an SU(N_c) matrix (Creutz, 1985). (see Exercise 5.2(5).) The key observation here is that all the Us from the Wilson loop and the $U^\dagger$s from $(-S_{\rm g})^l$ should be paired in the leading order of $1/g^2$ in Eq. (5.36). This means that the area inside the Wilson loop is tiled with a minimum number of plaquettes, as shown in Fig. 5.5. Structures other than the minimal surface are higher orders in $1/g^2$.

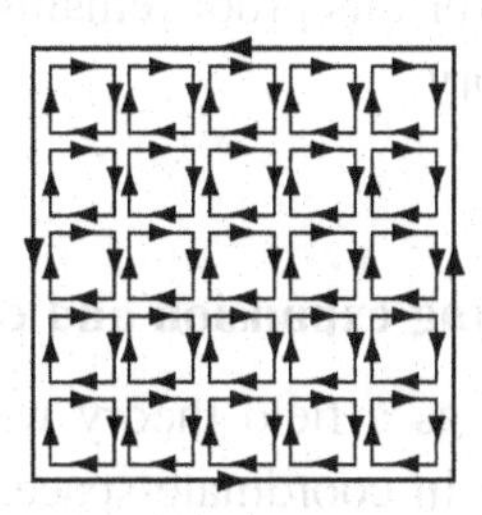

Fig. 5.5. A minimum surface in which the Wilson loop is tiled by the fundamental plaquettes in the strong coupling limit.

In the evaluation of the numerator of Eq. (5.36), each plaquette has a contribution, $1/g^2$. Also, each integration on the link gives a factor $1/N_c$, and the contraction of the color indices gives a factor N_c on each site. On the other hand, Z (the denominator of Eq. (5.36)) is unity in the leading order.

Therefore, one arrives at the following formula in the lowest order of the strong coupling expansion:

$$\frac{1}{N_c}\langle W(C)\rangle \xrightarrow[g^2\to\infty]{} \frac{1}{N_c}\cdot\left(\frac{1}{g^2}\right)^{N_{\rm plaq}}\cdot\left(\frac{1}{N_c}\right)^{N_{\rm link}}\cdot N_c^{N_{\rm site}} \tag{5.38}$$

$$= \left(\frac{1}{N_c g^2}\right)^{R\mathcal{T}/a^2} = \exp\left(-\frac{\ln N_c g^2}{a^2}R\mathcal{T}\right), \tag{5.39}$$

where we have used the relation $N_{\rm link} - N_{\rm site} + 1 = N_{\rm plaq}$ and $N_{\rm plaq}a^2 = R\mathcal{T}$ (the area inside C). Since it shows the area law, the confinement is proved in the strong coupling. The linear rising potential thus obtained is give by

$$V(R) = KR, \quad \text{with} \quad Ka^2 = \ln(N_c g^2). \tag{5.40}$$

If we consider higher orders of the strong coupling expansion, "rough" surfaces should be taken into account. Nevertheless, the confining feature is stable for small perturbations in $1/g^2$. In fact, there exists a theorem that, for sufficiently large g, the strong coupling expansion converges and exhibits confinement for all compact gauge groups in all space-time dimensions (Osterwalder and Seiler, 1978).

An obvious question here is whether the confining feature still survives in the weak coupling regime of QCD. For compact QED (quantum electrodynamics formulated in terms of the U(1) link variable), the confining phase in the strong coupling undergoes a phase change to a non-confining Coulomb phase in the weak coupling (Guth, 1980). On the other hand, in QCD in four space-time dimensions with $N_c = 3$, the confining phase is expected to persist in the weak coupling regime without phase transition. This receives strong support from numerical simulations on the lattice. However, a rigorous and analytic proof of this feature is still missing, and the search for this proof remains one of the most challenging problems in quantum field theory.

5.4 Weak coupling expansion and continuum limit

Lattice QCD may be regarded as a field theory with an ultraviolet (UV) regularization, i.e. the lattice cutoff in coordinate-space. To compare the predictions of lattice QCD with observables such as the hadron masses and the critical temperature, we need to take the continuum limit, $a \to 0$. If we formulate lattice perturbation theory in terms of the coupling constant, g, there appear an infinite number of vertices generated by the expansion of $U_\mu(n) = 1 + igaA_\mu(n) + \cdots$ in the gauge action, S_g, and in the measure, $[dU]$. Nevertheless, the theory is known to be renormalizable, i.e. all the UV divergences in the $a \to 0$ limit can be absorbed in the coupling constant and masses, and physical observables are independent of a (Reisz, 1989).

The lattice QCD with finite a may be alternatively regarded as an effective field theory defined at some length scale, a. In this case, $g(a)$ is interpreted as an effective coupling constant defined at the scale, a, where quantum fluctuations with wavelength shorter than a are integrated out. If a is small enough, the perturbation theory in terms of g is reliable because of the asymptotic freedom, $g(a \to 0) \to 0$, as will be shown shortly.

Since the lattice spacing is the only dimensionful parameter of the theory with massless quarks, any observable on the lattice, $\mathcal{O}$, can be written in the form:

$$\mathcal{O} = a^{-d} G(g(a)), \tag{5.41}$$

where d is the mass-dimension of $\mathcal{O}$ and G is a dimensionless function of g. If $\mathcal{O}$ is a physical quantity, such as the hadron mass or the string tension, it is a-independent and satisfies the following equation:

$$a\frac{d\mathcal{O}}{da} = \left(a\frac{\partial}{\partial a} - \beta_{\text{LAT}} \frac{\partial}{\partial g} \right) \mathcal{O} = 0; \tag{5.42}$$

$$\beta_{\text{LAT}}(g) = -a\frac{dg(a)}{da} = -\beta_0 g^3 - \beta_1 g^5 + \cdots. \tag{5.43}$$

By integrating Eq. (5.42), we obtain

$$G(g) = \exp\left(-d \int^g \frac{dg'}{\beta_{\text{LAT}}(g')} \right). \tag{5.44}$$

When a is extremely small, we may apply lattice perturbation theory for the short-distance part of the heavy quark potential to determine β_{LAT}. The first two terms on the right-hand side of Eq. (5.43) are known to be independent of how the theory is regularized (the momentum cutoff, the Pauli–Villers regularization, the lattice cutoff, the dimensional regularization, and so on). This is called the regularization-scheme independence of $\beta_{0,1}$ (see Exercise 5.5; also see Sect. 5.3 of Muta (1998)). Because of this, one may identify $\beta_{0,1}$ in Eq. (5.43) with $\beta_{0,1}$ given in Eq. (2.29).

By integrating Eq. (5.43), we obtain

$$a = \Lambda_{\rm LAT}^{-1} \cdot \exp\left(-\frac{1}{2\beta_0 g^2}\right) \cdot (\beta_0 g^2)^{-\frac{\beta_1}{2\beta_0^2}} \cdot (1 + O(g^2)). \tag{5.45}$$

This implies the asymptotic freedom in which the effective coupling on the lattice is a decreasing function of a, $g(a \to 0) \to 0$; $\Lambda_{\rm LAT}$ is the lattice scale parameter to be determined from experimental input. Expressing $g(a)$ in terms of a and $\Lambda_{\rm LAT}$ is also useful:

$$\frac{1}{g^2(a)} = \beta_0 \ln\left(\frac{1}{a^2\Lambda_{\rm LAT}^2}\right) + \frac{\beta_1}{\beta_0} \ln\ln\left(\frac{1}{a^2\Lambda_{\rm LAT}^2}\right) + \cdots. \tag{5.46}$$

By performing perturbative calculations of the same physical quantity in different regularization schemes, one can formulate a relation between $\Lambda_{\rm LAT}$ and $\Lambda_{\overline{\rm MS}}$. The latter is known from various experiments in high-energy processes. An alternative way of extracting $\Lambda_{\rm LAT}$ is to carry out a direct numerical simulation of a certain physical quantity (such as the string tension, hadron masses, mass splitting of hadrons, and so on) and to compare the result with experimental values.

For example, the string tension, which has mass-dimension two ($d = 2$), should behave as follows:

$$Ka^2 = C_K \exp\left(-\frac{1}{\beta_0 g^2}\right) (\beta_0 g^2)^{-\beta_1/\beta_0^2}, \tag{5.47}$$

where C_K is a dimensionless numerical constant independent of g. Thus, the functional form of the physical quantities for $g \sim 0$ is severely constrained. This is called the *asymptotic scaling*, which can be used to check whether the system is close enough to the continuum limit. Figure 5.6 shows a schematic illustration of the smooth crossover of the dimensionless string tension, Ka^2, from the strong coupling regime given by Eq. (5.40) to the asymptotic scaling regime given by Eq. (5.47).

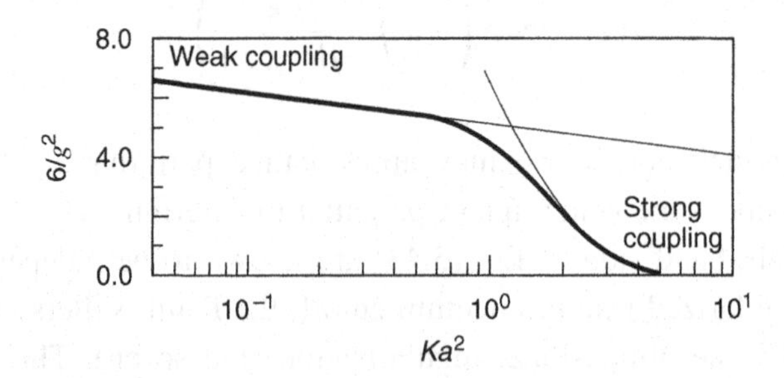

Fig. 5.6. The crossover behavior of the dimensionless string tension, Ka^2, from the strong coupling regime, $1/g^2 \sim 0$, to the weak coupling (asymptotic scaling) regime, $1/g^2 \sim \infty$.

5.5 Monte Carlo simulations

Suppose we have a lattice with N_s (N_t) sites in each spatial (temporal) direction. Then the total number of links is $N_s^3 \times N_t \times 4$. The total number of gluon integrations, $\int [dU]$, for a moderate lattice size, $N_s = N_t = 32$, is given by

$$(N_s^3 \times N_t \times 4)_{\text{links}} \times 8_{\text{color}} \sim 3 \times 10^7. \tag{5.48}$$

This is a hopelessly large number of dimensions for standard methods of numerical integration, such as the Simpson method, the Gaussian quadrature, etc. In this case, Monte Carlo (MC) integration, which is a statistical way of evaluating the integral, plays a powerful role. For a rapidly varying integrand, the MC integration should be supplemented by the importance sampling (IS) to achieve better accuracy, in which the rapidly varying part is sampled more often than the slowly varying part.

MC integration with importance sampling in the quenched approximation (Eq. (5.26) with Det $F = 1$) consists of two steps.

(i) Generate a chain of gauge configurations,

$$U^{(1)} \to U^{(2)} \to \cdots \to U^{(N)}, \tag{5.49}$$

where $U^{(i)}$ is a set of link variables on the lattice, which should be arranged to appear with a probability $W[U] = Z^{-1} \exp(-S_g(U))$.

(ii) Use the gauge configurations generated in Eq. (5.49) to calculate the expectation value of an arbitrary operator $A(U)$,

$$\langle A \rangle = \frac{1}{N} \sum_{n=1}^{N} A^{(n)} \pm \sqrt{\frac{\sigma^2}{N}}, \tag{5.50}$$

where $A^{(n)} = A(U^{(n)})$, and a statistical variance,

$$\sigma^2 = \frac{1}{N} \sum_{n=1}^{N} (A^{(n)} - \langle A \rangle)^2. \tag{5.51}$$

Generating a new sample, $U \to U'$, in (i) (which is called the updating) can be performed by the Markov process, defined as follows:

$$W'[U'] = \sum_U W[U] P(U \to U'); \tag{5.52}$$

$$\text{(a)} \quad \sum_{U'} P(U \to U') = 1; \tag{5.53}$$

$$\text{(b)} \quad P(U \to U') > 0 \text{ (strong ergodicity)}; \tag{5.54}$$

$$\text{(c)} \quad \sum_U W\,[U] = 1. \tag{5.55}$$

Here, $P(U \to U')$ is the probability that U' is accepted after U. We expect a unique equilibrium distribution, $W_{\rm eq}$, to be obtained after applying many successive updating processes to some initial distribution, W_0, and hence $W_{\rm eq}$ becomes a fixed point: $W_{\rm eq}[U'] = \sum_U W_{\rm eq}[U]P(U \to U')$. A sufficient condition for P to lead such an equilibrium distribution is the detailed balance:

$$W_{\rm eq}[U]P(U \to U') = W_{\rm eq}[U']P(U' \to U). \tag{5.56}$$

Typical methods for updating are the Metropolis algorithm (Metropolis *et al.*, 1953) and the heat-bath algorithm (Creutz, 1985):

$$P(U \to U') = \begin{cases} \min(1, {\rm e}^{-(S_{\rm g}(U')-S_{\rm g}(U))}) & \text{Metropolis} \\ {\rm e}^{-S_{\rm g}(U')} & \text{heat-bath,} \end{cases} \tag{5.57}$$

where we have dropped the unimportant normalization factor. It is easy to see that both algorithms satisfy the detailed balance, Eq. (5.56), and lead to the equilibrium "Boltzmann" distribution, $W_{\rm eq}[U] \propto {\rm e}^{-S_{\rm g}(U)}$. In both algorithms, the updating is carried out step by step. We start with a configuration, U, and change a single-link variable according to the rules above. Then we go to the next-link variable, repeat the procedure, and so on. One "sweep" corresponds to the change of all the link variables. After many sweeps, the gauge configuration becomes a member of the Boltzmann distribution.

For the full QCD simulation with dynamical quarks, one has to expend extra effort in order to deal with the fermionic determinant, Det$F(U)$, in Eq. (5.26). Various approaches, such as the pseudo-fermion method and the hybrid MC method, have been developed. For further details of these algorithms in full QCD simulations, see Chap. 7 of Montvay and Münster (1997).

Readers who wish to learn more about the basics and applications of lattice QCD should consult lecture notes (Ukawa, 1995; Gupta, 1999; Di Pierro, 2000) and books (Creutz, 1985; Montvay and Münster, 1997; Smit, 2002). For those who want to practice QCD MC simulations, some basic programs and tool kits may be downloaded.[1]

In the following, we present some high precision results using MC simulations in the quenched approximation.

Potential between heavy quark and anti-quark

The data in Fig. 5.7 shows a dimensionless $\rm Q\bar{Q}$ potential,

$$[V(R) - V(R_0)] \times R_0, \tag{5.58}$$

[1] QCDF90 may be downloaded from http://cpc.cs.qub.ac.uk/cpc/; the *Lattice Tool Kit in Fortran 90* may be downloaded from http://nio-mon.riise.hiroshima-u.ac.jp/LTK/.

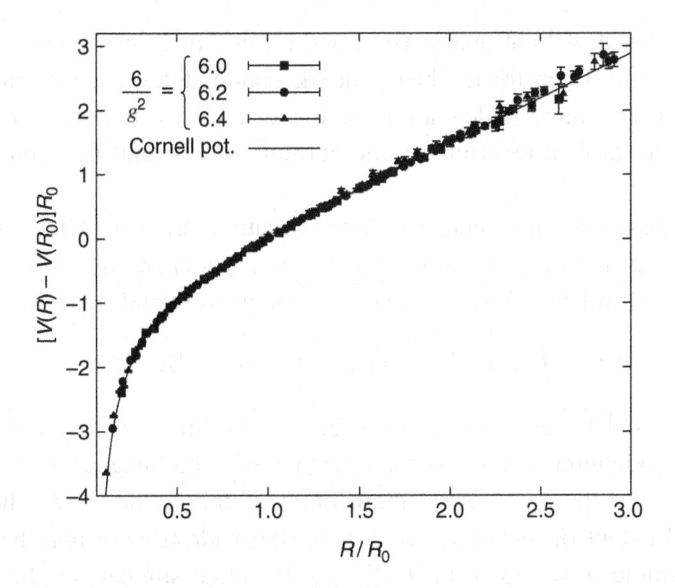

Fig. 5.7. A dimensionless $Q\bar{Q}$ potential as a function of a dimensionless quark–anti-quark separation, R/R_0, with R_0 being the Sommer scale defined by Eq. (5.59). Different symbols correspond to different lattice couplings, $g(a)$, and hence different lattice spacings. The solid line shows an empirical Cornell potential (linear + Coulomb type). Adapted from Bali (2001).

as a function of R/R_0 obtained by using the formula in Eq. (5.35) with MC simulations of the Wilson loop (Bali, 2001). Note that R is the distance between heavy quarks and R_0 is called the Sommer scale, defined as

$$R^2 \left. \frac{dV(R)}{dR} \right|_{R=R_0} = 1.65. \tag{5.59}$$

Simulations with different lattice couplings, $g(a)$, correspond to different lattice spacings, a. The latter can be fixed, for example by taking a phenomenological value, $R_0 \simeq 0.5\,\text{fm}$, obtained from the bottomonium phenomenology. In Fig. 5.7, $a = 0.094\,\text{fm}$ (squares: $6/g^2 = 6.0$), $a = 0.069\,\text{fm}$ (circles: $6/g^2 = 6.2$) and $a = 0.051\,\text{fm}$ (triangles: $6/g^2 = 6.4$).

The result of the MC simulations in Fig. 5.7 clearly shows that the heavy quark potential has a linear confining part at long distance and an attractive Coulombic interaction at short distance. The MC results agree not only qualitatively, but also quantitatively, with an empirical linear + Coulomb potential (the Cornell potential), shown by the solid line, $V(r) = Kr - b/r + \text{const.}$, with $b = 0.295$. The MC results in Fig. 5.7 show no appreciable dependence on a.

Mass spectrum of light hadrons

On the lattice, the masses of mesons and baryons may be calculated directly without recourse to the inter-quark potentials. This is particularly important for hadrons composed of light quarks (u, d and s), in which internal motion can be highly relativistic. Even for the heavy quark bound states, such as charmonium and bottomonium, direct calculations of the masses are important not only in themselves, but also in determining the current masses and the strong coupling constant.

Hadron masses in the quenched approximation are calculated as follows. Consider a local mesonic operator, $\mathcal{M}(x) = \bar{q}(x)\gamma q(x)$, where γ is an arbitrary γ-matrix. The correlation function of such operators in Euclidean space is given by

$$D(\tau) = \int d^3x \, \langle \mathcal{M}(\tau, \boldsymbol{x}) \mathcal{M}^\dagger(0) \rangle \xrightarrow[\tau \to \infty]{} |Z|^2 \, \mathrm{e}^{-m\tau}. \tag{5.60}$$

Here, m (Z) is the mass (the pole residue) of the lightest bound state, which has the same quantum number as the operator $\mathcal{M}$. The integration, $\int d^3x$, gives a projection onto the zero spatial momentum of the bound state. Therefore, if the temporal extent of the lattice is infinite, the hadron mass may be extracted from the formula $m = -(1/\tau) \ln D(\tau)|_{\tau \to \infty}$. In actual simulations, however, the space-time lattice volume is finite, $0 \le \tau \le N_\mathrm{t} a$ and $0 \le |\boldsymbol{x}| \le N_\mathrm{s} a$. Then we need to impose certain boundary conditions, such as the ones in Eqs. (4.7) and (4.8). Accordingly, the exponential on the right-hand side of Eq. (5.60) becomes $\exp[-m\tau] + \exp[-m(N_\mathrm{t} a - \tau)]$. (See Chapter 7.)

Shown in Fig. 5.8 is a mass spectrum of low-lying mesons and baryons composed of u, d and s quarks calculated by the MC simulations in the quenched

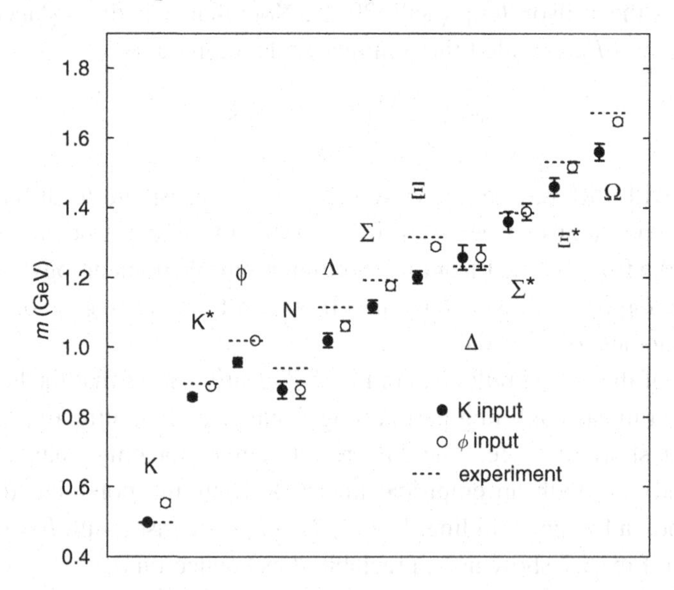

Fig. 5.8. Low-lying masses of mesons and baryons composed of u, d and s quarks. Horizontal bars are the experimental masses, while the black and white circles are the lattice QCD data in the quenched approximation. Adapted from Aoki *et al.* (2000).

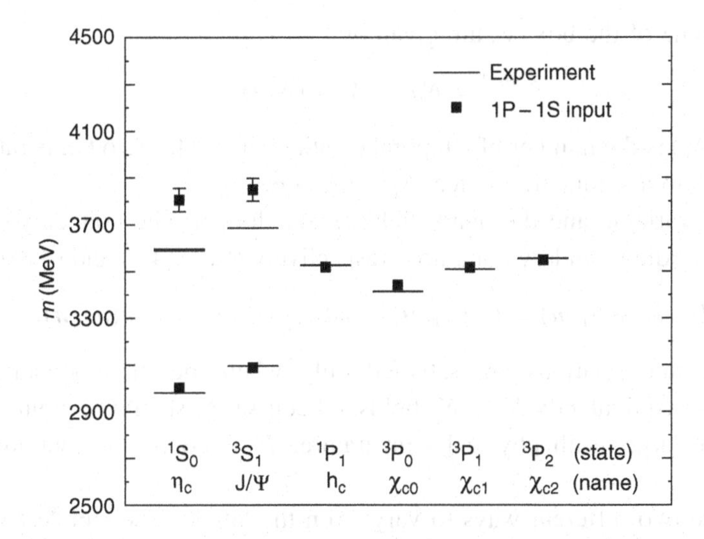

Fig. 5.9. Low-lying masses of charmonium (see Table 7.1). Horizontal bars are the experimental masses; black squares are the lattice QCD data in the quenched approximation. Each state has a classification in terms of its spin, s, orbital angular momentum, L, and total angular momentum, J, as $^{2s+1}L_J$. Adapted from Okamoto *et al.* (2002).

approximation (Aoki *et al.*, 2000, 2003). Extrapolation to the continuum limit is taken by using the data for $a = 0.05$ fm to 0.1 fm. (The experimental mass of the ρ-meson is used to set the lattice scale.) The u-d quark mass is fixed by the experimental pion mass. Black (white) circles are the lattice data where the s quark mass is fixed by the experimental kaon (ϕ-meson) mass. Simulation data and the experiments agree within 11%, even in the quenched approximation.

Shown in Fig. 5.9 is a mass spectrum of the $c\bar{c}$ bound states (charmonium) calculated using the MC simulations in the quenched approximation (Okamoto *et al.*, 2002). Extrapolation to the continuum limit is taken by using the data for $a = 0.07$ fm to 0.2 fm. (Splitting between the spin-averaged masses of 1S and 1P states is used to set the lattice scale.) The c quark mass is fixed by the spin-averaged mass of the 1S states. Lattice data and experimental data agree reasonably well, at least in the ground states in each of the quantum numbers.

5.6 Lattice QCD at finite T

As we have discussed in Chapter 4, field theories at finite T can be defined in a slab with a finite temporal extent (Fig. 4.2). Let us consider a hyper-cubic lattice with the lattice spacing a, as shown in Fig. 5.2. Then, the temperature, T, and spatial volume of the box, V, are given by

$$T^{-1} = N_t a, \quad V = (N_s a)^3, \tag{5.61}$$

where N_t (N_s) is the number of temporal (spatial) sites. The zero temperature limit corresponds to a symmetric lattice, $N_t = N_s \to \infty$.

The link variable and the quark field satisfy the periodic boundary condition and anti-periodic boundary condition, respectively (Eqs. (4.7) and (4.8)):

$$U_\mu(n_4 + N_t, \boldsymbol{n}) = U_\mu(n_4, \boldsymbol{n}), \quad \psi(n_4 + N_t, \boldsymbol{n}) = -\psi(n_4, \boldsymbol{n}). \tag{5.62}$$

Thermodynamic relations are satisfied only when the thermodynamic limit, $V^{1/3} \gg T$, or equivalently $N_s \gg N_t$, holds. Of course, N_t should be simultaneously taken to be large, with physical temperature, T, fixed in the continuum limit ($a \to 0$).

There are two different ways to vary T on the lattice. The simplest way is to change N_t, with a and N_s fixed. This method, however, can only change T by a discrete amount. Another way is to change a with N_s and N_t fixed, assuming that N_s is large enough. Changing a is equivalent to changing the lattice coupling, $g(a)$: a small (large) $g(a)$ corresponds to a small (large) a and hence to a large (small) T. In the leading order of the weak coupling expansion, Eq. (5.45) and Eq. (5.61) lead to

$$T \simeq \frac{\Lambda_{\rm LAT}}{N_t} \exp\left(\frac{1}{2\beta_0 g^2(a)}\right), \tag{5.63}$$

which shows that a small variation in g^2 is amplified exponentially in T. In this method, one can change T continuously. One should note that a decrease in g (or a) leads not only to the increase in T, but also to a decrease in V if $N_{s,t}$ are fixed, as seen from Eq. (5.61). Therefore, a necessary condition for the thermodynamic limit, $V^{1/3} \gg T$, will eventually be lost for small a.

The methods explained above are sufficient for studying the thermal average of various operators at fixed T and V. However, when we try to calculate the energy density, ε, the pressure, P, and the entropy density, s, we need to take derivatives of $\ln Z$ with respect to T or V, as shown in Eq. (3.16). To carry out the derivative explicitly, the introduction of an anisotropic lattice, where the lattice spacing in the temporal direction, a_t, is different from that in the spatial direction, a_s, is useful:

$$T^{-1} = N_t a_t, \quad V = (N_s a_s)^3, \quad \zeta \equiv \frac{a_s}{a_t} \neq 1. \tag{5.64}$$

Here, ζ is called an anisotropy parameter. The cost, however, is that the actions Eq. (5.9) and Eq. (5.13) need to be modified to incorporate the anisotropy; namely, the number of coupling constants (g_t, g_s, κ_t and κ_s) are doubled.

Another approach which may be taken to obtain bulk thermodynamic quantities within the isotropic lattice is the integral method. The pressure itself is not written as an expectation value of some operator; however, its derivative with respect to a set of coupling parameters, $\vec{\eta}$ (such as g, κ), may be written as thermal expectation values. Therefore, the pressure can be reconstructed as a line integral over a path parametrized by $\vec{\eta}$:

$$P = \frac{T}{V}\ln Z = \frac{T}{V}\int_{\eta_0}^{\eta} d\vec{\eta}\,\frac{\partial \ln Z}{\partial \vec{\eta}} + P_0$$
$$= -\frac{T}{V}\int_{\eta_0}^{\eta} d\vec{\eta}\,\left\langle \frac{\partial(S_g + S_q)}{\partial \vec{\eta}} \right\rangle + P_0. \tag{5.65}$$

Assuming that the integrand is a smooth function of $\vec{\eta}$ (which is valid as long as the lattice size is finite), the line integral does not depend on the path chosen to connect the initial point, η_0, and the end point, η. The pressure obtained as an integral of the entropy discussed in Section 3.7 is an example of the integral method with $\eta = T$. Once we obtain P in this method, the energy density and entropy density are simply obtained by using the thermodynamic relations $s = \partial P/\partial T$ and $\varepsilon + P = sT$, as long as the system is close to the thermodynamic limit, $N_s \gg N_t$.

Note that P and ε, simulated in the quenched approximation (Fig. 3.5) and in full QCD (Fig. 3.6), are calculated according to the integral method. The figures show that the energy density increases rapidly at a temperature of around 273 MeV for $N_f = 0$ and around 175 MeV for $N_f = 2$. This is clear evidence of the phase change from hadronic matter to the quark–gluon plasma associated with the liberation of color degrees of freedom.

5.7 Confinement–deconfinement transition in $N_f = 0$ QCD

For SU(N_c) gauge theory without dynamical quarks ($N_f = 0$), the QCD phase transition can be characterized by a discrete symmetry, Z(N_c) (Polyakov, 1978). This symmetry is unbroken in the confining phase at low T, whereas it is spontaneously broken in the deconfined phase at high T. For full QCD with N_f light flavors, Z(N_c) is no longer a good symmetry, but a continuous $\mathrm{SU}_L(N_f) \times \mathrm{SU}_R(N_f)$ chiral symmetry plays a crucial role, as will be discussed in Chapter 6.

In this section, we take $N_f = 0$ QCD and discuss the confinement–deconfinement transition based on the Z(N_c) symmetry. To warm up, let us first consider QED in continuous space-time at finite T and examine its gauge structure. The key observation is that, although the gauge field, $A_\mu(\tau, \boldsymbol{x})$, is periodic in the temporal direction, the gauge transformation does not have to be periodic. For example, consider the following aperiodic gauge transformation:

$$V(\tau + 1/T, \boldsymbol{x}) = e^{i\theta} V(\tau, \boldsymbol{x}), \tag{5.66}$$

where θ is a space-time-independent constant. The periodicity of A_μ is preserved under this gauge transformation as follows:

$$\begin{aligned} gA_\mu^V(\tau+1/T,\boldsymbol{x}) &= V(\tau+1/T,\boldsymbol{x})(gA_\mu(\tau+1/T,\boldsymbol{x})+i\partial_\mu)V^\dagger(\tau+1/T,\boldsymbol{x}) \\ &= gA_\mu^V(\tau,\boldsymbol{x})+\partial_\mu\theta = gA_\mu^V(\tau,\boldsymbol{x}); \end{aligned} \tag{5.67}$$

namely, the QED action is invariant even under this aperiodic gauge transformation. On the other hand, the following non-local quantity, called the Polyakov line, is not invariant under Eq. (5.66):

$$L(\boldsymbol{x}) = \exp\left[ig\int_0^{1/T} d\tau A_4(\tau,\boldsymbol{x})\right] \tag{5.68}$$

$$\rightarrow V(1/T,\boldsymbol{x})L(\boldsymbol{x})V^\dagger(0,\boldsymbol{x}) = \mathrm{e}^{i\theta}L(\boldsymbol{x}). \tag{5.69}$$

In the same way, let us look for the structure of the aperiodic gauge transformation of the following type in SU(N_c) gauge theory:

$$V(\tau+1/T,\boldsymbol{x}) = \mathbf{z}V(\tau,\boldsymbol{x}). \tag{5.70}$$

Since V is an element of SU(N_c), $\mathbf{z}$ must satisfy

$$\mathbf{z}\mathbf{z}^\dagger = 1 \text{ and } \det\mathbf{z} = 1. \tag{5.71}$$

The condition that the gauge field, A_μ, must be periodic implies that

$$\begin{aligned} gA_\mu^V(\tau+1/T,\boldsymbol{x}) &= V(\tau+1/T,\boldsymbol{x})(gA_\mu(\tau+1/T,\boldsymbol{x})+i\partial_\mu)V^\dagger(\tau+1/T,\boldsymbol{x}) \\ &= g\mathbf{z}A_\mu^V(\tau,\boldsymbol{x})\mathbf{z}^\dagger + i\mathbf{z}\partial_\mu\mathbf{z}^\dagger = gA_\mu^V(\tau,\boldsymbol{x}). \end{aligned} \tag{5.72}$$

The last equality is satisfied when

$$\mathbf{z}G\mathbf{z}^\dagger = G \quad \text{and} \quad \mathbf{z}\partial_\mu\mathbf{z}^\dagger = 0, \tag{5.73}$$

where G is an arbitrary element of the group SU(N_c). Note that $\mathbf{z}$, satisfying the conditions Eq. (5.71) and Eq. (5.73), has the form

$$\mathbf{z} = \mathrm{e}^{2\pi i n/N_c}\cdot\mathbf{1} \equiv z\cdot\mathbf{1} \qquad (n = 0, 1, 2, \ldots, N_c-1), \tag{5.74}$$

where $zz^* = 1$ and $z^{N_c} = 1$. Such $\mathbf{z}$s form a discrete subgroup of SU(N_c) and commute with all the elements of SU(N_c). This is called the "center" of SU(N_c), and is written as Z(N_c).

Under the aperiodic gauge transformation, Eq. (5.70), the gauge field is periodic, and hence the gauge action is invariant. However, the Polyakov line defined below is not invariant:

$$\begin{aligned} L(\boldsymbol{x}) &= \frac{1}{N_c}\,\mathrm{tr}\,\mathrm{P}\exp\left[ig\int_0^{1/T} A_4(\tau,\boldsymbol{x})\,d\tau\right] \equiv \mathrm{tr}\,\Omega(\boldsymbol{x}) \\ &\rightarrow \mathrm{tr}\left[V(1/T,\boldsymbol{x})\Omega(\boldsymbol{x})V^\dagger(0,\boldsymbol{x})\right] = zL(\boldsymbol{x}). \end{aligned} \tag{5.75}$$

The transformation V in Eq. (5.70) and its action on the gauge field can be written explicitly on the lattice. First, let us consider the gauge transformation which is aperiodic at the temporal boundary $V(n_4 = N_t, \boldsymbol{n}) = z$, with $V(n_4 \neq N_t, \boldsymbol{n}) = 1$. Then, applying suitable periodic gauge transformations on top of this, a transformation of the link variables at a fixed time-slice, $n_4 = l$, results (see Fig. 5.10):

$$U_4(n_4 = l, \boldsymbol{n}) \to zU_4(n_4 = l, \boldsymbol{n}). \tag{5.76}$$

Since the gauge action, S_g, on the lattice always contains the product of U_4 and U_{-4}, it is invariant under Eq. (5.76), as it should be. On the other hand, the lattice version of the Polyakov line transforms as

$$L(\boldsymbol{x}) = \frac{1}{N_c} \operatorname{tr} \prod_{n_4=0}^{N_t-1} U_4(n_4, \boldsymbol{x}) \to zL(\boldsymbol{x}). \tag{5.77}$$

Thus, $L(\boldsymbol{x})$ acts as an order parameter associated with the $Z(N_c)$ symmetry of the pure gauge action. A finite expectation value of $L(\boldsymbol{x})$ indicates the spontaneous breaking of $Z(N_c)$ symmetry.

Now, what is the physical meaning of $L(\boldsymbol{x})$? It can be interpreted as a partition function when an infinitely heavy quark is placed in the system at point $\boldsymbol{x}$. This heavy quark acts as a probe to determine whether the system (without dynamical quarks) is in the confined phase or in the deconfined phase. To see this explicitly, let us remember that the equation of motion for the heavy quark field, $\hat{\Psi}(\tau, \boldsymbol{x}) \equiv \mathrm{e}^{-m_Q \tau} \hat{\psi}(\tau, \boldsymbol{x})$ (which has only the upper component of the Dirac spinor), is given by

$$\left(i\frac{\partial}{\partial \tau} + gA_4(\tau, \boldsymbol{x}) \right) \hat{\psi}(\tau, \boldsymbol{x}) = 0. \tag{5.78}$$

The solution of this static Dirac equation is easily found to be

$$\hat{\psi}(\tau, \boldsymbol{x}) = \Omega(\boldsymbol{x}) \hat{\psi}(0, \boldsymbol{x}). \tag{5.79}$$

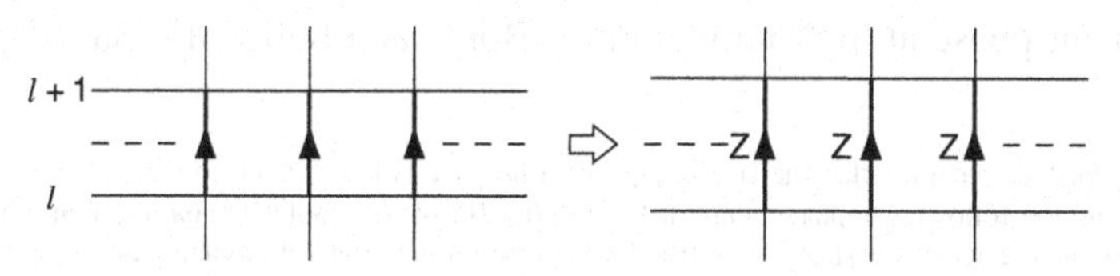

Fig. 5.10. Aperiodic gauge transformation acting on the temporal link variables.

Table 5.1. *Confinement–deconfinement phase transition in* $\mathrm{SU}(N_c)$ *gauge theory without dynamical quarks* ($N_f = 0$).

	Confined phase	Deconfined phase
T	$T < T_c$	$T > T_c$
F_Q	∞	finite
$\langle L \rangle$	0	finite
$Z(N_c)$ symmetry	unbroken	spontaneously broken

Let us now consider a partition function with a heavy quark placed at a spatial point $\boldsymbol{x}$. It may be defined by using $\hat{H}_{\mathrm{g}}$ (= the Yang–Mills part of the QCD Hamiltonian) and $\hat{H}_{\mathrm{Q}}$ (= $\hat{H}_{\mathrm{g}}$+ a static heavy quark):[2]

$$
\begin{aligned}
Z_{\mathrm{Q}}/Z_{\mathrm{g}} &= \mathrm{e}^{-(F_{\mathrm{Q}}(T,V)-F_{\mathrm{g}}(T,V))/T} \\
&\equiv \frac{1}{N_c}\sum_{a=1}^{N_c}\sum_{\mathrm{n}} \langle \mathrm{n}|\hat{\psi}^a(0,\boldsymbol{x})\, \mathrm{e}^{-\hat{H}_{\mathrm{Q}}/T}\hat{\psi}^{\dagger a}(0,\boldsymbol{x})|\mathrm{n}\rangle/Z_{\mathrm{g}} \\
&= \frac{1}{N_c}\sum_{a=1}^{N_c}\sum_{\mathrm{n}} \langle \mathrm{n}|\mathrm{e}^{-\hat{H}_{\mathrm{Q}}/T}\hat{\psi}^a(1/T,\boldsymbol{x})\hat{\psi}^{\dagger a}(0,\boldsymbol{x})|\mathrm{n}\rangle/Z_{\mathrm{g}} \\
&= \frac{1}{N_c}\,\mathrm{Tr}\left[\mathrm{e}^{-\hat{H}_{\mathrm{g}}/T}\ \mathrm{tr}\ \Omega(\boldsymbol{x})\right]\Big/ Z_{\mathrm{g}} = \langle L(\boldsymbol{x})\rangle.
\end{aligned} \tag{5.80}
$$

Here, $|\mathrm{n}\rangle$ is the complete set of eigenstates of $\hat{H}_{\mathrm{g}}$, and Z_{g} (F_{g}) is the partition function (free energy) of the gauge fields without the heavy quark. The color average factor, $1/N_c$, is introduced in the definition of Z_{Q} for convenience.

In the confining phase, the free energy of a single quark is infinite, and therefore $\langle L(\boldsymbol{x})\rangle = 0$. On the other hand, in the deconfined phase, F_{Q} is finite and $\langle L(\boldsymbol{x})\rangle$ is finite. Thus, $\langle L(\boldsymbol{x})\rangle$ serves as an order parameter for confinement–deconfinement transition. Refer to Table 5.1 for a summary of F_{Q} and $\langle L(\boldsymbol{x})\rangle$ in each phase. The existence of the deconfinement transition at high temperature was discussed within the context of the strong coupling lattice gauge theory (Polyakov, 1978; Susskind, 1979). It was later proved rigorously that for $\mathrm{SU}(N_c)$ and $\mathrm{U}(N_c)$ lattice gauge theory in the spatial dimension $d \geq 3$, there exists a phase transition to the deconfinement phase at high temperature (Borgs and Seiler, 1983a, b).

[2] Even with light dynamical quarks, the free energy of a heavy quark is related to the Polyakov line. This is seen by making the following replacements in Eq. (5.80): $\hat{H}_{\mathrm{g}} \to \hat{H}$ (the QCD Hamiltonian with dynamical quarks having the eigenstates $|\mathrm{n}\rangle$), $Z_{\mathrm{g}} \to Z$ (the QCD partition function with dynamical quarks) and $F_{\mathrm{g}} \to F$ (the QCD free energy with dynamical quarks). In this case, however, $\hat{H}$ does not have $Z(N_c)$ symmetry (Section 5.9) and $\langle L(\boldsymbol{x})\rangle$ does not serve as an order parameter.

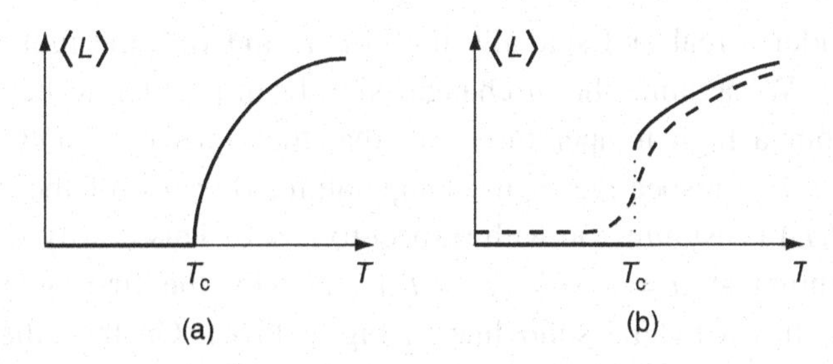

Fig. 5.11. Schematic behavior of the Polyakov line $\langle L \rangle$ for several cases. (a) Second order transition ($c = h = 0$). (b) Solid line: first order transition ($c \neq 0$, $h = 0$); dashed line: smooth crossover ($c \neq 0$, $h > h_c$).

Figure 5.11 presents the schematic behavior of $|\langle L \rangle|$ as a function of T. Two cases for $N_f = 0$ QCD may be imagined: the continuous (or second order) phase transition, where the order parameter is a continuous but non-differentiable function at the critical point; and the first order phase transition, where the order parameter itself is discontinuous at the critical point. Which of the two cases are realized depends on the number of colors: second order for $N_c = 2$ and first order for $N_c = 3$. The reason for this difference is discussed in Section 5.8.

5.8 Order of the phase transition for $N_f = 0$

First, we give an heuristic argument concerning the reason why the confinement–deconfinement transition is of first order for $N_c = 3$ (Yaffe and Svetitsky, 1982; Svetitsky, 1986). Let us define an effective action for the Polyakov line as follows:

$$Z = \int [dA]\ \mathrm{e}^{-S_g(A)} \int [dL] \prod_x \delta\left(L(x) - \mathrm{tr}\ \mathrm{P}\ \mathrm{e}^{ig\int_0^{1/T} A_4(\tau, x) d\tau}\right) \tag{5.81}$$

$$\equiv \int [dL]\ \mathrm{e}^{-S_{\mathrm{eff}}(L)}, \tag{5.82}$$

where the effective action must have Z(3) symmetry, $S_{\mathrm{eff}}(zL) = S_{\mathrm{eff}}(L)$. Assuming that S_{eff} can be expanded as a Taylor series in L and its derivatives, we have a general effective action (cf. Table 6.1),

$$S_{\mathrm{eff}}(L) \simeq \int d^3x \left[\frac{1}{2}(\nabla L^*)(\nabla L) + V(L)\right], \tag{5.83}$$

$$V(L) = \frac{a}{2} L^* L - \frac{c}{3}\ \mathrm{Re}(L^3) + \frac{b}{4}(L^* L)^2 + O(L^5), \tag{5.84}$$

where $|L|^2$ and L^3 are invariant under Z(3) because $zz^* = 1$ and $z^3 = 1$. The coefficients, a, b and c, can, in principle, be fixed by performing the A integration numerically, but they are left undetermined in the following discussion.

Let us consider a real and spatially uniform L and find the minimum of the potential, $V(L)$. We assume that a changes sign from positive to negative as we decrease T from a high temperature, and that the signs of b and c are fixed ($b > 0$ and $c > 0$), irrespective of the temperature. Because of the existence of the L^3 term, $V(L)$ is asymmetric with respect to $L = 0$. This is why we encounter degenerate minima at $a = c^2/4b$ ($T = T_c$) and why the first order transition is realized, as shown by the solid line in Fig. 5.11(b). On the other hand, for $N_c = 2$, only even powers of L are allowed. Therefore, as long as the sign in front of L^4 does not change across the critical temperature, we obtain the second order transition, as shown by the solid line in Fig. 5.11(a). (There is, however, a possibility that the first order transition is driven by the sign change of the coefficient in front of L^4. In this case, an L^6 term is necessary to stabilize the system.)

The existence of the deconfinement phase transition in MC simulations was first demonstrated for the $N_c = 2$ case in Kuti *et al.* (1981) and McLerran and Svetitsky (1981a) and also in Engels *et al.* (1981). Later, the first order (second order) deconfinement transition for $N_c = 3$ ($N_c = 2$) was firmly established in MC simulations with finite scaling analysis (Fukugita *et al.*, 1989, 1990). Determing the order of the transition is indeed more involved than just confirming the existence of the phase transition. Coexistence of two phases and discontinuity of the order parameter at T_c are direct signals for the first order transition. However, on a lattice with finite volume, these signals are blurred. In this case, the finite size scaling analysis, in which the lattice-volume dependence of observables is studied, provides detailed information on the order of the transition (Fukugita *et al.*, 1989, 1990; Ukawa, 1995).

5.9 Effect of dynamical quarks

In Section 5.8 we neglected the effect of dynamical quarks. Once they are introduced, the exact $Z(N_c)$ of the lattice action is explicitly broken. This is because the Dirac operator contains a time-like link, $U_4(n)$, not associated with its hermitian conjugate, Eq. (5.20). Therefore, $U_4(n) \to zU_4(n)$ is no longer a symmetry. This symmetry breaking becomes small for large quark mass. Indeed, the lowest order symmetry breaking contribution from the dynamical quarks in the hopping parameter expansion provides the following correction to $V(L)$ in Eq. (5.84):

$$V(L) \to V(L) - h\,\mathrm{Re}(L); \tag{5.85}$$

$h \propto \kappa^{N_t}$, where κ is the hopping parameter. The extra term is analogous to the external magnetic field applied to a spin system (Banks and Ukawa, 1983).

It requires only simple algebra to show that there exists a critical h_c, where the first order transition exists for $h < h_c$ and becomes a smooth crossover for $h > h_c$. (We have assumed $c > 0$, $b > 0$ and $h > 0$.) Typical crossover behavior of the Polyakov line for $h > h_c$ is shown by the dashed line in Fig. 5.11(b).

Since $Z(N_c)$ symmetry is badly violated by the light dynamical quarks, where κ is large, one cannot even talk about the confinement–deconfinement transition for small quark masses. This is physically reasonable because the existence of the $q\bar{q}$ pairs screens the color charge, even at zero temperature. Also, the potential between heavy quarks cannot rise linearly at large distances due to the string breaking by the $q\bar{q}$ pairs. For dynamical quarks with small masses, it is more appropriate to talk about dynamical breaking of chiral symmetry and its restoration at finite T. This will be extensively discussed in Chapter 6.

5.10 Effect of finite chemical potential

In this section, we consider the quark chemical potential, μ, in the lattice partition function, Z. As we have seen in Eq. (4.6), μ is introduced in the Euclidean Lagrangian in the continuum limit as $\mathcal{L}_q = \bar{q}F(U;\mu)q$, with

$$\begin{aligned} F(U;\mu) &\equiv -i\gamma\cdot D + m + i\mu\gamma_4 = \Gamma\cdot D + m - \mu\Gamma_4 \\ &= \Gamma_4(\partial_4 - i(gA_4 - i\mu)) + \boldsymbol{\Gamma}\cdot\boldsymbol{D} + m, \end{aligned} \tag{5.86}$$

where γ (Γ) denotes the anti-hermitian (hermitian) γ-matrices. Note that $F(U;\mu)$ has the following property (Exercise 5.6):

$$[F(U;\mu)]^\dagger = \gamma_5 F(U;-\mu)\gamma_5, \quad [\text{Det } F(U;\mu)]^* = \text{Det } F(U;-\mu). \tag{5.87}$$

The final equality in Eq. (5.86) implies that the chemical potential behaves as if it is an extra imaginary-valued gauge potential in the temporal direction: $gA_4 \to gA_4 - i\mu$. Therefore, μ on the lattice is naturally introduced by the following replacement in Eqs. (5.13), (5.14) and (5.20) (Hasenfratz and Karsch, 1983):

$$U_{\pm 4}(n) = \mathrm{e}^{\pm iagA_4(n)} \to \mathrm{e}^{\pm a\mu} U_{\pm 4}(n). \tag{5.88}$$

Although it is formally simple to introduce the chemical potential, the Monte Carlo simulation of QCD with finite μ has been a long standing challenge since the first attempt (Nakamura, 1984). The reason is that the fermionic determinant, Det $F(U;\mu)$, in Eq. (5.26) for $N_c = 3$ is complex (Eq. (5.87)) and allows a considerable cancellation among different gauge configurations (the complex phase problem). This requires the number of gauge configurations to increase exponentially as the spatial volume increases: $N \sim \mathrm{e}^{cV/T}$, where c is a positive constant. A similar situation appears in MC simulations of fermions in condensed matter physics and in nuclear physics. (For a simple example, see Exercise 5.7.)

For $\mu = 0$, Det $F(U; \mu = 0)$ is real, but it can be either positive or negative. Therefore, it still leads to the sign problem. However, for an even number of flavors with degenerate quark masses, the sign problem does not appear because we have (Det $F(U; \mu = 0))^{N_f}$. Also, for massive quarks, $m \gg \mu$, the determinant barely changes sign and is harmless. These nice features are lost for finite μ.

There have been many approaches proposed so far to solve QCD at finite μ. An example is the reweighting method, in which the factor Det $F(U; \mu)\exp(-S_{\rm g}(U))$ is reorganized in such a way as to enhance the signal effectively. Another approach, though limited to small μ, is to construct a Taylor expansion of Det $F(U; \mu)$ in terms of μ and to calculate the expansion coefficients at $\mu = 0$. Introducing the imaginary chemical potential, $\mu_{\rm I} = -i\mu$ (which is free from the complex phase problem), and then making an analytic continuation back to the real μ is another possible approach. More details on these approaches may be found in Muroya *et al.* (2003) and Nakamura *et al.* (2004). An interesting role played by μ in the chiral phase transition will be discussed in Section 6.13.5.

Exercises

5.1 Properties of the Wilson line.

(1) Remember that the time evolution operator, $U(t, t')$, in the interaction picture obeys the equation $i\partial_t U(t, t') = H_{\rm I}(t)U(t, t')$. Solve this equation to show that the solution is written in the form given in Eq. (5.2).

(2) Using the analogy between $U(t, t')$ and the Wilson line, $U_P(x, y; A)$, prove Eqs. (5.3) and (5.4).

(3) Prove Eq. (5.5) by discretizing $U_P(x, y; A)$ into a product of short Wilson lines along the path $z_\mu(s)$. Try to form an alternative proof by utilizing the fact that Eq. (5.5) is a unique solution of

$$\lambda_\mu(s) D_\mu(A^V) U_P(x, y; A^V) = 0.$$

5.2 Haar measure. Let us consider an integral over group elements, $g \in G$, so that the integral is invariant under the "shift" of the integration variable on the group manifold (the invariant integral) (Creutz, 1985; Gilmore, 1994). The left and right Haar measures associated with the integral are defined by

$$\int dg_{\rm L}\ f(g'g) = \int dg_{\rm L}\ f(g), \quad \int dg_{\rm R}\ f(gg') = \int dg_{\rm R}\ f(g),$$

respectively, where g' is an arbitrary element of the group G.

(1) If G is a compact Lie group in which the group elements are parametrized by variables varying over a closed interval, (i) the Haar measure is unique up to a multiplicative constant and (ii) the left measure and the right measure are the same: $dg_{\mathrm{L}} = dg_{\mathrm{R}} \equiv dg$. The multiplcative constant is fixed, for example by $\int dg\ 1 = 1$. Assuming uniqueness (i), prove property (ii).

(2) Let n be the dimension of the manifold of the compact Lie group, G, and parametrize the group elements as $\mathrm{g}(\theta_i)$ $(i = 1, \ldots, n)$. Then the invariant integral may be written as an n-dimensional integral:

$$\int dg\ f(\mathrm{g}) = \mathcal{N} \int d\theta\ J(\theta)\ f(\mathrm{g}(\theta)),$$

where $\mathcal{N}$ is a normalization constant. Using the multiplicative law of the group elements, $\mathrm{g}(\theta''(\theta, \theta')) = \mathrm{g}(\theta)\mathrm{g}(\theta')$, show that the Jacobian J is given by

$$J(\theta) = \left| \det_{i,j}\ \partial\theta''(\theta, \theta')/\partial\theta' \right|^{-1}_{\theta'=0}.$$

(3) Consider a metric tensor on the group G defined as $M_{ij} = \mathrm{tr}(L_i L_j) = \mathrm{tr}(R_i R_j)$, with $L_i = \mathrm{g}^{-1}(\partial_i \mathrm{g})$, $R_i = (\partial_i \mathrm{g})\mathrm{g}^{-1}$ and $\partial_i = \partial/\partial\theta_i$. By studying the relation between $J(\theta)$ and $\det M$, show that the alternative representation of the invariant integral is

$$\int dg\ f(\mathrm{g}) = \mathcal{N} \int d\theta\ |\det_{i,j}\ M|^{1/2}\ f(\mathrm{g}(\theta)),$$

where $\mathcal{N}$ is another normalization constant.

(4) Derive the Haar measure for the U(1) and SU(2) groups.

(5) Prove the relations in Eq. (5.37) for the SU(N) group by utilizing the invariant property of the integral. Derive general formulas for the SU(N) integral by consulting Chap. 8 of Creutz (1985).

5.3 Dirac operators on the lattice.

(1) Show that Wilson's Dirac operator, D_{W}, for $r \neq 0$ does not anti-commute with γ_5: $\{\gamma_5, D_{\mathrm{W}}\} \neq 0$. Explain that this implies that $\bar{\psi} D_{\mathrm{W}} \psi$ breaks chiral symmetry explicitly for finite a.

(2) Show that the Ginsparg–Wilson relation, Eq. (5.23), and γ_5-hermiticity, $\gamma_5 D^{\dagger}_{\mathrm{GW}} \gamma_5 = D_{\mathrm{GW}}$, are satisfied by D_{GW} in Eq. (5.24).

(3) By analyzing

$$m_0(p) = m_0 - \frac{1}{a} \sum_{\mu>0} (1 - \cos(p_\mu a))$$

in a similar way to Eq. (5.18), show that $m_0(p) \simeq m_0$ ($^\forall p_\mu \to 0$) and $m_0(p) \simeq m_0 - 2N_\pi/a$ ($^\exists p_\mu \to \pi/a$). Putting them into the definition of $D_{\rm GW}$, show that there is only one massless mode in the fermion spectrum and that a further 15 modes have positive and large masses $\frac{2}{a}(2N_\pi - m_0 a)$ as long as $0 < m_0 a < 2$.

5.4 Heavy quark propagator in the Minkowski space. By making an ansatz $S(x', x) = S_{\rm t}(x'_0, x_0) S_{\rm s}(\boldsymbol{x}', \boldsymbol{x})$, derive an equation for $S_{\rm t}(x'_0, x_0)$. Solve the resultant equation with the boundary condition that the positive (negative) energy quark propagates forward (backward) in time.

5.5 Regularization-scheme independence of β_0 and β_1. If we calculate a dimensionless physical quantity, such as the short distant part of the heavy quark potential, $V(R)$, multiplied by R in two different regularization schemes (for example, the dimensional regularization and the lattice regularization), we obtain the identity

$$g_{\overline{\rm MS}}(1/a) = g_{\rm LAT}(a) Z(g_{\rm LAT}(a))$$

with

$$Z(g_{\rm LAT}(a)) = 1 + d_1 g^2_{\rm LAT}(a) + d_2 g^4_{\rm LAT}(a) + \cdots .$$

Using the definition of β and the expansion $\beta = -\beta_0 g^3 - \beta_1 g^5 + \cdots$ in each regularization scheme, show that $\beta_0^{\overline{\rm MS}} = \beta_0^{\rm LAT}$ and $\beta_1^{\overline{\rm MS}} = \beta_1^{\rm LAT}$.

5.6 Fermion determinant with chemical potential.

(1) Prove the relations in Eq. (5.87).

(2) Show that Det $F(U; \mu)$ can be made real only for $N_c = 2$ by using the identity $[U_\nu]^* = \sigma_2 U_\nu \sigma_2$, where σ_2 is the 2×2 Pauli matrix in the color-space.

5.7 The sign problem. Consider a simple partition function of the form

$$Z = \sum_{\{\phi(\boldsymbol{x}) = \pm 1\}} \text{sign}(\phi)\, \mathrm{e}^{-S(\phi)},$$

where $\phi(\boldsymbol{x})$ is a field variable which takes only the values ± 1 at each point and S is assumed to be positive. The expectation value of an operator, $\mathcal{O}(\phi)$, is then written as $\langle \mathcal{O} \rangle = \langle \mathcal{O}(\phi)\ \text{sign}(\phi) \rangle_0 / \langle \text{sign}(\phi) \rangle_0$, where $\langle \cdot \rangle_0$ is an average with respect to the partition function without the sign factor, Z_0.

(1) Show that the denominator is given by

$$\langle \text{sign}(\phi) \rangle_0 = \mathrm{e}^{-(f - f_0)V/T},$$

where f (f_0) is the free-energy density corresponding to Z (Z_0), V is the spatial volume of the system and T is the temperature. Since $f > f_0$ holds, the denominator becomes exponentially small in the thermodynamic limit, $V \to \infty$.

(2) To achieve good statistical accuracy in the MC simulations, the variance of sign(ϕ) around its mean, $\langle \text{sign}(\phi) \rangle_0$, must be small enough; see Eq. (5.50). Show that this condition is given in terms of N (the number of MC samples) by

$$N \gg \mathrm{e}^{2(f-f_0)V/T},$$

which becomes extremely large for large spatial volume. This is the core of the sign problem and the complex phase problem.

6
Chiral phase transition

As we have seen in Section 2.4.1, the QCD Lagrangian with massless quarks is invariant under the $SU_L(N_f) \times SU_R(N_f)$ chiral rotation, while the operator $\bar{q}q = \bar{q}_L q_R + \bar{q}_R q_L$ is not invariant. Therefore, the thermal expectation value, $\langle \bar{q}q \rangle$, is a measure (but not necessary a unique measure) of the dynamical breaking of chiral symmetry at finite T:

$$\langle \bar{q}q \rangle = 0 : \text{the Wigner phase,} \tag{6.1}$$

$$\langle \bar{q}q \rangle \neq 0 : \text{the Nambu–Goldstone (NG) phase.} \tag{6.2}$$

As T increases, the $q\bar{q}$ pairing is dissociated by thermal fluctuations, and eventually the transition from the NG phase to the Wigner phase will take place. This is analogous to the phase transition in metallic superconductors, where the order parameter is the electron pairing $\langle e_\uparrow e_\downarrow \rangle$. In fact, the notion of dynamical breaking of chiral symmetry was originally introduced in analogy with BCS superconductivity (Nambu and Jona-Lasinio, 1961a,b). Also, the quark mass in $m\bar{q}q$, which plays a similar role to the external magnetic field, breaks chiral symmetry explicitly.

There are several interesting questions to be answered. (i) What is the critical temperature, T_c, of chiral phase transition? (ii) What will be the order of the chiral transition? (iii) What will be the observable phenomena associated with the chiral transition? We examine (i) and (ii) in this chapter, and (iii) in Chapter 7.

6.1 $\langle \bar{q}q \rangle$ in hot/dense matter

Let us start with the QCD partition function:

$$Z = \mathrm{Tr}\left[e^{-\hat{K}_{\mathrm{QCD}}/T} \right] = e^{-\Omega(T,V,\mu)/T} = e^{P(T,\mu)V/T}, \tag{6.3}$$

$$\hat{K}_{\mathrm{QCD}} = \hat{H}_{\mathrm{QCD}}(m_q = 0) + \sum_{q=\mathrm{u,d,s},\ldots} \int d^3x\, \bar{q}(m_q - \mu_q \gamma_0) q, \tag{6.4}$$

where m_q (μ_q) is the quark mass (the quark chemical potential) for each flavor, and we have

$$\langle \bar{q}q \rangle = -\frac{\partial P(T, \mu)}{\partial m_q}. \tag{6.5}$$

6.1.1 High-temperature expansion

Let us evaluate the right-hand side of Eq. (6.5) for $\mu_q = 0$ and $m_q \simeq 0$ in two extreme cases: the high-T limit and the low-T limit.

When T is high enough, the system may be approximated by the Stefan–Boltzmann gas of free quarks and gluons. Then, the total pressure at $\mu_q = 0$ is written as a sum, $P(T) = P_{\text{gluon}}(T) + P_{\text{quark}}(T)$. The quark mass, m_q, enters only in $P_{\text{quark}}(T)$, which has the form $\sum_q P_q(T; m_q)$, with

$$P_q(T; m_q) = 4N_c \int \frac{d^3k}{(2\pi)^3} T \ln\left(1 + e^{-E_q(k)/T}\right) \tag{6.6}$$

$$\simeq 4N_c \frac{7}{8}\left[\frac{\pi^2}{90}T^4 - \frac{1}{42}m_q^2 T^2 - \frac{1}{56\pi^2}m_q^4\left(\ln\left(\frac{m_q^2}{(\pi T)^2}\right) + C\right) + \cdots\right], \tag{6.7}$$

where $E_q(k) = (k^2 + m_q^2)^{1/2}$ and $C = 2\gamma - 3/2 \simeq -0.346$, with $\gamma \simeq 0.577$ being the Euler constant. To obtain Eq. (6.7), we have used the expansion in terms of m_q/T given in Exercise 3.4(3) for fermions.

Due to the absence of the linear term of m_q in Eq. (6.7), $\langle \bar{q}q \rangle$ vanishes for $m_q \to 0$ (the chiral limit) at high temperature. Although we have only shown this for a free gas of quarks and gluons, it is straightforward to see that the conclusion is the same even for interacting quarks and gluons, as long as perturbation theory is employed. The reason is that the quark–gluon vertex does not change chirality (namely, the transition between q_L and q_R is not allowed), and thus the expectation value of $\bar{q}q = \bar{q}_L q_R + \bar{q}_R q_L$ vanishes in any finite order of the perturbation if $m_q = 0$.

6.1.2 Low-temperature expansion

When T is low enough and $\mu_q = 0$, the system is composed of a weakly interacting gas of pions, and the total pressure may be decomposed as $P(T) = P_\pi(T) + P_{\text{vac}}$. We have $\langle \bar{q}q \rangle_{\text{vac}} = -\partial P_{\text{vac}}/\partial m_q$ by definition. On the other hand, $P_\pi(T)$ at low

T can be evaluated using chiral perturbation theory (an Euclidean version of Eq. (2.62)) (Gerber and Leutwyler, 1989):

$$e^{P_\pi(T)V/T} = \int [dU]\, e^{-\int_0^{1/T} d^4x\, (\mathcal{L}^{(2)}(U)+\mathcal{L}^{(4)}(U)+\mathcal{L}^{(6)}(U)+\cdots)}. \tag{6.8}$$

Using the GOR relations, Eqs. (2.51) and (2.52), we can change the variable from m_q to m_π, to obtain

$$\frac{\langle \bar{q}q \rangle}{\langle \bar{q}q \rangle_{\rm vac}} = 1 + \frac{1}{f_\pi^2} \frac{\partial P_\pi(T)}{\partial m_\pi^2}\bigg|_{m_\pi \to 0} \tag{6.9}$$

$$= 1 - \frac{T^2}{8f_\pi^2} - \frac{1}{6}\left(\frac{T^2}{8f_\pi^2}\right)^2 - \frac{16}{9}\left(\frac{T^2}{8f_\pi^2}\right)^3 \ln\left(\frac{\Lambda_q}{T}\right) + O(T^8), \tag{6.10}$$

where $\Lambda_q (= 470 \pm 110\,\mathrm{MeV})$ is a parameter extracted from the experimental pion–pion scattering length in the $I = 0$ and D-wave channel (Gerber and Leutwyler, 1989). The $O(T^2)$ contribution in Eq. (6.10) comes solely from the non-interacting gas of the pions. Therefore, the coefficient may be alternatively evaluated in a similar way as Eq. (6.7), with the formula given in Exercise 3.4(3) for bosons.

The low-T formula, Eq. (6.10), together with the result of high-T expansion strongly indicate that the chiral condensate decreases as T increases and eventually melts away at sufficiently high T, so restoring chiral symmetry. There is also a rigorous proof in $N_c = 2$ lattice gauge theory with massless dynamical fermions that the restoration of chiral symmetry takes place at sufficiently high temperature (Tomboulis and Yaffe, 1984, 1985).

In Section 6.2, we study the Nambu–Jona-Lasinio (NJL) model introduced in Chapter 2 (Eq. (2.60)) as a low-energy model of QCD to extract the physics behind the chiral restoration at finite T.

6.2 The NJL model

The simplest version of the NJL model (Nambu and Jona-Lasinio, 1961a, b; Hatsuda and Kunihiro, 1994) for two-flavor ($N_f = 2$) quarks in the Euclidean space-time is

$$\mathcal{L}_{\rm NJL} = \bar{q}(-i\gamma_\mu \partial_\mu + m)q - \frac{G^2}{2\Lambda^2}[(\bar{q}q)^2 + (\bar{q}i\gamma_5 \boldsymbol{\tau} q)^2], \tag{6.11}$$

where ${}^t q(x) = (\mathrm{u}(x), \mathrm{d}(x))$ and $m = \mathrm{diag}(m_\mathrm{u}, m_\mathrm{d}) = m \cdot \mathbf{1}$, where we assume the isospin symmetry, $m_\mathrm{u} = m_\mathrm{d}$, for simplicity. Note that G is a dimensionless coupling constant responsible for $q\bar{q}$ attraction in the scalar ($(I, J^P) = (0, 0^+)$) and pseudo-scalar ($(I, J^P) = (1, 0^-)$) channels; Λ^{-1} is a characteristic length scale below which the $q\bar{q}$ interaction is approximated as point-like in space-time, as

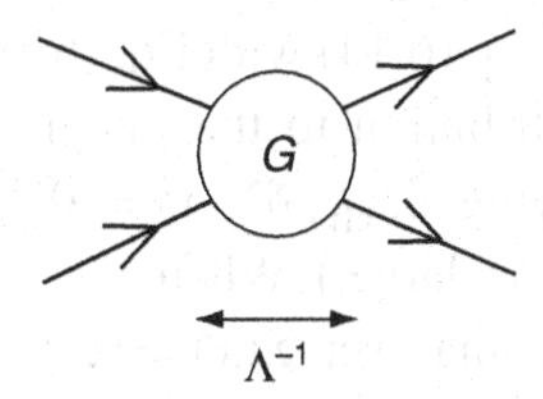

Fig. 6.1. The four-fermi interaction in the NJL model. The size of the interaction in the coordinate-space is $O(\Lambda^{-1}) \sim 0.2\,\text{fm}$.

shown in Fig. 6.1. We assume G^2 to be $O(1/N_c)$ so that the kinetic term and the interaction term are both $O(N_c)$.

Equation (6.11) has a global $\text{SU}_\text{L}(2) \times \text{SU}_\text{R}(2) \times \text{U}_\text{B}(1)$ symmetry, but breaks a $\text{U}_\text{A}(1)$ symmetry. The latter aspect is an explicit manifestation of the $\text{U}_\text{A}(1)$ axial anomaly discussed in Section 2.4.1. (See Exercise 6.1.) The partition function of the model at finite T and zero chemical potential may be given by

$$Z_{\text{NJL}} = \int [d\bar{q}\; dq]\; \mathrm{e}^{-\int_0^{1/T} d\tau \int d^3x\; \mathcal{L}_{\text{NJL}}} \tag{6.12}$$

$$= \int [d\bar{q}\; dq][d\Sigma]\; \mathrm{e}^{-\int_0^{1/T} d\tau \int d^3x\; \left[\bar{q}(-i\gamma\cdot\partial+m+G\Sigma)q+\frac{\Lambda^2}{2}\Sigma\Sigma^\dagger\right]} \tag{6.13}$$

$$\equiv \int [d\Sigma]\; \mathrm{e}^{-S_{\text{eff}}(\Sigma;T)}, \tag{6.14}$$

where $\Sigma(x)(=\sigma(x)+i\gamma_5\boldsymbol{\tau}\cdot\boldsymbol{\pi}(x))$ is a bosonic field with a 2×2 matrix structure in the isospin space and $[d\Sigma] = [d\sigma\; d\boldsymbol{\pi}]$.

To obtain Eq. (6.13) from Eq. (6.12), we have used a matrix generalization of the Gaussian trick, or the Hubbard–Stratnovich transformation (Appendix C):

$$\mathrm{e}^{\frac{1}{2}y^2} = \int_{-\infty}^{+\infty} \frac{dz}{\sqrt{2\pi}}\, \mathrm{e}^{-\frac{1}{2}z^2 \pm zy}. \tag{6.15}$$

This enables us to transform the four-fermi interaction into a bilinear form of the quark fields by the replacement $y \leftrightarrow (\bar{q}q, \bar{q}i\gamma_5\boldsymbol{\tau}q)$ with $z \leftrightarrow (\sigma, \boldsymbol{\pi})$. (See Exercise 6.2.)

Carrying out the Grassmann integration, $[d\bar{q}\; dq]$, in Eq. (6.13) (Appendix C.3), and using the identity $\det A = \exp(\text{Tr}\ln A)$ for a general matrix A, we obtain an explicit form of the effective action in Eq. (6.14) as follows:

$$S_{\text{eff}}(\Sigma;\, T) = -\text{Tr}\,\ln\,(-i\gamma\cdot\partial + m + G\Sigma) + \int_0^{1/T} d\tau \int d^3x \left(\frac{\Lambda^2}{2}\Sigma(x)\Sigma(x)^\dagger\right). \tag{6.16}$$

Here, Tr stands for the trace over color, flavor, spin and space-time coordinates.

Let us estimate the integral Eq. (6.14) with Eq. (6.16) in the mean-field approximation, where the main contribution to the integral is assumed to come from the stationary solution satisfying $\delta S_{\rm eff}/\delta\Sigma(x) = 0$. (In the present model, this assumption is justified if N_c is large.) When the stationary solution is space-time-independent and real, we may set $\Sigma(x) = \Sigma^\dagger(x) = \sigma$. Then, the stationary condition is equivalent to

$$\partial f_{\rm eff}/\partial\sigma = 0, \qquad \text{with } S_{\rm eff}(\sigma; T) \equiv f_{\rm eff}(\sigma; T)V/T. \tag{6.17}$$

In this case, the "Tr" ln part of Eq. (6.16) is simply a fermion contribution with a constant mass, $M = m + G\sigma$, which can be evaluated in the same way as the black body formula, Eq. (4.19):

$$f_{\rm eff}(\sigma; T) = \frac{\Lambda^2}{2}\sigma^2 + \int \frac{d^3k}{(2\pi)^3}\left[\frac{-d_q E(k)}{2} - d_q T \ln\left(1 + {\rm e}^{-E(k)/T}\right)\right], \tag{6.18}$$

where $E(k) = \sqrt{\boldsymbol{k}^2 + (m+G\sigma)^2}$, and d_q $(= 2_{\rm spin} \times 2_{q\bar{q}} \times N_c \times N_f = 24)$ is the quark degeneracy factor introduced in Eq. (3.47).

The first term on the right-hand side of Eq. (6.18) is the interaction energy originating from the four-fermion term in Eq. (6.11). The first term in the integrand of Eq. (6.18) is simply the zero-point fermion energy, $-E/2$, multiplied by the degeneracy factor, d_q, for the quark and anti-quark. It can be interpreted alternatively as the total energy of quarks in the Dirac sea, where $-E$ and $d_q/2$ are the quark energy and the degeneracy factor for the quark in the negative-energy sea, respectively. The last term of Eq. (6.18) is related to the entropy term $(-Ts)$ of thermally excited quarks. Thus, the free energy has the expected structure $f_{\rm eff} = \varepsilon - Ts$. We define $\bar{\sigma}$ as a true minimum of $f_{\rm eff}(\sigma; T)$.

Equation (6.17), $\partial f_{\rm eff}/\partial\sigma = 0$, is called the gap equation in analogy with a similar equation in BCS superconductivity. The dynamical (constituent) quark mass, M_q, and the chiral condensate, $\langle\bar{q}q\rangle$, in the chiral limit $(m = 0)$ are related to $\bar{\sigma}$ as follows:

$$M_q = G\bar{\sigma}, \quad \langle\bar{\rm u}{\rm u} + \bar{\rm d}{\rm d}\rangle = -\frac{\Lambda^2}{G}\bar{\sigma}. \tag{6.19}$$

6.2.1 Dynamical symmetry breaking at $T = 0$

Let us first study how the dynamical breaking of chiral symmetry takes place in the NJL model at $T = 0$ and $m = 0$ on the basis of Eq. (6.18).

Since the model is formulated at low energies below the scale Λ, we limit the momentum integral in the region, $|\boldsymbol{k}| \leq \Lambda$. Furthermore, we scale all the dimensionful quantities by Λ and rewrite as $f_{\rm eff}/\Lambda^4 \to f_{\rm eff}$ in this and the following subsection. Then we have

$$f_{\rm eff}(\sigma;0) = -\frac{d_q}{16\pi^2} + \frac{1}{2}\left(\frac{1}{G^2} - \frac{1}{G_{\rm c}^2}\right)(G\sigma)^2 + \frac{d_q}{64\pi^2}(G\sigma)^4 \ln\left(\frac{4}{(G\sigma)^2}\right) + O(\sigma^6), \tag{6.20}$$

where we have expanded the exact expression of the free energy around $\sigma \sim 0$, and

$$G_{\rm c} = \pi\sqrt{\frac{8}{d_q}}. \tag{6.21}$$

The coefficient of the σ^2 term in Eq. (6.20) changes sign at $G = G_{\rm c}$. This drives a second order phase transition from the Wigner phase to the NG phase:

$$G \le G_{\rm c} \to \bar{\sigma} = 0: \text{ the Wigner phase,} \tag{6.22}$$

$$G > G_{\rm c} \to \bar{\sigma} \ne 0: \text{ the NG phase.} \tag{6.23}$$

The gap equation obtained from Eq. (6.20) is given by

$$\frac{G_{\rm c}^2}{G^2} \simeq 1 - \frac{1}{2}(G\sigma)^2 \ln\left(\frac{4}{(G\sigma)^2 {\rm e}}\right). \tag{6.24}$$

The solution of this equation is written in terms of the Lambert function, $W(z)$, which satisfies $W{\rm e}^W = z$ (Corless *et al.*, 1996). Its asymptotic form near the critical point leads to (Exercise 6.3)

$$\bar{\sigma} \propto \sqrt{\frac{G^2 - G_{\rm c}^2}{-\ln(G^2 - G_{\rm c}^2)}} \qquad (G \searrow G_{\rm c}). \tag{6.25}$$

A schematic figure of the solution, $\bar{\sigma}$, as a function of G is shown by the solid line in Fig. 6.2(a).

6.2.2 Symmetry restoration at $T \ne 0$

Consider the case $G > G_c$, in which the system is in the NG phase at $T = 0$. Then, from the discussions in Section 6.1, we expect the restoration of chiral symmetry as T is increased.

If the phase transition is of second order, the condensate at finite temperature, $\bar{\sigma}(T)$, is a continuous function of T, and approaches zero near the critical point, $T \sim T_{\rm c}$. Anticipating this, we expand the free energy, Eq. (6.18), not only by σ/Λ

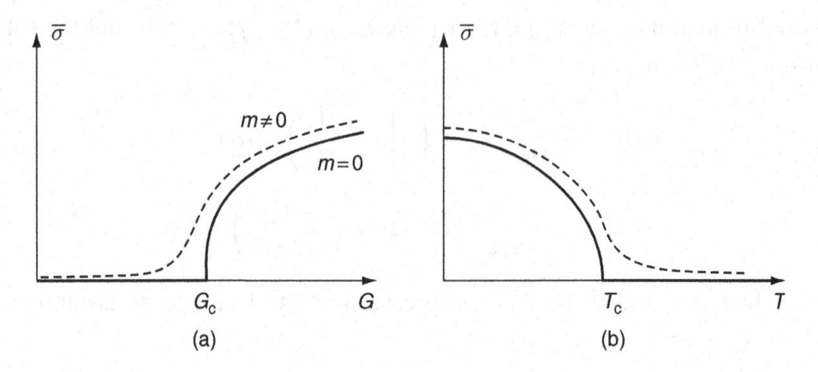

Fig. 6.2. Chiral condensate, $\bar{\sigma}$, as (a) a function of G at $T = 0$ and (b) as a function of T for $G > G_c$. Solid lines: $m = 0$; dashed lines: small, but non-vanishing, m.

but also by σ/T.[1] Using the high-temperature expansion given in Exercise 3.4(3), and noting that dimensionful quantities are all scaled by Λ as in Section 6.2.1, we find that

$$\begin{aligned} f_{\text{eff}}(\sigma; T) = &-\frac{d_q}{16\pi^2} - d_q \frac{7}{8}\frac{\pi^2}{90} T^4 \\ &+ \frac{d_q}{48}(T^2 - T_c^2)(G\sigma)^2 \\ &+ \frac{d_q}{64\pi^2}(G\sigma)^4 \left[\ln\left(\frac{1}{(\pi T)^2} \right) + C \right] + O(\sigma^6), \end{aligned} \tag{6.26}$$

where

$$T_c = \sqrt{\frac{24}{d_q}\left(\frac{1}{G_c^2} - \frac{1}{G^2} \right)}. \tag{6.27}$$

The term proportional to T^4 on the right-hand side of Eq. (6.26) represents the Stefan–Boltzmann value for massless quarks. The coefficient of the σ^2 term changes sign at $T = T_c$, while that of the σ^4 term is positive for $T \ll 1$. This is the behavior expected for a second order phase transition.

Note that $\bar{\sigma}(T)$ and the associated chiral condensate near T_c can be extracted from Eq. (6.26) and Eq. (6.19). In particular, we have

$$\langle \bar{q}q \rangle \propto \begin{cases} 0 & (T \geq T_c), \\ -(T^2 - T_c^2)^{1/2} & (T < T_c). \end{cases} \tag{6.28}$$

A schematic figure of $\bar{\sigma}$ at finite T is shown by the solid line in Fig. 6.2(b).

[1] The effect of the momentum cutoff on the entropy part of Eq. (6.18) is $O(e^{-\Lambda/T})$ and is exponentially suppressed for $\Lambda \gg T$.

As in the case for the BCS theory of superconductivity, a direct connection between T_c and the mass-gap at zero temperature, $M_0 = G\bar{\sigma}(T=0)$, is derived from the gap equation, $\partial f_{\rm eff}(\sigma; T)/\partial\sigma = 0$, evaluated at $T=0$ and $T=T_c$. For $\Lambda \gg \sigma, T$, we have (Exercise 6.4(1))

$$T_c \simeq \frac{\sqrt{3}}{\pi} M_0 = 0.55\ M_0. \tag{6.29}$$

Taking the standard value for the dynamical mass, $M(T=0) \sim 300$ to 350 MeV, we obtain $T_c \sim 165$ to 190 MeV, which is consistent with the numbers obtained in the lattice QCD simulations.

As mentioned in Section 6.1, the non-vanishing quark mass, m, plays a similar role to an external magnetic field. In fact, the second order transition is smeared out, as illustrated by the dashed lines in Figs. 6.2(a) and (b), which can be demonstrated explicitly in the NJL model (Exercise 6.4(2)).

6.3 Mean-field theory and the Landau function

By expanding $f_{\rm eff}(\sigma; T)$ in terms of σ near the critical point, we can make a general consideration without going into too much detail about the underlying dynamics. There are three interesting cases (all of which appear in QCD as shown later): the second order phase transition, the first order transition and the tricritical behavior. In this section, we will treat the three cases in a mean-field theory by neglecting the fluctuation of the order parameter around its mean. What happens beyond the mean-field theory will be discussed in later sections. For a pedagogical introduction to the phase transitions and critical phenomena in condensed matter physics, see Goldenfeld (1992).

6.3.1 Order of the phase transition

Consider the partition function in the thermodynamic limit,

$$Z = \mathrm{e}^{-\Omega(K)} = \mathrm{e}^{P(K)V},$$

where the factor $1/T$ is absorbed in the definition of Ω and P; $K = \{K_l\}$ is a set of generalized parameters such as the temperature, chemical potential, coupling constants, external fields and so on. We consider the cases where the pressure, $P(K)$, is a continuous function everywhere in the parameter space $\{K_l\}$, but is not necessarily analytic.[2] Phase boundaries are defined as a point, line, surface, etc., on which $P(K)$ is non-analytic in any one of the couplings, K_l. Depending

[2] The continuity and convexity of $P(K)$ as a function of K can be proved explicitly in many of the examples we encounter in the following (Goldenfeld, 1992).

on the nature of the discontinuity of $\partial P(K)/\partial K_l$ across the phase boundary, the phase transition is defined either as first order or continuous:

$$\text{first order,} \quad \text{if} \quad \frac{\partial P(K)}{\partial K_l} \quad \text{is discontinuous,} \tag{6.30}$$

$$\text{continuous,} \quad \text{if} \quad \frac{\partial P(K)}{\partial K_l} \quad \text{is continuous.} \tag{6.31}$$

For historical reasons, the continuous phase transition is sometimes called the second order phase transition, and we will follow this tradition in the following.[3]

It is useful to introduce an order parameter field, $\sigma(x)$, so that the partition function is given by

$$Z = \int [d\sigma]\, \mathrm{e}^{-S_{\text{eff}}(\sigma(x);K)}, \tag{6.32}$$

where $S_{\text{eff}}(\sigma(x); K)$ is called the Landau functional. Note that σ does not have to be an elementary field of the original Hamiltonian. For example, in the NJL model given in Section 6.2, $q(x)$ and $\bar{q}(x)$ are the elementary fields, while $\Sigma(x)$ is an auxiliary field introduced in the functional integral.

In the mean-field theory of phase transitions, this integral is assumed to be dominated by a minimum of $S_{\text{eff}}(\sigma(x); K)$, and the fluctuation around the minimum is neglected. Furthermore, for a uniform system, $S_{\text{eff}}(\sigma(x); K)$ may be replaced by $\mathcal{L}_{\text{eff}}(\sigma; K)V$, which is a function of the space-time-independent order parameter, σ. We will call $\mathcal{L}_{\text{eff}}(\sigma; K)V$ the Landau function. In the following, we study the generic feature of the phase transition by simply expanding $\mathcal{L}_{\text{eff}}(\sigma; K)$ into a power series of σ as follows:

$$\mathcal{L}_{\text{eff}}(\sigma; K) = \sum_n a_n(K)\sigma^n, \tag{6.33}$$

where $a_n(K)$ is a function of the generalized parameters K.[4]

In Sections 6.3.2–6.3.4, we will discuss the three typical transitions described by Eq. (6.33). Their explicit forms and their relevance to spin systems and QCD are summarized in Table 6.1.

[3] In the old Ehrenfest's classification of phase transitions, the nth order transition corresponds to the case where the first discontinuity appears for $\partial^n P(K)/\partial K_l^n$. Such a definition is, however, not quite appropriate from a modern point of view since some derivatives diverge instead of exhibiting discontinuity. A typical example is the specific heat of the two-dimensional Ising model.

[4] We do not consider the non-analytic terms in $\mathcal{L}_{\text{eff}}(\sigma; K)$ and $S_{\text{eff}}(\sigma(x); K)$ because they are the coarse-grained quantities obtained after integrating out the "hard" (short-wavelength) degrees of freedom with an infrared cutoff, Λ. In turn, this cutoff serves as an ultraviolet cutoff when we integrate out the "soft" (long-wavelength) modes in Eq. (6.32).

Table 6.1. *Landau functions, $\mathcal{L}_{\text{eff}}$, for the three typical cases: the second order transition, the first order phase transition and the case with tricritical behavior.*

Corresponding examples in spin systems with three spatial dimensions and in QCD are also shown. For spin systems, M ($\tilde{M}$) denotes magnetization (staggered magnetization); H ($\tilde{H}$) is the external magnetic (staggered magnetic) field. For QCD, $\langle \bar{q}q \rangle$ (q=u,d,s) are the quark condensates and $\langle L \rangle$ is the Polyakov loop. We define $m_{\text{ud}} = m_{\text{u}} = m_{\text{d}}$ (two degenerate light flavors) and $m_{\text{uds}} = m_{\text{u}} = m_{\text{d}} = m_{\text{s}}$ (three degenerate light flavors); m_{Q} is the heavy quark mass and μ is the quark chemical potential. Notation such as $(a, b) \leftrightarrow (T, H)$ implies that a and b are the linear combinations of T and H near the critical point.

$\mathcal{L}_{\text{eff}}$	Spin system in $d = 3$	QCD
Fig. 6.3(a) Second order, $\frac{a}{2}\sigma^2 + \frac{b}{4}\sigma^4 - h\sigma$, controlled by (a, h)	Ising model $\begin{cases} \sigma \sim M \\ (a, h) \leftrightarrow (T, H) \end{cases}$	$N_c = 3$, $N_f = 2$ $\begin{cases} \sigma \sim \langle \bar{\text{u}}\text{u} + \bar{\text{d}}\text{d} \rangle \\ (a, h) \leftrightarrow (T, m_{\text{ud}}) \end{cases}$ $N_c = 2$, $N_f = 0$ $\begin{cases} \sigma \sim \langle L \rangle \\ (a, h) \leftrightarrow (T, 1/m_{\text{Q}}) \end{cases}$
Fig. 6.3(b) First order, $\frac{a}{2}\sigma^2 - \frac{c}{3}\sigma^3 + \frac{b}{4}\sigma^4 - h\sigma$, controlled by (a, h)	Z(3) Potts model $\begin{cases} \sigma \sim M \\ (a, h) \leftrightarrow (T, H) \end{cases}$	$N_c = 3$, $N_f = 3$ $\begin{cases} \sigma \sim \langle \bar{\text{u}}\text{u} + \bar{\text{d}}\text{d} + \bar{\text{s}}\text{s} \rangle \\ (a, h) \leftrightarrow (T, m_{\text{uds}}) \end{cases}$ $N_c = 3$, $N_f = 0$ $\begin{cases} \sigma \sim \langle L \rangle \\ (a, h) \leftrightarrow (T, 1/m_{\text{Q}}) \end{cases}$
Fig. 6.5 Tricritical behavior, $\frac{a}{2}\sigma^2 + \frac{b}{4}\sigma^4 + \frac{c}{6}\sigma^6 - h\sigma$, controlled by (a, b, h)	model for metamagnet $\begin{cases} \sigma \sim \tilde{M} \\ (a, b, h) \leftrightarrow (T, H, \tilde{H}) \end{cases}$	$N_c = 3$, $N_f = 2+1$ $\begin{cases} \sigma \sim \langle \bar{\text{u}}\text{u} + \bar{\text{d}}\text{d} \rangle \\ (a, b, h) \leftrightarrow (T, m_{\text{s}}, m_{\text{ud}}) \\ (a, b, h) \leftrightarrow (T, \mu, m_{\text{ud}}) \end{cases}$

6.3.2 Second order phase transition

Consider a Landau function, Eq. (6.33), truncated at $n = 4$, without the $n = 3$ term:

$$\mathcal{L}_{\text{eff}} = \frac{1}{2}a\sigma^2 + \frac{1}{4}b\sigma^4 - h\sigma, \tag{6.34}$$

where we assume that

$$a = a_t t \equiv a_t \frac{T - T_c}{T_c} \quad (a_t > 0), \tag{6.35}$$

$$b > 0, \ h \geq 0. \tag{6.36}$$

The Landau function, $\mathcal{L}_{\text{eff}}$, for different values of T, with $h = 0$, is shown in the upper part of Fig. 6.3(a).

The stationary condition for Eq. (6.34), $\partial \mathcal{L}_{\text{eff}} / \partial \sigma = 0$, is

$$a\sigma + b\sigma^3 = h. \tag{6.37}$$

Its solutions, corresponding to the global minimum of $\mathcal{L}_{\text{eff}}$ for $h = 0$, are given by

$$\bar{\sigma}|_{h=0} = \begin{cases} 0 & (T \geq T_c) \\ \pm\left(-\frac{a}{b}\right)^{1/2} & (T < T_c). \end{cases} \tag{6.38}$$

The behavior of $\bar{\sigma}$ as a function of T is shown in the lower part of Fig. 6.3(a).

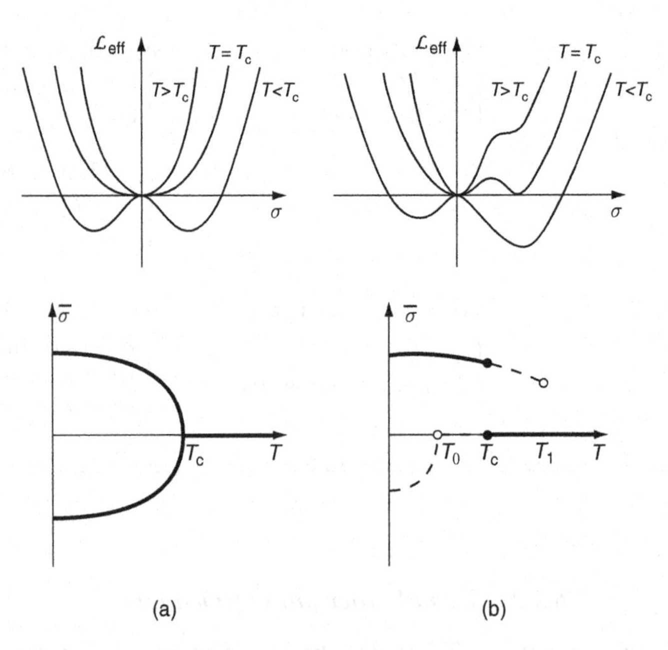

Fig. 6.3. (a) The Landau function, $\mathcal{L}_{\text{eff}}$, for the second order phase transition (upper part) and the behavior of its global and local minima corresponding to Eq. (6.34) for $h = 0$ (lower part). (b) The Landau function for the first order phase transition induced by cubic interaction (upper part) and the behavior of its global and local minima, corresponding to Eq. (6.47) with $h = 0$ (lower part). The dashed lines indicate metastable states.

The specific heat of the system is given by

$$C_{\rm V}(T, h=0) = -T \left. \frac{\partial^2 \mathcal{L}_{\rm eff}(\bar{\sigma}; T)}{\partial T^2} \right|_{h=0} = \begin{cases} 0 & (T \geq T_{\rm c}) \\ \frac{a_t^2}{2b} \frac{T}{T_{\rm c}^2} & (T < T_{\rm c}). \end{cases} \tag{6.39}$$

The effect of a finite external field, h, generally smears the second order phase transition and changes it to a smooth crossover from the low-T phase to the high-T phase. For $h \neq 0$ (we take $h > 0$ without loss of generality), we obtain

$$\bar{\sigma}(T = T_{\rm c}, h) = \left(\frac{h}{b} \right)^{1/3}, \tag{6.40}$$

where $T = T_{\rm c}$ corresponds to $a = 0$ in Eq. (6.35). Then the static magnetic susceptibility is given by

$$\chi_{\rm T}(T, h)\big|_{h=0} = \left. \frac{\partial \bar{\sigma}}{\partial h} \right|_{h=0} = \begin{cases} \frac{1}{a} \sim |T - T_{\rm c}|^{-1} & (T \geq T_{\rm c}) \\ \frac{1}{-2a} \sim |T - T_{\rm c}|^{-1} & (T < T_{\rm c}), \end{cases} \tag{6.41}$$

which is shown in Fig. 6.4 for $h = 0$ and $h \neq 0$.

If we define the critical exponents near the critical point as follows:

$$\bar{\sigma}(T \to T_{\rm c}^-, h = 0) \sim |T - T_{\rm c}|^{\beta}, \tag{6.42}$$

$$C_{\rm V}(T \to T_{\rm c}^{\pm}, h = 0) \sim |T - T_{\rm c}|^{-\alpha_{\pm}}, \tag{6.43}$$

$$\bar{\sigma}(T = T_{\rm c}, h \to 0) \sim h^{1/\delta}, \tag{6.44}$$

$$\chi_{\rm T}(T \to T_{\rm c}^{\pm}, h = 0) \sim |T - T_{\rm c}|^{-\gamma_{\pm}}, \tag{6.45}$$

we obtain

$$\alpha_{\pm} = 0, \quad \beta = 1/2, \quad \gamma_{\pm} = 1, \quad \delta = 3. \tag{6.46}$$

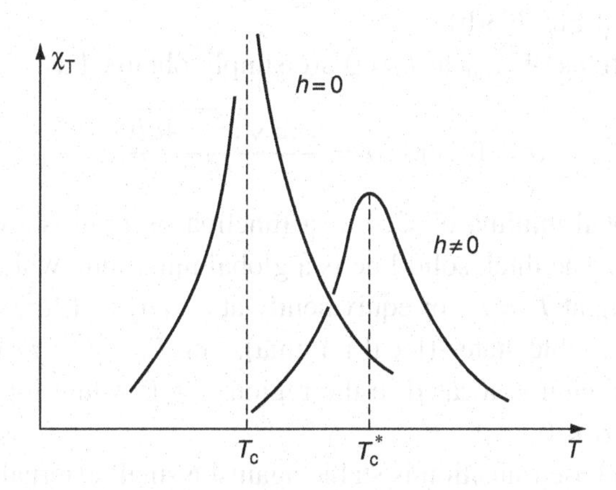

Fig. 6.4. The magnetic susceptibility for $h = 0$ and $h \neq 0$; $T_{\rm c}$ ($T_{\rm c}^*$) is the critical (pseudo-critical) temperature.

The critical exponents in the mean-field theory depend neither on the dimensionality of space nor on the details of the underlying dynamics.

Crucial conditions for obtaining Eq. (6.46) include the absence of a cubic term, σ^3, and the positivity of b. If one of these conditions is not met, other possibilities appear; these will be discussed in Sections 6.3.3 and 6.3.4.

The predictions of the Landau theory will be modified if fluctuations around the mean-field are taken into consideration. First, the fluctuation may change the second order transition either to first order or to no phase transition at all. Secondly, even in the case that the second order nature is not destroyed, critical exponents deviate from the mean-field estimate in Eq. (6.46). The deviation depends on the dimensionality of space and the internal symmetry of the system. This is because the nature of the fluctuation essentially depends on these aspects.

6.3.3 First order transition driven by cubic interaction

Consider a Landau function, Eq. (6.33), with all the terms up to $n=4$,

$$\mathcal{L}_{\text{eff}} = \frac{1}{2}a\sigma^2 - \frac{1}{3}c\sigma^3 + \frac{1}{4}b\sigma^4 - h\sigma, \tag{6.47}$$

where we assume that

$$a = a_t t \equiv a_t \frac{T - T_0}{T_0} \quad (a_t > 0), \tag{6.48}$$

$$b > 0, \quad c > 0, \quad h \geq 0. \tag{6.49}$$

Note that $\mathcal{L}_{\text{eff}}$, for different values of a, is shown for $h=0$ in the upper part of Fig. 6.3(b). In the present case, the critical temperature, T_c, corresponds not to T_0 but to the point where the two minima have the same free energy, as shown in the upper part of Fig. 6.3(b).

Stationary solutions of $\mathcal{L}_{\text{eff}}$ at $h=0$ are simply obtained as

$$\bar{\sigma} = 0 \quad \text{and} \quad \bar{\sigma} = \frac{c \pm \sqrt{c^2 - 4ab}}{2b}. \tag{6.50}$$

The local and global minima of $\mathcal{L}_{\text{eff}}$ as a function of T are shown in the lower part of Fig. 6.3(b). The thick solid line is a global minimum, which experiences a discontinuous jump at $T = T_c$, or equivalently at $a = a_c \equiv (2c^2/9b)$. The dashed lines indicate metastable states (local minima). For $T_0 < T < T_1$ $(0 < a < a_{\text{ms}} \equiv \frac{c^2}{4b})$, the local minimum is located in the region $\sigma \geq 0$, while for $T < T_0$ $(a < 0)$ it is in the region $\sigma < 0$.

The first order phase transition is stable against a small external field (small h). However, it will be washed out eventually because the stationary condition, $a\sigma + b\sigma^3 - c\sigma^2 = h$ with $b > 0$, has only one solution for large h.

6.3.4 Tricritical behavior with sextet interaction

Consider a Landau function, Eq. (6.33), truncated at $n = 6$ without the $n = 3, 5$ terms,

$$\mathcal{L}_{\text{eff}} = \frac{1}{2}a\sigma^2 + \frac{1}{4}b\sigma^4 + \frac{1}{6}c\sigma^6 - h\sigma, \tag{6.51}$$

where we assume $c > 0$ so that $\mathcal{L}_{\text{eff}}$ is bounded from below for large $|\sigma|$. Then both a and b can change sign and may be parametrized as

$$a = a_t t + a_s s, \quad b = b_t t + b_s s, \tag{6.52}$$

$$t = \frac{T - T_c}{T_c}, \quad s = \frac{S - S_c}{S_c}. \tag{6.53}$$

The point $(a, b) = (0, 0)$ is called the tricritical point (TCP) for reasons given later. The behavior of the system around the TCP is governed by a reduced temperature, t, and another independent parameter, s. This situation is realized in various systems, for example the ^{3}He–^{4}He mixture, the multi-component fluid mixture, some metamagnets, and so on (Lawrie and Sarnach, 1984). In QCD, linear combinations of the temperature and the baryon chemical potential take the roles of t and s. One can also consider linear combinations of the temperature and the strange-quark mass to be appropriate for t and s.

The phase structure in the (a, b)-plane and the associated form of $\mathcal{L}_{\text{eff}}(\sigma)$ for $h = 0$ are shown in Fig. 6.5(a). The TCP is located at the origin, $a = b = 0$. The stationary solutions of $\mathcal{L}_{\text{eff}}$ for $h = 0$ are given by

$$\bar{\sigma} = 0, \quad \bar{\sigma}^2 = \frac{-b \pm \sqrt{b^2 - 4ac}}{2c}. \tag{6.54}$$

Then, two critical lines smoothly connected at TCP are obtained, as shown in Fig. 6.5(a); the second order critical line (the thick solid line in $b > 0$) and the first order critical line (the thick solid line in $b < 0$). They are characterized as follows:

$$\text{second order line} : b > 0, \ a = 0, \tag{6.55}$$

$$\text{first order line} : b < 0, \ a = \frac{3b^2}{16c}. \tag{6.56}$$

Around the first order critical line, there is a region where metastable states (local minima of $\mathcal{L}_{\text{eff}}$) exist; It is bounded by the two dashed lines, $(a = 0, b < 0)$ and $(a = \frac{b^2}{4c}, b < 0)$. The existence of the metastable state is a generic feature of the first order transition, where several local minima compete with each other.

Let us introduce an external field, $h \neq 0$. As we have seen already in Section 6.3.2, the second order transition is unstable for small h. The first order

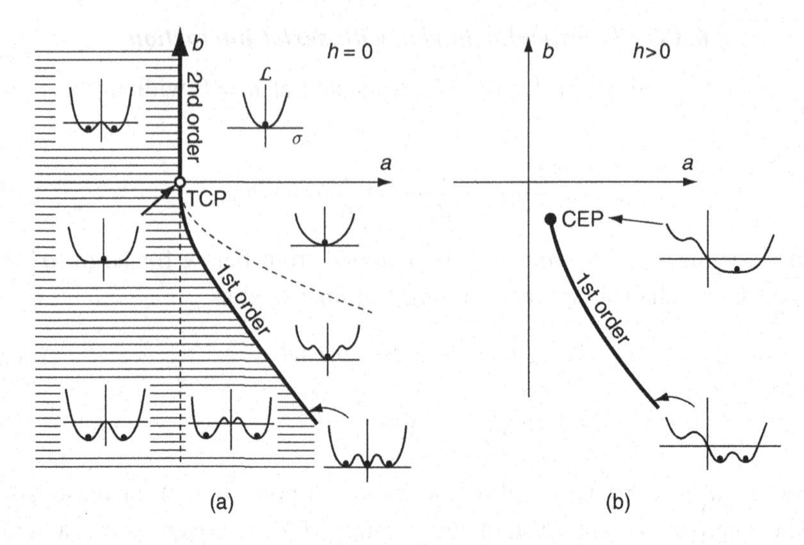

Fig. 6.5. (a) Phase diagram, depicting tricritical behavior ($h = 0$). The thick solid lines correspond to either first or second order transitions. There are metastable states around the first order critical line, bounded by the dashed lines. The hatched region corresponds to the symmetry-broken phase. (b) Case for $h \neq 0$, where the first-order critical line terminates at the critical end point.

transition is relatively stable for h, but can be wiped out if $h(> 0)$ is large enough or $b(< 0)$ is small enough. Namely, the second order critical line disappears and the first order critical line shifts for $h \neq 0$, as shown in Fig. 6.5(b). The point where the first order line terminates is called the critical end point (CEP). Note that $\bar{\sigma}$ exhibits a second order behavior and is continuous, but non-analytic, across the CEP (Exercise 6.5). In terms of $\mathcal{L}_{\text{eff}}$, two minima located in $\sigma > 0$ collapse to form a global minimum at the CEP. As $|h|$ increases, the CEP moves further and further away from the TCP.

In the three-dimensional parameter space governed by a, b and h, the TCP is identified as $(a, b, h) = (0, 0, 0)$. For $(a, b, h) = (0, b > 0, 0)$, there is a standard second order critical line, as we know from Fig. 6.5(a). For $b < 0$, the CEP forms two second order critical lines starting from the TCP: one is along the direction of negative b and positive h, and the other is along the direction of negative b and negative h.

This situation is shown schematically in Fig. 6.6. The cross-section of the figure at $h = 0$ corresponds to Fig. 6.5(a), and that at $h \neq 0$ corresponds to Fig. 6.5(b). The shaded surface (the wing) in Fig. 6.6 is a first order critical surface across which the first order transition occurs. Also, three second order critical lines merge together at $(a, b, h) = (0, 0, 0)$; this is why it is called the tricritical point.

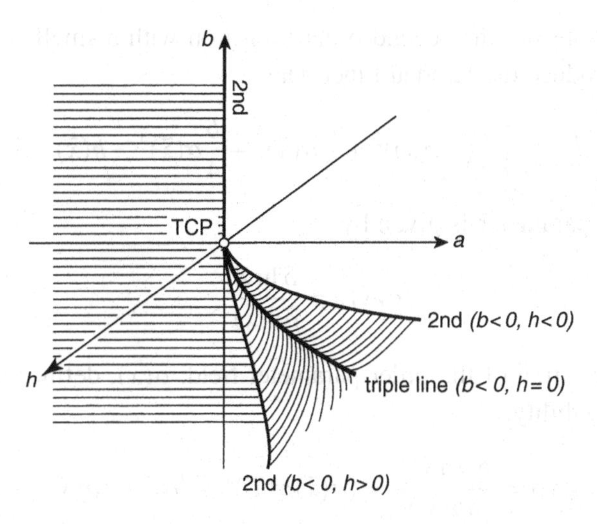

Fig. 6.6. The phase diagram in the three-dimensional (a, b, h) space. A "wing" develops from the TCP in the $b < 0$ region. The center of the wing, at $h = 0$, corresponds to a triple line (a line of triple points on which the three different phases coexist).

The edge of the wing (second order critical lines formed by the CEP) in (a, b, h)-space can be identified by the condition that $\mathcal{L}_{\text{eff}}$ is "flat":

$$\frac{\partial^n \mathcal{L}_{\text{eff}}}{\partial \sigma^n} = 0 \quad (n = 1, 2, 3), \tag{6.57}$$

which may be rewritten as

$$a = 5c\sigma^4, \quad b = -\frac{10}{3}c\sigma^2, \quad h = \frac{8}{3}c\sigma^5. \tag{6.58}$$

Eliminating σ, we have

$$\pm h = \frac{8c}{3}\left(\frac{a}{5c}\right)^{5/4} = \frac{8c}{3}\left(\frac{-3b}{10c}\right)^{5/2} \quad (a \geq 0, b \leq 0). \tag{6.59}$$

One may also define the n-critical point by introducing terms up to σ^n in the Landau function.

6.4 Spatial non-uniformity and correlations

Let us apply a small but spatially non-uniform external field, $h(\vec{x})$, to a system in d spatial dimensions. Then the order parameter acquires spatial dependence, $\sigma(\vec{x})$, with $\vec{x} = (x_1, x_2, \ldots, x_d)$. In this case, we need to study the Landau functional,

$S_{\rm eff}(\sigma(\vec{x}); K)$. Consider the second order transition with a small spatial inhomogeneity and introduce the Landau functional

$$S_{\rm eff} = \int d^d x \left[\frac{1}{2}(\nabla\sigma(\vec{x}))^2 + \frac{a}{2}\sigma(\vec{x})^2 + \frac{b}{4}\sigma(\vec{x})^4 - h(\vec{x})\sigma(\vec{x}) \right]. \tag{6.60}$$

The local order parameter is given by

$$\langle\sigma(\vec{x})\rangle = \frac{\delta \ln Z}{\delta h(\vec{x})}. \tag{6.61}$$

The spatial correlation of the order-parameter field, $\sigma(\vec{x})$, defines the dynamical magnetic susceptibility,

$$\chi(\vec{x}, \vec{x}') = \frac{\delta\langle\sigma(\vec{x})\rangle}{\delta h(\vec{x}')} = \langle\sigma(\vec{x})\sigma(\vec{x}')\rangle - \langle\sigma(\vec{x})\rangle\langle\sigma(\vec{x}')\rangle. \tag{6.62}$$

which is related to the static susceptibility for the uniform case, Eq. (6.41), as $\chi(\vec{x}, \vec{x}) = V \cdot \chi_{_{\rm T}}(T, h)$, where V is the spatial volume in d dimensions.

In the mean-field approximation, $\langle\sigma(\vec{x})\rangle$ can be determined by the stationary condition, $\delta S_{\rm eff}/\delta\sigma(\vec{x}) = 0$. For a small external field, $h \to 0$, we arrive at a linear equation for $\chi(\vec{x}, \vec{x}')$:

$$\left[-\nabla_x^2 + \xi^{-2}\right]\chi(\vec{x} - \vec{x}') = \delta(\vec{x} - \vec{x}'), \tag{6.63}$$

with the correlation length given by

$$\xi = \left(a + 3b\bar{\sigma}^2\right)^{-1/2} = \begin{cases} \xi_+(T \geq T_{\rm c}) = a^{-1/2} \\ \xi_-(T < T_{\rm c}) = (-2a)^{-1/2}, \end{cases} \tag{6.64}$$

where $\bar{\sigma}$ is the global minimum of the Landau function in the homogeneous case, Eq. (6.38).

The Fourier transform of $\chi(\vec{x} - \vec{x}')$ is given by

$$\chi(\vec{k}) = \frac{1}{\vec{k}^2 + \xi_\pm^{-2}} \xrightarrow[T \to T_{\rm c}]{} \frac{1}{\vec{k}^2}. \tag{6.65}$$

If we define the two more critical exponents:

$$\xi_\pm(T \to T_{\rm c}^\pm) \propto |T - T_{\rm c}|^{-\nu_\pm}, \tag{6.66}$$

$$\chi(\vec{x} - \vec{x}')\big|_{T = T_{\rm c}} \propto \frac{1}{|\vec{x} - \vec{x}'|^{d-2+\eta}}, \tag{6.67}$$

the mean-field theory predicts

$$\nu_\pm = 1/2, \quad \eta = 0. \tag{6.68}$$

For $d = 3$, $\chi(\vec{x} - \vec{x}')$ has a Yukawa form,

$$\chi(\vec{x} - \vec{x}')\big|_{d=3} = \frac{1}{4\pi} \frac{1}{|\vec{x} - \vec{x}'|} \mathrm{e}^{-|\vec{x}-\vec{x}'|/\xi_{\pm}}, \tag{6.69}$$

which implies that the spatial correlation becomes long-ranged near the critical point of the second order phase transition because ξ (the correlation length) becomes large. This is not the case for the first order transition, where the correlation length stays finite at the critical point.

6.5 Critical fluctuation and the Ginzburg region

In Section 6.4, we saw how a spatial fluctuation of the order parameter, $\sigma(\vec{x})$, becomes long-ranged or "soft" as the system approaches the critical point of the second order phase transition. Since such soft fluctuations (soft modes) are easily excited thermally, they may affect the partition function and the nature of the second order phase transition. In the context of QCD, the pion and its chiral partner (usually called the σ-meson) correspond to these soft modes.

The effect of the fluctuations depends on the number of spatial dimensions, d, the number of internal degrees of freedom and the symmetry of the underlying theory. Therefore, the behavior in mean-field theory will be modified once the fluctuations are taken into account. The results may be summarized as follows.

- For $d \le d_{\mathrm{LC}}$ (the lower critical dimension), dynamical symmetry breaking cannot occur even if we take the limit $T \to 0$. (Mermin–Wagner–Coleman's theorem Mermin and Wagner, 1966; Coleman, 1973). This is because the fluctuation generally becomes large as d decreases. For small d, the direction of the fluctuation is limited in space. This tends to enhance its effect to destroy the long-range order. On the other hand, for large d the effect of the fluctuation is diluted in many spatial directions and becomes weak. The dimension at which the long-range order disappears is called the lower critical dimension, d_{LC}. For the Ising model, which has only a discrete symmetry, $d_{\mathrm{LC}} = 1$. For models with continuous symmetry, such as the Heisenberg model, $d_{\mathrm{LC}} = 2$.
- For $d > d_{\mathrm{LC}}$, a phase transition at a non-zero value of T_{c} takes place. However, the order of the phase transition is affected by the thermal fluctuations. Even if the mean-field theory predicts the second order phase transition, it may turn into first order due to fluctuations (fluctuation-induced first order transition). Typical examples are the case with multiple coupling constants among the order parameters and the case with the coupling to gauge fields. The former example in QCD is the chiral transition with massless three-flavors (Pisarski and Wilczek, 1984), which will be discussed later, and the latter example in QCD is the super-to-normal transition of the color superconductor coupled to thermal gluons in analogy with the metallic superconductors coupled to thermal photons (Halperin *et al.*, 1974; Amit, 1984).

- Even when the effect of fluctuations is mild and the second order nature is maintained for $d > d_{\rm LC}$, the critical exponents may be significantly modified when $d < d_{\rm UC}$, where $d_{\rm UC}$ is the upper critical dimension. The naive perturbation theory breaks down due to infrared divergences below $d_{\rm UC}$. As we will see, the model given in Eq. (6.60) takes $d_{\rm UC} = 4$. The renormalization group method, combined with the ϵ-expansion (expansion by $\epsilon = d_{\rm UC} - d$), is one of the systematic techniques used to treat the situation, which will be discussed in Section 6.6 in more detail.

To see how to obtain the upper critical dimension, $d_{\rm UC}$, let us formulate a naive perturbation theory by taking Eq. (6.60) as an example:

$$Z = \int [d\sigma]\, \mathrm{e}^{-S_{\rm eff}}, \tag{6.70}$$

$$S_{\rm eff} = \int d^d x \left[\frac{1}{2}(\nabla_x \sigma)^2 + \frac{1}{2} a\sigma^2 + \frac{1}{4} b\sigma^4 \right]. \tag{6.71}$$

We consider the case $a > 0$ ($T > T_{\rm c}$) for simplicity. We introduce a dimensionless length scale, $\vec{r}$, a dimensionless field variable, ρ, and a dimensionless coupling, b', defined relative to the correlation length, $\xi = 1/\sqrt{a}$:

$$\vec{x} = \xi \vec{r}, \quad \sigma^2 = \xi^{2-d} \rho^2, \quad b' = b\xi^{4-d}. \tag{6.72}$$

Then we have

$$S_{\rm eff} = \int d^d r \left[\frac{1}{2}(\nabla_r \rho)^2 + \frac{1}{2}\rho^2 + \frac{1}{4} b'\rho^4 \right]. \tag{6.73}$$

The perturbative expansion is set up by treating the quartic term in ρ as an interaction. Then the dimensionless expansion parameter is b'. Since ξ becomes large near the critical point, we obtain

$$b' = b\xi^{4-d} \to \begin{cases} 0 & (d > 4) \\ \infty & (d < 4). \end{cases} \tag{6.74}$$

Namely, the naive perturbation theory breaks down for a spatial dimension less than 4, whereas it is a sensible expansion (at least in the sense of the asymptotic expansion) for $d > 4$; i.e., $d_{\rm UC} = 4$ in this model.

The physical reason for the breakdown of the perturbation theory is clear if one goes back to the action before rescaling. No matter how small the coupling b is, the soft fluctuation with "mass" $a = 1/\xi^2$ causes severe infrared (IR) divergences in higher orders near the critical point.

By using b', we may estimate the region of temperature in which the fluctuation becomes important. The condition $b' > O(1)$ is rewritten as

$$|t| = \frac{|T - T_{\rm c}|}{T_{\rm c}} < \frac{b^2}{a_t} \quad (\text{for } d = 3), \tag{6.75}$$

where we have used the definition $a = a_t t$. This characterizes the critical region (the Ginzburg region) where the fluctuations become significant (Ginzburg, 1961). The size of the critical region depends highly on the system: a weak coupling BCS superconductor in $d = 3$ has quite a narrow critical region, whereas spin systems and the superfluid ^{4}He have wide critical regions (Goldenfeld, 1992).

6.6 Renormalization group and ϵ-expansion

As we have seen in Section 6.5 for the Landau functional with quadratic and quartic terms, Eq. (6.71), severe IR divergence is encountered near the critical point of the second order phase transition for $d < 4$. The renormalization group (RG) method, together with the dimensional regularization, supply us with a convenient way to handle this problem (Wilson and Kogut, 1974; Zinn-Justin, 2002).

In the dimensional regularization, the spatial dimension, d, is analytically continued to non-integer values with a continuous parameter ϵ:

$$d = 4 - \epsilon, \tag{6.76}$$

where ϵ acts as a "small" expansion parameter, which will eventually be taken to be unity.

In the following, we will outline the basic idea of the RG method based on the ϵ-expansion and then apply the idea to study the second order phase transition of the ϕ^4 model. The method may also be used to detect a signature of first order transitions. The chiral phase transition in QCD will be discussed in Section 6.13.

6.6.1 *Renormalization in $4 - \epsilon$ dimensions*

Let us introduce a one-component ϕ^4 model in Euclidean d dimensions:

$$\begin{aligned} Z &= \int [d\phi_{\rm B}]\, \mathrm{e}^{-S_{\rm eff}}, \\ S_{\rm eff} &= \int d^d x \left[\frac{1}{2}(\partial_\mu \phi_{\rm B}(x))^2 + \frac{a_{\rm B}}{2}\phi_{\rm B}^2(x) + \frac{b_{\rm B}}{4!}\phi_{\rm B}^4(x) - h_{\rm B}(x)\phi_{\rm B}(x) \right]. \end{aligned} \tag{6.77}$$

This is equivalent to what was discussed in Eq. (6.60), with the replacements, $\sigma(x) \to \phi_{\rm B}(x)$, $a \to a_{\rm B}$, $b \to b_{\rm B}/6$ and $h \to h_{\rm B}$. The subscript B stands for bare (unrenormalized) parameters and field.

The basic idea of the renormalization is to reorganize the perturbation theory so that the "large" contribution from high-momentum (short-distance) fluctuations is absorbed in the redefinition of the parameters of the theory: For this purpose, we introduce the following redefinition:

$$a_{\mathrm{B}} = z_a\, a, \quad b_{\mathrm{B}} = z_b\, b\, \kappa^{\epsilon}, \quad h_{\mathrm{B}}(x) = z_h\, h(x), \tag{6.78}$$

where $z_{a,b,h}$ are called renormalization constants.[5] The scale, κ, introduced in Eq. (6.78) is an arbitrary, dimensionful parameter, which makes the renormalized coupling, b, dimensionless in d dimensions.

Since Eq. (6.78) is merely a redefinition of the couplings, the partition function is invariant,

$$Z[h_{\mathrm{B}}, a_{\mathrm{B}}, b_{\mathrm{B}}] = Z[h, a, b|\kappa]. \tag{6.79}$$

Our goal is to change a badly behaved perturbation series in terms of the bare couplings $(h_{\mathrm{B}}, a_{\mathrm{B}}, b_{\mathrm{B}})$ into a better behaved perturbation series in terms of the renormalized couplings (h, a, b).

The field $\phi_{\mathrm{B}}(x)$ in Eq. (6.77), which is conjugate to the bare external field, $h_{\mathrm{B}}(x)$, is simply the integration variable. We may introduce a renormalized field, $\phi(x)$, conjugate to the renormalized external field, $h(x)$, as follows:

$$\phi_{\mathrm{B}}(x) = z_h^{-1}\phi(x). \tag{6.80}$$

Note that z_h^{-1} is often defined as $\sqrt{z_\phi}$ in the literature.

Since κ is arbitrary, we obtain

$$\kappa \frac{d}{d\kappa} Z[h, a, b|\kappa] = 0, \tag{6.81}$$

which is the master equation for the renormalization group (Brown, 1995). The reason it is called the "group" is that the procedure used to integrate out the high-momentum fluctuations in order to construct an effective low-energy theory forms a group. Strictly speaking, it is a semi-group, since the procedure corresponds to a coarse graining and is not invertible.

Since the renormalized couplings, a, b and h, are defined as being associated with the scale, κ, they are implicit functions of κ. Therefore, the master equation may be rewritten as

$$\left(\mathcal{D} + \beta_h \int d^d x\, h(x) \frac{\delta}{\delta h(x)}\right) Z[h, a, b|\kappa] = 0, \tag{6.82}$$

$$\mathcal{D} \equiv \kappa \frac{\partial}{\partial \kappa} + a\beta_a \frac{\partial}{\partial a} + \beta_b \frac{\partial}{\partial b}, \tag{6.83}$$

[5] Strictly speaking, we need another renormalization constant, a constant term in S_{eff} or "the cosmological constant." However, it does not appear explicitly in the following discussions.

where

$$\kappa \frac{da}{d\kappa} = a\beta_a(b, \epsilon), \quad \kappa \frac{db}{d\kappa} = \beta_b(b, \epsilon), \tag{6.84}$$

$$\kappa \frac{dh(x)}{d\kappa} = h(x)\beta_h(b, \epsilon), \quad \kappa \frac{d\phi(x)}{d\kappa} = -\phi(x)\beta_h(b, \epsilon). \tag{6.85}$$

Factors β_i $(i = a, b, h)$ in Eqs. (6.84) and (6.85) are "velocities" which control the renormalization-group (RG) flow of the couplings as functions of κ. Note that the bare couplings do not know anything about κ; $d(a_B, b_B, h_B)/d\kappa = 0$.

A few remarks are in order here. In the dimensional regularization combined with the minimal subtraction scheme, renormalized couplings absorb only the leading singularities proportional to $1/\epsilon^n$ for $\epsilon \sim 0$. In this case, the dimensionless renormalization constants, z_i, are shown to be independent of a (Muta, 1998) (the mass independent renormalization)[6]:

$$z_i(b, \epsilon) = 1 + \sum_{n=1}^{\infty} \frac{z_i^{(n)}(b)}{\epsilon^n}, \qquad (i = a, b, h). \tag{6.86}$$

Therefore, the "velocity" parameters, β_i, are also a-independent in this scheme. Furthermore, by using the κ-independence of the bare couplings, together with the definition of the βs, it may be shown that (Exercise 6.6)

$$\beta_i(b, \epsilon) = -\epsilon b \delta_{ib} + \beta_i(b) \quad (i = a, b, h). \tag{6.87}$$

Namely, the ϵ appears only in the first term of $\beta_b(b, \epsilon)$. In the following discussion, we will follow this simple scheme exclusively.

6.6.2 Running couplings

Let us introduce running couplings and running fields under the scale transformation of κ:

$$\begin{aligned} \bar{b}(s) &= b(e^s\kappa), \quad \bar{a}(s) = a(e^s\kappa), \\ \bar{h}(x; s) &= h(x; e^s\kappa), \quad \bar{\phi}(x; s) = \phi(x; e^s\kappa). \end{aligned} \tag{6.88}$$

[6] The dimensionful coupling, a, can enter into z_i only as a combination of the form $\ln(\kappa^2/a)$. However, such terms, for example in the form $(\ln \kappa^2)/\epsilon$, are always canceled by the counter-terms in sub-diagrams and do not appear.

Then the flow equations, Eqs. (6.84) and (6.85), together with the β-functions, Eq. (6.87), may be rewritten as follows:

$$\frac{d\bar{b}(s)}{ds} = \beta_b(\bar{b}(s), \epsilon), \tag{6.89}$$

$$\bar{a}(s) = a \exp\left(\int_0^s ds' \beta_a(\bar{b}(s'))\right), \tag{6.90}$$

$$\bar{h}(x; s) = h(x) \exp\left(\int_0^s ds' \beta_h(\bar{b}(s'))\right), \tag{6.91}$$

$$\bar{\phi}(x; s) = \phi(x) \exp\left(-\int_0^s ds' \beta_h(\bar{b}(s'))\right). \tag{6.92}$$

We define the initial conditions at $s = 0$ as $b \equiv \bar{b}(0)$, $a \equiv \bar{a}(0)$, $h(x) \equiv \bar{h}(x; 0)$ and $\phi(x) \equiv \bar{\phi}(x; 0)$.

If we can compute β-functions in perturbation theory, we obtain a non-perturbative behavior of the coupling constants by integration from the above formulas. The procedure may be interpreted as constructing an envelops of a set of perturbative solutions (Ei *et al.*, 2000).

6.6.3 Vertex functions

The partition function, Z, is a generating functional of the Green's functions:

$$\begin{aligned} G^{(n)}(x_1, \ldots, x_n) &= \langle \phi(x_1) \cdots \phi(x_n) \rangle \\ &= \frac{1}{Z[h]} \left(\frac{\delta}{\delta h(x_1)} \cdots \frac{\delta}{\delta h(x_n)} \right) Z[h] \bigg|_{h=0}. \end{aligned} \tag{6.93}$$

On the other hand, the grand potential, $\Omega[h] = -\ln Z[h]$, is a generating functional of connected Green's functions because of the linked cluster theorem (Exercise 4.3):

$$\begin{aligned} G_{\mathrm{c}}^{(n)}(x_1, \ldots, x_n) &= \langle \phi(x_1) \cdots \phi(x_n) \rangle_{\mathrm{c}} \\ &= -\left(\frac{\delta}{\delta h(x_1)} \cdots \frac{\delta}{\delta h(x_n)} \right) \Omega[h] \bigg|_{h=0}. \end{aligned} \tag{6.94}$$

It is also convenient to introduce the vertex function, Γ, as a Legendre transform of Ω:

$$\Gamma[\varphi] \equiv \Omega[h] + \int d^d x \; h(x)\varphi(x), \tag{6.95}$$

$$\varphi(x) \equiv \langle \phi(x) \rangle = -\frac{\delta \Omega[h]}{\delta h(x)}.$$

By definition, the variation of the vertex function with respect to the "classical field," $\varphi(x)$, is given by

$$h(x) = \frac{\delta \Gamma[\varphi]}{\delta \varphi(x)}. \tag{6.96}$$

The vertex function can be formally expanded by $\varphi(x)$ as follows:

$$\Gamma[\varphi] = \sum_{n=0}^{\infty} \frac{1}{n!} \int d^d x_1 \cdots d^d x_n \ \varphi(x_1) \ldots \varphi(x_n) \ \Gamma^{(n)}(x_1, \ldots, x_n). \tag{6.97}$$

Then the expansion coefficients, $\Gamma^{(n)}(x_1, \ldots, x_n)$, become one-particle irreducible (1PI) n-point vertices. Here, "1PI" means that corresponding Feynman diagrams cannot be separated into two parts by cutting any one of their internal lines. Here, "vertices" implies that all the external lines are amputated (Exercise 6.7). For example, the two-point vertex is given schematically by

$$\Gamma^{(2)} = \frac{\delta^2 \Gamma[\varphi]}{\delta \varphi \delta \varphi} = \frac{\delta h}{\delta \varphi} \tag{6.98}$$

$$= [\langle \phi \phi \rangle - \langle \phi \rangle \langle \phi \rangle]^{-1} = [G_{\mathrm{c}}^{(2)}]^{-1} = \frac{G_{\mathrm{c}}^{(2)}}{G_{\mathrm{c}}^{(2)} G_{\mathrm{c}}^{(2)}}. \tag{6.99}$$

We see that $\Gamma^{(2)}$ is simply an inverse of the connected Green's function, which is rewritten as a two-point function with two amputations of the external lines as shown in the last equality; $\Gamma^{(2)}$ and $\Gamma^{(4)}$ for the ϕ^4 model are shown in Fig. 6.7.

6.6.4 RG equation for vertex function

Renormalization group (RG) equations for general Green's functions are the consequence of the RG invariance of Z, i.e. Eq. (6.81), or equivalently Eq. (6.82). By combining Eq. (6.82) with Eq. (6.93), we immediately obtain

$$(\mathcal{D} + n\beta_h) G^{(n)}(x_1, \ldots, x_n; a, b|\kappa) = 0. \tag{6.100}$$

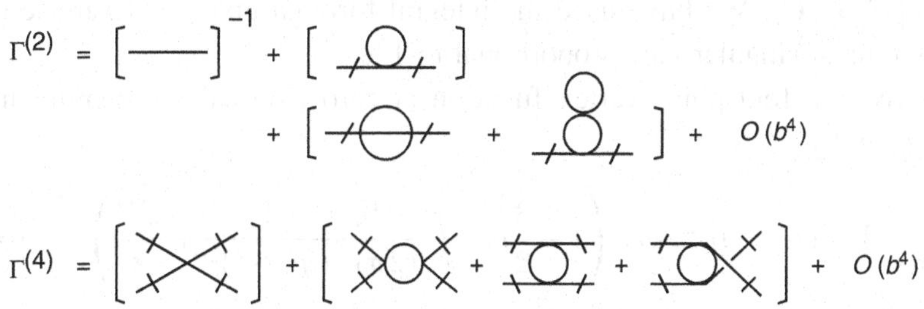

Fig. 6.7. An example of 1PI two-point and four-point vertices up to $O(b^2)$ in the ϕ^4 model. Each slash indicates the amputation of the external lines.

Also, Eq. (6.81) and the definition of Γ imply that $\Gamma[\varphi, a, b|\kappa]$ is RG-invariant; $\kappa d\Gamma/d\kappa = 0$. This is rewritten as

$$\left(\mathcal{D} - \beta_h \int d^d x \; \varphi(x) \frac{\delta}{\delta\varphi(x)}\right) \Gamma[\varphi; a, b|\kappa] = 0, \tag{6.101}$$

$$(\mathcal{D} - n\beta_h)\Gamma^{(n)}(x_1, \ldots, x_n; a, b|\kappa) = 0. \tag{6.102}$$

Furthermore, the RG equation for the "magnetic" field, h, is obtained from Eq. (6.102) and Eq. (6.96) as

$$\left[\mathcal{D} - \beta_h \int d^d x \left(1 + \varphi(x) \frac{\delta}{\delta\varphi(x)}\right)\right] h(x) = 0. \tag{6.103}$$

6.7 Perturbative evaluation of β_i

Perturbation theory in terms of the renormalized coupling constant, b, may be formulated by rewriting the action, $S_{\rm eff}$, in Eq. (6.77) as follows:

$$S_{\rm eff} = \int d^d x \left[\frac{1}{2}(\partial_\mu \phi(x))^2 + \frac{a}{2}\phi^2(x) + \frac{b\kappa^\epsilon}{4!}\phi^4(x) - h(x)\phi(x)\right] + S_{\rm CT}, \tag{6.104}$$

where $S_{\rm CT}$ are the counter-terms defined by identifying the above Landau functional, $S_{\rm eff}$, with that given in Eq. (6.77) and using Eq. (6.78).

To calculate the β_i in the leading order, the two-point vertex function in the momentum-space with zero external momentum in one loop can be evaluated from Fig. 6.7 as

$$\Gamma^{(2)}(p=0) \simeq a + \frac{1}{2} b\kappa^\epsilon \int \frac{d^d q}{(2\pi)^d} \frac{1}{q^2 + a} + a\frac{z_a^{(1)}}{\epsilon} \tag{6.105}$$

$$= a\left(1 - \frac{b}{16\pi^2 \epsilon} + \frac{z_a^{(1)}}{\epsilon}\right) + \text{finite terms}. \tag{6.106}$$

Here, the 1/2 factor in front of the integral is a combinatorial factor: $1/2 = (1/4!) \times {}_4C_2 \times {}_2C_1$. We have used the integral formula given in Exercise 6.8 to extract the most singular part proportional to $1/\epsilon$.

Similarly, the four-point vertex function at zero external momentum in one loop is given by

$$\Gamma^{(4)}(p_i = 0) \simeq b\kappa^\epsilon \left(1 - \frac{3}{2} b\kappa^\epsilon \int \frac{d^d q}{(2\pi)^d} \frac{1}{(q^2 + a)^2} + \frac{z_b^{(1)}}{\epsilon}\right) \tag{6.107}$$

$$= b\kappa^\epsilon \left(1 - \frac{3b}{16\pi^2 \epsilon} + \frac{z_b^{(1)}}{\epsilon}\right) + \text{finite terms}. \tag{6.108}$$

Here, the 3/2 factor in front of the integral is $3/2 = (1/2) \times (1/4!)^2 \times ({}_4C_2)^2 \times 2 \times 4!$ For an alternative way of obtaining Eqs. (6.105) and (6.107) without being bothered by the combinatorial consideration, see Exercise 6.9.

A large correction for $\epsilon \to 0$ is absorbed in $z_{a,b}$, and we find that

$$a_{\mathrm{B}} \simeq \left(1 + \frac{b}{16\pi^2}\frac{1}{\epsilon}\right) a, \quad b_{\mathrm{B}} \simeq \left(1 + \frac{3b}{16\pi^2}\frac{1}{\epsilon}\right) b\kappa^{\epsilon}. \tag{6.109}$$

Also, $h_{\mathrm{B}} = h$ in the one-loop order. This leads to the one-loop evaluation of β_i as

$$\beta_a(b) = \frac{1}{16\pi^2} b, \quad \beta_b(b) = \frac{3}{16\pi^2} b^2, \quad \beta_h(b) = 0. \tag{6.110}$$

6.8 Renormalization group equation and fixed point

The vertex function, Γ, as well as the partition function, Z, are independent of κ. This is simply because the bare theory does not know anything about the arbitrary scale, κ. In other words, all the coupling constants in the theory conspire to hide the κ-dependence of Z and Γ.

If we combine this fact with a dimensional analysis, we can compare the Green's functions with a common κ but with different momenta and couplings. Then we may study the UV and IR behavior of the Green's functions from the RG flow of the coupling constants. In high-energy processes of elementary particle collisions, our main interest lies in the UV behavior of the scattering amplitude. On the other hand, in the study of the critical phenomena associated with the second order phase transition, the long-wavelength (IR) behavior of the various observables is of interest. In both cases, the RG analysis plays a crucial role.

6.8.1 Dimensional analysis and solution of RG equation

Let us first count the dimension of the basic couplings and the field in d dimensions:

$$[a] = \mathrm{E}^2, \quad [b] = \mathrm{E}^0, \quad [h] = \mathrm{E}^{\frac{d}{2}+1}, \quad [\phi] = \mathrm{E}^{\frac{d}{2}-1}, \tag{6.111}$$

where E is a scale with the dimension of energy. Since $[\Gamma[\varphi]] = [W[h]] = \mathrm{E}^0$, we have

$$[G^{(n)}(x)] = \mathrm{E}^{(\frac{d}{2}-1)n}, \quad [\Gamma^{(n)}(x)] = \mathrm{E}^{(\frac{d}{2}+1)n}. \tag{6.112}$$

Here and below, we abbreviate $(x_1, \ldots, x_n)$ by x, for simplicity.

Now we define the Green's function and the vertex function in the momentum-space by making a Fourier transform:

$$(2\pi)^d \delta^{(d)}(k_1+\cdots+k_n)\Gamma^{(n)}(k)$$
$$\equiv \int d^d x_1 \cdots d^d x_n e^{-i(k_1 x_1+\cdots+k_n x_n)}\Gamma^{(n)}(x), \tag{6.113}$$

where Γ in the coordinate-space (the momentum-space) is specified only by its argument, x (k), for simplicity. Then we obtain

$$[G^{(n)}(k)] = \mathrm{E}^{d-(\frac{d}{2}+1)n}, \quad [\Gamma^{(n)}(k)] = \mathrm{E}^{d-(\frac{d}{2}-1)n}. \tag{6.114}$$

The dimensions obtained in the above analysis are called the canonical dimensions. For later use, we define the canonical dimensions of $h(x)$, $G^{(n)}(k)$ and $\Gamma^{(n)}(k)$ as

$$d_h = \frac{d}{2}+1, \tag{6.115}$$

$$d_G = d-\left(\frac{d}{2}+1\right)n, \quad d_\Gamma = d-\left(\frac{d}{2}-1\right)n. \tag{6.116}$$

Suppose we make a scaling of all the dimensionful quantities according to the rule $\mathrm{E} \to \lambda \mathrm{E}$. Then the dimensional counting yields

$$\Gamma[\varphi, a, b\ |\kappa] = \Gamma\left[\frac{\varphi}{\lambda^{d/2-1}}, \frac{a}{\lambda^2}, b\ \left|\frac{\kappa}{\lambda}\right.\right], \tag{6.117}$$

$$\Gamma^{(n)}(k; a, b|\kappa) = \lambda^{d_\Gamma}\Gamma^{(n)}\left(\frac{k}{\lambda}; \frac{a}{\lambda^2}, b\ \left|\frac{\kappa}{\lambda}\right.\right). \tag{6.118}$$

Similar formulas to Eq. (6.118) hold for $G^{(n)}(k)$ by the replacement $d_\Gamma \to d_G$.

Now let us apply this dimensional analysis to obtain a useful solution of the RG equation, Eq. (6.102). Instead of taking the partial differential form of the RG equation, we go back to its original form,

$$\Gamma[\varphi(x;\kappa), a(\kappa), b(\kappa)|\kappa] = \Gamma[\varphi(x;\kappa'), a(\kappa'), b(\kappa')|\kappa']. \tag{6.119}$$

By choosing $\kappa' = e^s\kappa (\equiv \lambda\kappa)$ and then applying the dimensional analysis, Eq. (6.117), we arrive at

$$\Gamma[\varphi(x), a, b|\kappa] = \Gamma[\bar{\varphi}(x;s)\, e^{-s(d/2-1)}, \bar{a}(s)\, e^{-2s}, \bar{b}(s)|\kappa]. \tag{6.120}$$

Differentiating both sides of Eq. (6.119) with respect to $\varphi(x;\kappa)$ and using Eq. (6.92), we have

$$\Gamma^{(n)}(k; a, b|\kappa)$$
$$= e^{sd_\Gamma - n\int_0^s \beta_h(\bar{b}(s'))ds'} \cdot \Gamma^{(n)}(k\, e^{-s}; \bar{a}(s)\, e^{-2s}, \bar{b}(s)|\kappa). \tag{6.121}$$

As mentioned earlier, the renormalization group leads to a relation for the vertex functions with different κ. Then the dimensional analysis turns it into a relation with the same κ but with different momenta and couplings. This is clearly seen in Eq. (6.121).

Similar to Eq. (6.121), the solution of the RG equation for the "magnetic" field, Eq. (6.103), is given by

$$\begin{aligned} &h(\varphi(x), a, b|\kappa) \\ &= \mathrm{e}^{sd_h - \int_0^s \beta_h(\bar{b}(s'))ds'} \cdot h(\bar{\varphi}(x;s)\,\mathrm{e}^{-s(d/2-1)}, \bar{a}(s)\,\mathrm{e}^{-2s}, \bar{b}(s)|\kappa). \quad (6.122) \end{aligned}$$

6.8.2 Renormalization group flow

Equation (6.121) can be written in a slightly different form by setting $k = \mathrm{e}^s p$:

$$\Gamma^{(n)}(\mathrm{e}^s p; a, b|\kappa) = \mathrm{e}^{sd_\Gamma - n\int_0^s \beta_h ds'} \cdot \Gamma^{(n)}(p; \bar{a}(s)\,\mathrm{e}^{-2s}, \bar{b}(s)|\kappa), \qquad (6.123)$$

where $s \to +\infty$ $(-\infty)$ corresponds to the UV (IR) limit. Physics in these limits is essentially controlled by the running couplings, $\bar{b}(s)$ and $\bar{a}(s)\,\mathrm{e}^{-2s}$, on the right-hand side.

Let us first consider the behavior of $\bar{b}(s)$. Suppose there is a coupling, b^*, at which the "velocity" vanishes, $\beta_b(b^*, \epsilon) = 0$. Such b^* terms are called the fixed points (FPs) of the RG flow. For example, in the lowest order in the ϕ^4 model, Eq. (6.110) yields

$$\beta_b(b, \epsilon) = -\epsilon b + \beta_b(b) = -\epsilon b + A b^2, \qquad (6.124)$$

with $A = 3/16\pi^2$. Then there arise two fixed points for $\epsilon > 0$ $(d < 4)$, namely

$$b^* = 0, \quad b^* = A^{-1}\epsilon. \qquad (6.125)$$

The former is called the Gaussian FP, b^*_{G}, and the latter is called the Wilson–Fisher FP, b^*_{WF}.

As shown in Fig. 6.8(a), β_b is negative (positive) for $0 < \bar{b} < b^*_{\mathrm{WF}}$ $(b^*_{\mathrm{WF}} < \bar{b})$. Combining this with the flow equation, $d\bar{b}/ds = \beta_b(\bar{b}, \epsilon)$, immediately implies that the Wilson–Fisher FP is "attractive" for $s \to -\infty$ (IR limit) and "repulsive" for $s \to +\infty$ (UV limit). The running coupling, $\bar{b}(s)$, as a function of s is shown in Fig. 6.8(b); $\bar{b}(s)$ with any initial conditions (as long as they are positive) flows into b^*_{WF} in the IR limit.

For $\epsilon = 0$ $(d = 4)$, two fixed points in Eq. (6.125) merge into a single FP at $b^* = 0$. This is IR-attractive and UV-repulsive. The behavior of $\beta(\bar{b})$ and $\bar{b}(s)$ in this case are shown in Fig. 6.9. This situation is called the "IR asymptotic freedom," in the sense that the running coupling approaches zero in the IR region. If the sign of $\beta(\bar{b})$ is opposite from that in Fig. 6.9(a) in $d = 4$, it is "UV asymptotic

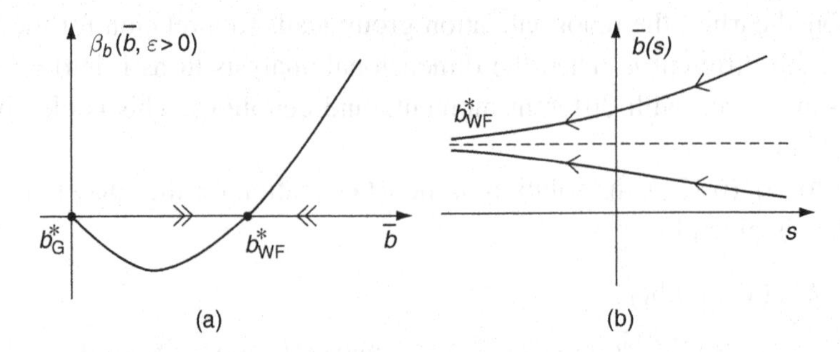

Fig. 6.8. (a) β_b as a function of the running coupling, $\bar{b}$, for the ϕ^4 model with $d < 4$. The double arrows show the flow direction of $\bar{b}$ toward the IR limit. (b) Corresponding running coupling as a function of the scaling parameter s for $d < 4$. The arrows show the flow direction toward the IR limit.

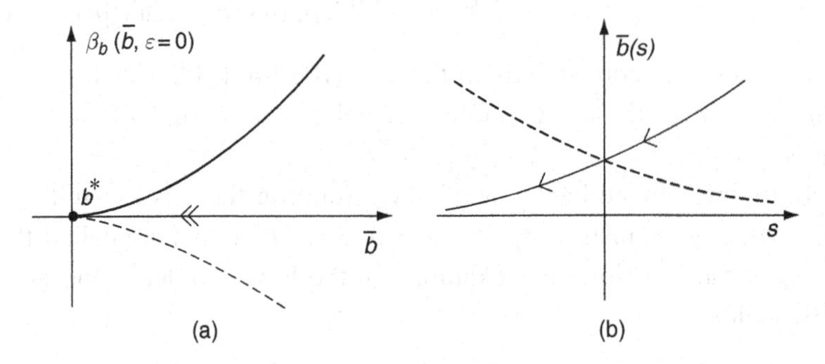

Fig. 6.9. (a) The solid line is β_b as a function of $\bar{b}$ for the ϕ^4 model with $d = 4$. The double arrows show the flow direction of $\bar{b}$ toward the IR limit. The dashed line represents the schematic behavior of β_b for UV asymptotic free theory such as the Yang–Mills theory. (b) Corresponding running coupling as a function of the scaling parameter s for $d = 4$. The arrows show the flow direction toward the IR limit.

freedom," in which perturbation theory in terms of $\bar{b}(s)$ becomes better and better in the UV region. It has been proven that *only* Yang–Mills theories have such a property (Coleman and Gross, 1973).

We now turn to the RG flow in the two-dimensional ($\bar{a}\,\mathrm{e}^{-2s}$, $\bar{b}$)-plane. Assuming that the flow of $\bar{a}(s)$ does not lead to a qualitative change of the exponential behavior, e^{-2s}, which is always the case in the ϵ-expansion, we may write the flow pattern for $d < 4$ as shown in Fig. 6.10. The Gaussian FP is unstable in any direction in this two-dimensional plane. On the other hand, the Wilson–Fisher FP is attractive in the horizontal direction and repulsive in the vertical direction; i.e. it is a saddle point on this plane.

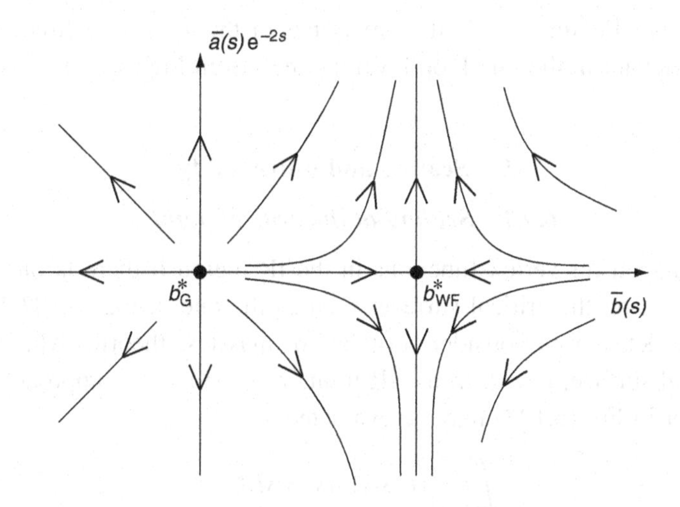

Fig. 6.10. RG flow of the ϕ^4 model with $d < 4$ toward the IR limit, $s \to -\infty$, in the $(\bar{a}\,e^{-2s},\ \bar{b})$-plane. The Gaussian FP is an unstable point; the Wilson–Fisher FP is a saddle point.

In Fig. 6.10, the line characterized by $\bar{a} = 0$ is called the critical line along which the correlation length, $\xi \sim |\bar{a}|^{-1/2}$, diverges; b^*_{WF} is a stable FP on the critical line toward which the RG trajectories flow. If there exist multiple couplings, $\bar{b}_l(s)$ $(l = 1, 2, \ldots)$, $\bar{a} = 0$ defines a critical hyper-surface. There may be several stable FPs on the critical surface which have their own basin of attraction of the RG flow and dictate the physics in each basin, as will be shown in the following sections. There is also the case that no stable FPs are found and RG flow has runaway trajectories. This is a sign of the first order phase transition, as will be discussed in Section 6.11.

It is appropriate here to define relevant, irrelevant and marginal couplings with respect to a given fixed point. The relevant coupling is the one which is amplified by the action of the RG, while the irrelevant coupling approaches zero. The marginal coupling remains fixed under the RG action. To see these couplings explicitly, let us consider the Gaussian FP in Fig. 6.10. Around this FP, $\bar{a}(s)\,e^{-2s}$ grows as $s \to -\infty$ because of the rapidly increasing factor, e^{-2s}, in Eq. (6.121), which originates from the scaling factor, $1/\lambda^2$, in Eq. (6.118). Therefore, the coupling a is a relevant coupling, and the corresponding operator, ϕ^2, is said to be a relevant operator. The coupling, b, is also a relevant operator near the Gaussian FP for $4 - d = \epsilon > 0$ since $\bar{b} \propto e^{-\epsilon s}$.

If we have sextet or higher order interactions, $\sum_{n=3}^{\infty} c_n \phi^{2n}$ in Eq. (6.77), $c_{n\geq 3}$ become irrelevant operators for $d > 3$ since they are scaled by positive powers of $\lambda = e^s$ in Eq. (6.118). This is why we have not considered these irrelevant couplings in Eq. (6.77). An example of the marginal coupling is b with respect to the Gaussian FP for $d = 4$. It is marginal in the zero-loop (tree) level, but becomes irrelevant in the one-loop level, as seen from Fig. 6.9(b).

6.9 Scaling and universality

6.9.1 Scaling at the critical point

What will happen for vertex functions in the IR region if there is one IR stable fixed point, b^*, on the critical surface, such as the one in Fig. 6.10? To answer this question, let us first consider "critical" or massless theories which are right on the critical surface, $a = 0$. In the IR limit, $s \to -\infty$, any b approaches b^* and the pre-factor in Eq. (6.123) may be evaluated as

$$\int_0^s \beta_h(\bar{b}(s'))\, ds' \sim s\beta_h^*, \tag{6.126}$$

where $\beta_h(b^*)$ is abbreviated to β_h^* for simplicity. Then, we immediately find that

$$\Gamma^{(n)}(\mathrm{e}^s p; 0, b|\kappa) \to \mathrm{e}^{s(d_\Gamma - n\beta_h^*)}\Gamma^{(n)}(p; 0, b^*|\kappa). \tag{6.127}$$

This simple exercise tells us several important facts, as follows.

(i) Long-wavelength physics at the critical point ($a = 0$) is governed solely by the IR fixed point, b^*. All the different theories on the critical surface with different values of the coupling, b, have the same critical properties dictated by b^*, as long as they belong to the same basin of attraction of b^*. This is an explicit manifestation of the "universality."

(ii) Even for the critical theories in which there are no dimensionful parameters, a naive scaling factor with canonical dimension d_Γ is modified by a correction, $n\beta_h^*$. This extra exponent is called the "anomalous dimension." It originates from the fact that there always exists an implicit dimensionful scale, κ.

(iii) Since b^* is a solution of $\beta_b(b, \epsilon) = -\epsilon b + \beta_b(b)$, and the power series expansion of $\beta_b(b)$ starts from b^2, b^* can be written as a systematic power series by a "small" parameter, ϵ, starting from $O(\epsilon)$. For $d = 3$, it is at most an asymptotic series, but it may be improved by the standard resummation techniques, such as the Borel and Padé methods.

To clarify point (ii) further, let us consider the two-point vertex function at $s \to -\infty$:

$$\Gamma^{(2)}(\mathrm{e}^s p; 0, b^*|\kappa) = \mathrm{e}^{s(2-2\beta_h^*)}\Gamma^{(2)}(p; 0, b^*|\kappa). \tag{6.128}$$

Since the only dimensionful parameters are p and κ, the above formula implies that

$$\Gamma^{(2)}(p; 0, b^*|\kappa) \propto \kappa^{2\beta_h^*} \cdot p^{2-2\beta_h^*}, \tag{6.129}$$

which shows explicitly that the origin of the anomalous dimension, $-2\beta_h^*$, is related to the existence of κ.

The two-point vertex function is related to the connected two-point Green's function as $G_c^{(2)}(p) = [\Gamma^{(2)}(p)]^{-1}$. Then, by making the Fourier transform of $1/p^{2-2\beta_h^*}$ in d dimensions, we have

$$G_c^{(2)}(x) \sim \frac{1}{x^{d-2+\eta}}, \qquad \eta = 2\beta_h^*, \tag{6.130}$$

at $T = T_c$; η is one of the critical exponents defined in Eq. (6.67) in Section 6.4. It is now identified with the anomalous dimension at the IR fixed point.

6.9.2 Scaling near the critical point

Let us now consider the case where T is slightly above T_c or, equivalently, a is small but positive, $a/\kappa^2 \ll 1$. In Eq. (6.121) for fixed κ, s is still arbitrary. So, we choose s so that the dimensionless ratio $\bar{a}(s)\,e^{-2s}/\kappa^2 = 1$ is satisfied. In Fig. 6.10, these conditions lead to an RG trajectory which starts at a point very close to the critical line (the x-axis) and passes by the Wilson–Fisher FP. Then the RG equation for $\bar{a}(s)$ in Eq. (6.90) yields

$$e^s \simeq \left(\frac{a}{\kappa^2}\right)^{\nu_+} \quad \text{with} \quad \nu_+ = (2 - \beta_a^*)^{-1}. \tag{6.131}$$

By taking $\kappa = 1$ in an appropriate unit for notational simplicity, we obtain

$$\Gamma^{(n)}(k; a, b|1) \xrightarrow[a\to 0]{} a^{\nu_+(d_\Gamma - n\beta_h^*)} g_+(ka^{-\nu_+}, b^*), \tag{6.132}$$

where $g_+(ka^{-\nu_+}, b^*) \equiv \Gamma^{(n)}(ka^{-\nu_+}; 1, b^*|1)$.

Since only the combination $ka^{-\nu_+}$ enters in the vertex function, and its Fourier transform should be a function of xa^{ν_+}, ξ_+ (the correlation length above T_c) is given by

$$\xi_+ \sim a^{-\nu_+} \sim (T - T_c)^{-\nu_+}. \tag{6.133}$$

Thus, the behavior of the correlation length is solely determined by the β-function at the Wilson–Fisher FP.

As given in Eq. (6.62), the dynamical magnetic susceptibility above T_c is defined by the behavior of the two-point Green's function. Also, it is related to the two-point vertex function, as shown in Eq. (6.99). Taking $n = 2$ and $k = 0$ in Eq. (6.132), we have

$$\Gamma^{(2)}(k; a, b|1) \to a^{2\nu_+(1-\beta_h^*)} g_+(0, b^*) \sim (T - T_c)^{\gamma_+}, \tag{6.134}$$

which leads to

$$\gamma_+ = \frac{2-2\beta_h^*}{2-\beta_a^*} = \nu_+(2-\eta). \tag{6.135}$$

6.10 Magnetic equation of state

To study the critical exponents below T_c, one should first determine the equilibrium value of φ as a solution of Eq. (6.122) with $h = \text{const}$. For this purpose, it is convenient to take the condition $\bar{\varphi}/(\kappa\, e^s)^{d/2-1} = 1$ together with $\varphi/\kappa^{d/2-1} \ll 1$. These two, combined with the RG equation for $\bar{\varphi}$ in Eq. (6.92), lead to

$$e^s \sim \left(\frac{\varphi}{\kappa^{d/2-1}}\right)^{\frac{2}{d-2+2\beta_h^*}}, \tag{6.136}$$

which gives the magnetic equation of state (EOS):

$$h(\varphi, a, b|1) \simeq \varphi^\delta f(a\varphi^{-1/\beta}, b^*), \tag{6.137}$$

with

$$\delta = \frac{d+2-2\beta_h^*}{d-2+2\beta_h^*} = \frac{d+2-\eta}{d-2+\eta}, \tag{6.138}$$

$$\beta = \frac{1}{2}\frac{d-2+2\beta_h^*}{2-\beta_a^*} = \frac{1}{2}\nu(d-2+\eta). \tag{6.139}$$

Let us consider the case for $T < T_c$ with $h = 0$. Then the non-trivial solution of Eq. (6.137) corresponds to a zero of $f(z)$, which leads to

$$\varphi \sim (-a)^\beta \sim (T_c - T)^\beta. \tag{6.140}$$

Therefore, β in Eq. (6.139) is the exponent defined in Eq. (6.42). For $T = T_c$ ($a = 0$), Eq. (6.137) simply yields $h \sim \varphi^\delta$. Therefore, δ in Eq. (6.138) is the same as that defined in Eq. (6.44). Note that Landau's mean-field theory yields $\delta = 3$ and $\beta = 1/2$ with $f(z) = b/6 + z$ (Exercise 6.10(1)).

Since the magnetic EOS obtained above is valid both below and above T_c, we can derive a formula for the static magnetic susceptibilities, $\gamma_\pm$, by calculating $\chi_T(T, 0) = (\partial\varphi/\partial h)_{h=0} \sim |T - T_c|^{-\gamma_\pm}$ (Exercise 6.10(2)):

$$\gamma_- = \gamma_+ = \beta(\delta - 1). \tag{6.141}$$

It should be noted that the first equality in Eq. (6.141) is valid only for the one-component theory. If there is a continuous symmetry, such as in the case for $O(N)$ ϕ^4 model, γ_- is not defined because the susceptibility always diverges due to the contribution from the Nambu–Goldstone bosons.

Table 6.2. *Scaling relations obtained in the RG approach and the definitions of the critical exponents for the one-component ϕ^4 model.*

Scaling relations	Critical exponents	
$\alpha = 2 - d\nu$	$C_V \sim \lvert T - T_c \rvert^{-\alpha}$	$T \sim T_c,\ h = 0$
$\beta = \frac{1}{2}\nu(d - 2 + \eta)$	$\varphi \sim \lvert T - T_c \rvert^{-\beta}$	$T \nearrow T_c,\ h = 0$
$\gamma = \nu(2 - \eta)$	$\chi_T \sim \lvert T - T_c \rvert^{-\gamma}$	$T \sim T_c,\ h = 0$
$\delta = (d + 2 - \eta)(d - 2 + \eta)^{-1}$	$\varphi \sim h^{1/\delta}$	$T = T_c,\ h \sim 0$
$(\alpha + 2\beta + \gamma = 2)$		
$\nu = (2 - \beta_a(b^*))^{-1}$	$\xi \sim \lvert T - T_c \rvert^{-\nu}$	$T \sim T_c,\ h = 0$
$\eta = 2\beta_h(b^*)$	$\chi \sim \lvert \vec{x} - \vec{x}' \rvert^{-(d-2+\eta)}$	$T = T_c,\ h = 0$

When $\varphi \neq 0$ is realized for $T < T_c$ as a solution of the magnetic EOS, the vertex function, $\Gamma[\varphi]$, may be expanded around the equilibrium value:

$$\Gamma_\varphi^{(n)}(x; a, b|\kappa) = \left(\frac{\delta}{\delta\varphi(x_1)} \cdots \frac{\delta}{\delta\varphi(x_n)} \right) \Gamma[\varphi(x), a, b|\kappa] \Big|_{\varphi(x)=\varphi} . \tag{6.142}$$

Using $\varphi \sim (-a)^\beta$ and following the same procedure as in the derivation of the magnetic EOS, we obtain

$$\Gamma_\varphi^{(n)}(k; a, b|1) \sim (-a)^{2\nu_+ d - n\beta} g_-(k(-a)^{-\nu_+}, b^*). \tag{6.143}$$

Therefore,

$$\nu_- = \nu_+, \quad \gamma_- = \gamma_+ = \nu_+(2 - \eta). \tag{6.144}$$

Again, the first equalities $\nu_- = \nu_+$ and $\gamma_- = \gamma_+$ are valid only for the one-component theory.

As a final critical exponent, let us briefly discuss $\alpha_\pm$, characterizing the singularity of the specific heat. By using the definition of the specific heat together with the behavior of $\Gamma^{(0)}$ below and above T_c given in Eqs. (6.132) and (6.143), we obtain

$$C_V \propto \frac{\partial^2 \Gamma^{(0)}}{\partial a^2} \sim |a|^{-(2-d\nu_+)}. \tag{6.145}$$

Therefore, we have

$$\alpha_\pm = 2 - d\nu_+. \tag{6.146}$$

Table 6.2 summarizes the scaling relations obtained so far using the RG method together with the definition of the critical exponents. All the exponents are related through the Wilson–Fisher FP.

6.11 Stability of the fixed point

When there are several dimensionless couplings, $\boldsymbol{b} = (b_1, \ldots, b_n)$, the RG flow on a multi-dimensional critical hyper-surface defined by $a = 0$ should be considered. The flow is dictated by

$$\frac{d\bar{\boldsymbol{b}}(s)}{ds} = \boldsymbol{\beta}(\boldsymbol{b}(s), \epsilon). \tag{6.147}$$

Assuming that we found a FP solution of $\boldsymbol{\beta} = \mathbf{0}$,[7] let us study the stability of the FP. By linearizing $\boldsymbol{\beta}$ around the FP,

$$\frac{d\bar{\boldsymbol{b}}(s)}{ds} \sim \boldsymbol{\Omega} \cdot \bar{\boldsymbol{b}}(s), \quad (\boldsymbol{\Omega})_{ll'} = \left.\frac{\partial \beta_l}{\partial \bar{b}_{l'}}\right|_{\bar{\boldsymbol{b}} = b^*}, \tag{6.148}$$

where $\boldsymbol{\Omega}$ is an $n \times n$ stability matrix which is not necessarily symmetric. Consider the special case that $\boldsymbol{\Omega}$ has n independent eigenvectors, $\bar{B}_l(s)$, with complex eigenvalues, ω_l. Then one can diagonalize $\boldsymbol{\Omega}$ as $\mathbf{P}\boldsymbol{\Omega}\mathbf{P}^{-1} = \mathrm{diag}(\omega_1, \ldots, \omega_n)$, which leads to

$$\bar{B}_l(s) = (\mathbf{P} \cdot \bar{\boldsymbol{b}}(s))_l = \mathrm{e}^{s\omega_l}. \tag{6.149}$$

Therefore, in the IR limit ($s \to -\infty$), the fixed point, b^*, is stable (unstable) on the critical hyper-surface toward the direction of Re $\omega_l > 0$ (Re $\omega_l < 0$).

There are cases where no stable FP can be found and the flow pattern has runaway trajectories. This indicates the possibility of the existence of the fluctuation-induced first order phase transition (Iacobson and Amit, 1981; Amit, 1984). We will encounter such a case in QCD for $N_f \geq 3$.

6.12 Critical exponents for the O(N)-symmetric ϕ^4 model

In this section, we present the critical exponents of the O(N)-symmetric ϕ^4 model in the ϵ-expansion, where O(N) is a rotational symmetry in N-dimensional internal space. This model is a generalization of that given in Eq. (6.77) to the N-component field, $\vec{\phi} = (\phi_0, \phi_1, \ldots, \phi_{N-1})$, with an O($N$)-symmetric interaction, $(\vec{\phi}^2)^2$:

$$S_{\mathrm{eff}} = \int d^d x \left[\frac{1}{2}(\partial\vec{\phi})^2 + \frac{a}{2}\vec{\phi}^2 + \frac{b}{4!}\vec{\phi}^4 \right]. \tag{6.150}$$

The redefinition of the couplings and fields can be performed in a similar manner to Eq. (6.78).

[7] Unlike the case of a single coupling, other flow patterns, such as the limit cycle and ergodic behavior, are, in principle, possible. We do not, however, consider these cases here.

We recapitulate here the result of β-functions at one loop (Exercise 6.9(3)):

$$\beta_b(g,\epsilon) = g\left[-\epsilon + \frac{N+8}{6}g + O(g^2)\right], \tag{6.151}$$

$$\beta_a(g) = \frac{N+2}{6}g + O(g^2), \tag{6.152}$$

$$\beta_h(g) = O(g^2), \tag{6.153}$$

where g is defined as

$$g \equiv b\frac{S_d}{(2\pi)^d}, \quad S_d \equiv \frac{2\pi^{d/2}}{\Gamma(d/2)}, \tag{6.154}$$

and S_d is the surface area of a unit sphere in Euclidean d dimensions (Exercise 6.8); $(2\pi)^d$ originates from the measure of the momentum integration.

From Eq. (6.151), we immediately obtain

$$g^* = \frac{6}{N+8}\epsilon. \tag{6.155}$$

Then it is straightforward to calculate $\nu = (2-\beta_a(g^*))^{-1}$ and $\eta = 2\beta_h(g^*)$. All the other exponents are simply obtained by referring to the scaling relations in Table 6.2. Since g^* is $O(\epsilon)$, the β-functions around the fixed point, as well as the critical exponents, have a systematic expansion in terms of ϵ. The exponents at $O(\epsilon)$ obtained in this way are summarized in the third column of Table 6.3 together with the prediction of the mean-field (MF) theory in the second column. As we have discussed before, the critical exponents of the mean-field theory do not depend on the dimensionality of space, d, or on the internal degrees of freedom, N. On the other hand, the ϵ-expansion introduces corrections which depend on both d and N.

The higher order corrections in the ϵ-expansion show that it is, at most, an asymptotic expansion. This implies that the series should be resummed to obtain sensible results for $\epsilon \to 1$. The fourth column in Table 6.3 presents the resummed results for the O(4) ϕ^4 model with the use of a series calculated up to seven loops (Zinn-Justin, 2001). These results agree remarkably well with the results of Monte Carlo numerical simulations shown in the fifth column (Hasenbusch, 2001). Further details concerning the theoretical calculations of the critical exponents of various models, and their comparison to numerical and experimental data, are given in Pelissetto and Vicari (2002).

The result for $N=4$ in Table 6.3 is particularly important for us. This is because QCD with two massless flavors has chiral symmetry, $\mathrm{SU_L}(2)\times\mathrm{SU_R}(2)\simeq \mathrm{O}(4)$, and is expected to be in the same universality class as the O(4) ϕ^4 model in $d=3$.

Table 6.3. *Critical exponents for the* O(N)*-symmetric* ϕ^4 *model.*

MF and MC refer to the mean-field method and the Monte Carlo method, respectively. The numbers in parentheses are the standard errors in the last digits; ω is an eigenvalue of the stability matrix, $\boldsymbol{\Omega}$, at the Wilson–Fisher FP (see Section 6.11).

Exponents	MF	ϵ-exp. at $O(\epsilon)$	ϵ-exp. ($N=4$) resummed	MC ($N=4$) $d=3$
α	0	$-\frac{N-4}{2(N+8)}\epsilon$	−0.211(24)	−0.247(6)
β	$\frac{1}{2}$	$\frac{1}{2}-\frac{3}{2(N+8)}\epsilon$	0.382(4)	0.388(1)
γ	1	$1+\frac{N+2}{2(N+8)}\epsilon$	1.447(16)	1.471(4)
δ	3	$3+\epsilon$	4.792(19)	4.789(5)
ν	$\frac{1}{2}$	$\frac{1}{2}+\frac{N+2}{4(N+8)}\epsilon$	0.737(8)	0.749(2)
η	0	0	0.0360(40)	0.0365(10)
ω	–	$+\epsilon$	0.795(30)	0.765

6.13 Chiral phase transition of QCD at finite T

Let us now consider an effective theory of QCD near the critical point of the chiral phase transition.

For massless N_f flavors, the classical QCD Lagrangian has a large symmetry,

$$G = SU_L(N_f) \times SU_R(N_f) \times U_A(1) \times U_B(1) \times SU_c(3). \tag{6.156}$$

Let us focus on the dynamical breaking of chiral symmetry and its restoration, assuming that the $U_B(1) \times SU_c(3)$ part is not broken. This is an assumption which turns out to be invalid if color superconductivity takes place at high baryon density. We will discuss this in Chapter 9.

We now introduce a chiral order parameter, which is an $N_f \times N_f$ matrix and is singlet under a $U_B(1) \times SU_c(3)$ rotation:

$$\Phi_{ij} = \frac{1}{2}\bar{q}^j(1-\gamma_5)q^i = \bar{q}^j_R q^i_L, \tag{6.157}$$

where i and j are flavor indices. This order parameter transforms as

$$\Phi \to e^{i\alpha} V_L \Phi V_R^\dagger, \tag{6.158}$$

under an $SU_L(N_f) \times SU_R(N_f) \times U_A(1)$ rotation; V_L (V_R) is an element of $SU_L(N_f)$ ($SU_R(N_f)$) and α is the $U_A(1)$ rotational angle. Transformations of the left- and right-handed quarks under the same rotation are given by

$$q_L \to e^{-i\frac{\alpha}{2}} V_L q_L, \quad q_R \to e^{+i\frac{\alpha}{2}} V_R q_R. \tag{6.159}$$

If the dynamical breaking of chiral symmetry takes place, the thermal average of Φ is non-vanishing.

A decomposition of Φ useful for later purposes is given by

$$\Phi = \sum_{a=0}^{N_f^2-1} \Phi^a \frac{\lambda^a}{\sqrt{2}}, \quad \text{with} \quad \Phi^a = S^a + iP^a. \tag{6.160}$$

Here, $t^a = \lambda^a/2\,(a = 1, 2, \ldots, N_f^2 - 1)$ (Appendix B.3) and $t^0 \equiv \sqrt{1/(2N_f)}$ constitute the generators of $\mathrm{U}(N_f)$ in the fundamental representation. Note that S^a (P^a) are the hermitian fields with N_f^2 components with even (odd) parity. For $N_f = 2$, $P^{1,2,3}$ correspond to the pionic fields, $\pi^{1,2,3}$.

6.13.1 Landau functional of QCD

Motivated by the universality argument, we construct a Landau functional, $S_{\text{eff}} = \int d^d x\, \mathcal{L}_{\text{eff}}$, in terms of the order-parameter field, Φ. The general structure of $\mathcal{L}_{\text{eff}}$, which has the same symmetries as the QCD Lagrangian and is expanded by Φ near the critical point, is given by (Pisarski and Wilczek, 1984)

$$\begin{aligned}
\mathcal{L}_{\text{eff}} &= \frac{1}{2} \operatorname{tr} \partial\Phi^\dagger \partial\Phi + \frac{a}{2} \operatorname{tr} \Phi^\dagger \Phi \\
&\quad + \frac{b_1}{4!} \left(\operatorname{tr} \Phi^\dagger \Phi\right)^2 + \frac{b_2}{4!} \operatorname{tr} \left(\Phi^\dagger \Phi\right)^2 \\
&\quad - \frac{c}{2} \left(\det\Phi + \det\Phi^\dagger\right) \\
&\quad - \frac{1}{2} \operatorname{tr} h(\Phi + \Phi^\dagger).
\end{aligned} \tag{6.161}$$

Here, "tr" and "det" are taken over the flavor indices. The first four terms on the right-hand side of Eq. (6.161) have $\mathrm{SU_L}(N_f) \times \mathrm{SU_R}(N_f) \times \mathrm{U_A}(1)$ symmetry. The fifth term with determinant structure consists of the lowest dimensional operator which preserves $\mathrm{SU_L}(N_f) \times \mathrm{SU_R}(N_f)$ symmetry but breaks $\mathrm{U_A}(1)$ symmetry. This term is a manifestation of the axial anomaly in QCD discussed in Chapter 2, and was first introduced in Kobayashi and Maskawa (1970) and 't Hooft (1986). The last term of Eq. (6.161) originates from the quark masses, $h \propto \mathrm{diag}(m_\mathrm{u}, m_\mathrm{d}, m_\mathrm{s}, \ldots)$, and it breaks not only $\mathrm{SU_L}(N_f) \times \mathrm{SU_R}(N_f)$, but also $\mathrm{U_A}(1)$ explicitly.

The Φ fields are would-be soft modes which are expected to have divergent correlation length at the critical point if the system exhibits the second order phase transition. The hard modes, which do not have a divergent correlation length, are supposed to be integrated out in the path integral and affect only the coefficients

a, b_1, b_2, c and h. Whether there exist Lorentz-vector soft modes (such as the vector mesons) or not is a non-trivial issue (Harada and Yamawaki, 2003). We assume here that the soft modes are only Lorentz scalars.

6.13.2 Massless QCD without axial anomaly

To study the phase structure obtained from $\mathcal{L}_{\rm eff}$ step by step, let us first consider Eq. (6.161) with $c = 0$ and $h = 0$; i.e. the case where $\mathcal{L}_{\rm eff}$ has $\mathrm{SU_L}(N_f) \times \mathrm{SU_R}(N_f) \times \mathrm{U_A}(1)$ symmetry.

For large values of Φ, $\mathcal{L}_{\rm eff}$ is bounded from below if $b_1 + \frac{b_2}{N_f} > 0$ and $b_2 > 0$ are satisfied (Exercise 6.11). Under these conditions, the phase transition driven by the sign change of a is of second order in the mean-field theory. This is easily seen by rewriting Eq. (6.161) with $c = h = 0$ in terms of S^0 $(= \Phi^0)$ and comparing the result with Eq. (6.34).

The order of the transition, however, may change once the thermal fluctuation of Φ is taken into account. To see this, let us look at the β-functions up to the next-to-leading order given in terms of the scaled couplings, $g_{1,2} \equiv b_{1,2} S_d/(2\pi)^d$:

$$\beta_1 = -\epsilon g_1 + \frac{N_f^2 + 4}{3} g_1^2 + \frac{4N_f}{3} g_1 g_2 + g_2^2, \tag{6.162}$$

$$\beta_2 = -\epsilon g_2 + 2 g_1 g_2 + \frac{2N_f}{3} g_2^2. \tag{6.163}$$

These are obtained from the one-loop effective action at the critical point, $a = 0$, with the four-point vertices proportional to b_1 and b_2 (Exercise 6.9(4)).

Depending on the number of flavors, Eqs. (6.162) and (6.163) have different RG flow.

$N_f = 1$ case

In this case, Eq. (6.161) with $c = h = 0$ is equivalent to the O(2) symmetric ϕ^4 model discussed in Section 6.12 with a single coupling constant, $b \equiv b_1 + b_2$. It is easy to discover that the fixed point, $g^* = 3\epsilon/5$, is IR-stable. Namely, the phase transition is of second order and the critical exponents are those given in Table 6.3 with $N = 2$.

$N_f \geq 2$ case

In this case, b_1 and b_2 are independent coupling constants, and Eq. (6.161) with $c = h = 0$ has a symmetry $\mathrm{SU_L}(N_f) \times \mathrm{SU_R}(N_f) \times \mathrm{U_A}(1)$. There are two solutions of $\beta_1 = \beta_2 = 0$:

$$g^* = (0, 0) \text{ and } g^* = \left(\frac{3}{N_f^2 + 4}\epsilon, 0 \right).$$

As we have discussed in Section 6.11, eigenvalues, ω_l, of the stability matrix, $\Omega_{ll'}(=\partial\beta_l/\partial g_{l'})$, determine whether these fixed points are IR-stable. After simple algebra, we obtain

$$(\omega_1, \omega_2) = \begin{cases} (-\epsilon, -\epsilon) & \text{for } g^* = (0,0) \\ \left(\epsilon, -\frac{N_f^2-2}{N_f^2+4}\epsilon\right) & \text{for } g^* = \left(\frac{3}{N_f^2+4}\epsilon, 0\right). \end{cases} \tag{6.164}$$

Since there is always a negative eigenvalue for $N_f \geq 2$, we find no IR-stable fixed point on the critical surface, which indicates that the phase transition is fluctuation-induced first order, as discussed in Section 6.11. The RG flow in the two-dimensional critical surface in this case is shown in Fig. 6.11. Irrespective of the starting point of the flow in Fig. 6.11, the coupling constants are sent to the unbounded region (either $b_2 < 0$ or $b_1 + b_2/N_f < 0$).

6.13.3 Massless QCD with axial anomaly

Let us consider the case for $c \neq 0$ and $h = 0$, where $\mathcal{L}_{\text{eff}}$ has $\text{SU}_\text{L}(N_f) \times \text{SU}_\text{R}(N_f)$ symmetry. This situation is closer to the real world since $\text{U}_\text{A}(1)$ symmetry is always broken by the axial anomaly irrespective of the temperature.[8]

Table 6.4 summarizes the orders of the phase transition for massless quarks, $h = 0$. Since the number of flavors plays a crucial role, let us discuss the physics behind Table 6.4 for different N_f.

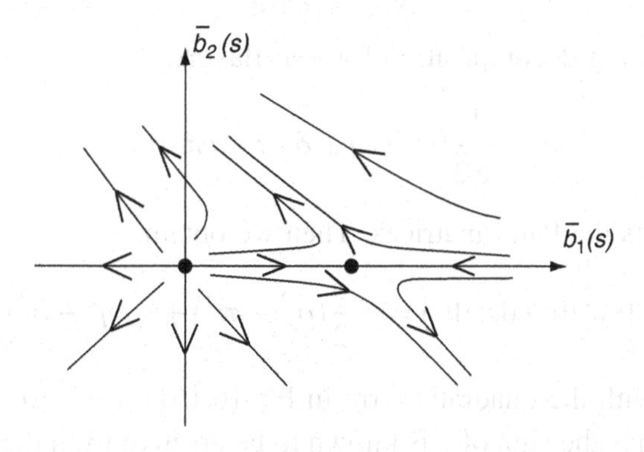

Fig. 6.11. Schematic RG flow of the $\text{SU}_\text{L}(N_f) \times \text{SU}_\text{R}(N_f) \times \text{U}_\text{A}(1)$ model. The two filled circles correspond to the fixed points $g^* = (0,0)$ and $g^* = \left(\frac{3}{N_f^2+4}\epsilon, 0\right)$.

[8] In principle, c is temperature-dependent and may become much smaller than g_1 and g_2 near the critical point. If so, the analyses in Section 6.13.2 are valid except for in the region very close to the critical point.

Table 6.4. *Summary of the chiral phase transition of massless QCD ($h = 0$) without axial anomaly ($c = 0$) and with axial anomaly ($c \neq 0$). Symmetries characterizing the second order transitions are shown in the square brackets.*

	Massless QCD without anomaly ($h = 0, c = 0$)	Massless QCD with anomaly ($h = 0, c \neq 0$)
$N_f = 1$	second order [O(2)]	none
$N_f = 2$	first order	second order [O(4)]
$N_f = 3$	first order	first order[a]
$N_f \geq 4$	first order	first order

[a] First order transition induced by the cubic term originating from the axial anomaly.

$N_f = 1$ case

Using the decomposition Eq. (6.160) for one flavor with notation $S^0 + iP^0 = \sigma + i\eta$, we obtain

$$-\frac{c}{2}\left(\det\Phi + \det\Phi^\dagger\right) = -c\sigma. \tag{6.165}$$

This has the same structure as the quark mass term in Eq. (6.161) (or the external magnetic field for the spin system) and breaks chiral symmetry explicitly. Therefore, the second order phase transition for $c = 0$ becomes a smooth crossover for $c \neq 0$, as shown in Fig. 6.2.

$N_f = 2$ case

We use the following decomposition for two flavors:

$$\Phi = \frac{1}{\sqrt{2}}(\sigma + i\eta + \boldsymbol{\delta}\cdot\boldsymbol{\tau} + i\boldsymbol{\pi}\cdot\boldsymbol{\tau}), \tag{6.166}$$

where $\boldsymbol{\tau}$ represents the Pauli matrices. Then we obtain

$$-\frac{c}{2}\left(\det\Phi + \det\Phi^\dagger\right) = -\frac{c}{2}(\sigma^2 + \boldsymbol{\pi}^2) + \frac{c}{2}(\eta^2 + \boldsymbol{\delta}^2). \tag{6.167}$$

This, combined with the quadratic term in Eq. (6.161), leads to $\frac{a-c}{2}(\sigma^2 + \boldsymbol{\pi}^2) + \frac{a+c}{2}(\eta^2 + \boldsymbol{\delta}^2)$. Since the sign of c is known to be positive from the particle spectra at $T = 0$ (Exercise 6.12), σ and $\boldsymbol{\pi}$ are almost massless near the critical point ($a - c \sim 0$), while η and $\boldsymbol{\delta}$ stay massive. Thus, we end up with an O(4)-symmetric ϕ^4 model,

$$\mathcal{L}_{\text{eff}} = \frac{1}{2}(\partial\vec{\phi})^2 + \frac{a-c}{2}\vec{\phi}^2 + \frac{b_1 + b_2/2}{4!}(\vec{\phi}^2)^2, \tag{6.168}$$

with $\vec{\phi} = (\phi_0, \phi_1, \phi_2, \phi_3) = (\sigma, \boldsymbol{\pi})$. As discussed in Section 6.12, this model shows the second order transition with the critical exponents given in Table 6.3 for $N = 4$.

$N_f = 3$ case

The determinant term produces cubic interactions:

$$-\frac{c}{2}\left(\det\Phi + \det\Phi^\dagger\right) = -\frac{c}{3\sqrt{3}}\sigma^3 + \cdots . \tag{6.169}$$

This is exactly the case discussed in Section 6.3.3, and the phase transition becomes first order even in the mean-field theory.

$N_f \geq 4$ case

The determinant term produces a quartic term for $N_f = 4$ and a term higher order than quartic for $N_f > 4$. In the former case, the new term is, in principle, relevant for the critical properties, but IR-stable fixed points in (b_1, b_2, c) space do not appear (Paterson, 1981). In the latter case, the new term is not relevant, and the result with $c = 0$ in Section 6.13.2 holds. Therefore, the fluctuation-induced first order transition is expected to occur for all N_f greater than or equal to four.

6.13.4 Effect of light quark masses

So far, we have discussed a hypothetical world where all the quarks are massless. To forge a closer connection with the real world, let us consider u, d and s quarks and take into account their current masses, i.e. $c \neq 0$ and $h \neq 0$.

Shown in Fig. 6.12 is the nature of the finite T phase transition in the $(m_{\rm ud} m_{\rm s})$-plane, where we have assumed the isospin symmetry to be $m_{\rm ud} \equiv m_{\rm u} = m_{\rm d}$. The four corners of Fig. 6.12 correspond to the following four limiting cases:

$$(m_{\rm ud}, m_{\rm s}) = \begin{cases} (\infty, \infty) & N_f = 0 \text{ (no quarks)} \\ (\infty, 0) & N_f = 1 \text{ (massless 1 flavor)} \\ (0, \infty) & N_f = 2 \text{ (massless 2 flavors)} \\ (0, 0) & N_f = 3 \text{ (massless 3 flavors).} \end{cases} \tag{6.170}$$

As we have discussed in Section 6.3.3, the first order transition does not disappear as long as the external field is weak. The first order region in the lower left (upper right) corner of Fig. 6.12 corresponds to such a case with m_q $(1/m_q)$ as an external field.

The first order regions in the figure are separated by the crossover region. The boundaries separating these regions are characterized by the second order

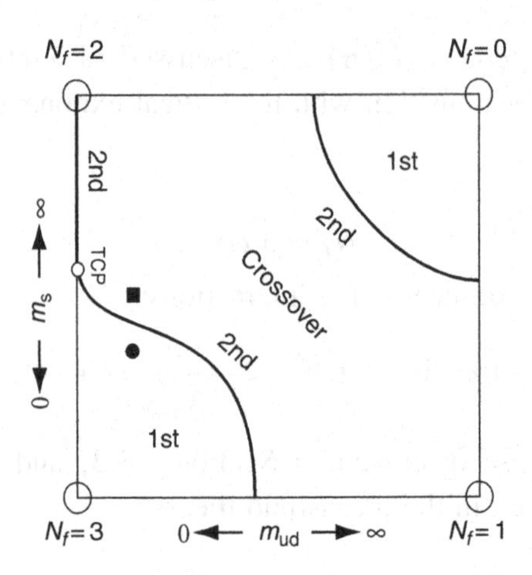

Fig. 6.12. Finite T phase transition of QCD in the (m_{ud}, m_s)-plane. "TCP" denotes the tricritical point. The filled circle and square denote possible candidates in the real world.

transition, which is in the same universality class as the Ising model with Z(2) symmetry. (See Exercise 6.13(1).)

The point separating the first order and second order transitions on the $m_{ud} = 0$ line is an example of the tricritical point discussed in Section 6.3.4. Let us assume that m_s is heavy around this point. Then we may write down the functional only in terms of the light modes $(\sigma, \boldsymbol{\pi})$. Since the system has O(4) symmetry in the limit $m_{ud} \to 0$, the relevant functional is given by

$$\mathcal{L}_{\text{eff}} = \frac{1}{2}(\partial\vec{\phi})^2 + \frac{a(m_s, T)}{2}\vec{\phi}^2 + \frac{b(m_s, T)}{4!}(\vec{\phi}^2)^2 + \frac{c}{6!}(\vec{\phi}^2)^3 - h\phi_0, \quad (6.171)$$

where $h \propto m_{ud}$. The tricritical point is defined as $a = b = 0$ with $c > 0$. Three second order lines meet at the tricritical point, and the wing develops to positive and negative values of m_{ud} from the tricritical point. Using the result from Eq. (6.59) in Section 6.3.4, the leading behavior of the wing near the tricritical point is given by $m_{ud} \sim (m_s^{\text{tri}} - m_s)^{5/2}$ (Exercise 6.13(2)).

The location of the physical quark masses in the (m_{ud}, m_s)-plane is not known exactly. It could be in the first order region, as indicated by the filled circle in Fig. 6.12, or it could be in the crossover region, as indicated by the filled square. Lattice QCD simulations with dynamical quarks provide some evidence that the real world belongs to the crossover region; further confirmation with smaller quark masses and larger volume is one of the most important issues in lattice QCD. Table 6.5 summarizes the order of the phase transition, the relevant symmetry of the Landau functional and the critical temperature from lattice QCD simulations for different numbers of flavors.

Table 6.5. *Summary of QCD phase transition for different numbers of flavors with $N_c = 3$.*

Lattice data for T_c are taken from Laermann and Philipsen (2003).

N_f	0	2	2+1	3
m_{ud}	∞	0	$\sim$5 MeV	0
m_s	∞	∞	$\sim$100 MeV	0
Order	first	second	first or crossover	first
Symmetry	$Z(3)$	O(4)	$\sim SU_L(3) \times SU_R(3)$	$SU_L(3) \times SU_R(3)$
T_c (lattice)	$\sim$270 MeV	$\sim$170 MeV		$\sim$150 MeV

6.13.5 Effect of finite chemical potential

Introducing the quark chemical potential (μ) enriches the phase diagram of QCD. Let us fix the strange quark mass, m_s, to be a relatively large value corresponding to the crossover region (i.e. the filled square in Fig. 6.12) and consider a phase diagram in the three-dimensional (T, μ, m_{ud}) space.

When m_{ud} is small, the Landau functional that describes the system takes a similar form to the O(4) ϕ^4 model given by Eq. (6.171):

$$\mathcal{L}_{\text{eff}} = \frac{1}{2}(\partial\vec{\phi})^2 + \frac{a(\mu, T)}{2}\vec{\phi}^2 + \frac{b(\mu, T)}{4!}(\vec{\phi}^2)^2 + \frac{c}{6!}(\vec{\phi}^2)^3 - h\phi_0, \quad (6.172)$$

where $h \propto m_{ud}$ and c is assumed to be positive. Since we have two parameters, μ and T, characterizing a and b, we may, in principle, have a tricritical point (TCP) at $a = b = h = 0$. Indeed the TCP at finite T and μ was first recognized by Asakawa and Yazaki (1989) on the basis of the NJL model discussed in Section 6.2, and was later studied in a more general context (see, for example, Halasz *et al.* (1998), Stephanov *et al.* (1998) and Hatta and Ikeda (2003)).

Shown in Fig. 6.13 is the schematic phase structure expected from Eq. (6.172) with $m_s > m_s^{\text{tri}}$. For $m_{ud} = 0$, the second order transition at low μ turns into the first order transition at large μ. For small, non-zero, m_{ud}, the second order transition is smoothed out, and the TCP becomes the CEP (the critical end point). CEPs for different values of m_{ud} form the edge of a wing starting from the TCP. We have already seen all these features in our general discussion in Section 6.3.4, in particular Fig. 6.6. Shown in Fig. 6.14 is a schematic QCD phase diagram in the (T, μ)-plane for $m_{ud} = 0$ (the line with the TCP) and for $m_{ud} > 0$ (the line with the CEP). It corresponds to the cross-section of Fig. 6.13 with fixed values of m_{ud}. Finding the precise locations of the TCP and the CEP is one of the most challenging problems in lattice QCD studies.

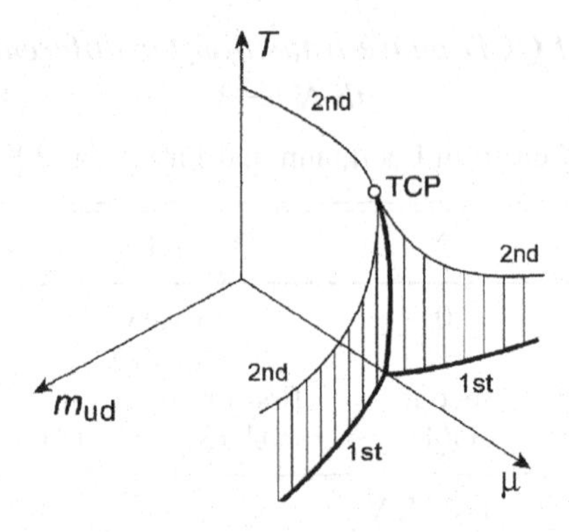

Fig. 6.13. Schematic QCD phase diagram formed from the Landau functional in the (T, μ, m_{ud}) space.

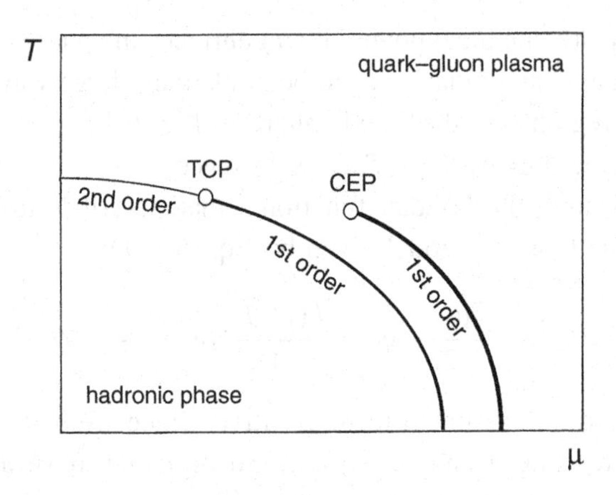

Fig. 6.14. Schematic phase diagram in temperature (T) and quark chemical potential (μ) space. The line with the tricritical point, TCP, corresponds to $m_{ud} = 0$, and the line with the critical end point, CEP, corresponds to $m_{ud} > 0$. For the strange quark mass, $m_s > m_s^{tri}$ is assumed.

Exercises

6.1 NJL Lagrangian with $N_f = 2$. Write down the most general form of the four-fermi interaction which has an $SU_L(2) \times SU_R(2)$ symmetry. Decompose it into a $U_A(1)$ symmetric part and a $U_A(1)$ breaking part. What kind of conditions are necessary to obtain the four-fermi interaction in Eq. (6.11) from the general interaction?

6.2 Auxiliary field method. The functional integral of the self-interacting fermion system and the boson system can be converted into a simple form by introducing auxiliary fields. Typical tricks are to use the following decomposition:

$$1 = \int [d\chi]\, e^{-(\chi-\Psi)^2} = \int [d\chi]\, e^{i(\chi-\Psi)^2},$$

$$1 = \int [d\chi]\, \delta(\chi - \Psi) = \int [d\chi d\lambda]\, e^{i\lambda(\chi-\Psi)},$$

where numerical constants are absorbed in the definition of the functional measure. Apply these to the NJL model, where $\Psi \sim \bar{q}q$, and to the ϕ^4 model, where $\Psi \sim \phi^2$.

6.3 Lambert function $W(z)$. The Lambert function is a solution of $We^W = z$ and has a kth branch, $W_k(z)$. For real and positive z, we find $W_{-1}(z \to 0) \simeq \mathrm{Ln}(-z/\mathrm{Ln}(z)) - i\pi + O(\pi/\mathrm{Ln}(z), \mathrm{Ln}(-\mathrm{Ln}(z))/\mathrm{Ln}(z))$ (Corless *et al.*, 1996). Starting from the gap equation, Eq. (6.24), and making a change of variables, find the relation between $\bar{\sigma}$ and W_{-1} and then derive Eq. (6.25).

6.4 Gap equation and its solution at finite T.

(1) T_c is defined as a temperature where the non-trivial solution of the gap equation vanishes. By comparing T_c and the non-trivial solution at zero temperature, $\sigma(T=0)$, derive an approximate relation, Eq. (6.29). (Assuming that Λ is large compared with other scales helps to simplify the analysis.)

(2) Derive the free energy of the NJL model for $m \neq 0$ and show that it indeed has a term proportional to $-m\sigma$. Show that this term leads to the dashed lines in Figs. 6.2(a) and (b).

6.5 Critical behavior near the TCP and CEP. On the basis of Eq. (6.51), study the critical exponents and susceptibilities in the vicinity of the TCP and CEP within the mean-field theory. Consult Lawrie and Sarnach (1984) and Hatta and Ikeda (2003) for further details that go beyond the mean-field theory.

6.6 β-functions in the ϵ-expansion. Derive Eq. (6.87) by noting that the bare parameters (a_B, b_B, h_B) are κ-independent and that the renormalized parameters (a, b, h) are finite in the limit $\epsilon \to 0$.

6.7 Vertex function and 1PI diagrams.

(1) Starting from the definition in Eq. (6.95), and making the change of variables, derive the relation

$$\Gamma[\varphi] = \int d^d x\, \mathcal{L}(\varphi) + \tilde{\Gamma}[\varphi]$$

$$= -\ln \int [d\phi]\, e^{-\int d^d x \left[\mathcal{L}(\varphi) + \frac{1}{2}\phi \mathcal{D}^{-1}(\varphi)\phi + \mathcal{L}_{\mathrm{int}}(\phi;\varphi) - \frac{\delta\tilde{\Gamma}[\varphi]}{\delta\varphi}\phi\right]}, \quad (6.173)$$

where $\mathcal{L}$ is defined from $S_{\rm eff}[\phi] \equiv \int d^d x \, (\mathcal{L}(\phi) - h\phi)$. The following power series expansion in terms of ϕ has been used: $\mathcal{L}(\phi+\varphi) = \mathcal{L}(\varphi) + (\delta\mathcal{L}(\varphi)/\delta\varphi)\phi + (1/2)\phi\mathcal{D}^{-1}\phi + \mathcal{L}_{\rm int}(\phi;\varphi)$.

(2) Show that the term $(\delta\tilde{\Gamma}[\varphi]/\delta\varphi)\phi$ on the right-hand side acts to subtract out all the one-particle reducible diagrams. Then the resultant vertex function, $\Gamma[\varphi]$, contains only the one-particle irreducible (1PI) diagrams that cannot be separated into two by cutting any one of the propagators.

6.8 Basic integral in d Euclidean dimensions.

(1) Prove the following formula for an integral I_l in d-dimensional Euclidean space:

$$I_l = \int \frac{d^d q}{(2\pi)^d} \frac{1}{(q^2+a)^l} = \int \frac{d\Omega_d}{(2\pi)^d} \int_0^\infty dr \frac{r^{d-1}}{(r^2+a)^l}$$
$$= \frac{S_d}{(2\pi)^d} \frac{1}{2} B\left(\frac{d}{2}, l-\frac{d}{2}\right) a^{\frac{d-2l}{2}} = \frac{S_d}{(2\pi)^d} \frac{\Gamma(d/2)\Gamma(l-d/2)}{2\Gamma(l)} a^{\frac{d-2l}{2}},$$

where $S_d = 2\pi^{d/2}/\Gamma(d/2)$ is the area of a unit sphere in d-dimensional space and the beta function is given by

$$B(x,y) = \int_0^1 dt \, t^{x-1}(1-t)^{y-1} = \int_0^\infty dt \, \frac{t^{x-1}}{(1+t)^{x+y}} = \frac{\Gamma(x)\Gamma(y)}{\Gamma(x+y)}.$$

(2) Show that the leading contribution to the integral for small ϵ is given by

$$I_1 = -\frac{1}{8\pi^2}\frac{a}{\epsilon} + \text{finite terms},$$
$$I_2 = \frac{1}{8\pi^2}\frac{1}{\epsilon} + \text{finite terms}.$$

It is useful to use the following relations for the Γ-function:

$$\Gamma(z+1) = z\Gamma(z),$$
$$\Gamma(\epsilon) = \frac{1}{\epsilon} - \gamma + \frac{1}{2}\left(\gamma^2 + \frac{\pi^2}{6}\right)\epsilon + \cdots,$$
$$\Gamma(-l+\epsilon) = \frac{(-1)^l}{l!}\left[\frac{1}{\epsilon} + \left(1 + \frac{1}{2} + \cdots + \frac{1}{l} - \gamma\right) + O(\epsilon)\right],$$

where $\gamma \simeq 0.577$ is the Euler constant (see Exercise 3.4(3)).

6.9 One-loop formula for the vertex function, $\Gamma[\varphi]$.

(1) Using the definition of the vertex function given in Section 6.6.3, show that the following formula is valid in the one-loop approximation:

$$\Gamma[\varphi] \simeq S_{\text{eff}}(h=0) + \frac{1}{2}\ \text{Tr}\ \ln \frac{\delta^2 S_{\text{ren}}(h=0)}{\delta\varphi(x)\delta\varphi(y)},$$

where S_{eff} and S_{ren} ($\equiv S_{\text{eff}} - S_{\text{CT}}$) are defined in Eq. (6.104).

(2) Using the above formula, derive $\Gamma^{(n)}$ for the one-component ϕ^4 model and reproduce Eqs. (6.105) and (6.107).

(3) Generalize the above results to the O(N) ϕ^4 model and derive the β-functions at one loop given in Eqs. (6.151), (6.152) and (6.153).

(4) Generalize the above results to the $\text{SU}_{\text{L}}(N_f) \times \text{SU}_{\text{R}}(N_f) \times \text{U}_{\text{A}}(1)$ model given in Eq. (6.161) and derive the β-functions in Eqs. (6.162) and (6.163).

6.10 Magnetic equation of state.

(1) Derive the magnetic EOS, Eq. (6.137), in the mean-field theory with the use of Eq. (6.34), and show that $\delta = 3$, $\beta = 1/2$ and $f(z) = b/6 + z$.

(2) Derive Eq. (6.141) using the magnetic EOS, Eq. (6.137), below and above T_{c}.

6.11 Stability of the effective potential. Derive the stability condition of $\mathcal{L}_{\text{eff}}$ for $c = h = 0$ by rewriting $\mathcal{L}_{\text{eff}}$ in terms of S^a and P^a.

6.12 Axial anomaly and the meson mass spectra. Write down Eq. (6.161) explicitly in terms of σ, η, $\boldsymbol{\delta}$ and $\boldsymbol{\pi}$. Assuming $h = 0$ and $a - c < 0$, derive the mass spectra of these particles in terms of $\langle\sigma\rangle$, b_1, b_2 and c. Assuming that the pion, $\boldsymbol{\pi}$, is the massless Nambu–Goldstone boson, determine the sign of c. Carry out a similar analysis for $h \neq 0$ and derive the mass spectra of σ, η, $\boldsymbol{\delta}$ and $\boldsymbol{\pi}$. For more details, see 't Hooft (1986).

6.13 Phase structure in the (m_{ud}, m_{s})-plane.

(1) By taking the line $m = m_{\text{ud}} = m_{\text{s}}$ in Fig. 6.12 and studying the Landau functional for the chiral field, Eq. (6.161), with $c \neq 0$ and small m, show that the boundary between the first order region and the crossover region is the second order line with Z(2) universality. Construct a similar analysis for large m by taking the Landau functional for the Polyakov line, Eqs. (5.83) and (5.85).

(2) Study the structure around the tricritical point (TCP) in Fig. 6.12 using Eq. (6.171).

7
Hadronic states in a hot environment

The purpose of this chapter is to discuss how the properties of hadrons, in particular the quark–anti-quark bound states, are modified when they exist in hot hadronic plasma and in quark–gluon plasma.

In Section 7.1, we will consider heavy quarkonia, such as J/ψ and Υ, as impurities inside the plasma composed of light quarks and gluons. These impurities can provide various information on the plasma properties, especially the confinement–deconfinement phase transition of the system.

In Section 7.2, we will discuss the light mesons, such as ρ, ω and ϕ, as well as π and σ, in the plasma. Unlike for heavy quarkonia, light mesons are part of the plasma and thus reflect a collective nature of the system that is intimately related to chiral symmetry and its restoration.

In Section 7.3, we discuss a lattice QCD approach that may be taken to study hadronic modes in a hot QCD environment on the basis of the maximum entropy method (MEM). This provides clues on how to extract spectral properties of the hot plasma from first principle calculations.

Section 7.4 provides emission rates of photons and dileptons from hot matter in terms of spectral functions.

7.1 Heavy quarkonia in hot plasma

7.1.1 $\mathrm{Q\bar{Q}}$ *spectra at* $T = 0$

Consider heavy quark bound states, such as charmonium and bottomonium. Since the masses of the charm and bottom quarks are much larger than the QCD scale parameter, $\Lambda_{\mathrm{QCD}} \sim 200$ MeV, the non-relativistic Schrödinger equation is a good place to start to analyze their properties:

$$\left(-\frac{\nabla^2}{2(m_{\mathrm{Q}}/2)} + V(r)\right)\Psi(\boldsymbol{r}) = E\Psi(\boldsymbol{r}), \tag{7.1}$$

Table 7.1. *Well established low-lying heavy quarkonia.*

J and P denote the total angular momentum and the parity; L and S are the orbital angular momentum and the spin of the $Q\bar{Q}$ pair. Masses, decay widths and some of the electromagnetic (EM) branching ratios are also given (Eidelman *et al.*, 2004).

	J^P	L	S	Mass M (GeV)	Total width Γ_{tot} (MeV)	EM branching ratios
$\eta_c(1s)$	0^-	0	0	2.98	~ 16	$B(\gamma\gamma) \sim 0.046\%$
$\eta_c(2s)$	0^-	0	0	3.65	< 55	
$J/\psi(1s)$	1^-	0	1	3.097	~ 0.09	$B(e^+e^-) \sim B(\mu^+\mu^-) \sim 6\%$
$\psi(2s)$	1^-	0	1	3.686	~ 0.28	$B(e^+e^-) \sim B(\mu^+\mu^-) \sim 0.75\%$
$\chi_{c0}(1p)$	0^+	1	1	3.42	~ 11	$B(\gamma\, J/\psi) \sim 1\%$
$\chi_{c1}(1p)$	1^+	1	1	3.51	~ 0.9	$B(\gamma\, J/\psi) \sim 32\%$
$\chi_{c2}(1p)$	2^+	1	1	3.56	~ 2.1	$B(\gamma\, J/\psi) \sim 20\%$
$\Upsilon(1s)$	1^-	0	1	9.46	~ 53	$B(e^+e^-) \sim B(\mu^+\mu^-) \sim 2.4\%$
$\Upsilon(2s)$	1^-	0	1	10.02	~ 43	$B(e^+e^-) \sim B(\mu^+\mu^-) \sim 1.3\%$
$\Upsilon(3s)$	1^-	0	1	10.36	~ 26	$B(\mu^+\mu^-) \sim 1.8\%$
$\chi_{b0}(1p)$	0^+	1	1	9.86		
$\chi_{b1}(1p)$	1^+	1	1	9.89		
$\chi_{b2}(1p)$	2^+	1	1	9.91		

where $m_Q/2$ is a reduced mass and E is the binding energy. The inter-quark potential as a function of r has the following characteristic form:

$$V(r) = Kr - \frac{4}{3}\frac{\alpha_s}{r} + \frac{32\pi\alpha_s}{9}\frac{\mathbf{s}_1 \cdot \mathbf{s}_2}{m_Q^2}\delta(\boldsymbol{r}) + \cdots, \tag{7.2}$$

where the first, second and third terms are the linear confining potential, the color Coulomb interaction and the color magnetic spin–spin interaction, respectively. The terms denoted by "$\cdots$" include the tensor, the spin-orbit and other higher order relativistic corrections.

The typical value of the string tension, K, is known to be $K \simeq 0.9\,\text{GeV fm}^{-1}$, as given in Eq. (2.58). The argument, κ, for the running coupling constant, $\alpha_s(\kappa)$, may be chosen to be m_Q: then the actual value of $\alpha_s(\kappa = m_Q)$ can be obtained from Fig. 2.2. An appropriate heavy quark mass, m_Q, that may be used in the Schrödinger equation, Eq.(7.1), is the pole mass given in Table 2.2: $m_c = 1.5$ to $1.8\,\text{GeV}$ and $m_b = 4.6$ to $5.1\,\text{GeV}$.

The experimental properties of the heavy quark bound states are summarized in Table 7.1. The corresponding mass spectra of low-lying charmonia and bottomonia are shown in Fig. 7.1. If the excited state masses are larger than the $D\bar{D}$ ($B\bar{B}$) threshold shown by the dash-dotted lines in Fig. 7.1, the excited states have large widths due to fall-apart decay. On the other hand, if they are below the threshold,

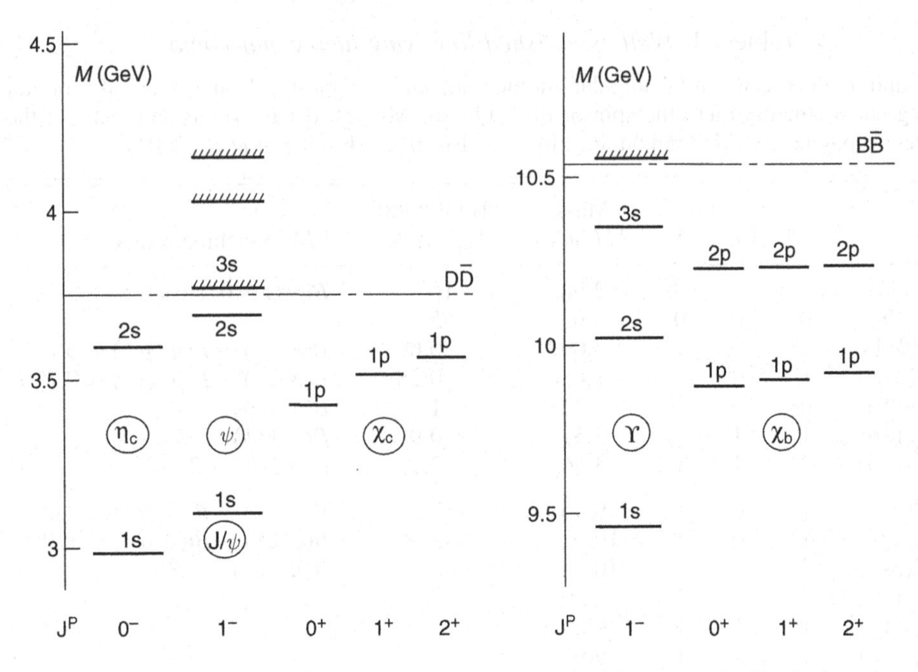

Fig. 7.1. Mass spectra of low-lying charmonia and bottomonia. Dash-dotted lines show the $D\bar{D}$ and $B\bar{B}$ thresholds.

annihilations of the heavy quarks are necessary, and the widths are suppressed according to the Okubo–Zweig–Iizuka (OZI) rule.

7.1.2 $Q\bar{Q}$ at $T \neq 0$

Let us consider what will happen if we put a heavy quarkonium inside hot QCD plasma. To simplify the discussion, we limit ourselves to a temperature satisfying the condition $T \ll m_c$, so that the plasma is composed of gluons and light flavors (u, d and s and their anti-particles) but not of the heavy flavors. Under this assumption, we may treat the heavy quarkonium as an impurity and study its properties as if it were a two-body problem of $Q\bar{Q}$. Similar to the idea of the Born–Oppenheimer approximation for molecules, the "fast" motion of gluons and light quarks in the plasma may be renormalized into T-dependent parameters of the Schrödinger equation for slowly moving heavy quarks. In particular, changes in Eq. (7.1) would result:

$$V(r) \to V_{\text{eff}}(r; T), \quad m_Q \to m_Q(T). \tag{7.3}$$

Here we have implicitly assumed that the heavy $Q\bar{Q}$ pair is in the color-singlet state and that $V_{\text{eff}}(r; T)$ is a potential acting in this channel. The color-singlet pair is of direct physical significance because the annihilation into dileptons goes through this channel in the leading order of α_s: $\mathrm{Q}+\bar{\mathrm{Q}} \to \gamma^* \to l^+l^-$.

In the deconfined plasma, the color-singlet $\mathrm{Q}\bar{\mathrm{Q}}$ pair and the color-octet $\mathrm{Q}\bar{\mathrm{Q}}$ pair can mix with each other by emitting or absorbing color-octet thermal gluons. To accommodate such a coupling, we need to introduce the wave functions both in the singlet and octet channels and consider a coupled channel Schrödinger equation, which will not be considered below.

The $V_{\text{eff}}(r; T)$ in the color-singlet channel may have several modifications from $V(r)$, as illustrated in Fig. 7.2, and discussed in the following.

(i) The string tension, K, will decrease as T approaches T_c, and it eventually vanishes above T_c, as expected from the string model discussed in Section 3.3. Associated with this, the binding between Q and $\bar{\mathrm{Q}}$ becomes weak. This may be detected as a shift of the J/ψ peak in the dilepton spectra for $T < T_c$ (Hashimoto *et al.*, 1986).

(ii) Because of the thermal pairs of light quarks ($q\bar{q}$) in the hot environment, the QCD string between the heavy quarks is going to be broken at long distances, and the heavy $\mathrm{Q}\bar{\mathrm{Q}}$ pair breaks up into heavy–light pairs, $\mathrm{Q}\bar{q}$ and $\bar{\mathrm{Q}}q$. Then, $V_{\text{eff}}(r; T)$ becomes flat above some critical distance, $r > r_c(T)$. This may lead to a change in the decay properties of heavy quarkonia for $T < T_c$ (Vogt and Jackson, 1988).

(iii) For $T > T_c$, the confining potential disappears. Furthermore, the short-range part of the gluonic interaction is Debye screened and a Yukawa potential arises, as we have discussed in Chapter 4:

$$V_{\text{eff}}(r; T) \to -\frac{4}{3}\frac{\alpha_s}{r}\, \mathrm{e}^{-r/\lambda_D}, \qquad \omega_D = 1/\lambda_D. \tag{7.4}$$

Since the Yukawa potential does not always support bound states (unlike the Coulomb potential), dissociation of heavy quarkonia takes place if ω_D is sufficiently large at high T (Matsui and Satz, 1986).

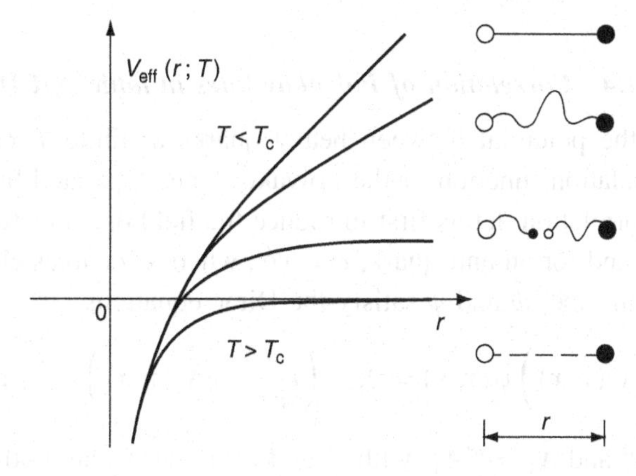

Fig. 7.2. Schematic behavior of the heavy quark potential at finite T.

7.1.3 Charmonium suppression at high T

Let us consider situation (iii) in Section 7.1.2 in more detail. In order to estimate the temperature above which the heavy quarkonium fails to be a bound state, let us compare the Bohr radius, r_B, of the $Q\bar{Q}$ pair with the Debye screening length, $\lambda_D = \omega_D^{-1}$. If the former is smaller than the latter, the short-range part of the potential is effectively Coulombic and can support bound states, even for $K = 0$. If the former is larger than the latter, the Debye screening prevents the formation of the bound states. We have already derived the QCD Debye mass in the lowest order perturbation theory in Chapter 4 (Eq.(4.87)): $\omega_D = gT\sqrt{N_c/3 + N_f/6}$. The Bohr radius, r_B, of a hydrogen atom is $1/(m_e\alpha)$, where m_e is the electron mass and α is the QED fine structure constant. Replacing m_e by the reduced mass, $m_Q/2$, and α by $4\alpha_s/3$, we obtain $r_B = 3/(2m_Q\alpha_s)$. Thus, the condition that the bound state disappears, $r_B > \lambda_D$, is rewritten as

$$T > 0.15 \times m_Q\sqrt{\alpha_s} \sim \begin{cases} 0.16 \text{ GeV} & \text{(charm)} \\ 0.46 \text{ GeV} & \text{(bottom)}, \end{cases} \tag{7.5}$$

where we take $N_f = 3$ and have used the central values of the pole mass from Table 2.2: $m_c = 1.65$ GeV and $m_b = 4.85$ GeV. For the coupling, we have taken $\alpha_s(\kappa = 2\pi T) \sim 0.4$ for $T \sim 200$ MeV.

The above estimate is, at most, qualitative. Nevertheless, it is interesting to see that charmonium suppression takes place at a temperature close to T_c obtained from the lattice QCD simulations shown in Eq. (3.63). On the other hand, the bottomonium may survive to higher temperature. To construct a more quantitative discussion of the fate of the heavy quarkonia at finite T, we require non-perturbative knowledge of $V_{\text{eff}}(r; T)$ calculated, for example, in lattice QCD simulations; this is discussed in Section 7.1.4. An alternative way of calculating the quarkonium spectrum directly on the lattice will be discussed in Section 7.3.

7.1.4 Correlation of Polyakov lines in lattice QCD

On the lattice, the potential between heavy quarks at finite T can be defined through the correlation function of the Polyakov lines separated by a distance r. To define this correlation, let us first introduce the field operator for an infinitely heavy quark, $\hat{\psi}$, and for an anti-quark, $\hat{\phi} = \hat{\psi}^C$, where C denotes charge conjugation. In Euclidean time, $\hat{\phi}$ and $\hat{\psi}$ satisfy the Dirac equations

$$\left(i\frac{\partial}{\partial\tau} + gA_4(\tau, \boldsymbol{x})\right)\hat{\psi}(\tau, \boldsymbol{x}) = 0, \quad \left(i\frac{\partial}{\partial\tau} + g\bar{A}_4(\tau, \boldsymbol{x})\right)\hat{\phi}(\tau, \boldsymbol{x}) = 0, \tag{7.6}$$

where $A_4 = t^a A_4^a$ and $\bar{A}_4 = \bar{t}^a A_4^a$, with t^a and $\bar{t}^a = -(t^a)^*$, are both generators of $SU(N_c)$ (Exercise 7.1). The solution of this static Dirac equation is easily found, as before (see Eq. (5.79)): $\hat{\psi}(\tau, \boldsymbol{x}) = \Omega(\boldsymbol{x})\hat{\psi}(0, \boldsymbol{x})$ and $\hat{\phi}(\tau, \boldsymbol{x}) = \bar{\Omega}(\boldsymbol{x})\hat{\phi}(0, \boldsymbol{x})$ with $\bar{\Omega} = \Omega^*$. Thus, the Polyakov lines for the quark and the anti-quark are defined as $L(\boldsymbol{x}) = (1/N_c)$ tr $\Omega(\boldsymbol{x})$ and $\bar{L}(\boldsymbol{x}) = (1/N_c)$ tr $\bar{\Omega}(\boldsymbol{x}) = L^\dagger(\boldsymbol{x})$, respectively.

Now, consider the partition function of Q and $\bar{\text{Q}}$ with an independent sum of color orientations (McLerran and Svetitsky, 1981b):

$$\begin{aligned}\frac{Z_{\text{Q}\bar{\text{Q}}}}{Z} &= \mathrm{e}^{-(F_{\text{Q}\bar{\text{Q}}}-F)/T} \\ &= \frac{1}{V^2}\frac{1}{N_c^2}\sum_{a,b=1}^{N_c}\sum_{\text{n}}\langle \text{n}|\hat{\phi}_b(\boldsymbol{y})\hat{\psi}_a(\boldsymbol{x})\mathrm{e}^{-H_{\text{Q}\bar{\text{Q}}}/T}\hat{\psi}_a^\dagger(\boldsymbol{x})\hat{\phi}_b^\dagger(\boldsymbol{y})|\text{n}\rangle/Z \\ &= \langle L(\boldsymbol{x})L^\dagger(\boldsymbol{y})\rangle. \end{aligned} \tag{7.7}$$

Here the state vectors, $\{|\text{n}\rangle\}$, form a set without heavy quarks. Since there are $N_c \times N_c$ possible color combinations of $\text{Q}\bar{\text{Q}}$, we take an average over the color orientations; V is the spatial volume of the system. To arrive at the final expression, we performed the same manipulation as that in Eq. (5.80) and its footnote. Equation (7.7) is simply a spatial correlation of the Polyakov lines. It is gauge-invariant, since the summation over color indices, a and b, is taken independently in Eq. (7.7).

Note that Q at $\boldsymbol{x}$ and $\bar{\text{Q}}$ at $\boldsymbol{y}$ may be decomposed into singlet and adjoint combinations in the color-space as $\mathbf{N_c} \otimes \bar{\mathbf{N}}_\mathbf{c} = \mathbf{1} \oplus (\mathbf{N_c^2} - \mathbf{1})$. Then the partition function, $Z_{\text{Q}\bar{\text{Q}}}$, for $N_c = 3$ is written as an average over the singlet channel and that in the octet channel (Brown and Weisberger, 1979):

$$\mathrm{e}^{-F_{\text{Q}\bar{\text{Q}}}/T} = \frac{1}{9}\mathrm{e}^{-F^{(1)}/T} + \frac{8}{9}\mathrm{e}^{-F^{(8)}/T}. \tag{7.8}$$

The left-hand side is gauge-invariant for arbitrary $\boldsymbol{x}$ and $\boldsymbol{y}$ by construction. However, the decomposition in the right-hand side is gauge-dependent and thus should be taken with care (Exercise 7.2).

In perturbation theory at high temperature, we have $F^{(1)} = -8F^{(8)} = -\frac{4}{3}\frac{\alpha_s}{r}\mathrm{e}^{-r/\lambda_\text{D}}$ (see Exercise 7.3 for the coefficient 4/3). Therefore, the leading contribution to $F_{\text{Q}\bar{\text{Q}}}$ at high T is given by

$$\frac{F_{\text{Q}\bar{\text{Q}}}}{T} \xrightarrow[T\to\infty]{} -\frac{1}{16}\left(\frac{F^{(1)}}{T}\right)^2. \tag{7.9}$$

The reason why the right-hand side starts from a term proportional to α_s^2 is that the color-singlet Polyakov lines can interact with each other only from the two-gluon exchange in perturbation theory.

The lattice measurements of the free energy, $F_{\text{Q}\bar{\text{Q}}}$, in QCD simulations with dynamical quarks are shown in Fig. 7.3 (Karsch *et al.*, 2001). Some features

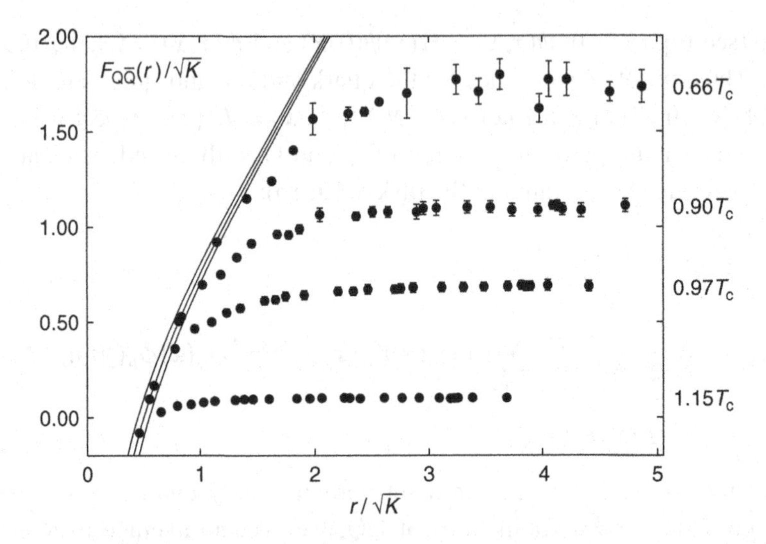

Fig. 7.3. The color-averaged free energy of a heavy quark pair at finite T as a function of the separation, r, in lattice QCD simulations with three dynamical quarks. The free energy is normalized to the phenomenological Cornell potential at $r = 1/(4T)$. The quark mass adopted in this figure corresponds to the pion mass, $m_\pi \simeq 1.8\sqrt{K} \sim 760\,\text{MeV}$, which is about five times heavier than the physical mass. Adapted from Karsch *et al.* (2001).

discussed in Section 7.1.2, such as the decreasing string tension, K, and the string breaking (flattening of the potential) are indeed observed. However, one cannot naively compare $F_{Q\bar{Q}}$ (the color-averaged free energy) with $V_{\text{eff}}(r; T)$ (the in-medium potential in the color-singlet channel). Further work will be necessary in order to extract useful information to allow us to calculate the heavy quarkonium properties from $F_{Q\bar{Q}}$.

7.2 Light quarkonia in a hot medium

7.2.1 $q\bar{q}$ spectra at $T = 0$

Unlike the case of the heavy mesons discussed in Section 7.1, the light mesons composed of u, d and s quarks cannot be described by a simple non-relativistic approach, as is obvious from Table 2.1: the current quark mass m_{ud} (m_{s}) is much smaller than (comparable to) Λ_{QCD}.

The typical mesons composed of light quarks are shown in Table 7.2. The major constituents of the masses do not originate from the current quark masses but from the non-perturbative QCD interactions. Figure 7.4 presents the meson mass spectra that have the same quantum numbers except for the parity. Universal mass splitting can be observed between different parity states; this is closely related to the dynamical breaking of chiral symmetry in the QCD vacuum.

Table 7.2. *Low-lying neutral mesons composed of light quarks.*

J, P and I denote the total angular momentum, the parity and the isospin, respectively. Masses, decay widths and some of the electromagnetic (EM) branching ratios are given (Eidelman *et al.*, 2004).

	J^P	I	M (MeV)	$\Gamma_{\rm tot}$ (MeV)	EM branching ratios
π^0	0^-	1	134.98	7.7×10^{-6}	$\Gamma(2\gamma)\simeq7.6\,{\rm eV}$
η	0^-	0	547.8	1.2×10^{-3}	$\Gamma(2\gamma)\simeq0.46\,{\rm keV}$
$\eta'(958)$	0^-	0	957.8	0.2	$\Gamma(2\gamma)\simeq4.3\,{\rm keV}$
$f_0(600)$ or σ	0^+	0	400–1200	600–1000	$\Gamma(2\gamma)\sim5\,{\rm keV}$
$\rho^0(770)$	1^-	1	769	151	$B({\rm e^+e^-})\simeq B(\mu^+\mu^-)\simeq4.5\times10^{-5}$
$\omega(782)$	1^-	0	783	8.4	$B({\rm e^+e^-})\sim B(\mu^+\mu^-)\sim8\times10^{-5}$
$\phi(1020)$	1^-	0	1019	4.3	$B({\rm e^+e^-})\simeq B(\mu^+\mu^-)\sim3\times10^{-4}$
$a_1(1260)$	1^+	1	1230	250–600	
$f_1(1285)$	1^+	0	1282	24	
$f_1(1420)$	1^+	0	1426	56	

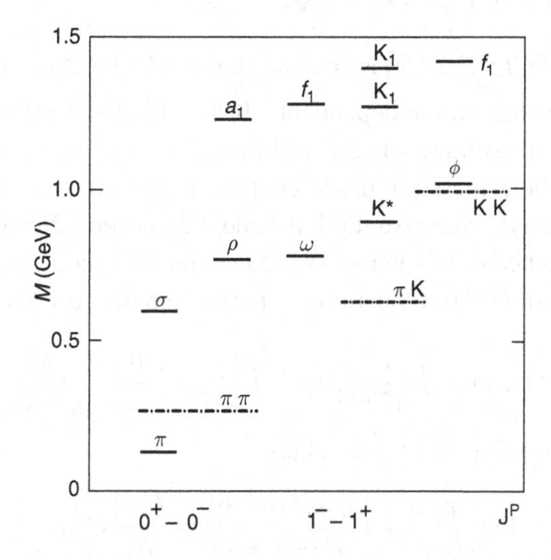

Fig. 7.4. Mass spectra of typical low-lying mesons. Possible chiral multiplets are paired, for example the π-meson versus the σ-meson, and the ρ-meson versus the a_1-meson. Dash-dotted lines indicate the $\pi\pi$, πK and KK thresholds.

In the finite-T medium, with the condition $m_{\rm u,d} \ll T \ll m_{\rm c}$, there are many thermal excitations of light $q\bar{q}$ pairs. Namely, the light quarkonium constitutes part of the medium, unlike the heavy quarkonium, which is an impurity. This leads to the idea that the study of the light quarkonia at finite T is the study of collective excitations of the finite T medium.

There exist several methods that calculate the collective properties of the hot QCD medium; for example, effective theories, such as the Nambu–Jona-Lasinio model, the low-T expansion with the chiral perturbation theory, the in-medium QCD sum rules and so on. Among others, a promising approach is the first principle QCD simulations combined with the maximal entropy method, which will be discussed in Section 7.3.

7.2.2 Nambu–Goldstone theorem at finite T

Consider N_f massless quarks and define a retarded correlation of the flavor axial current, $J^a_{5\mu} = \bar{q}\gamma_\mu\gamma_5 t^a q$, and the pseudo-scalar density, $P^b = \bar{q}i\gamma_5 t^b q$ (a and b in the following denote flavor indices running from 1 to $N_f^2 - 1$):

$$\Pi^{ab}_\mu(t, \boldsymbol{x}) \equiv \theta(t)\langle [J^a_{5\mu}(t, \boldsymbol{x}), P^b(0)]\rangle. \tag{7.10}$$

Taking the total divergence of the left-hand-side and integrating over the four-volume, we obtain

$$\int d^4x\ \partial^\mu \Pi^{ab}_\mu(t, \boldsymbol{x}) = \langle [Q^a_5, P^b(0)]\rangle = -i\frac{\langle \bar{q}q\rangle}{N_f}\delta^{ab}. \tag{7.11}$$

Here, $Q^a_5(t) = \int d^3x\, J^a_{5,\mu=0}(t, \boldsymbol{x})$ is an axial charge which is conserved for massless quarks and is therefore time-independent.[1] If the chiral symmetry is dynamically broken, the thermal average on the right-hand side is non-zero. This implies that there should be a massless mode coupled to both the axial current and the pseudo-scalar density; otherwise the left-hand side is zero after integration.

To see the existence of this massless mode explicitly, let us introduce the spectral representation of Π^{ab}_μ (see Section 4.4 for the general spectral decomposition):

$$\Pi^{ab}_\mu(t, \boldsymbol{x}) = \int \frac{d^4q}{(2\pi)^4} \mathrm{e}^{-iqx} \int_{-\infty}^{+\infty} d\omega' \frac{\rho^{ab}_\mu(\omega', \boldsymbol{q})}{\omega' - q_0 - i\delta}. \tag{7.12}$$

Substituting this into Eq. (7.11), we obtain

$$\lim_{q_0 \to 0} \int_{-\infty}^{+\infty} d\omega' \frac{q_0 \rho^{ab}_0(\omega', \boldsymbol{0})}{\omega' - q_0 - i\delta} = \frac{\langle \bar{q}q\rangle}{N_f}\delta^{ab}. \tag{7.13}$$

The left-hand side is non-vanishing only when the spectral function has the following singularity at zero energy:

$$\rho^{ab}_0(\omega, \boldsymbol{0}) = C\delta(\omega)\delta^{ab} + \rho^{ab}_{\text{regular}}, \tag{7.14}$$

where $C = -\langle \bar{q}q\rangle / N_f$. The delta function singularity corresponds to the $N_f^2 - 1$ Nambu–Goldstone massless pions in hot matter.

[1] The basic commutation relation is $\delta(x_0 - y_0)[q^\dagger Aq(x), q^\dagger Bq(y)] = \delta^4(x - y)q^\dagger [A, B]q$, where A and B are any matrices characterizing the internal degrees of freedom.

Although the theorem predicts massless excitation with no width at zero momentum, the dispersion relation, $\omega = \omega(\boldsymbol{q})$, for finite $\boldsymbol{q}$ can be different from its vacuum relation, $\omega = |\boldsymbol{q}|$; $\omega(\boldsymbol{q} \neq 0)$ may even become complex (the thermal width). Once we introduce small quark masses, the pion is assigned a finite mass and it also has a finite thermal width, even at $\boldsymbol{q} = 0$.

7.2.3 Virial expansion and the quark condensate

If the temperature of the system is low compared to the pion mass, $T < m_\pi$, the dominant thermal excitations in the hot plasma are the pions, as seen in Fig. 3.3. Furthermore, the pion gas at low T is dilute, so that the expansion in terms of the pion number density (the virial expansion) is a sensible way to estimate physical quantities. In particular, the leading order virial expansion for thermal average of an arbitrary operator, $\hat{O}$, is given by

$$\langle \hat{O} \rangle \simeq \langle \hat{O} \rangle_{\rm vac} + \sum_a \int \frac{d^3k}{(2\pi)^3 2\varepsilon_k} \langle \pi^a(k) | \hat{O} | \pi^b(k) \rangle n_{\rm B}(k;T) \tag{7.15}$$

$$\simeq \langle \hat{O} \rangle_{\rm vac} - \frac{1}{f_\pi^2} \int \frac{d^3k}{(2\pi)^3 2\varepsilon_k} \sum_a \langle 0 | [Q_5^a, [Q_5^a, \hat{O}]] | 0 \rangle n_{\rm B}(k;T). \tag{7.16}$$

In obtaining Eq. (7.16), the soft pion theorem (Exercise 2.9) is used to evaluate the matrix element. Note also that we use the covariant normalization for the one-pion state, which is why we have a covariant volume element, $d^3k/((2\pi)^3 2\varepsilon_k)$ (Exercise 7.4). For the realistic case, $m_{\rm u} \simeq m_{\rm d} \ll m_{\rm s}$, a takes values from 1 to 3, while for hypothetical N_f degenerate flavors, a takes values from 1 to $N_f^2 - 1$. We have neglected the finite quark mass in the matrix element, but we have retained it in the Boltzmann factor, because the latter is exponentially sensitive to the quark mass.

Applying the lowest order virial expansion to the quark condensate, we obtain

$$\frac{\langle \bar{q}q \rangle}{\langle \bar{q}q \rangle_{\rm vac}} \simeq 1 - \frac{N_f^2 - 1}{12 N_f} \frac{T^2}{f_\pi^2} B_1(m_\pi/T). \tag{7.17}$$

Here, B_1 is defined by

$$\int \frac{d^3k}{(2\pi)^3 2\varepsilon_k} n_{\rm B}(k;T) = \frac{T^2}{24} B_1(m_\pi/T), \tag{7.18}$$

in which $\varepsilon_k = (k^2 + m_\pi^2)^{1/2}$, $B_1(0) = 1$ and $B_1(\infty) = 0$. If we take massless pions for $N_f = 2$, we recover the first correction of $O(T^2)$ in Eq. (6.10) in Chapter 6.

7.2.4 Pions at low T

Let us now consider the dispersion relation of the pion at low temperature with reference to the virial expansion. Information concerning the pion properties inside the hot plasma is all buried in its in-medium propagator,

$$D_\pi(\omega, \boldsymbol{p}) = \left[\omega^2 - \varepsilon_p^2 + \Sigma(\omega, \boldsymbol{p})\right]^{-1}, \tag{7.19}$$

where $\varepsilon_p = (\boldsymbol{p}^2 + m_\pi^2)^{1/2}$ and Σ is the pion self-energy in the medium which has both real and imaginary parts.

Consider the case $\Sigma \ll \varepsilon_p$, which should be satisfied when $T \ll m_\pi$. In the lowest order of the virial expansion, the dispersion relation of the pion is given by

$$\omega = \varepsilon_p - \frac{\Sigma(\varepsilon_p, \boldsymbol{p})}{2\varepsilon_p} \tag{7.20}$$

$$= \varepsilon_p - \frac{1}{2\varepsilon_p} \sum_{\pi'=\pi^{0,\pm}} \int \frac{d^3k}{(2\pi)^3 2\varepsilon_k} \mathcal{F}_{\pi\pi'}(s) n_{\mathrm{B}}(k; T). \tag{7.21}$$

Here $\mathcal{F}_{\pi\pi'}(s)$ is the forward scattering amplitude of a pion (π) with momentum $\boldsymbol{p}$ and another pion inside the medium (π') with momentum $\boldsymbol{k}$. Also, $s = (p_\mu + k_\mu)^2$. Note that $\mathcal{F}$ is related to the invariant $\pi\pi$ scattering amplitude, $\mathcal{M}$, as follows: $\mathcal{F}_{\pi\pi'}(s) = \mathcal{M}_{\pi\pi' \to \pi\pi'}(s, t = 0)$; see Appendix F.

First we consider the modification of the real part of the dispersion relation. This takes place in the leading order of $\mathcal{F}_{\pi\pi'}(s)$ in the chiral expansion, which leads to $\sum_{\pi'} \mathcal{F}_{\pi\pi'}(s) = -m_\pi^2/f_\pi^2 + O(s^2, m_\pi^4)$ (Exercise 7.5). Then, the *mass shift* of the pion in the leading order may be given by

$$\frac{m_\pi(T)}{m_\pi} = 1 + \frac{1}{48}\frac{T^2}{f_\pi^2} B_1(m_\pi/T). \tag{7.22}$$

The mass shift is positive, but by no more than a small percentage, even at $T \simeq m_\pi$. For degenerate N_f-flavors, the coefficient 48 in Eq. (7.22) should be replaced by 24 N_f. For higher order calculations, see, for example, Toublan (1997).

Next we consider the damping rate of the pion in the medium. This is related to the imaginary part of the self-energy. By defining $\omega = \mathrm{Re}\ \omega(\boldsymbol{p}) - i\gamma(\boldsymbol{p})/2$, and taking the imaginary part of Eq. (7.20), we obtain (Goity and Leutwyler, 1989; Leutwyler and Smilga, 1990)

$$\gamma(\boldsymbol{p}) = \sum_{\pi'} \int \frac{d^3k}{(2\pi)^3} \bar{v}_{\mathrm{rel}}\ \sigma_{\pi\pi'}^{\mathrm{tot}}(s)\ n_{\mathrm{B}}(k; T) \tag{7.23}$$

$$= \sum_{\pi'} \int \frac{d^3k}{(2\pi)^3 2\varepsilon_p \varepsilon_k} \sqrt{s(s - 4m_\pi^2)}\ \sigma_{\pi\pi'}^{\mathrm{tot}}(s)\ n_{\mathrm{B}}(k; T). \tag{7.24}$$

Here we have used the optical theorem, which relates the forward scattering amplitude, $\mathcal{F}$, to the total cross-section, $\sigma^{\rm tot}$ (Appendix F.2):

$$\mathrm{Im}\mathcal{F}_{\pi\pi'}(s) = 2\varepsilon_p\varepsilon_k\,\bar{v}_{\rm rel}\,\sigma^{\rm tot}_{\pi\pi'}(s) = \sqrt{s(s-4m_\pi^2)}\,\sigma^{\rm tot}_{\pi\pi'}(s), \tag{7.25}$$

where $\bar{v}_{\rm rel}(=\sqrt{(p_\mu k^\mu)^2 - m_\pi^4}/\varepsilon_p\varepsilon_k)$ is a generalized relative velocity.

Let us consider the case in the chiral limit, $m_\pi = 0$. The leading order total cross-section in this case is known to be $\sum_{\pi'}\sigma^{\rm tot}_{\pi\pi'}(s) = 5s/48\pi f_\pi^4$. Then it is easy to see that the width vanishes at zero momentum, $\gamma(\boldsymbol{p}\to 0)|_{m_\pi=0} = 0$. This is consistent with what we have seen in the general derivation of the Nambu–Goldstone theorem in Section 7.2.2: there is no absorption of the pion wave at zero momentum in the chiral limit.

Let us now consider the case with finite pion mass. The mean free path, $\ell_\pi(\boldsymbol{p})$, of the pion wave may be defined as the distance that the wave travels during the damping time, $\gamma^{-1}(\boldsymbol{p})$. Since the group velocity of the wave is simply $v_{\rm g} = \partial\omega/\partial|\boldsymbol{p}| \simeq |\boldsymbol{p}|/\omega$ in the leading order, we have

$$\ell_\pi(\boldsymbol{p}) = \frac{|\boldsymbol{p}|}{\omega\gamma(\boldsymbol{p})}. \tag{7.26}$$

In Fig. 7.5, the mean free paths for $T = 120\,\mathrm{MeV}$ and $150\,\mathrm{MeV}$ are shown as a function of the pion momentum. This figure is based on a formula for $\gamma(\boldsymbol{p})$ derived on the basis of the kinetic theory combined with the lowest order two-body $\pi\pi$ scattering amplitude. It reduces to the same form as Eq. (7.24) when the pion gas is dilute (Goity and Leutwyler, 1989; Schenk, 1993). As T increases,

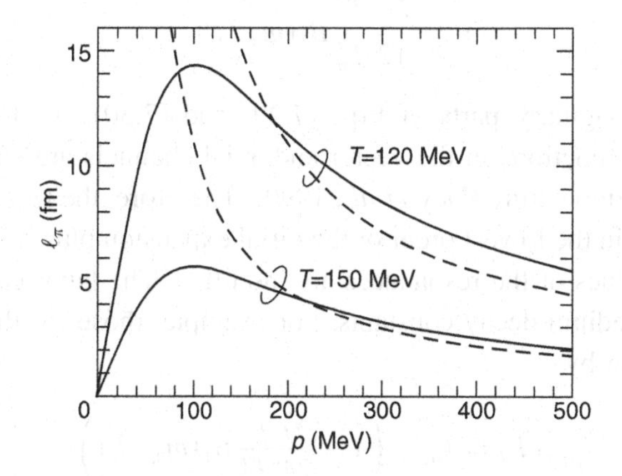

Fig. 7.5. Mean free path of the pion in the hot pion gas. The solid lines denote $m_\pi = 140$ MeV and the dashed lines denote $m_\pi = 0$. Adapted from Goity and Leutwyler (1989).

resonances heavier than the pion are also excited (Fig. 3.3). Therefore, the solid curves should be considered as an upper bound of the mean free path.

7.2.5 Vector mesons at low T

So far, we have discussed the pion at low T. We generalize this to vector mesons such as ρ, ω, ϕ and their axial-vector partners. As long as the chiral symmetry is dynamically broken in the vacuum, vector and axial-vector mesons have different spectra, as shown in Fig. 7.4. However, as T increases, the thermal pions mix vector and axial-vector mesons and thus tend to restore the broken symmetry.

To observe the above-mentioned mixing explicitly, let us consider the retarded correlation functions of the vector and axial-vector currents in degenerate N_f-flavors:

$$\Pi^{\rm V} \equiv i\langle {\rm R}J^a_\mu(x)J^b_\nu(y)\rangle, \qquad \Pi^{\rm A} \equiv i\langle {\rm R}J^a_{5\mu}(x)J^b_{5\nu}(y)\rangle. \tag{7.27}$$

Taking the lowest order virial expansion of the above correlation functions, and using the soft pion theorem twice (Exercise 2.9(2)) with the commutation relation,

$$\sum_c [Q^c_5, [Q^c_5, {\rm R}J^a_\mu(x)J^b_\mu(y)]] = 2N_f\left({\rm R}J^a_\mu(x)J^b_\nu(y) - {\rm R}J^a_{5\mu}(x)J^b_{5\nu}(y)\right), \tag{7.28}$$

we arrive at the following formulas:

$$\Pi^{\rm V} \simeq (1-\theta)\Pi^{\rm V}_{\rm vac} + \theta\Pi^{\rm A}_{\rm vac}, \tag{7.29}$$

$$\Pi^{\rm A} \simeq (1-\theta)\Pi^{\rm A}_{\rm vac} + \theta\Pi^{\rm V}_{\rm vac}, \tag{7.30}$$

$$\theta = \frac{N_f}{12}\frac{T^2}{f_\pi^2}B_1(m_\pi/T). \tag{7.31}$$

By taking the imaginary parts of Eqs. (7.29) and (7.30), we find that the in-medium spectral functions in the vector and axial channels are simply a mixture of those at zero temperature (Dey *et al.*, 1990). Therefore, the resonance positions are not modified in the lowest order of the virial expansion plus soft pion theorem, whereas the residues of the resonances are modified. The latter can be characterized by the in-medium decay constants. For example, those for the ρ-meson and the pion are given by

$$f_{\rho,\pi}(T) = f_{\rho,\pi}\cdot\left(1 - \frac{N_f}{24}\frac{T^2}{f_\pi^2}B_1(m_\pi/T)\right). \tag{7.32}$$

The mixing among different channels without shift of the resonance positions is a general feature at low temperature applicable to all hadrons (Exercise

7.6). Of course, the situation will change as T increases, where the interaction among hadrons, and even the quark structure of hadrons, are not negligible (Pisarski, 1982; Hatsuda and Kunihiro, 1985, 1994). For example, the scalar (σ) and pseudo-scalar (π) mesons are going to degenerate and become massless at the second order critical point of the massless $N_f = 2$ QCD discussed in Chapter 6. The fate of vector mesons, such as ρ, ω and ϕ, around the critical point is also of particular interest and is actively studied in various approaches (Brown and Rho, 1991, 1996; Hatsuda *et al.*, 1993; Harada and Yamawaki, 2003). A key issue is whether the chiral symmetry restoration is associated with the softening of the vector-meson spectra or not; it is related to the dilepton emission rate from the hot plasma, as discussed in Sections 7.4 and 15.5.

7.3 In-medium hadrons from lattice QCD

The first principle lattice QCD simulations provide us with an extremely useful tool with which to study finite T QCD, as we have seen in Chapters 2, 5 and 6. The method also provides useful information on hadron properties below and above the QCD phase transition, as we will see in this section.

Let us first consider the spectral function, $\rho(\omega, \boldsymbol{p})$, defined in Eq. (4.55). This is related to the imaginary part of the retarded Green's function, $\mathcal{G}^{\mathrm{R}}(t, \boldsymbol{x})$, in Eq. (4.49): $\hat{O}_{1,2}$ should be chosen to produce the hadronic states we are interested in. We assume a correlation of bosonic operators for simplicity. Then, for the vector mesons, $\hat{O}_1$ is chosen as: $\bar{\mathrm{c}}\gamma_\mu \mathrm{c}$ for J/ψ; $\bar{\mathrm{s}}\gamma_\mu \mathrm{s}$ for the ϕ-meson; $\frac{1}{2}(\bar{\mathrm{u}}\gamma_\mu \mathrm{u} - \bar{\mathrm{d}}\gamma_\mu \mathrm{d})$ for the ρ^0-meson; and $\frac{1}{2}(\bar{\mathrm{u}}\gamma_\mu \mathrm{u} + \bar{\mathrm{d}}\gamma_\mu \mathrm{d})$ for the ω-meson. Note that $\hat{O}_2$ has the same structure, with the Lorentz index μ replaced by ν. Both the real-time (retarded) and imaginary-time (Matsubara) correlations can be reconstructed from $\rho(\omega, \boldsymbol{p})$ through the spectral representations Eq. (4.53) and Eq. (4.57).

Let us now introduce an imaginary-time correlation in a mixed representation:

$$G(\tau, \boldsymbol{p}) = \int d^3x \; \mathcal{G}(\tau, \boldsymbol{x}) \, \mathrm{e}^{-i\boldsymbol{p}\cdot\boldsymbol{x}}, \tag{7.33}$$

where the Matsubara correlation, $\mathcal{G}(\tau, \boldsymbol{x})$, is defined in Eq. (4.50). On using Eq. (4.57) and carrying out the Fourier transform with an identity ($\omega_n = 2n\pi T$),

$$T \sum_n \frac{\mathrm{e}^{-i\omega_n \tau}}{x - i\omega_n} = \frac{\mathrm{e}^{-x\tau}}{1 - \mathrm{e}^{-x/T}} \quad (0 \le \tau < \beta), \tag{7.34}$$

we arrive at

$$G(\tau, \boldsymbol{p}) = \int_{-\infty}^{+\infty} \frac{e^{-\tau\omega}}{1 - e^{-\beta\omega}} \rho(\omega, \boldsymbol{p}) \, d\omega \qquad (0 \leq \tau < \beta). \tag{7.35}$$

Equation (7.35) is always convergent and does not require subtraction for $\tau \neq 0$ as long as $\rho(\omega \to \infty, \boldsymbol{p})$ does not grow exponentially.

For $\hat{O}_1 = \hat{O}_1^\dagger$, $\rho(\omega \geq 0, \boldsymbol{p}) \geq 0$, and it has a symmetry under the change of variables for bosonic correlation, $\rho(-\omega, -\boldsymbol{p}) = -\rho(\omega, \boldsymbol{p}) = -\rho(\omega, -\boldsymbol{p})$, where the parity is assumed to be unbroken. Then we arrive at

$$G(\tau, \boldsymbol{p}) = \int_{0}^{+\infty} K(\tau, \omega) \, \rho(\omega, \boldsymbol{p}) \, d\omega \quad (0 \leq \tau < \beta), \tag{7.36}$$

$$K(\tau, \omega) = \frac{e^{-\tau\omega} + e^{-(\beta-\tau)\omega}}{1 - e^{-\beta\omega}}. \tag{7.37}$$

Mathematically, K is an integral kernel which reduces to the Laplace kernel at $T = 0$. Physically, K is the free boson propagator in Euclidean time, τ, with an energy ω. The factor $\rho(\omega, \boldsymbol{p})$ denotes a spectral distribution as a function of the energy (Exercise 7.7).

Lattice Monte Carlo simulations provide $G(\tau, \boldsymbol{p})$ on the left-hand side of Eq. (7.36) over a discrete set of τ and $\boldsymbol{p}$ (Hashimoto *et al.*, 1993). From such numerical data, we want to extract the spectral function, ρ, which is a continuous function of ω. This is a typical ill-posed problem, where the number of data points is much smaller than the number of degrees of freedom to be reconstructed. The standard likelihood analysis (χ^2-fitting) is obviously inapplicable here, since many degenerate solutions appear in minimizing χ^2.

The maximum entropy method (MEM) is the approach taken to circumvent this difficulty, using Bayesian probability theory (Box and Tiao, 1992; Wu, 1997). In MEM, *a priori* assumptions or parametrizations of the spectral functions need not be made. Nevertheless, for any given lattice data, a unique solution is obtained, if it exists. Furthermore, one can carry out an error analysis on the obtained spectral function and evaluate its statistical significance. Basic concepts and techniques of MEM applied to lattice QCD are reviewed in Asakawa *et al.* (2001). We show in Fig. 7.6 how the MEM is used to extract the spectral function of the J/ψ channel below and above the critical temperature of the deconfinement transition in the quenched approximation. Unlike the naive expectation based on the Debye screening above T_c, the J/ψ peak around 3 GeV still exists, at least up to $T \sim 1.6T_c$ (Fig. 7.6(a)), and then it disappears above $T \sim 1.8T_c$ (Fig. 7.6(b)). This may suggest that the plasma is strongly interacting, so it can hold bound states, even through it is deconfined. Further studies, including a simulation with dynamical quarks, are necessary to reveal the true nature of the plasma just above T_c.

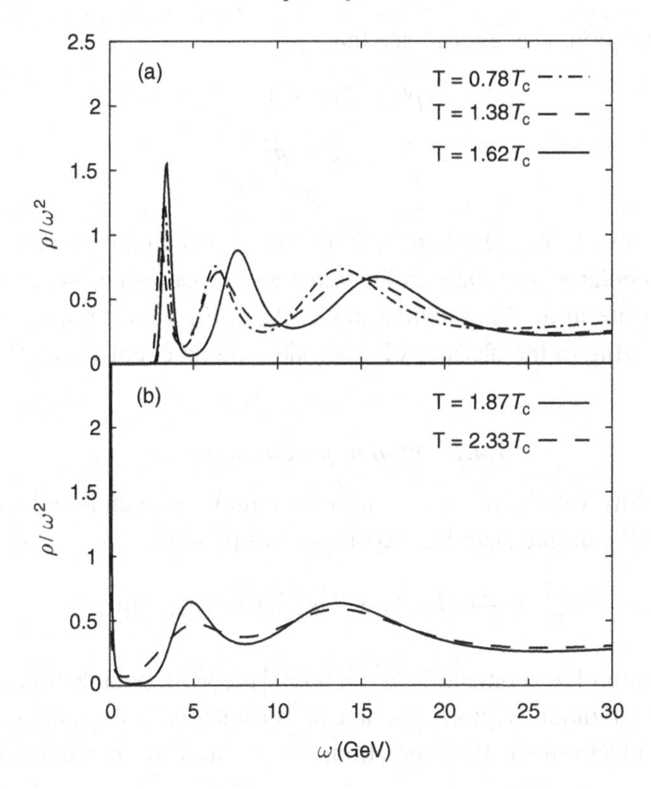

Fig. 7.6. The dimensionless spectral function, $\rho(\omega, \boldsymbol{p}=0)/\omega^2$, in the J/$\psi$ channel as a function of ω for several different temperatures. Since $\boldsymbol{p}=0$, we have $\rho = \rho_{_\mathrm{T}} = \rho_{_\mathrm{L}}$, where T and L stand for the transverse and longitudinal parts, respectively (see Section 7.4 for the definition). Adapted from Hatsuda and Asakawa (2004).

7.4 Photons and dileptons from hot/dense matter

The spectral function, $\rho(\omega, \boldsymbol{p})$, in the vector channel has a direct relevance to experimental observables, such as the production rates of real photons and dileptons from the hot and/or dense system (Feinberg, 1976). Consider the QCD electromagnetic current,

$$j_\mu^{\mathrm{em}} = \frac{2}{3}\bar{\mathrm{u}}\gamma_\mu \mathrm{u} - \frac{1}{3}\bar{\mathrm{d}}\gamma_\mu \mathrm{d} - \frac{1}{3}\bar{\mathrm{s}}\gamma_\mu \mathrm{s} + \cdots, \tag{7.38}$$

and choose $\hat{O}_1 = j_\mu^{\mathrm{em}}$ and $\hat{O}_2 = j_\nu^{\mathrm{em}}$ in Eq. (4.55). The resultant spectral function, $\rho_{\mu\nu}$, may be decomposed as follows:

$$\rho_{\mu\nu}(\omega, \boldsymbol{p}) = \rho_{_\mathrm{T}}(\omega, \boldsymbol{p})(P_\mathrm{T})_{\mu\nu} + \rho_{_\mathrm{L}}(\omega, \boldsymbol{p})(P_\mathrm{L})_{\mu\nu}, \tag{7.39}$$

where P_T (P_L) is the projection operator to the transverse (longitudinal) component in the Minkowski space (see the footnote to Eqs. (4.68)–(4.71) in Section 4.5).

From the definition, it is easy to see that

$$-\rho^{\mu}_{\mu} = 2\rho_{\rm T} + \rho_{\rm L}, \tag{7.40}$$

$$\rho_{\rm L} = \frac{\omega^2 - \boldsymbol{p}^2}{\boldsymbol{p}^2}\rho_{00}. \tag{7.41}$$

The coefficients on the right-hand side of Eq. (7.40) simply reflect the fact that there are two polarization states in the transverse mode, whereas there is only one polarization state in the longitudinal mode. Also, Eq. (7.41) implies that $\rho_{\rm L}(\omega = |\boldsymbol{p}|, \boldsymbol{p}) = 0$ owing to the absence of a massless mode coupled to $j_0^{\rm em}$ in QCD.

7.4.1 Photon production rate

From Fig. 7.7(a), we obtain the transition amplitude from an initial hadronic state $|{\rm i}\rangle$ to the final hadronic state $|{\rm f}\rangle$ + a single real photon:

$$S_{\rm fi}^{(\lambda)} = -ie\int d^4x\, {\rm e}^{ipx}\, \varepsilon_{\mu}^{(\lambda)}(p)\langle {\rm f}|j^{\mu}_{\rm em}(x)|{\rm i}\rangle, \tag{7.42}$$

where the photon has momentum $p^{\mu} = (\omega = |\boldsymbol{p}|, \boldsymbol{p})$, a polarization, λ, with $\varepsilon_{\mu}^{(\lambda)}$ being the polarization vector. The above formula is only valid in the lowest order of the electromagnetic interaction, but is valid in all orders of the strong interaction.

The production rate of real photons from a thermalized initial state is defined as the number of photons emitted per unit time and unit spatial volume, $R_{\gamma} = d^4N_{\gamma}/d^4x$. We have

$$R_{\gamma} = \frac{1}{\int d^4x}\int \frac{d^3p}{2\omega(2\pi)^3}\frac{1}{Z}\sum_{\rm f,i,\lambda} {\rm e}^{-(E_{\rm i}-\mu N_{\rm i})/T}|S_{\rm fi}^{(\lambda)}|^2. \tag{7.43}$$

All possible final states are summed and the thermal average (with temperature, T, and chemical potential, μ) is taken for the initial state. Putting Eq. (7.42) into

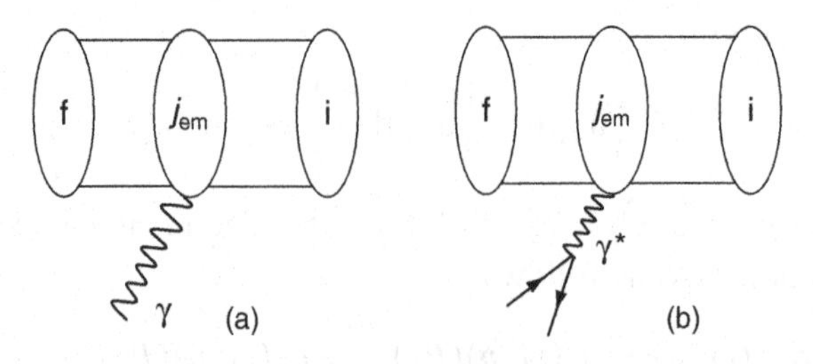

Fig. 7.7. (a) Transition from a hadronic initial state to a final state by emitting an on-shell photon. (b) Similar process except for the emission of a virtual photon decaying into dileptons.

Eq. (7.43) and using $\sum_\lambda \varepsilon_\mu^{(\lambda)*} \varepsilon_\nu^{(\lambda)} = -g_{\mu\nu}$ together with the translation invariance of the system, we end up with the following simple formula:

$$\omega \frac{d^3 R_\gamma}{d^3 p} = -\frac{\alpha}{2\pi} \frac{\rho_\mu^\mu(\omega = |\boldsymbol{p}|, \boldsymbol{p})}{\mathrm{e}^{\omega/T} - 1} = \frac{\alpha}{\pi} \frac{\rho_{\mathrm{T}}(\omega = |\boldsymbol{p}|, \boldsymbol{p})}{\mathrm{e}^{\omega/T} - 1}, \tag{7.44}$$

where $\alpha = e^2/(4\pi) \sim 1/137$. (Equation (7.41) is used in the second equality.) This is essentially the optical theorem (Appendix F.2), in which the photon emission rate is related to the imaginary part of the electromagnetic correlation function at $\omega = |\boldsymbol{p}|$. It is natural that only the transverse part of the spectral function, ρ_{T}, appears in the final expression for the production of the on-shell photon.

7.4.2 Dilepton production rate

A general formula for dilepton production rate is obtained in a similar way to the case for real photons (Feinberg, 1976; Weldon, 1990). The only difference is that, in the present case, the emitted photon is off-shell. From Fig. 7.7(b), we find the transition matrix element, from the initial hadronic state, $|\mathrm{i}\rangle$, to the final hadronic state, $|\mathrm{f}\rangle$, plus a lepton pair, l^+l^- (such as $\mathrm{e}^+\mathrm{e}^-$ and $\mu^+\mu^-$), as follows:

$$S_{\mathrm{fi}}(p_1, p_2) = -i\frac{e^2}{p^2}[\bar{u}(p_1)\gamma_\mu v(p_2)] \int d^4x \, \mathrm{e}^{ipx} \, \langle \mathrm{f}|j_{\mathrm{em}}^\mu(x)|\mathrm{i}\rangle. \tag{7.45}$$

Here, $p_1^\mu = (E_1, \boldsymbol{p}_1)$ and $p_2^\mu = (E_2, \boldsymbol{p}_2)$ are the four-momenta of a lepton and its anti-particle; $q^\mu = (\omega, \boldsymbol{p}) = p_1^\mu + p_2^\mu$ is the total momentum of the lepton pair, which is time-like, $p^2 \equiv \omega^2 - \boldsymbol{p}^2 > 0$. The factor $1/p^2$ is from the virtual photon propagator, and $e\bar{u}\gamma_\mu v$ is the electromagnetic current of the outgoing lepton pair.

The production rate of the dilepton from a thermalized initial state is defined as follows:

$$R_{l^+l^-} = \frac{1}{\int d^4x} \int \frac{d^3p_1}{2E_1(2\pi)^3} \int \frac{d^3p_2}{2E_2(2\pi)^3} \frac{1}{Z} \sum_{\mathrm{f,i}} \mathrm{e}^{-(E_{\mathrm{i}} - \mu N_{\mathrm{i}})/T} |S_{\mathrm{fi}}|^2 \tag{7.46}$$

$$= \int d^4p \, \frac{d^4 R_{l^+l^-}}{d^4p}. \tag{7.47}$$

Now we put Eq. (7.45) into Eq. (7.46) and use the following relation:

$$\int \frac{d^3p_1}{2E_1(2\pi)^3} \int \frac{d^3p_2}{2E_1(2\pi)^3} L_{\mu\nu}(p_1, p_2) \, (2\pi)^4 \delta^4(p - p_1 - p_2)$$

$$= \frac{1}{6\pi}(p_\mu p_\nu - p^2 g_{\mu\nu}) F(m_l^2/p^2). \tag{7.48}$$

Here, F is a kinematical factor defined by $F(x) = (1+2x)(1-4x)^{1/2}\theta(1-4x)$, which is derived in Exercise 14.5, and $L_{\mu\nu}$ is a leptonic tensor,

$$L^{\mu\nu}(p_1, p_2) = \sum_{s,r}[\bar{u}_s(\boldsymbol{p}_1)\gamma^\mu v_r(\boldsymbol{p}_2)][\bar{u}_s(\boldsymbol{p}_1)\gamma^\nu v_r(\boldsymbol{p}_2)]^*$$

$$= 4\left[p_1^\mu p_2^\nu + p_1^\nu p_2^\mu - (p_1 \cdot p_2 - m_l^2)g^{\mu\nu}\right], \tag{7.49}$$

where the summation is taken over spin (s and r); see Appendix B. We arrive at the final formula, as follows:

$$\frac{d^4 R_{l^+l^-}}{d^4 p} = \frac{-\alpha^2}{3\pi^2 p^2}\frac{\rho^\mu_\mu(\omega, \boldsymbol{p})}{e^{\omega/T}-1}F(m_l^2/p^2)$$

$$= \frac{\alpha^2}{3\pi^2 p^2}\frac{(2\rho_{\mathrm{T}}+\rho_{\mathrm{L}})(\omega, \boldsymbol{p})}{e^{\omega/T}-1}F(m_l^2/p^2). \tag{7.50}$$

This is again regarded as an optical theorem (Appendix F.2), in which the dilepton emission rate is related to the imaginary part of the electromagnetic correlation function with off-shell four-momentum (Exercise 7.8). Unlike the case of production of the on-shell photon, both ρ_{T} and ρ_{L} contribute to dilepton production.

Note that Eqs. (7.44) and (7.50) are valid, irrespective of the nature of the initial state: either the hadronic phase or the quark–gluon plasma. The spectral functions calculated on the lattice, as in Section 7.3, are a useful first principle input to the production rate of photons and dileptons. In actual heavy ion collisions, the temperature and chemical potential vary in space and time. Therefore, we need to combine the "local" production rate obtained in this section with the hydrodynamical evolution to predict realistic spectra. This approach will be discussed in Section 14.3.3.

Exercises

7.1 Dirac equation for charge-conjugate field. Derive the Dirac equation for the field, $\hat{\phi}$, in Eq. (7.6). Use the definition of the charge conjugation, $\hat{\phi} = \hat{\psi}^{\mathrm{C}} = C\hat{\psi}^*$, with $C = i\gamma^2\gamma^0$.

7.2 Gauge-dependent potential at finite T. By rearranging the color summation over (a, b) in Eq. (7.7) into color-singlet and adjoint combinations, express $e^{-F^{(1)}/T}$ in terms of the spatial correlation of $\Omega(\boldsymbol{x})$. Show that the result is gauge-dependent.

7.3 Projection operators in color-space.

(1) Consider the two spin 1/2 particles with total spin operator, $\mathbf{S} = \mathbf{s}_1 + \mathbf{s}_2$. Show that the matrix element of $\mathbf{s}_1 \cdot \mathbf{s}_2$ is given by $(\mathbf{s}_1 \cdot \mathbf{s}_2)_{S=0} = -3/4$ for the spin singlet state and $(\mathbf{s}_1 \cdot \mathbf{s}_2)_{S=1} = 1/4$ for the spin triplet state.

Then derive the projection operator, P, to the singlet state and the triplet state:

$$\mathrm{P}_{S=0} = \frac{1}{4}(1 - 4\mathbf{s}_1 \cdot \mathbf{s}_2), \qquad \mathrm{P}_{S=1} = \frac{1}{4}(3 + 4\mathbf{s}_1 \cdot \mathbf{s}_2).$$

(2) Let us generalize the above formulas for a particle in the fundamental representation, $\mathbf{N_c}$, of SU(N_c) and an anti-particle in the fundamental representation, $\bar{\mathbf{N}}_c$, of SU(N_c). The total color operator is defined as $T^a = t_1^a + \bar{t}_2^a$ $(a = 1, 2, \ldots, N_c^2 - 1)$, with $\bar{t}^a = -t^{a*}$. Since $\mathbf{N_c} \otimes \bar{\mathbf{N}}_c = \mathbf{1} \oplus (\mathbf{N_c^2} - \mathbf{1})$, the color states of two particles belong either to color-singlet or color-adjoint representations. Considering $(T^a)^2$ and its trace, derive the following relation:

$$t_1^a \bar{t}_2^a = \begin{cases} -\frac{N_c^2 - 1}{2N_c} & \text{(singlet)} \\ \frac{1}{2N_c} & \text{(adjoint)}. \end{cases}$$

Then derive the projection operators:

$$\mathrm{P}_{\text{singlet}} = \frac{1}{N_c^2}(1 - 2N_c t_1^a \bar{t}_2^a), \qquad \mathrm{P}_{\text{adjoint}} = \frac{1}{N_c^2}(N_c^2 - 1 + 2N_c t_1^a \bar{t}_2^a).$$

Generalize this to the two particles that belong to the fundamental representation, $\mathbf{N_c}$. (See the Appendices of Brown and Weisberger (1979) and Nadkarni, (1986a, b) for more details.)

7.4 Leading order virial expansion. Confirm Eq. (7.15) by taking into account the covariant normalization of the state vector (see Exercise 2.9), $\langle \pi^a(k) | \pi^b(k) \rangle = 2\varepsilon_k \delta^{ab} V$, where V is a volume of the three-dimensional space. Note that the sum of discrete momenta and the momentum integration are related: $\sum_k / V = \int d^3k/(2\pi)^3$.

7.5 $\pi\pi$ scattering amplitude.

(1) Show that the general $\pi\pi$ scattering amplitude in $N_f = 2$ is given by

$$\mathcal{M}_{ab \to cd}(s, t, u) = A(s, t, u)\delta_{ab}\delta_{cd} + A(t, s, u)\delta_{ac}\delta_{bd} + A(u, t, s)\delta_{ad}\delta_{bc},$$

where s, t and u are the Mandelstam variables, with $s + t + u = 4m_\pi^2$, and a, b, c and d are the isospin indices.

(2) Using Eq. (2.63), show that $A(s, t, u) = (s - m_\pi^2)/f_\pi^2$ in the leading order of the chiral expansion. Then, taking the forward limit $t = 0$, show that

$$\sum_{\pi'} \mathcal{F}_{\pi\pi'}(s) = \sum_b \mathcal{M}_{ab \to ab}(s, 0, 4m_\pi^2 - s) = -\frac{m_\pi^2}{f_\pi^2}.$$

(3) Generalize the above formulas for the case of degenerate N_f flavors.

7.6 **Mixing of hadronic correlations at finite T.** Derive the mixing formula, similar to Eqs. (7.29) and (7.30), in other channels such as the scalar and pseudo-scalar channels. Generalize the analysis to baryonic correlations by consulting Leutwyler and Smilga (1990).

7.7 **Hadronic spectral functions.** To obtain a rough idea about the shape, $G(\tau, p)$, as a function of τ in Eq. (7.36), consider a pole plus a continuum form of the spectral function:

$$\rho(\omega, \boldsymbol{p}=0) = \omega^n \cdot \left[a\ \delta(\omega^2 - m^2) + b\ \theta(\omega - \omega_0)\right],$$

where n is a positive integer, m is the pole position and ω_0 is a continuum threshold. Note that a and b are the constants which characterize the pole residue and the continuum height, respectively. This simple parametrization is commonly used in the QCD sum rules (Colangelo and Khodjamirian, 2001). Derive the corresponding correlation function, $G(\tau, \boldsymbol{p}=0)$, and study its τ-dependence. Which region of τ is dominated by the pole and which region is dominated by the continuum?

7.8 ***R*-ratio and the spectral function.** Using the optical theorem (Appendix F.2), show that the R-ratio in the e^+e^- annihilation in the vacuum is related to the ρ^μ_μ as follows:

$$R(s) \equiv \frac{\sigma(e^+e^- \to \text{hadrons})}{\sigma(e^+e^- \to \mu^+\mu^-)} = -\frac{4\pi^2}{s}\rho^\mu_\mu(s),$$

where $s \equiv \omega^2 - \boldsymbol{p}^2$.

II

Quark–Gluon Plasma in Astrophysics

8

QGP in the early Universe

In this chapter, we discuss the implications of the quark–gluon plasma (QGP) and the quantum chromodynamics (QCD) phase transition on the history of the early Universe. First, we summarize the basics of Big Bang cosmology and the general theory of relativity. Then we discuss the thermal history and the composition of matter in the Universe (Kolb and Turner, 1989; Peebles, 1993). The QCD phase transition, which took place at 10^{-5} to 10^{-4} s after the Big Bang, will be explained in some detail by using the bag equation of state coupled with the Einstein equation.

8.1 Observational evidence for the Big Bang

It is now widely accepted that the Universe started out as an initial hot stage some 10^{10} years ago and that it is continuously expanding, even today (Fig. 8.1). There are three key observations supporting this Big Bang scenario:

(i) Hubble's law (Hubble, 1929), which states that the velocity, v, of a distant galaxy or star relative to the Earth is proportional to its distance, l, from the Earth:

$$v = H_0 \times l, \tag{8.1}$$

$$H_0 = 100h\,\text{km s}^{-1}\,\text{Mpc}^{-1} = (9.78h^{-1} \times 10^9\ \text{year})^{-1},$$

where 1 pc (parsec) $= 3.09 \times 10^{13}$ km $= 3.26$ light years (Appendix A.2). The factor H_0 is called the Hubble constant, and results from the Hubble Space Telescope (HST) Key Project yield (Freedman *et al.*, 2001)

$$h = 0.72 \pm 0.08. \tag{8.2}$$

The determination of this number, which has been based on the observation of various distant objects up to 400 Mpc, is shown in Fig. 8.2. If we accept the "cosmological principle" (that no observer occupies a preferred position in the Universe), Eq. (8.1) applies to any two observers in the Universe, and thus indicates that the Universe is expanding (Fig. 8.1).

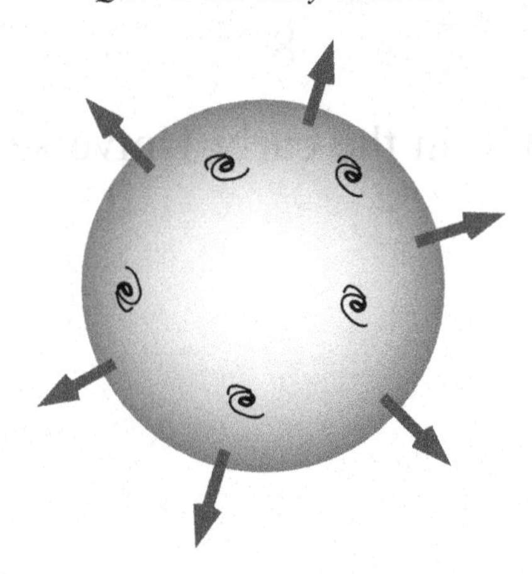

Fig. 8.1. Expanding Universe, modeled by a balloon with galaxies (or, more precisely, clusters of galaxies) on its surface.

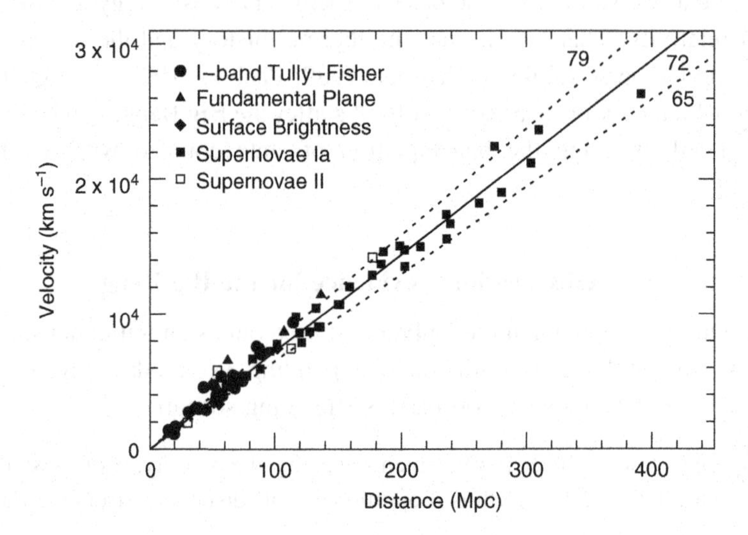

Fig. 8.2. Determination of Hubble's law from the measurement of the velocity vs. distance of distant galaxies and supernovae by Hubble Space Telescope (HST). The area bounded by straight dotted lines corresponds to the Hubble constant, $H_0 = 65\text{–}79$ km s^{-1} Mpc^{-1}. Adapted from Freedman *et al.* (2001).

(ii) The present Universe is filled with cosmic microwave background (CMB) radiation, which was first observed by Penzias and Wilson (1965). The Cosmic Background Explorer (COBE) has measured a very accurate black body CMB spectrum with a temperature (Fixsen *et al.*, 1996; Mather, 1999)

$$T_{\rm CMB} = 2.725 \pm 0.002\,\text{K}. \tag{8.3}$$

Photon intensity as a function of frequency observed by COBE is shown in Fig. 8.3. The CMB is a remnant of hot thermal era in the early Universe.

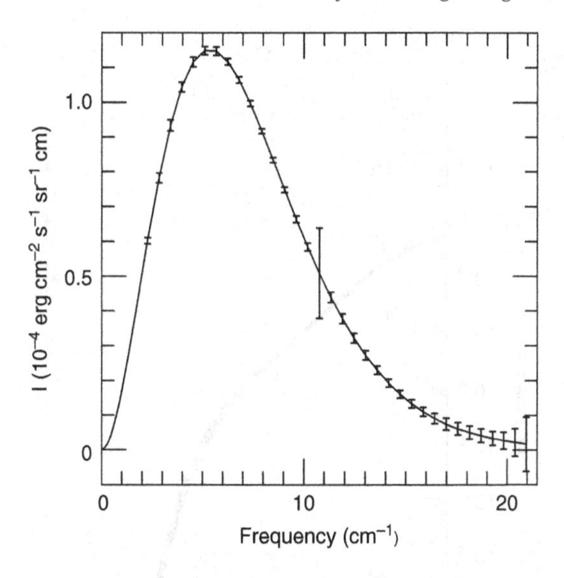

Fig. 8.3. Cosmic microwave background (CMB) spectrum taken by Cosmic Background Explorer (COBE). The solid curve is the black body radiation (the Planck formula) of $T = 2.73$ K. Error bars are magnified by a factor of 400. Therefore, the actual errors are within the thickness of the solid line. Adapted from Turner and Tyson (1999).

(iii) The observed mass fraction of primordial ^{4}He relative to the total baryon masses is given by (Tytler, 2000)

$$Y_{\rm p} = \frac{4n_{\rm He}}{n_{\rm B}} \simeq 0.25, \tag{8.4}$$

where $n_{\rm He}$ is the primordial number density of ^{4}He and $n_{\rm B}$ is the net baryon number density, bound or free. The number 0.25 is reasonably consistent with the scenario of primordial nucleosynthesis, in which light elements are produced in a hot environment a few minutes after the Big Bang (Alpher, Bethe and Gamow, 1948; Hayashi, 1950). Many sophisticated calculations of the Big Bang nucleosynthesis, compared with observations shown in Fig. 8.4, can even pin down the total baryon mass density of the Universe: $\rho_{\rm B} \sim (3\text{–}5) \times 10^{-31}$ g cm^{-3} (Schramm and Turner, 1998; Tytler *et al.*, 2000).

The Wilkinson Microwave Anisotropy Probe (WMAP) launched in 2001 has made remarkable progress in measuring the CMB spectrum, and its anisotropy as shown in Fig. 8.5. By fitting the data, many of the cosmological parameters are constrained, which advances the field of precision cosmology. Some of the best fit cosmological parameters obtained from WMAP are shown in Table 8.1 (Bennett *et al.*, 2003).

The expansion of the Universe and the existence of an early hot era, which are both suggested by these observations, are well described by the solution of the

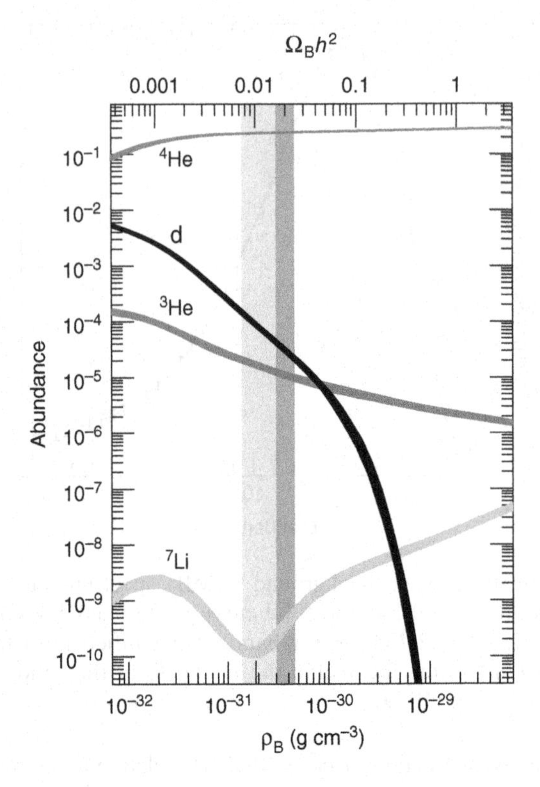

Fig. 8.4. Calculated abundances of ^{4}He (mass fraction), d, ^{3}He and ^{7}Li (number relative to H) as a function of the baryon mass density. The wider band is a constraint for ρ_B from all four elements, while the narrower band is from deuteron, d. Adapted from Turner and Tyson (1999).

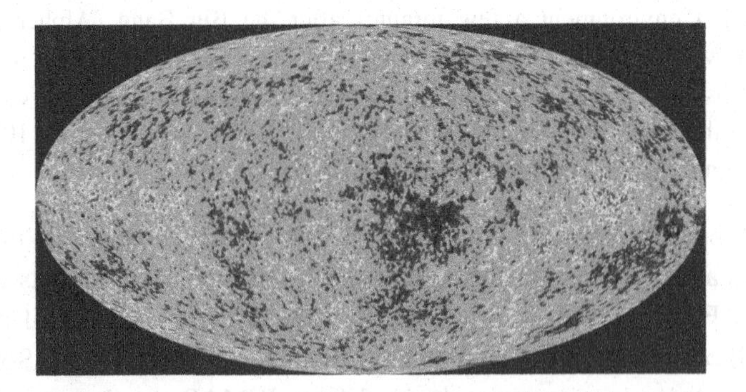

Fig. 8.5. WMAP sky map of the CMB anisotropy in Mollweide projection in galactic coordinates (Courtesy of NASA/WMAP Science Team). The contrast between the darkest area and the brightest area ranges from -2×10^{-4} K to 2×10^{-4} K.

Table 8.1. *Some of the cosmological parameters determined by WMAP (Bennett et al., 2003).*

Description	Symbol	Value	± Uncertainty
Total density	$\Omega_{\rm tot}$	1.02	± 0.02
Dark energy density	Ω_Λ	0.73	± 0.04
Baryon density	$\Omega_{\rm B}$	0.044	± 0.004
Matter density	$\Omega_{\rm B}+\Omega_{\rm DM}$	0.27	± 0.04
Baryon-to-photon ratio	η	6.1×10^{-10}	$^{+0.3}_{-0.2}\times 10^{-10}$
Red shift of decoupling	$z_{\rm dec}$	1089	± 1
Age of decoupling (year)	$t_{\rm dec}$	379×10^3	$^{+8}_{-7}\times 10^3$
Hubble constant	h	0.71	$^{+0.04}_{-0.03}$
Age of the Universe (year)	t_0	13.7×10^9	$\pm 0.2\times 10^9$

Einstein equation coupled with matter and radiation. In the following sections, we first discuss the thermal history of the Universe and then examine how the QGP enters into the scenario when the age of the Universe was less than 10^{-4} s.

8.2 Homogeneous and isotropic space

8.2.1 Robertson–Walker metric

Observations show that the Universe is approximately homogeneous and isotropic if averaged over a scale much greater than the size of galaxies and clusters of galaxies. Then, to describe the global structure of the Universe, one may take the Robertson–Walker (RW) metric by assuming spatial uniformity (Appendix D.3):

$$\begin{aligned} ds^2 &= dt^2 - a^2(t)\left[\frac{dr^2}{1-Kr^2} + r^2(d\theta^2 + \sin^2\theta d\phi^2)\right] \\ &\equiv dt^2 - a^2(t)\, d\sigma^2. \end{aligned} \tag{8.5}$$

Here (t, r, θ, ϕ) are the co-moving coordinates, namely the observers at rest in this coordinate system are always at rest. Consequently, t is the proper time for these observers.

The scale factor, $a(t)$, parametrizes the expansion of the Universe. Its t-dependence can be determined by solving the Einstein equation. The spatial part of the RW metric is simply a metric for the three-dimensional surface with a constant curvature which is homogeneous and isotropic; i.e. there is no special point on this surface. The three-dimensional scalar curvature calculated from this metric is given by (Exercise 8.1)

$${}^3R = \frac{6K}{a^2(t)}. \tag{8.6}$$

Therefore, the parameter K dictates the sign of the spatial curvature: $K=+1$ (closed space with positive curvature); $K=0$ (flat space); $K=-1$ (open space with negative curvature). Note that a simple rescaling of the coordinate, r, can change the magnitude of K and $a(t)$ with ${}^3R$ unchanged. Thus we choose K to take either $+1$, -1 or 0.

8.2.2 *Hubble's law and red shift*

Hubble's law is easily obtained from the RW metric. Consider two nearby points, O and P, which are spatially separated and have co-moving coordinates, $(t,0,0,0)$ and $(t,r,0,0)$, respectively. The physical distance between O and P is given by

$$l(t)=\int_{O}^{P} a(t)\, d\sigma = a(t)\int_0^r \frac{dr}{\sqrt{1-Kr^2}}, \tag{8.7}$$

where $ds^2=-a^2(t)\,d\sigma^2$. The integral in the above equation depends only on r and does not depend on t. Therefore, we obtain a relation between the velocity, $v(t)=dl(t)/dt\equiv \dot{l}(t)$, and the spatial distance, $l(t)$:

$$v(t)=\left(\frac{\dot{a}}{a}\right) l(t)\equiv H(t)l(t). \tag{8.8}$$

The Hubble constant, H_0, defined in Eq. (8.1), is simply $H_0=H(t_0)$, where t_0 is the present time. In the following, we will include the subscript 0 for the quantities at $t=t_0$.

Equation (8.8) is valid only when the distance, l, is not large and the expansion of the Universe during the traveling of light from point P to O is negligible. For P distant from O, we need to consider the geodesic, $ds^2=0$, for the light propagation. Consider a source emitting light with a frequency ν. Let us assume that the photon emitted at $\mathrm{P}=(t,r,0,0)$ reaches $\mathrm{O}=(t_0,0,0,0)$, as shown in Fig. 8.6, while the photon emitted at $\mathrm{P}'=(t+\delta t,r,0,0)$, with $\delta t=1/\nu$, reaches $\mathrm{O}'=(t_0+\delta t_0,0,0,0)$. Because the light propagates on the geodesic, we obtain

$$\int_t^{t_0}\frac{dt}{a(t)}=\int_0^r\frac{dr}{\sqrt{1-Kr^2}}=\int_{t+\delta t}^{t_0+\delta t_0}\frac{dt}{a(t)}. \tag{8.9}$$

For light with large enough frequency (i.e. small enough δt),

$$\frac{\delta t}{a(t)}=\frac{\delta t_0}{a_0}, \tag{8.10}$$

or, equivalently,

$$\frac{a_0}{a(t)}=\frac{\nu}{\nu_0}=\frac{\lambda_0}{\lambda}\equiv 1+z, \tag{8.11}$$

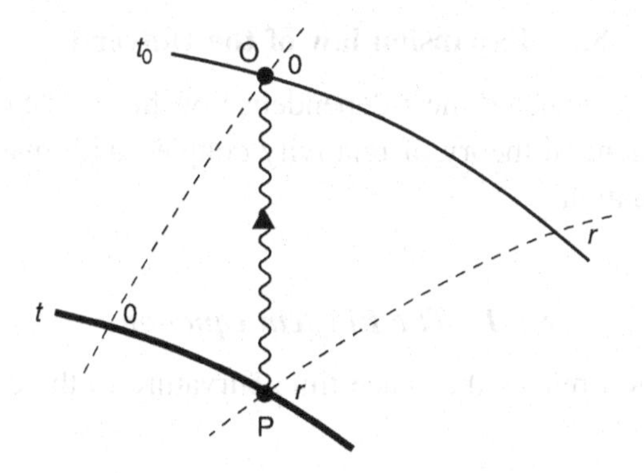

Fig. 8.6. Light traveling from point P to point O under the effect of the red shift due to the expansion of the Universe.

where λ (λ_0) is the wavelength of the emitted (received) light; z is the red shift of the light emitted in the past at time t ($< t_0$) and observed at t_0 (the Doppler effect).

If the Universe is expanding, $a_0 > a(t)$, and hence $z > 1$. After some algebra, corrections to the simple Hubble law, Eq. (8.1), can be derived for small z:

$$z - \frac{1}{2}(1+q_0)z^2 + \cdots = H_0 l_0, \tag{8.12}$$

where q_0 $(= -\ddot{a}_0/(a_0 H_0^2))$ is the deceleration parameter. If z is very small, $z \simeq v$ and Eq. (8.1) is recovered.

8.2.3 Horizon distance

The horizon distance is defined as a length scale which specifies the causally connected region. Suppose that light emitted at $t = 0$ with the co-moving coordinate, r_{H}, is observed at time t with a co-moving coordinate, $r = 0$. Then,

$$\int_0^t \frac{dt}{a(t)} = \int_0^{r_{\mathrm{H}}(t)} \frac{dr}{\sqrt{1-Kr^2}}. \tag{8.13}$$

Therefore, the horizon distance, $l_{\mathrm{H}}(t)$, is given by

$$\begin{aligned} l_{\mathrm{H}}(t) &\equiv a(t) \int_0^{r_{\mathrm{H}}(t)} \frac{dr}{\sqrt{1-Kr^2}} \\ &= a(t) \int_0^t \frac{dt}{a(t)} = a(t) \int_0^{a(t)} \frac{da}{a^2 H}. \end{aligned} \tag{8.14}$$

As we will see in Eq. (8.31), $a(t)$ is proportional to $\sqrt{t}$ in the radiation-dominant era of the early Universe. Then, we have $l_{\mathrm{H}}(t) \propto t$ in this era.

8.3 Expansion law of the Universe

So far, we have not calculated the t-dependence of the scale factor, $a(t)$. This is determined by the general theory of relativity coupled with matter and radiation, i.e. the Einstein equation.

8.3.1 *The Einstein equation*

The Einstein equation relates the space-time curvature to the energy-momentum tensor (Appendix D.2):

$$G_{\mu\nu} \equiv R_{\mu\nu} - \frac{1}{2} R g_{\mu\nu} = 8\pi G \, T_{\mu\nu}. \tag{8.15}$$

Here, $R_{\mu\nu}$ and R are the Ricci tensor and scalar curvature, respectively; G is the gravitational constant; $T_{\mu\nu}$ is the total energy-momentum tensor, which contains not only terms originating from the matter and the radiation, but also the term $\Lambda g_{\mu\nu}$, where Λ is the cosmological constant. (See Exercise 8.2.)

The energy-momentum tensor must satisfy the covariant conservation law (Exercise 8.3):

$$\nabla_\mu T^\mu_{\ \nu} = 0, \tag{8.16}$$

where ∇ implies the covariant derivative defined in Appendix D.1. Let us introduce a parametrization of $T^\mu{}_\nu$ consistent with the spatial homogenity and isotropy:

$$T^\mu{}_\nu = \mathrm{diag}(\varepsilon(t), -P(t), -P(t), -P(t)). \tag{8.17}$$

Because of the homogenity, ε and P are functions of t only. Note that ε (P) is naturally interpreted as a local energy density (the local pressure) at a given time.

The left-hand side of Eq. (8.15) can be rewritten by using the explicit form of the RW metric, Eq. (8.5), as shown in Appendix D.3. In particular, $G^0{}_0 = 8\pi G\varepsilon$ leads to

$$H^2 = \left(\frac{\dot{a}}{a}\right)^2 = \frac{8\pi G}{3}\varepsilon - \frac{K}{a^2}. \tag{8.18}$$

This is an equation for the expansion velocity, $\dot{a}$, and is called the *Friedmann equation*. Combining $G^0{}_0 = 8\pi G\varepsilon$ and $G^i{}_i = -8\pi G P$, an independent equation for the acceleration $\ddot{a}$ is obtained, as follows:

$$\frac{\ddot{a}}{a} = -\frac{4\pi G}{3}(\varepsilon + 3P). \tag{8.19}$$

Another combination yields an equation which does not contain $\ddot{a}$:

$$\frac{d(\varepsilon a^3)}{dt} + P\frac{da^3}{dt} = 0. \tag{8.20}$$

This is a *balance equation* between the total energy in a co-moving volume element and the work done, $dE = -P\,dV$. Equation (8.20) can also be obtained from $\nabla_\mu T^\mu{}_0 = 0$. Only two of the above three equations are independent. In the following, we will take the Friedmann equation, Eq. (8.18), and the balance equation, Eq. (8.20), as the independent equations. Note that the balance equation can be rewritten in a form useful for later purposes:

$$\frac{d\varepsilon}{da} = -\frac{3}{a}(\varepsilon + P). \tag{8.21}$$

8.3.2 Critical density

Dividing both sides of the Friedmann equation, Eq. (8.18), by H^2, we obtain

$$k \equiv \frac{K}{H^2 a^2} = \frac{\varepsilon}{\varepsilon_c} - 1 \equiv \Omega - 1, \tag{8.22}$$

where the critical density defined by $\varepsilon_c = 3H^2/8\pi G$ sets a boundary between the closed Universe and the open Universe:

$$\begin{aligned} \Omega > 1 &\Longleftrightarrow k > 0 \qquad \text{(closed space)}, \\ \Omega = 1 &\Longleftrightarrow k = 0 \qquad \text{(flat space)}, \\ \Omega < 1 &\Longleftrightarrow k < 0 \qquad \text{(open space)}. \end{aligned} \tag{8.23}$$

The critical density of the present Universe, ε_{c0}, is given by

$$\varepsilon_{c0} = \frac{3H_0^2}{8\pi G} = 1.88h^2 \times 10^{-29}\ \text{g cm}^{-3}, \tag{8.24}$$

which is a tiny number. Observations, in particular the WMAP results in Table 8.1, indicate that Ω_0 is close to unity, indicating a flat Universe. Since $|1-\Omega|$ grows in time (as we will see in Section 8.3.3), it implies that Ω must have been extremely fine tuned to be unity in the very early stages of the Universe. A possible resolution of this flatness problem, together with other cosmological problems, are given by the hypothesis of the inflational Universe (Kazanas, 1980; Guth, 1981; Sato, 1981a, b; Peebles, 1993).

Note that Ω_0 may be decomposed into three components as following:

$$\Omega_0 = \Omega_{\rm B} + \Omega_{\rm DM} + \Omega_\Lambda, \tag{8.25}$$

where $\Omega_{\rm B} \simeq 0.04$ is the baryonic contribution, $\Omega_{\rm DM} \sim 0.23$ is the contribution from non-baryonic dark matter and $\Omega_\Lambda \sim 0.73$ is the contribution from the dark energy (see Table 8.1.) The latter two factors are the major components of the present Universe; unraveling their origin is one of the biggest challenges in current cosmology and particle physics.

8.3.3 Solution of the Friedmann equation

Let us consider a simplified equation of states to see how $a(t)$ behaves as a function of time. We consider three cases, as follows.

(i) *Radiation-dominant* (RD), $P = \varepsilon/3$; such an equation of state is realized for relativistic non-interacting particles.
(ii) *Matter-dominant* (MD), $P = 0$; this is realized for heavy non-relativistic particles.
(iii) *Vacuum* (VAC), $\varepsilon = -P$; this corresponds to the case where the uniform vacuum condensates exist.

In each case, the balance equation, Eq. (8.21), can be easily solved to yield

$$\text{(i)} \quad \varepsilon \propto a^{-4} \quad \text{(RD)}, \tag{8.26}$$

$$\text{(ii)} \quad \varepsilon \propto a^{-3} \quad \text{(MD)}, \tag{8.27}$$

$$\text{(iii)} \quad \varepsilon = \text{constant} \quad \text{(VAC)}. \tag{8.28}$$

Then the Friedmann equation, Eq. (8.18), can be cast into the form of a "classical particle" moving in a "potential," $V(a)$, with a "total energy," $E = -K$:

$$\left(\frac{da}{dt}\right)^2 + V(a) = -K, \tag{8.29}$$

$$V(a) = -\frac{8\pi G}{3}\frac{C}{a^{\alpha}}, \tag{8.30}$$

where we have (i) $C = \varepsilon a^4 = \varepsilon_0 a_0^4$ with $\alpha = 2$; (ii) $C = \varepsilon a^3 = \varepsilon_0 a_0^3$ with $\alpha = 1$; (iii) $C = \varepsilon = \varepsilon_0 \equiv \Lambda$ with $\alpha = -2$. The motion of the classical particle is "closed" or "open" depending on $E < 0$ $(K = -1)$ or $E > 0$ $(K = +1)$.

As long as there exist matter and radiation in the Universe, their energy densities in the early epoch ($a \sim 0$) dominate the curvature term and the vacuum contribution in the Friedmann equation. However, the vacuum term (the dark energy), if it exists, eventually dominates in the late stage of the Universe. Shown in Fig. 8.7 is the motion of $a(t)$, starting from $a = 0$, for three cases, $K = 0, \pm 1$. The potential, $V(a)$, in the figure is a "realistic" one, which is dominated by the radiation and matter for small a and by the dark energy for large a.

Particles in the Universe became more and more relativistic as the energy density increased. Therefore, in the very early stage of the Universe, the RD era should have been realized. In this case, Eq. (8.29) is easily solved as follows:

$$a(t) = \left[\frac{32\pi G}{3}(\varepsilon a^4)\right]^{1/4} \sqrt{t} \quad \text{(RD)}, \tag{8.31}$$

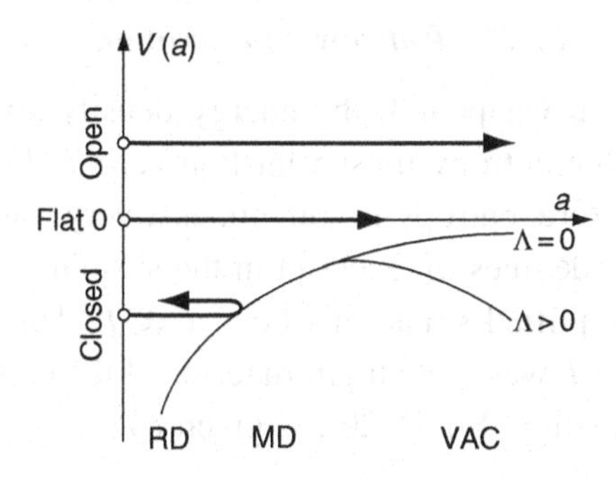

Fig. 8.7. Classical mechanical picture of the motion of $a(t)$.

where $\varepsilon a^4 =$ constant in the RD case. This equation also implies that

$$\varepsilon = \frac{3}{32\pi G}\,\frac{1}{t^2} \qquad \text{(RD)}. \tag{8.32}$$

As the Universe expanded and $a(t)$ increased, the RD era, with $a(t) \propto t^{1/2}$, turned into the MD era, with $a(t) \propto t^{2/3}$. Eventually, it will experience exponential growth, $a(t) \propto \exp(\sqrt{8\pi G\Lambda/3}\; t)$, in the VAC era for $\Lambda > 0$.

Suppose that the energy density in the RD era is described by the black body spectrum of massless particles with temperature T. Then we obtain (from results in Section 3.2)

$$\varepsilon = d_{\text{eff}} \frac{\pi^2}{30} T^4, \tag{8.33}$$

$$d_{\text{eff}} = \sum_i \left(d_{\text{B}}^i + \frac{7}{8} d_{\text{F}}^i \right), \tag{8.34}$$

where i labels the particle species (leptons, photons, quarks, gluons etc.). Then Eq. (8.32) reduces to

$$T = \left[\frac{45}{16\pi^3 G\, d_{\text{eff}}} \right]^{1/4} \frac{1}{\sqrt{t}} \qquad \text{(RD)}. \tag{8.35}$$

If the number of species is large, the Universe would have expanded quickly, and hence T decreased rapidly.

It should be noted that Eq. (8.34) is approximate, even in the RD case, in the sense that different species may have different temperatures depending on the time when a given species decouples from the system. In such a case, d_{eff} should be replaced by

$$d_{\text{eff}} = \sum_i \left[d_{\text{B}}^i \left(\frac{T_i}{T} \right)^4 + \frac{7}{8} d_{\text{F}}^i \left(\frac{T_i}{T} \right)^4 \right], \tag{8.36}$$

where T_i is the temperature for species, i, and T is the photon temperature.

8.3.4 Entropy conservation

As we have discussed in Chapter 3, the energy density and pressure can exist in a vacuum, whereas the entropy must vanish at zero T due to the third law of thermodynamics. Therefore, entropy is sometimes a very useful tool for counting the effective number of degrees of freedom in the system.

Let us consider a simplified situation where $\mu \ll T$. This is not unrealistic in the early Universe when T was quite high. In terms of the unit co-moving volume, $V = a^3$, the balance equation, Eq. (8.20), is given by

$$dE + PdV = 0, \tag{8.37}$$

where $E = \varepsilon V = \varepsilon a^3$. This, together with the thermodynamic relations $TS = E + PV$ (Eq. (3.5)) and $TdS = dE + PdV$ (Eq. (3.7)), immediately imply the conservation of the total entropy in the volume:

$$0 = dS = d(sa^3) \to s \propto \frac{1}{a^3}, \tag{8.38}$$

where s is the entropy density, $s = S/V$.

If the system contains independent "thermal" components with temperature T_i, not only is the total entropy conserved, $dS = d(\sum_i S_i) = 0$, but also each component is conserved, $dS_i = 0$. This is because the covariant conservation of the energy-momentum tensor in each component, $\nabla_\nu T_i^{\mu\nu} = 0$, gives each balance equation. The entropy density for the relativistic particles in this case is given by

$$s = d_{\text{eff}}^s \frac{2\pi^2}{45} T^3, \tag{8.39}$$

$$d_{\text{eff}}^s = \sum_i \left[d_{\text{B}}^i \left(\frac{T_i}{T} \right)^3 + \frac{7}{8} d_{\text{F}}^i \left(\frac{T_i}{T} \right)^3 \right]. \tag{8.40}$$

8.3.5 Age of the Universe

The RD era ended at about 10^{12} s ($\sim 3 \times 10^4$ year) after the Big Bang, a time which is negligible when compared with the present age of the Universe. Therefore, to estimate the age of the Universe, t_0, from the Einstein equation, we can safely neglect the contribution from the RD era. Let us start with the Friedmann equation, Eq. (8.18), divided by a_0:

$$\frac{d(a/a_0)}{dt} = \sqrt{\frac{8\pi G}{3} \varepsilon \left(\frac{a}{a_0} \right)^2 - \frac{K}{a_0^2}}. \tag{8.41}$$

This can be integrated to yield

$$t = H_0^{-1} \int_0^{(1+z)^{-1}} \frac{dx}{\sqrt{(1-\Omega_\Lambda)x^{-1} + \Omega_\Lambda x^2}}, \tag{8.42}$$

where we have simplified the expression by neglecting the contribution from the RD era and also by assuming that the Universe is quite flat at present, $\Omega_0 \simeq 1$. The first term inside the square root is from the matter contribution ($\Omega_B + \Omega_{DM}$) and the second term is from the vacuum contribution (Ω_Λ); these are defined in Eq. (8.25).

Then, t_0 is simply obtained by setting $z = 0$ in Eq. (8.42):

$$t_0 = \frac{2}{3} H_0^{-1} \Omega_\Lambda^{-1/2} \sinh^{-1} \sqrt{\frac{\Omega_\Lambda}{1-\Omega_\Lambda}}. \tag{8.43}$$

This is an increasing function of Ω_Λ: for $\Omega_\Lambda = 0$ and 1, we have $t_0 = \frac{2}{3} H_0^{-1}$ and ∞, respectively. If we adopt $\Omega_\Lambda \simeq 0.73$ and $h \simeq 0.71$ from Table 8.1, the age of the Universe is estimated to be

$$t_0 \simeq 13.7 \times 10^9 \text{ yr}. \tag{8.44}$$

Shown schematically in Fig. 8.8 is the time-dependence of the scale factor, $a(t)$, for several different combinations of K and Λ. The WMAP results in Table 8.1 suggest that the expansion of our Universe corresponds to $K = 0$ and $\Lambda > 0$, the dash-dotted line in Fig. 8.8.

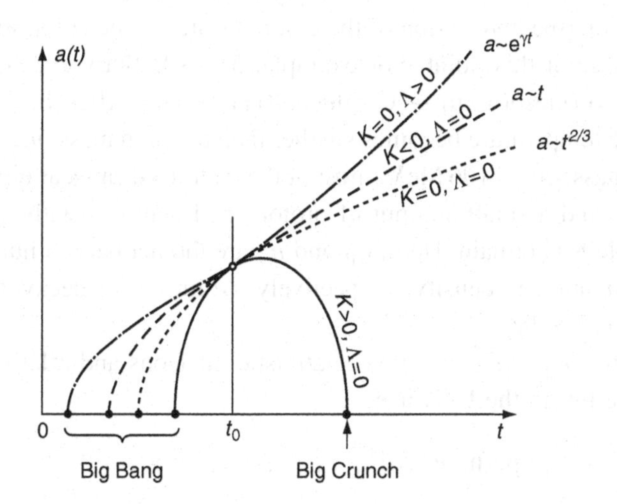

Fig. 8.8. Time-dependence of the scale factor, $a(t)$, with four different sets of the curvature parameter, K, and the cosmological constant, Λ, where $\gamma = \sqrt{8\pi G\Lambda/3}$. (See Section 8.3.3.) The gradient of each curve, $\dot{a}_0/a_0$, at the present time ($t_0 \simeq 13.7 \times 10^9$ yr) should be $H_0 \simeq 0.71 \times 100\,\text{km s}^{-1}\,\text{Mps}^{-1}$.

8.4 Thermal history of the Universe: from QGP to CMB

Let us briefly summarize the history of the Universe from the early epoch to the present time by decomposing it into several stages (see Fig. 1.7).

Early epoch: $T > m_c \simeq 1\,\text{GeV}$

Up until the point at which the temperature of the Universe cools to 1 GeV, several phase transitions in grand unified theories and in electroweak theory have already taken place. At the time when $T \sim m_c$, heavy particles, such as weak bosons ($W^{\pm}$, Z^0), the Higgs boson, the heavy quarks (t, b and c) and the heavy lepton (τ), have already decayed into lighter quarks, leptons and gauge bosons.

Quark–gluon plasma: $m_c > T > T_{QCD} \simeq 170\,\text{MeV}$

This is the region of initial temperature which the relativistic heavy ion collision experiments at RHIC and LHC may reach. Major thermal components of the Universe are given by

$$\text{quarks} : \text{u, d, s}, \bar{\text{u}}, \bar{\text{d}}, \bar{\text{s}};$$
$$\text{leptons} : \text{e}^-, \mu^-, \text{e}^+, \mu^+, \nu_{\text{e},\mu,\tau}, \bar{\nu}_{\text{e},\mu,\tau};$$
$$\text{gauge bosons} : \text{g (gluon)}, \gamma \text{ (photon)}.$$

QCD phase transition: $T \sim T_{QCD}$ $(t \simeq 10^{-5}\text{–}10^{-4}\,\text{s})$

The hadronization (the conversion of the colored states to the color-singlet hadron states) takes place at this point. For example, $3q \to \text{B}$ (baryon), $3\bar{q} \to \bar{\text{B}}$ (anti-baryon), $q\bar{q} \to \text{M}$ (meson), $\text{gg} \to \text{G}$ (glueball) and so on. After this conversion is finished and the temperature becomes smaller than the pion mass, $m_\pi \simeq 140\,\text{MeV}$, and the muon mass, $m_\mu = 106\,\text{MeV}$, most of the hadrons disappear either by decay or annihilation, and a small amount of protons and neutrons with $\eta = n_B/n_\gamma \simeq 6 \times 10^{-10}$ (Table 8.1) remain. Here, n_B and n_γ are the net baryon number density and the photon number density, respectively. Muons also decay into photons through $\mu^- + \mu^+ \to 2\gamma$.

Eventually, for $m_\mu > T > m_e$, non-relativistic nucleons and relativistic leptons and photons are left in the Universe:

$$\text{p, n, e}^-, \text{e}^+, \nu_{\text{e},\mu,\tau}, \bar{\nu}_{\text{e},\mu,\tau}, \gamma. \tag{8.45}$$

Neutrino decoupling: $T \sim 1\,\text{MeV}$ $(t \sim 0.1\text{–}1\,\text{s})$

As the Universe further expands and cools, the Hubble expansion rate exceeds the reaction rate of neutrinos through weak interactions. Then the neutrinos decouple from other components. This decoupling temperature can be estimated as follows. First, the typical neutrino reaction rate is given by

$$\Gamma_\nu = \langle n\sigma_\nu v\rangle \sim T^3 G_F^2 T^2 = G_F^2 T^5. \tag{8.46}$$

Here, v ($\simeq c$) is the relative velocity of the neutrino and other particles and $n \propto T^3$ is the number density of particles in the plasma to interact with the neutrino. The factor $\sigma_\nu \sim G_F^2 T^2$ is a typical cross-section of the neutrino in a thermal environment, with $G_F = (292.80\,\text{GeV})^{-2}$ being the Fermi constant. (The T-dependence of the cross-section can be obtained from simple dimensional analysis: $\sigma_\nu \sim G_F^2 \times (\text{energy})^2 \sim G_F^2 \times T^2$.) By taking the ratio of Γ_ν with the Hubble constant, H, in the RD era ($H = \sqrt{(8\pi/3)G\varepsilon} \propto \sqrt{GT^4}$), we obtain

$$\frac{\Gamma_\nu}{H} \sim \frac{G_F^2 T^5}{\sqrt{GT^4}} \sim \left(\frac{T}{1\,\text{MeV}}\right)^3, \tag{8.47}$$

where Newton's constant, $G = (1.2211 \times 10^{19}\,\text{GeV})^{-2}$ (Appendix A.2), is used. Thus, the neutrino decouples from other particles for T smaller than approximately 1 MeV.

Once the neutrino decoupling occurs, the original thermal distribution is frozen and the number of neutrinos in a unit co-moving volume is conserved. Let us consider a unit co-moving volume, a^3, and a volume in momentum-space, d^3p. Since the particle number after decoupling is conserved, we obtain

$$a^3 d^3p\, n_\nu(p, T) = (a')^3 d^3p'\, n_\nu(p', T'), \tag{8.48}$$

where $T < T' < 1$ MeV. The wave number is red-shifted according to the expansion, so that $p = (a'/a)p'$. Noting that the distribution function for a massless particle is a function of the ratio p/T, we thus obtain

$$n_\nu(p, T) = n_\nu(p', T') = n_\nu\left(p, T'\frac{a'}{a}\right), \tag{8.49}$$

which leads to

$$T = T_{\text{dec}}\frac{a_{\text{dec}}}{a} \propto \frac{1}{a}. \tag{8.50}$$

Although the distribution function has a thermal form even below the neutrino decoupling, it does not mean that the particles interact with each other.

For later purposes, we define T_ν as the neutrino "temperature" after the decoupling. Then, Eq. (8.50) and the conservation of entropy imply that

$$T_\nu \propto \frac{1}{a}, \quad s_\nu \propto n_\nu \propto \frac{1}{a^3}. \tag{8.51}$$

Photon reheating: $T < m_e = 0.51\,\mathrm{MeV}$ $(t \sim 10\,\mathrm{s})$

Shortly after the neutrino decoupling, the temperature drops below the electron mass and electrons and positrons (e^- and e^+) start to disappear via $e^- + e^+ \to 2\gamma$. Simultaneously, the number of photons increases, and the photon temperature (hereafter defined as T) starts to deviate from T_ν.

Let us evaluate the ratio T/T_ν at the end of this reheating. Before the e^+e^- annihilation, but after the neutrino decoupling, the entropy density for e^-, e^+ and γ is given by (see Section 3.2)

$$s_{\gamma+e^\pm} = \left(2 + \frac{7}{8} \times 2 \times 2\right) \frac{2\pi^2}{45} T^3 = \frac{11}{2} \cdot \frac{2\pi^2}{45} T^3. \tag{8.52}$$

On the other hand, after the e^+e^- annihilation, we have

$$s_\gamma = 2 \cdot \frac{2\pi^2}{45} T^3. \tag{8.53}$$

Due to separate entropy conservations in the electromagnetic and neutrino sectors, we obtain

$$\begin{aligned} (aT)_{\text{after}} &= \left(\frac{11}{4}\right)^{1/3} (aT)_{\text{before}}, \\ (aT_\nu)_{\text{after}} &= (aT_\nu)_{\text{before}} = (aT)_{\text{before}}, \end{aligned} \tag{8.54}$$

where we have used the fact that $T = T_\nu$ before the reheating. Thus we obtain

$$\left(\frac{T}{T_\nu}\right)_{\text{after}} = \left(\frac{11}{4}\right)^{1/3} \simeq 1.401. \tag{8.55}$$

At which temperature is this ratio indeed realized? To answer this question, we need to include the electron mass, m_e, to evaluate the entropy density in Eq. (8.52). Then we find that the above ratio is satisfied only when the temperature is smaller than about $m_e/10$ (Exercise 8.4).

Recombination: $T = T_{\text{rec}} \sim 4000\,\mathrm{K}$

After the e^+e^- annihilation has been completed, there still remains a tiny fraction of electrons, so that the net electric charge of the Universe is zero: $n_{e^-}/n_\gamma = n_p/n_\gamma \sim 10^{-9}$–$10^{-10}$. However, once the temperature falls below 4000 K, recombination of the electrons and the protons occurs quickly to form hydrogen ($e^- + p \to H + \gamma$), and the local charge neutralization of the Universe takes place. Note that roughly 25% of the baryonic matter is in the form of α particles, which are composed of two protons and two neutrons. The neutralization of α into ^{4}He atoms takes place around 8000 K because of the relatively high ionization energy of ^{4}He.

Photon decoupling: $T = T_{\rm dec} \sim 2700$ K

As the temperature falls further, the collision rate of photons with atoms through Thomson scattering becomes smaller than the Hubble expansion rate. Then the photon decouples from the matter, as previously occurred for neutrinos. After this decoupling, the Universe becomes optically transparent.

Cosmic background radiation: $T \ll T_{\rm dec}$

At the present time, the Universe is filled with black body photons and neutrinos, which are the remnant of the hot era in the early Universe. As we have already discussed in Section 8.1, accurate measurements of the CMB and its fluctuation became available thanks to COBE and WMAP. Although the cosmic neutrino background (CνB) radiation has not yet been observed, we expect the following temperature from the discussion given above:

$$T_{\nu 0} = \left(\frac{4}{11}\right)^{1/3} T_{\rm CMB} \simeq 1.95 \text{ K}. \tag{8.56}$$

Correspondingly, the number densities of CMB and CνB are

$$n_{\rm CMB} = 2\frac{\zeta(3)}{\pi^2} T_{\rm CMB}^3 \simeq 410 \text{ cm}^{-3}, \tag{8.57}$$

$$n_{{\rm C}\nu{\rm B}} = 6 \cdot \frac{1}{4}\frac{\zeta(3)}{\pi^2} T_{\nu 0}^3 = \frac{3}{11} n_{\rm CMB}, \tag{8.58}$$

where we have used the integral formula in Exercise 3.4. Note that the six degrees of freedom for CνB originate from the three generations of left-handed neutrinos and their anti-particles.

The energy density of CMB+CνB at the present time is given by

$$\begin{aligned} \varepsilon_{\gamma+\nu+\bar{\nu},0} &= \left[2 + \frac{7}{8} \times 2 \times 3 \times \left(\frac{4}{11}\right)^{4/3}\right] \frac{\pi^2}{30} T_{\rm CMB}^4 \\ &\simeq 7.80 \times 10^{-34} \text{ g cm}^{-3}, \end{aligned} \tag{8.59}$$

which is more than four orders of magnitude smaller than ε_{c0} in Eq. (8.24).

8.5 Primordial nucleosynthesis

In the epoch between the QCD phase transition ($T_{\rm QCD} \sim 170$ MeV) and the neutrino decoupling ($\sim$1 MeV), protons and neutrons are equilibrated by the weak interaction. Their relative abundance is dictated by the neutron–proton mass difference, $Q = m_n - m_p = 1.3\,\text{MeV}$:

$$\left(\frac{n_n}{n_p}\right)_{eq} = e^{-Q/T}, \tag{8.60}$$

where the chemical potentials of the electrons and the neutrinos are assumed to be negligible. At $T = T_n \simeq 0.8\,\text{MeV}$, the weak interaction rate, such as $p + e^- \leftrightarrow \nu_e + n$, becomes smaller than the Hubble expansion rate. Then the neutron-to-proton ratio is fixed and the neutron freeze-out takes place. After this freeze-out, neutrons decay ($n \to p + e^- + \bar{\nu}_e$) slowly with a lifetime of about 900 s. When T reaches $T_d \simeq 0.07\,\text{MeV}$, the formation of d (deuteron), ^{3}H, ^{3}He, ^{4}He, ^{7}Li and ^{7}Be takes place according to the following reaction chains (Fig. 8.9):

$$\begin{aligned}
&p + n \to d + \gamma, \\
&d + p \to {}^3\text{He} + \gamma, \quad d + d \to {}^3\text{H} + p, \quad d + d \to {}^3\text{He} + n, \\
&{}^3\text{H} + d \to {}^4\text{He} + n, \quad {}^3\text{H} + {}^4\text{He} \to {}^7\text{Li} + \gamma, \\
&{}^3\text{He} + n \to {}^3\text{H} + p, \quad {}^3\text{He} + d \to {}^4\text{He} + p, \quad {}^3\text{He} + {}^4\text{He} \to {}^7\text{Be} + \gamma, \\
&{}^7\text{Li} + p \to {}^4\text{He} + {}^4\text{He}, \quad {}^7\text{Be} + n \to {}^7\text{Li} + p.
\end{aligned} \tag{8.61}$$

Since most of the neutrons are eventually used to form ^{4}He, it is relatively easy to estimate the primordial abundance of this element at the time of the neutron freeze-out $(n_n/n_p)_{T=T_n} \simeq 1/6$. Until nucleosynthesis begins, this ratio is decreased a little by the neutron decay from 1/6 to 1/7. Then, assuming all the neutrons are exhausted to form ^{4}He, the mass fraction of the primordial ^{4}He is given by

$$Y_p \simeq \frac{4 \cdot n_n/2}{n_n + n_p} = \frac{2(n_n/n_p)}{1 + (n_n/n_p)} = 0.25, \tag{8.62}$$

which is consistent with the observed number Eq. (8.4). Detailed numerical calculations of the primordial nucleosynthesis show that one can predict the abundance of the light elements reasonably accurately as a function of the baryon density,

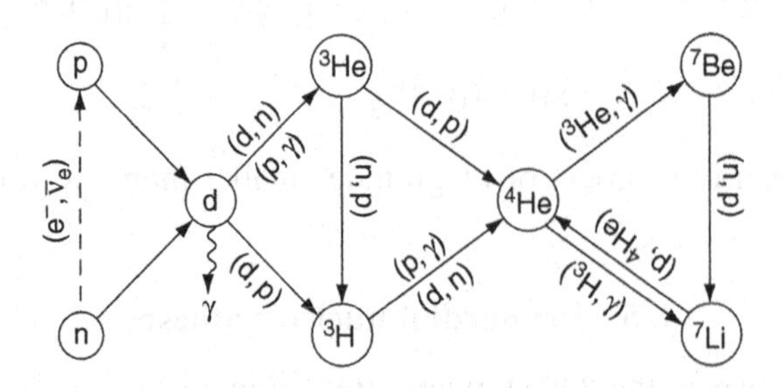

Fig. 8.9. Major nuclear reactions in the Big Bang nucleosynthesis (BBN).

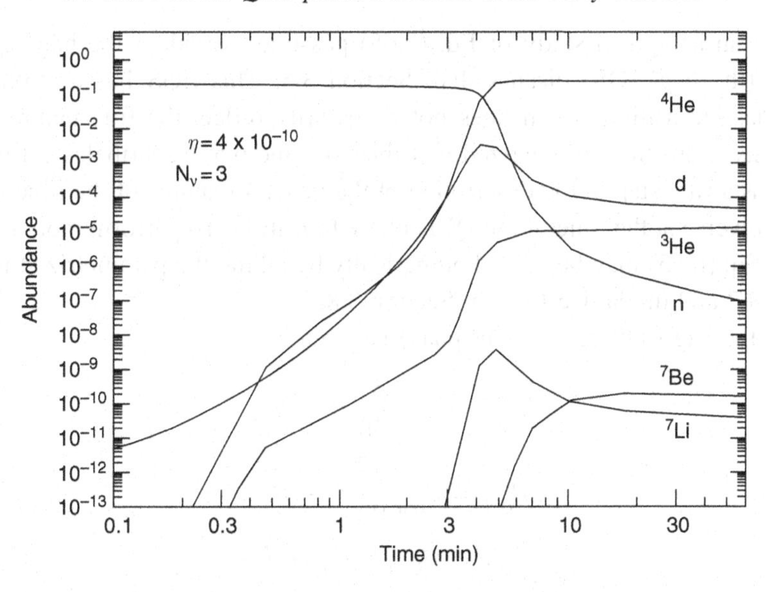

Fig. 8.10. The mass fraction of ^{4}He and the number abundance (relative to H) of the other light elements during the first hour after the Big Bang. Note that $\eta = n_B/n_\gamma$ and N_ν denote the baryon-to-photon ratio and the number of light neutrino species, respectively. Adapted from Schramm and Turner (1998).

n_B, or the baryon-to-photon ratio, $\eta \equiv n_B/n_\gamma$. The time history of the primordial nucleosynthesis in the first hour of the Universe is shown in Fig. 8.10. Most of the light elements are produced during the first three minutes (Weinberg, 1977).

As shown in Fig. 8.4, by comparing the resultant abundance and the observation of primordial light elements, we obtain

$$\rho_B = (3-5) \times 10^{-31} \text{ g cm}^{-3}, \tag{8.63}$$

$$\Omega_{B0} h^2 \simeq 0.02. \tag{8.64}$$

This indicates that in order to guarantee the flatness of the present day Universe, $(\Omega_0 \sim 1)$, the baryon cannot be a major component.

8.6 More on the QCD phase transition in the early Universe

Let us now return to the epoch of the QCD phase transition and examine the effect of phase transition on the time history of the Universe. Since the scale factor, a, is small, the Universe is radiation-dominant, and the curvature term proportional to K can be neglected. Then the balance equation, Eq. (8.21), and the Friedmann equation, Eq. (8.18) yield

$$\frac{\dot{a}}{a} = -\frac{\dot{\varepsilon}}{3(\varepsilon + P)} = \sqrt{\frac{8\pi G}{3}\varepsilon}. \tag{8.65}$$

To make an analytical study of Eq. (8.65) possible, we adopt the bag equation of state (the bag EOS) discussed in Section 3.5. This EOS has a strong first order phase transition which does not necessarily reflect the true nature of the QCD phase transition summarized in Tables 6.4 and 6.5. Nevertheless, it is good enough as a first step in the description of the general features of the time history of the Universe at the time of the QCD phase transition. To perform more realistic studies, Eq. (8.65) may be solved numerically by taking the parametrized EOS in Section 3.7 and the lattice EOS in Section 3.8.

Now, the bag EOS in the QGP phase is given by

$$\varepsilon_{\rm QGP} = d_{\rm QGP}\frac{\pi^2}{30}T^4 + B, \tag{8.66}$$

$$P_{\rm QGP} = d_{\rm QGP}\frac{\pi^2}{90}T^4 - B, \tag{8.67}$$

$$s_{\rm QGP} = (\varepsilon_{\rm QGP} + P_{\rm QGP})/T. \tag{8.68}$$

The effective number of degrees of freedom is given by

$$d_{\rm QGP} = 16 + \frac{21}{2}N_f + 14.25 = 51.25 \quad (N_f = 2), \tag{8.69}$$

in which 14.25 is a contribution from the photons and leptons ($e^{\pm}$, $\mu^{\pm}$, $\nu_{e,\mu,\tau}$, $\bar{\nu}_{e,\mu,\tau}$). We take $N_f = 2$ from now on for simplicity.

The bag EOS in the hadronic phase with $N_f = 2$ is composed of massless pions together with photons and leptons:

$$\varepsilon_{\rm H} = d_{\rm QGP}\frac{\pi^2}{30}T^4 = 3P_{\rm H} = \frac{3}{4}s_{\rm H}T, \tag{8.70}$$

where

$$d_{\rm H} = 3 + 14.25 = 17.25. \tag{8.71}$$

As we have discussed in Section 3.5, phase equilibrium is achieved when $P_{\rm QGP} = P_{\rm H}$ at $T = T_c$, which gives

$$B = (d_{\rm QGP} - d_{\rm H})\frac{\pi^2}{90}T_c^4 = L/4, \tag{8.72}$$

$$r \equiv \frac{s_{\rm QGP}(T_c)}{s_{\rm H}(T_c)} = \frac{d_{\rm QGP}}{d_{\rm H}} = \frac{51.25}{17.25} \simeq 3. \tag{8.73}$$

As the temperature decreases and reaches T_c, the Universe experiences the coexistence of the QGP phase and the hadronic phases for a certain period of time for the EOS with the first order phase transition. During this period, the temperature of the system is fixed at T_c because the liberation of the latent heat, L, compensates for the cooling of the system due to the expansion. The energy density of this mixed phase may be parametrized by the volume fraction factor, $f(t)$, as follows:

$$\varepsilon(t) = \varepsilon_{\rm H}(T_{\rm c})f(t) + \varepsilon_{\rm QGP}(T_{\rm c})(1 - f(t)), \tag{8.74}$$

where $f(t)$ takes the values 0 (1) at the start (end) of the co-existence. Note that the time scale of the expansion of the Universe at the epoch of the QCD phase transition is 10^{-5}–10^{-4} s, which is many orders of magnitude larger than the time scale of the strong interaction (10^{-23}–10^{-22} s). Therefore, the expansion of the Universe is an adiabatic process, and we can safely neglect, for example, the supercooling at the time of the phase transition. (For more discussions on this point, see, for example, Kajantie and Kurki-Suonio (1986) and Kapusta (2001).) This feature is already assumed in Eq. (8.74), where we use a fixed temperature, $T_{\rm c}$, for $\varepsilon_{\rm H,QGP}$ on the right-hand side. Even if the phase transition is not first order, but is second order or a smooth crossover, the analysis based on Eq. (8.74) still provides us with a good idea of the time history of the temperature, $T(t)$, and the scale factor, $a(t)$, as long as the entropy density experiences rapid change.

For later purposes, we define the typical time scale, λ, of the QCD phase transition as follows:

$$\lambda = \left(\frac{8\pi GB}{3}\right)^{-1/2} = (78\ \mu\text{s})\left(\frac{170\ \text{MeV}}{T_{\rm c}}\right)^2, \tag{8.75}$$

where $1\ \mu\text{s} = 10^{-6}$ s and we adopt 170 MeV as a typical value of $T_{\rm c}$.

8.6.1 $t < t_{\rm I}$ $(T > T_{\rm c})$

With the bag EOS, Eqs. (8.67) and (8.68), the second equality in Eq. (8.65) yields a differential equation for $\varepsilon(\equiv \varepsilon_{\rm QGP})$:

$$\frac{-d\varepsilon}{4\sqrt{\varepsilon}(\varepsilon - B)} = \sqrt{\frac{8\pi G}{3}}\, dt. \tag{8.76}$$

By introducing a variable $X = \sqrt{(r-1)/3r}(T_{\rm c}/T)^2$, this can be cast into a simple form, $dX/dt = (2/\lambda)\sqrt{1+X^2}$. Then we obtain

$$\left(\frac{T(t)}{T_{\rm c}}\right)^2 = \sqrt{\frac{r-1}{3r}}\frac{1}{\sinh\,(2t/\lambda)}. \tag{8.77}$$

On the other hand, the first equality in Eq. (8.65) yields

$$\frac{a(t)}{a_{\rm I}} = \frac{T_{\rm c}}{T(t)}, \tag{8.78}$$

where $a_{\rm I}$ is simply defined as the scale factor when the phase transition starts to take place: $a_{\rm I} = a(t = t_{\rm I})$ and $T(t = t_{\rm I}) = T_{\rm c}$.

Equation (8.77) shows that the temperature decreases as $1/\sqrt{t}$ at early times and then turns into the form $\exp(-2t/\lambda)$ for large t due to the existence of B. The onset time of the phase transition, t_I, is obtained from Eq. (8.77) as follows:

$$t_I = \frac{\lambda}{2} \ln \left(\sqrt{\frac{r-1}{3r}} + \sqrt{1 + \frac{r-1}{3r}} \right) \tag{8.79}$$

$$\simeq 18 \ \mu\text{s} \quad (\text{for } T_c = 170 \text{ MeV}). \tag{8.80}$$

8.6.2 $t_I < t < t_F$ $(T = T_c)$

During the mixed phase, Eq. (8.65) and Eq. (8.74) lead to

$$\frac{\dot{a}}{a} = \frac{\dot{f}}{3(\frac{r}{r-1} - f)} = \lambda^{-1} \sqrt{4(1-f) + \frac{3}{r-1}}. \tag{8.81}$$

This is solved analytically as follows (Exercise 8.5):

$$f(t) = 1 - \frac{1}{4(r-1)} \left[\tan^2 \left(\frac{3}{2\sqrt{r-1}} \frac{t - t_I}{\lambda} - \tan^{-1} \sqrt{4r-1} \right) - 3 \right], \tag{8.82}$$

$$\frac{a(t)}{a(t_I)} = (4r)^{1/3} \left[\sin \left(\frac{3}{2\sqrt{r-1}} \frac{t - t_I}{\lambda} + \sin^{-1} \frac{1}{\sqrt{4r}} \right) \right]^{2/3}. \tag{8.83}$$

The mixed phase ends at $f(t_F) = 1$, which yields

$$t_F - t_I = +\lambda \frac{2\sqrt{r-1}}{3} \left[\tan^{-1} \sqrt{4r-1} - \tan^{-1} \sqrt{3} \right] \tag{8.84}$$

$$\simeq 17 \ \mu\text{s} \quad (\text{for } T_c = 170 \text{ MeV}). \tag{8.85}$$

The scale factor at the end of the transition is given by

$$a(t_F) = r^{1/3} a(t_I) = 1.44 \ a(t_I), \tag{8.86}$$

which is obtained either from the solution given in Eq. (8.83) or from the conservation of the total entropy, $s(t_I) a^3(t_I) = s(t_F) a^3(t_F)$.

8.6.3 $t > t_F$ $(T < T_c)$

After the phase transition, the temperature of the Universe starts to decrease again, following the standard law in the radiation-dominant era. With the EOS for the hadronic phase in Eq. (8.70), the second equality of Eq. (8.65) for ε $(\equiv \varepsilon_H)$ becomes

$$\frac{-d\varepsilon}{4\sqrt{\varepsilon}\varepsilon} = \sqrt{\frac{8\pi G}{3}} \, dt. \tag{8.87}$$

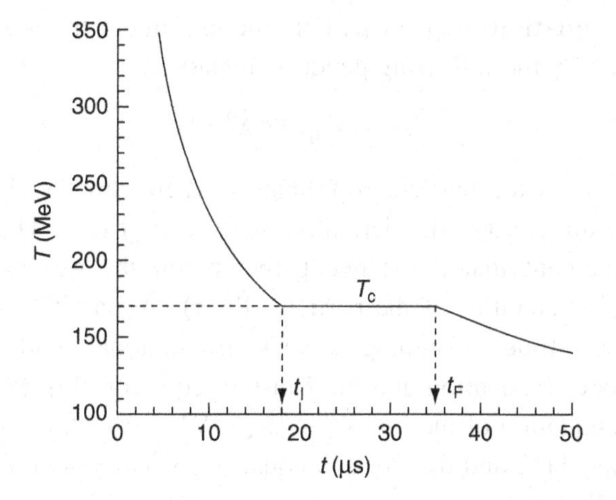

Fig. 8.11. Temperature, T, of the Universe as a function of the age, t, around the epoch of the QCD phase transition. Typical scales used are $T_c = 170\,\text{MeV}$ and $\lambda = 78\ \mu\text{s}$. Compare this figure with that in the relativistic heavy ion collisions, Fig. 13.7.

This may be rewritten as $dY/dt = \lambda^{-1}$ by introducing a change of variable, $Y = \frac{(r-1)}{12}(T_c/T)^2$, which gives

$$\left(\frac{T(t)}{T_c}\right)^2 = \frac{1}{1+\sqrt{\frac{12}{r-1}\frac{t-t_F}{\lambda}}}. \tag{8.88}$$

On the other hand, the first equality in Eq. (8.65) yields $a(t)/a_F = T_c/T(t)$, as before.

Shown in Fig. 8.11 is the time-dependence of the temperature of the Universe with the first order QCD phase transition discussed in this section. Extension to a more realistic case with smooth crossover between the hadronic phase and the quark–gluon plasma can be easily achieved by taking the parametrized EOS discussed in Section 3.7. As long as the energy density and the entropy density have increase rapidly around the (pseudo) critical point, the qualitative behavior of $T(t)$ is unchanged from that given in Fig. 8.11. (See Exercise 8.6.)

Exercises

8.1 **Three-dimensional curvature.** Derive the relation between the three-dimensional scalar curvature, ${}^3R$, and the scale parameter, a, given in Eq. (8.6). Remember that the four-dimensional scalar curvature, R, in Eq. (D.26) reduces to $-{}^3R$ if we neglect the t-dependence of $a(t)$. For more details, see Sec. 111 in Landau and Lifshitz (1988).

8.2 Geodesic equation. A point particle moving under the gravitational field is governed by the following geodesic equation:

$$\ddot{X}^\lambda + \Gamma^\lambda_{\mu\nu} \dot{X}^\mu \dot{X}^\nu = 0,$$

where $X^\lambda(\tau)$ is the particle coordinate as a function of the proper time, τ, and the dot denotes the derivative with respect to τ. This is equivalent to the statement that any tangent vectors on the geodesic line, which are the four-velocities of the particle, $\dot{X}^\lambda(\tau)$ are parallel to each other in curved space-time. Assuming a weak gravitational field, show that the above geodesic equation and the Einstein equation, Eq. (8.15), reduce to Newton's equation of motion $\ddot{X}^\lambda + \partial\phi_\mathrm{N}/\partial X^\lambda = 0$ with the scalar potential $\phi_\mathrm{N} \equiv (g_{00} - 1)/2$ and the Poisson equation, $\nabla^2\phi_\mathrm{N} = 4\pi G\varepsilon$, respectively.

8.3 Bianchi identity.

(1) Derive the identity, Eq. (D.11), and the Bianchi identity, Eq. (D.12), in Appendix D starting from the Jacobi identity acting on an arbitrary vector, A^α.

(2) Prove the symmetry property of the Ricci tensor, $R_{\mu\nu} = R_{\nu\mu}$, in Eq. (D.14).

(3) Using the Bianchi identity and making appropriate contractions, prove that $\nabla_\mu G^\mu{}_\nu = 0$, where $G_{\mu\nu}$ is defined by Eq. (8.15).

8.4 Photon reheating. Evaluate the entropy density, $s_{\gamma+\mathrm{e}^\pm}$, in Eq. (8.52) before the photon reheating by taking into account the electron mass, m_e. Assuming entropy conservation, relate the photon temperature, T, to the neutrino temperature, T_ν. Plot T/T_ν as a function of m_e/T and ascertain when the ratio Eq. (8.55) is approximately satisfied.

8.5 Mixed phase and QGP volume fraction. Solve Eq. (8.81) in the mixed phase and derive the time-dependence of the volume fraction, $f(t)$, given in Eq. (8.82) and the scale factor, $a(t)$, given in Eq. (8.83).

8.6 Realistic EOS and expansion of the Universe. Generalize the parametrized EOS, Eqs. (3.57) and (3.58) with Eqs. (3.60) and (3.61) by taking into account the contributions from leptons and photons. By using the EOS, solve the balance equation + Friedmann equation, Eq. (8.65), numerically. Plot T as a function of t, and compare the result with that of Fig. 8.11.

9

Compact stars

In this chapter, we present an elementary introduction to the physics of compact stars and its connection to the quark–hadron phase transition at high baryon density.

The existence of a giant nucleus of macroscopic size composed mainly of neutrons, the neutron star hypothesis, was proposed shortly after the discovery of the neutron by Chadwick in 1932. In particular, in 1934 Baade and Zwicky argued that such stars may be produced after a supernova explosion. Oppenheimer and Volkoff (1939) were the first to calculate the structure of neutron stars using Einstein's theory of relativity. More than three decades after this proposal was made, the first radio pulsar was discovered in 1967 by Bell and Hewish (Hewish *et al.*, 1968); this has now been identified as a rapidly rotating neutron star. The first binary system of rotating neutron stars (the binary pulsar) was discovered by Hulse and Taylor (1975), which provides us with a stringent test for the general theory of relativity. Furthermore, neutrinos from the supernova 1987A were detected by the KAMIOKANDE-II Collaboration led by Koshiba; this opens the door to the study of neutrino astronomy and the observational studies of the formation of compact stars (Hirata *et al.*, 1987; Koshiba, 1992).

Soon after the discovery of the asymptotic freedom of QCD (Gross and Wilczek, 1973; Politzer, 1973; 't Hooft, 1972, 1985), the possibility of the existence of a transition to quark matter in the core of high density neutron stars was realized (Collins and Perry, 1975; Baym and Chin, 1976). The existence of an exotic strange quark star, which is a star entirely made of deconfined u, d and s quarks, was also proposed (Witten, 1984) as a modern version of the early idea of the quark star (Itoh, 1970) and of stable strange matter (Bodmer, 1971). So far, there has been no convincing evidence of the existence of quark stars, but Nature may yet be strange enough to accommodate such compact objects in our Universe and we should prepare ourselves for their future discovery.

9.1 Characteristic features of neutron stars

A schematic view of various forms of a compact star, from a neutron star to a quark star, is presented in Fig. 9.1.

The typical radius of a neutron star is about 10 km, and its mass is comparable to the mass of our Sun, $M_\odot$ ($\simeq 2\times 10^{30}$ kg). Since neutrons cannot be bound by the strong interaction, the presence of the gravitational force is essential to hold the star in place. This implies that R increases as M decreases. From measurements of radio pulsars in binary systems, the masses of neutron stars are likely to be in the range $1\sim 2M_\odot$, as shown in Fig. 9.2. The surface temperature of neutron stars is likely to be less than 10^9 K one year after their birth. In the early stage, cooling occurs through neutrino emissions, while in the later stages it is dominated by surface photon emissions (Tsuruta, 1998). Some exotic cooling mechanisms may also be present, such as the effects of the pion and kaon condensations and of the quark matter in the core. See Fig. 9.3 for observations and theoretical predictions of the surface photon luminosity as a function of age of a neutron star.

For pulsars (rotating neutron stars), the measured rotational frequency is in the range of milliseconds to several seconds. The surface magnetic field is typically 10^{12} gauss for ordinary pulsars with rotational period $P\sim 1$ s and $dP/dt \sim 10^{-15}$. There are also stars with much larger (smaller) dP/dt and larger (smaller)

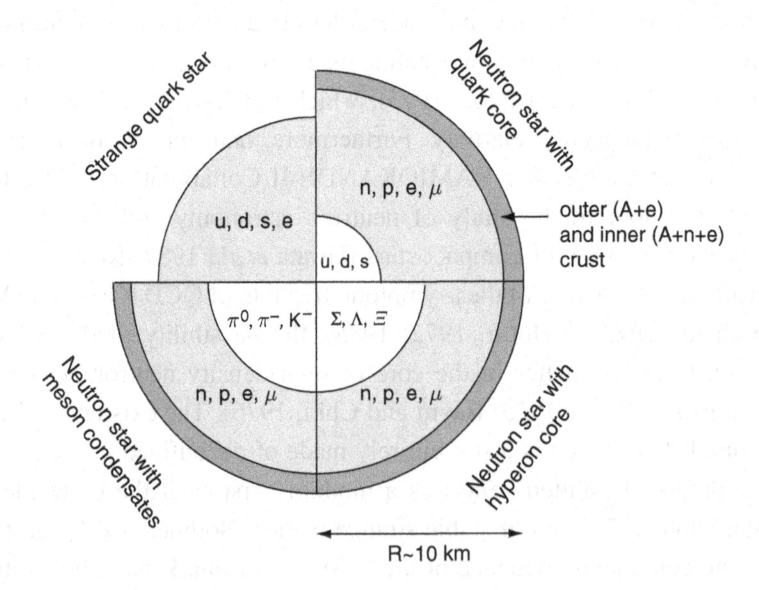

Fig. 9.1. Possible internal structures and compositions of four different types of compact stars.

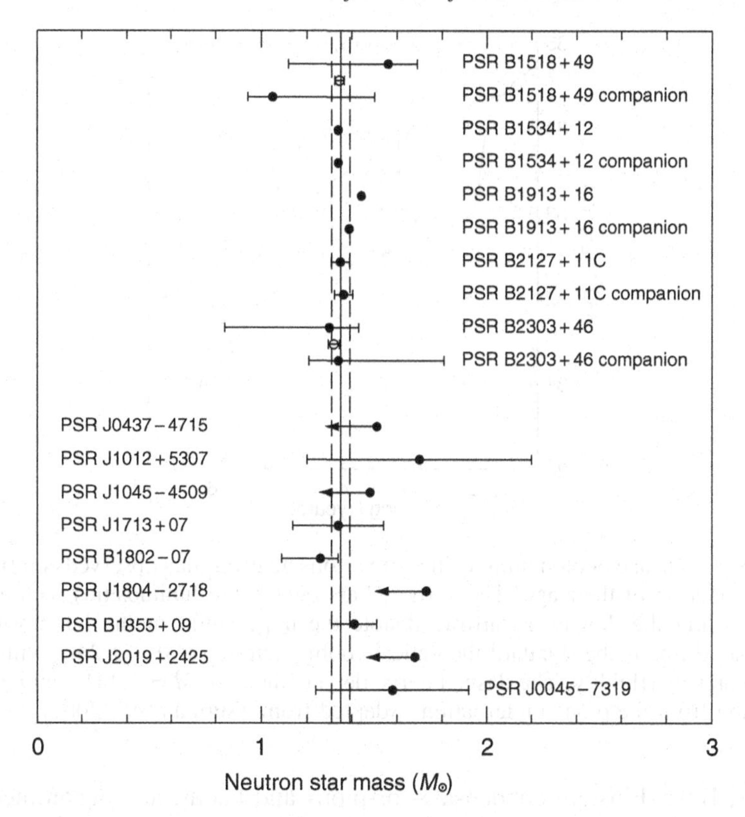

Fig. 9.2. Neutron star masses from radio pulsars in binary systems (double neutron stars, neutron star – white dwarf stars and neutron star – main sequence stars). The vertical solid and dashed lines denote $M/M_\odot = 1.35 \pm 0.04$. Taken from Thorsett and Chakrabarty (1999).

magnetic field, $\sim 10^{15}$ (10^9) gauss. A sudden spin up of the rotation associated with a subsequent relaxation to the normal rotation has been observed, and is called the glitch. This phenomenon may be related to the internal structure of neutron stars, in particular the superfluidity of the neutron liquid.

The outer crust of neutron stars is solid; it is composed of heavy nuclei forming a Coulomb lattice in a sea of degenerate electrons. As the pressure and density increase toward the inner region, electrons tend to be captured by nuclei and, at the same time, neutrons drip out from the nuclei, so that the system is composed of neutron-rich heavy nuclei in a Fermi sea of neutrons and electrons. Eventually, the nuclei dissociate into a neutron liquid and the system becomes a degenerate Fermi system composed of superfluid neutrons and a small fraction of superconducting protons and normal electrons. When the baryon density becomes more than a few times the normal nuclear density, one may expect exotic components such as

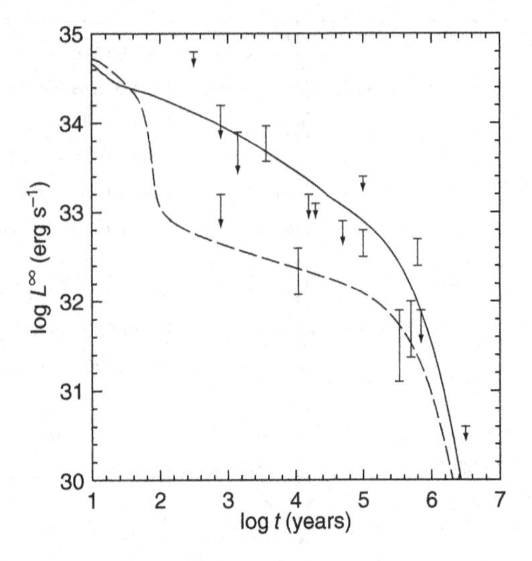

Fig. 9.3. Surface photon luminosities of various neutron stars observed at infinity as a function of their age. The vertical bars denote the luminosities with error bars, while the downward arrows denote the upper limit of the luminosities. The solid line is the standard theoretical cooling curve for $M = 1.2M_\odot$ with the nucleon superfluidity. The dashed curve depicts the case $M = 1.4M_\odot$ with extra cooling from the pion condensation. Adapted from Tsuruta *et al.* (2002).

hyperons, Bose–Einstein condensates of pions and kaons, and deconfined quark matter.

If the strange quark matter is the absolute ground state in QCD, quark stars, in which deconfined quarks exist all the way up to the star surface, are a possibility. Since the matter is assumed to be self-bound in this case, R decreases as M decreases unless general relativistic effects become too strong.

For both neutron stars and quark stars, there is a theoretical upper limit of the mass from general relativity. It is around $2M_\odot$ in both cases, although the precise value depends on the equation of state of the matter inside the star.

9.2 Newtonian compact stars

In this section, the physics underlying the existence of compact stars (white dwarfs and neutron stars) is discussed by considering Newtonian gravity combined with the equation of state for degenerate free fermions.

First, let us recapitulate the basic formula for non-relativistic degenerate fermions with spin 1/2 at zero temperature. The fermion number density and the Fermi momentum of the system are given by $\rho = N/V$ and $k_{\rm F}$, respectively, where N is the total baryon number and V is the volume of the system (which is later identified as the volume of the compact star).[1] The corresponding Fermi energy is given by (Exercise 9.1)

$$E_{\mathrm{F}} = \frac{\hbar^2 k_{\mathrm{F}}^2}{2m} = \frac{\hbar^2}{2m}\left(3\pi^2 \frac{N}{V}\right)^{2/3}. \tag{9.1}$$

The total energy of the degenerate system is obtained as an integral of the kinetic energy, $\hbar^2 k^2/2m$, up to the Fermi energy:

$$E_{\mathrm{tot}} = \frac{3}{5} N E_{\mathrm{F}}. \tag{9.2}$$

The corresponding pressure is given by

$$P \equiv -\frac{\partial E_{\mathrm{tot}}}{\partial V} = \frac{2}{5}\rho E_{\mathrm{F}}. \tag{9.3}$$

The white dwarfs (neutron stars) exist as a result of the balance between the pressure, P, of degenerate electrons (neutrons) and the gravitational pressure.

9.2.1 White dwarfs

Although the masses and radii of white dwarfs (WDs) vary, the characteristic values are

$$M \sim M_{\odot}, \quad R \sim 10^{-2} R_{\odot}, \tag{9.4}$$

where the mass and the radius of the Sun ($M_{\odot}$ and $R_{\odot}$) are given by (Appendix A.2)

$$M_{\odot} = 1.989 \times 10^{30}\ \mathrm{kg}, \quad R_{\odot} = 6.960 \times 10^{5}\ \mathrm{km}. \tag{9.5}$$

Matter inside a WD is composed of electrons and nuclei such as ^{4}He, ^{12}C and ^{16}O. The relative fraction of the nuclei depends on the density of the WD. The typical interior temperature of observed WDs is 10^{6-7} K. Sirius B was the first star identified as a WD. It is companion to a main sequence star, Sirius A. The mass and radius of Sirius B are known to be $M \simeq 1.03 M_{\odot}$ and $R \simeq 8.4 \times 10^{-3} R_{\odot}$, respectively.

To evaluate the electron density, consider a WD with $M = M_{\odot}$ and $R = 10^{-2} R_{\odot}$. The mass density of this star is given by

$$\varepsilon = \frac{M}{V} = \left(\frac{M}{V}\right)_{\odot} \times 10^6 = 1.41 \times 10^6\ \mathrm{g\ cm}^{-3}. \tag{9.6}$$

[1] Throughout Chapter 9, we use ρ for the *number* density and ε for the mass or energy density, following the standard convention in nuclear physics.

The corresponding electron number density, which is equal to the proton number density due to charge neutrality, is given by

$$\rho_e = \rho_p = \left(\frac{\varepsilon/2}{m_p}\right) = 0.42 \times 10^{30}\ \text{cm}^{-3}, \tag{9.7}$$

where we have assumed that the number of protons equals the number of neutrons and the mass density of the star is dominated by the nucleon mass, which is 2000 times heavier than the electron mass.

Since the electron density is so high, the averaged inter-electron distance, a, is an order of magnitude smaller than the Bohr radius in the oxygen nucleus:

$$\begin{aligned} a = \left(\frac{4\pi}{3}\rho_e\right)^{-1/3} &\simeq 0.83 \times 10^{-10}\ \text{cm} \\ < a_B(^{16}\text{O}) = \frac{a_B}{8} &= 6.6 \times 10^{-10}\ \text{cm}, \end{aligned} \tag{9.8}$$

where $a_B = \hbar/(m_e \alpha)$ is the Bohr radius of a hydrogen atom. This implies that most of the electrons in a WD are not bound to the nucleus. The nuclei themselves form a lattice structure to prevent mutual Coulomb repulsion, and the electrons form a Fermi sea.

The electrons can be well approximated by a degenerate Fermi gas, since the Fermi energy of the electron is much larger than the typical temperature of a WD:

$$\begin{aligned} E_F^e &= \frac{\hbar^2}{2m_e}(3\pi^2\rho_e)^{2/3} \simeq 200\ \text{keV} \\ &\gg T \sim 10^{6-7}\ \text{K} \simeq (0.1 - 1)\ \text{keV}. \end{aligned} \tag{9.9}$$

Now, what would be a general relation between M and R for WDs? Such a relation is obtained by the pressure balance. First, we consider the Newtonian gravitational energy of a star with a uniform density, which is given by

$$E_G(R) = -\frac{3}{5}\frac{GM^2}{R}. \tag{9.10}$$

On the other hand, the total energy of the matter composed of nuclei and electrons with electromagnetic interactions is given by

$$E_{\text{matt}}(R) = \sum_i m_i N_i + \frac{3}{5} N_e E_F^e + \cdots, \tag{9.11}$$

where i denotes possible species inside the WD such as the nucleus and the electron; m_i (N_i) is the mass (number) of each species.

The second term on the right-hand side of Eq. (9.11) is the energy of the degenerate electrons. The main contribution to the third term, denoted by "$\cdots$," is the Coulomb energy among the nuclei forming a lattice structure.

The energy $E_{\text{matt}}(R)$ for fixed R is dominated by the energy of the nuclei in the first term of Eq. (9.11), while the pressure of the matter, $P_{\text{matt}} = -\partial E_{\text{matt}}/\partial V$, is dominated by the degenerate electrons in the second term. In other words, the mass of the WD is determined by baryons, whereas the stability is dictated by electrons. This was first recognized by Fowler (1926), immediately following the development of Fermi–Dirac statistics in 1926.

Neglecting the binding energy of the nuclei, the proton–neutron mass difference and the Coulomb corrections, we may write

$$E_{\text{matt}}(R) \simeq 2N_{\text{e}}m_{\text{p}} + 1.1\frac{\hbar^2}{m_{\text{e}}}\frac{N_{\text{e}}^{5/3}}{R^2}, \tag{9.12}$$

where m_{p} is the proton mass and we have used charge neutrality and isospin symmetry, $N_{\text{n}} = N_{\text{p}} = N_{\text{e}}$.

Because the electron energy increases as $1/R^2$ for small R, and the gravitational energy decreases as $-1/R$ for large R, there is always a minimum of the total energy, $E_{\text{tot}}(R) = E_{\text{matt}} + E_{\text{G}}$, as a function of R for fixed N_{e}. The minimum is determined by the pressure balance:

$$\frac{\partial E_{\text{tot}}}{\partial R} = -1.1\frac{2\hbar^2}{m_{\text{e}}}\frac{N_{\text{e}}^{5/3}}{R^3} + \frac{3}{5}\frac{GM^2}{R^2} = 0. \tag{9.13}$$

This immediately leads to a mass–radius relation as follows:

$$M^{1/3}R = 1.2\left(\frac{1}{G}\right)\left(\frac{\hbar^2}{m_{\text{e}}}\right)\left(\frac{1}{m_{\text{p}}^{5/3}}\right) \simeq 0.8\times 10^{20}\ \text{g}^{1/3}\ \text{cm}, \tag{9.14}$$

where we have eliminated N_{e} by using $N_{\text{e}} \simeq M/(2m_{\text{p}})$. This is a remarkable relation in the sense that the properties of the macroscopic star given on the left-hand side are dictated by the fundamental constants in Nature, such as G and $\hbar$, on the right-hand side. If we adopt $M = M_{\odot} \simeq 2\times 10^{33}$ g, we have

$$R \simeq 0.6\times 10^4\ \text{km} \sim 10^{-2}R_{\odot}. \tag{9.15}$$

This is consistent with the observation given in Eq. (9.4).

9.2.2 *Neutron stars*

The typical mass and radius of a neutron star are given by

$$M \sim M_{\odot}, \quad R \sim (1\text{–}2)\times 10^{-5}R_{\odot}, \tag{9.16}$$

and the internal composition comprises mainly neutrons with a small admixture of protons and electrons. Typical internal temperatures are less than 10^9 K one year after their birth.

Because the radius is small ($\sim$10 km), the averaged mass density is quite high:

$$\varepsilon \sim \left(\frac{M}{V}\right)_{\odot} \times 10^{14-15} \sim 10^{14-15}\ \mathrm{g\ cm^{-3}}. \tag{9.17}$$

Assuming that the neutron star is composed only of neutrons, the neutron number density is given by

$$\rho_{\mathrm{n}} = \frac{\varepsilon}{m_{\mathrm{n}}} = 10^{38-39}\ \mathrm{cm^{-3}} = (0.1\text{–}1)\ \mathrm{fm^{-3}}. \tag{9.18}$$

This should be compared with the nuclear matter density, $\rho_{\mathrm{nm}} = 0.16\,\mathrm{fm}^{-3}$: the neutron star density is so high that all the nuclei dissolve into its constituents (except for the low-density outer surface). Namely, the neutron star is a giant, neutron-rich nucleus.

The Fermi energy of the neutron corresponding to the density interval in Eq. (9.18) is given by

$$E_{\mathrm{F}}^{\mathrm{n}} = \frac{\hbar^2}{2m_{\mathrm{n}}}(3\pi^2\rho_{\mathrm{n}})^{2/3} \sim (50\text{–}200)\ \mathrm{MeV}. \tag{9.19}$$

This is much larger than the internal temperature ($T < 10^9\,\mathrm{K} \sim 0.1\,\mathrm{MeV}$). Therefore, the neutrons are well approximated by a degenerate Fermi liquid.

The mass–radius relation for a neutron star is estimated in the same way as that for a WD if we assume uniform and degenerate neutron matter and Newtonian gravity. Taking into account the fact that both the energy and pressure of the system originate from the neutrons in the present case, an equation similar to Eq. (9.14) results:

$$M^{1/3}R \simeq 3.7\left(\frac{1}{G}\right)\left(\frac{\hbar^2}{m_{\mathrm{n}}}\right)\left(\frac{1}{m_{\mathrm{n}}^{5/3}}\right) \sim 10^{17}\ \mathrm{g^{1/3}\ cm}. \tag{9.20}$$

Using $M = M_{\odot}$, we find $R \simeq 10\,\mathrm{km}$, which is consistent with the observation given in Eq. (9.16).

9.3 General relativistic stars

In this section, we discuss the effect of general relativity on the structure of compact stars by looking at the Oppenheimer–Volkoff equation for static and spherically symmetric stars.

9.3.1 Maximum mass of compact stars

Equations (9.14) and (9.20) show that the mass, M, of a compact star grows without an upper bound as its radius, R, decreases. However, in reality, there does exist an upper limit of the mass, beyond which instability occurs for several reasons.

(i) As the central density of the star increases, degenerate fermions become relativistic and the pressure is reduced in comparison to the non-relativistic case. This causes an instability and means that the star has maximum mass under Newtonian gravity.
(ii) As the density increases, the composition of the matter changes such that the degenerate Fermi pressure decreases. This also causes an instability and sets a maximum mass.
(iii) General relativistic instability inevitably occurs at a high density, irrespective of the interior composition of the star.

Case (i) was first discussed by Chandrasekhar (1931) in relation to WDs; the corresponding maximum mass is called the "Chandrasekhar limit." The idea may be discussed as follows. Suppose we increase the baryon number density in the WD. Then the electron density also increases due to charge neutrality. (Here, we assume that $N_{\rm e} = N_{\rm p} = N_{\rm n}$. This will, however, be relaxed in case (ii).) Then the Fermi energy of the electrons increases and eventually becomes relativistic, i.e. $E_{\rm F} \to \hbar k_{\rm F}$. The total energy of degenerate relativistic electrons is easily calculated to be

$$E_{\rm matt} = \frac{3}{4} N_{\rm e} E_{\rm F} = 1.44\hbar \, \frac{N_{\rm e}^{4/3}}{R} \propto 1/R, \tag{9.21}$$

which is qualitatively different from the $1/R^2$ behavior of the non-relativistic electrons in Eq. (9.12). The Newtonian gravitational energy is still dictated by the nuclei as before: $E_{\rm G}(R) = -(3/5)G(2m_{\rm p}N_{\rm e})^2/R \propto -1/R$. Therefore, the pressure balance at high density (small R) is given by

$$4\pi R^2 P = \frac{N_{\rm e}}{R^2}\left(1.44\hbar \, N_{\rm e}^{1/3} - \frac{12}{5} G N_{\rm e} m_{\rm p}^2\right), \tag{9.22}$$

which becomes negative if the electron density exceeds a critical value,

$$N_{\rm e}^{\rm cr} \simeq \left(\frac{0.6\hbar}{G m_{\rm p}^2}\right)^{3/2} \sim 10^{57}, \tag{9.23}$$

$$M_{\rm max} \simeq 2 m_{\rm p} N_{\rm e}^{\rm cr} \simeq 2 \times (1.67 \times 10^{-27}\ {\rm kg}) \times 10^{57} \sim 1.7 M_{\odot}. \tag{9.24}$$

Thus we find that the maximum mass of a WD is of the same order as the solar mass.

The Chandrasekhar limit, Eq. (9.24), is realized only at asymptotically high density, where all the electrons become relativistic. In an actual WD, there is a more stringent constraint bound originating from (ii). As the electron Fermi energy increases, it eventually exceeds the threshold of an inverse β-decay:

$$e^- + Z \to (Z-1) + \nu_e, \tag{9.25}$$

where Z denotes a nucleus with an electric charge, Ze. Then the electrons are absorbed in the nuclei and the Fermi energy of the electrons is reduced. The typical Q-values for such processes in WDs are 13.37 MeV, 10.42 MeV and 4.64 MeV for ^{12}C, ^{16}O, and ^{28}Si, respectively. Because of the reduction of relativistic electron pressure due to this "neutronization" the WD becomes unstable above the central energy density, $\sim 10^{9-10}$ g cm^3. The corresponding maximum mass is $\sim M_\odot$, as shown in Fig. 9.4.

So far, we have discussed only the instabilities arising from cases (i) and (ii), which are relevant to WDs. For neutron stars, the general relativistic effect is more important, and the bound originating from case (iii) is most relevant to set the maximum mass and density. This will be discussed in Section 9.3.2.

9.3.2 Oppenheimer–Volkoff equation

To describe the static and spherically symmetric stars in general relativity, we start with the Schwarzschild metric (Appendix D.4):

$$ds^2 = e^{a(r)}dt^2 - e^{b(r)}dr^2 - r^2(d\theta^2 + \sin^2\theta\, d\phi^2), \tag{9.26}$$

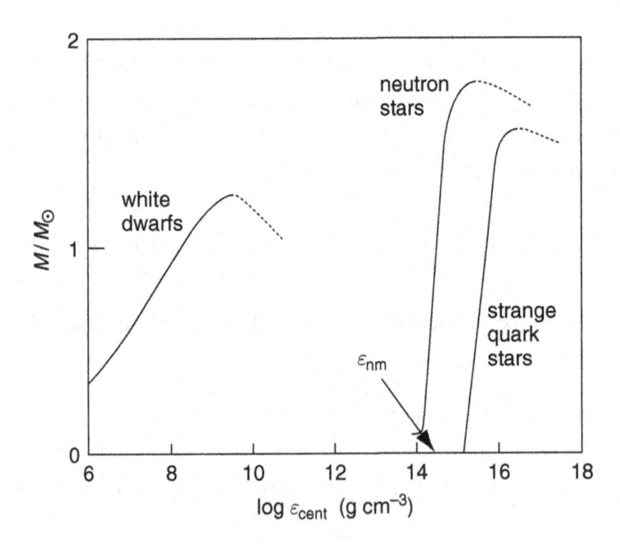

Fig. 9.4. Compact star mass, M, as a function of its central energy density, $\varepsilon_{\text{cent}}$. The energy density of the normal nuclear matter is shown by $\varepsilon_{\text{nm}}(\simeq 2.7 \times 10^{14}$ g cm$^{-3})$. The solid lines depict branches satisfying the necessary condition for stability; see Eq. (9.35).

together with the energy momentum tensor of a perfect fluid (Section 11.2),

$$T^{\mu\nu} = (P+\varepsilon)u^{\mu}u^{\nu} - Pg^{\mu\nu}, \tag{9.27}$$

where u^{μ} is the fluid four-velocity with the normalization $g_{\mu\nu}u^{\mu}u^{\nu} = 1$.

Then the Einstein equation reduces to (Appendix D.5):

$$\frac{d\mathcal{M}(r)}{dr} = 4\pi r^2 \varepsilon(r), \tag{9.28}$$

$$-\frac{dP(r)}{dr} = \frac{G\varepsilon\mathcal{M}}{r^2}\left(1-\frac{2G\mathcal{M}}{r}\right)^{-1}\left(1+\frac{P}{\varepsilon}\right)\left(1+\frac{4\pi r^3 P}{\mathcal{M}}\right), \tag{9.29}$$

where $\mathcal{M}(r)$ is the mass inside radius r. Equations (9.28) and (9.29) are the first order coupled differential equations originally derived by Oppenheimer and Volkoff (1939); they are called the "OV equations."

The OV equations can be solved together with the equation of state (EOS), $P = P(\varepsilon)$, and the initial conditions, $\mathcal{M}(r=0) = 0$ and $\varepsilon(r=0) = \varepsilon_{\text{cent}}$. The pressure decreases uniformly from the center of the star to the surface. The radius of the star, R, is defined at the point where $P(r=R) = 0$. The gravitational mass (or simply the mass) of the star, M, is given by

$$M = \int_0^R 4\pi r^2 \varepsilon(r)\, dr, \tag{9.30}$$

which contains the contribution from the matter and the gravitational field, and their mutual interactions.

We define the binding energy of a neutron star as the difference of M and the total mass of the nucleons in free space:

$$-\mathcal{B} \equiv M - m_{\text{N}}A, \tag{9.31}$$

where m_{N} is the nucleon mass and A is the total baryon number of the star;

$$A = \int j^0(r)\sqrt{-g}\; dr\; d\theta\; d\phi \tag{9.32}$$

$$= \int_0^R 4\pi r^2 \rho(r)\left(1-\frac{2G\mathcal{M}(r)}{r}\right)^{-1/2} dr, \tag{9.33}$$

where we have used the relations $\sqrt{-g} = \mathrm{e}^{(a(r)+b(r))/2} r^2 \sin\theta$, $j^{\mu}(r) \equiv u^{\mu}\rho(r)$, with $u^{\mu} = (g_{00}^{-1/2} = \mathrm{e}^{-a(r)/2}, 0, 0, 0)$, for the Schwarzschild metric. Using realistic EOSs of hadronic matter, we obtain $\mathcal{B}/A = 50\text{–}100$ MeV for neutron stars with $M = (1\text{–}1.5)M_{\odot}$ (see Sect. 3.14 in Glendenning (2000)).

A crucial difference between the Newtonian star and the relativistic star stems from the existence of the last three factors on the right-hand side of Eq. (9.29).

In particular, the factor $(1-2G\mathcal{M}(r)/r)^{-1}$ is related to the change of metric. Assuming that the limit $\varepsilon(r\to 0)$ is regular, we have $1-2G\mathcal{M}(r)/r|_{r\to 0}>0$ from Eq. (9.28). Then Eq. (9.29) tells us that $P(r)$ decreases monotonically as r increases from the center of the star. If $2G\mathcal{M}(r)/r$ approaches unity, the pressure gradient, dP/dr, becomes so large that P goes to zero quickly and vanishes at $r=R$, which defines the surface of the star. Therefore, $1-2G\mathcal{M}(r)/r$ is always positive inside the star.

The gravitational red shift of a photon emitted from the surface of the neutron star is given by (Exercise 9.2)

$$z=\frac{\nu_{\rm e}}{\nu_{\rm o}}-1=\left(\frac{g_{00}(\infty)}{g_{00}(R)}\right)^{1/2}-1=\left(1-\frac{2GM}{R}\right)^{-1/2}-1. \tag{9.34}$$

On inserting characteristic values for the mass and radius of the neutron star ($M\simeq M_\odot$ and $R\simeq 10\,\text{km}$), we obtain $z\simeq 0.2$; i.e. the effect of general relativity is not negligible for neutron stars. (The upper limit of z is 2, as will be shown in Section 9.3.3.) Although the metric changes substantially over the macroscopic distance from the center of the star to the surface, its change in the microscopic scale is negligibly small. Therefore, one can always take a local Lorentz frame to calculate the EOS, $P(r)=P(\varepsilon(r))$.

The existence of pressure on the right-hand side of Eq. (9.29) is a source of the limiting mass with a genuine relativistic origin. In the Newtonian star, the pressure acts only to support the system against gravitational collapse. In the relativistic star, however, the pressure can act to destabilize the star. In fact, as the mass of the star increases, the gravity compresses the matter and P increases. The right-hand side of Eq. (9.29) tells us that the increase of P further increases the pressure gradient and thus decreases the size of the star. Eventually the star becomes unstable, and suffers from gravitational collapse to become a black hole.

A necessary condition for the stability of compact stars can be shown to be as follows (see Harrison *et al.* (1965) and also Chap. 6 of Shapiro and Teukolsky (1983)):

$$\frac{dM}{d\varepsilon_{\rm cent}}>0. \tag{9.35}$$

Namely, the mass of the stable star should increase as the central energy density increases. This is a necessary, but not sufficient, condition for stability. Whether the star is truly stable or not should be examined by studying the vibrational modes of the star. If there is a tachyonic mode with an imaginary frequency, the star is unstable, even if Eq. (9.35) is satisfied. White dwarfs and neutron stars shown by the solid lines in Fig. 9.4 belong to the truly stable branch.

9.3.3 Schwarzschild's uniform density star

If we assume the existence of a star with uniform energy density, $\varepsilon(r) = \varepsilon_{\mathrm{cent}}$, we can solve the OV equations analytically. The assumption of uniform density (the incompressible fluid) is unphysical because the speed of sound violates causality, $c_s = dP/d\varepsilon = \infty$. Nevertheless, we can learn something about the general structure of the compact stars, as well as the relativistic upper bound on $2GM/R$ in an analytic manner.

If the density is uniform, Eq. (9.28) reduces to

$$\mathcal{M}(r) = \frac{4\pi}{3}\varepsilon_{\mathrm{cent}} r^3. \tag{9.36}$$

Then, Eq. (9.29) is given by

$$-\frac{dP}{(P+\varepsilon_{\mathrm{cent}})(3P+\varepsilon_{\mathrm{cent}})} = \frac{4\pi G}{3}\frac{r\,dr}{1-8\pi G\varepsilon_{\mathrm{cent}} r^2/3}, \tag{9.37}$$

which is easily solved as

$$\frac{P(r)}{\varepsilon_{\mathrm{cent}}} = \frac{\sqrt{1-2G(M/R)(r/R)^2}-\sqrt{1-2G(M/R)}}{3\sqrt{1-2G(M/R)}-\sqrt{1-2G(M/R)(r/R)^2}}. \tag{9.38}$$

The numerator is always positive, but the denominator can change sign for a certain value of r. Assuming that $P(r)$ is not singular and is positive everywhere inside the star, such a singularity can be hidden away only when the denominator at $r = 0$ is positive:

$$3\sqrt{1-\frac{2GM}{R}} - 1 > 0 \to \frac{2GM}{R} < \frac{8}{9}. \tag{9.39}$$

This upper limit turns out to be correct, even for a star with non-uniform energy density, as long as several reasonable conditions are met as follows: (i) $\varepsilon(r > R) = 0$; (ii) M is fixed; (iii) $2G\mathcal{M}(r)/r < 1$; and (iv) $d\varepsilon(r)/dr \leq 0$. For the proof, see Sect. 11.6 of Weinberg (1972). The above upper limit also implies that the red shift for the light emitted from the surface of the star satisfies $z < 2$ via Eq. (9.34).

9.4 Chemical composition of compact stars

We now discuss the possible forms of high density matter in chemical equilibrium inside the compact stars.

9.4.1 Neutron star matter and hyperon matter

Although the main ingredient of a neutron star is the neutron, there are other species mixed in the system because of the strong and weak interactions. First, pure neutron matter is unstable against the β-decay $n \to p + e^- + \bar{\nu}_e$. After the decay, the electron-neutrino leaves the star without experiencing many interactions,[2] while protons and electrons remain in the star and form a degenerate Fermi liquid together with the neutrons. The equilibrium configuration of n, p and e^-, which we call the standard neutron star matter, is determined by the following three conditions: chemical equilibrium, charge neutrality and the conservation of total baryon number. These may be given as

$$\mu_n = \mu_p + \mu_e \qquad (n \leftrightarrow p + e^-), \tag{9.40}$$

$$\rho_p = \rho_e, \tag{9.41}$$

$$\rho = \rho_n + \rho_p. \tag{9.42}$$

These are the general conditions which should be satisfied, no matter what kind of interactions take place, as long as they conserve charge and baryon number.

If we adopt the non-interacting Fermi gas model for a degenerate fermion system, the chemical potential, μ, and the fermion number density, ρ, are simply related to the Fermi momentum, k_F. Therefore, it is easy to find, for example the proton fraction in a neutron star for a given baryon density, ρ. Indeed, Eq. (9.40) is given by

$$\sqrt{k_n^2 + m_n^2} = \sqrt{k_p^2 + m_p^2} + \sqrt{k_e^2 + m_e^2}, \tag{9.43}$$

where $k_n = (3\pi^2 \rho_n)^{1/3}$ and $k_p = (3\pi^2 \rho_p)^{1/3} = (3\pi^2 \rho_e)^{1/3} = k_e$ (see Fig. 9.5(a)).

For simplicity, let us neglect the proton–neutron mass difference, $m_n - m_p \simeq 1.3\,\text{MeV}$, and the electron mass, $m_e \simeq 0.51\,\text{MeV}$, which are small compared with the Fermi energy. Then, the proton-to-neutron ratio is given by

$$\frac{\rho_p}{\rho_n} \simeq \frac{1}{8}\left[1 + \left(\frac{m_n^3}{3\pi^2 \rho_n}\right)^{2/3}\right]^{-3/2}. \tag{9.44}$$

This is a monotonically increasing function of ρ_n that approaches the asymptotic limit, 1/8, from below (Exercise 9.3). In the same approximation, the electron Fermi momentum is given by

$$k_e = k_p = \frac{m_n}{2} \frac{(3\pi^2 \rho_n / m_n^3)^{2/3}}{1 + (3\pi^2 \rho_n / m_n^3)^{1/3}}. \tag{9.45}$$

As the neutron density increases, the electron chemical potential increases and eventually exceeds the muon mass: $k_e > m_\mu = 106\,\text{MeV}$. Then, the Fermi

[2] This is true only for the cold neutron star. At the time of formation of the neutron star in the supernova explosion, the temperature of the system is high enough such that the neutrinos interact rather strongly with matter. This leads to the formation of the neutrino sphere, inside which neutrinos are effectively confined.

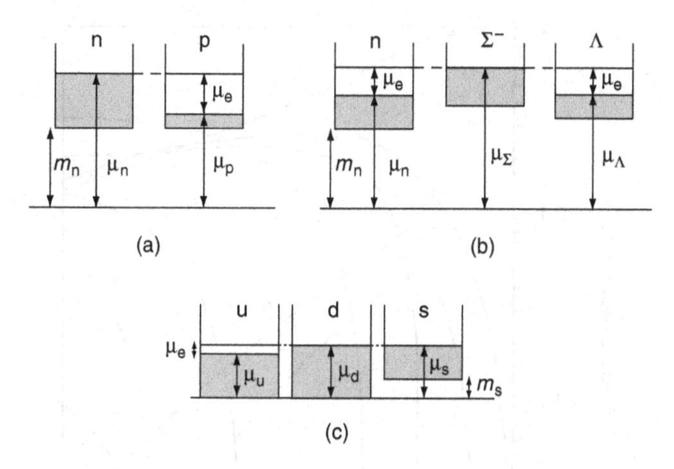

Fig. 9.5. Matter under chemical equilibrium and charge neutrality conditions in the Fermi gas model. Shaded areas show the occupied states. (a) The neutron star matter with n, p and e^-; (b) n, Σ^- and Λ in hyperon matter; (c) the composition of the uds quark matter with finite strange quark mass, m_s.

surface of the electrons becomes unstable against the decay $e^- \to \mu^- + \bar{\nu}_\mu + \nu_e$. Neutrinos produced in this reaction escape from the star, while muons stay and form a degenerate Fermi system. At the same time, the process $n \to p + \mu + \bar{\nu}_\mu$ also becomes available. Then the chemical equilibrium, charge neutrality and the conservation of baryon number for n, p, e^- and μ^- are given by

$$\mu_n = \mu_p + \mu_e, \quad \mu_e = \mu_\mu, \tag{9.46}$$

$$\rho_p = \rho_e + \rho_\mu, \tag{9.47}$$

$$\rho = \rho_n + \rho_p. \tag{9.48}$$

If we assume a non-interacting Fermi gas, these equations can determine the composition of all species once ρ_n is given.

As the baryon density increases further, not only the excited states of the nucleon but also baryons with strange quarks (hyperons) enter into the game. This is because the Fermi energy of the neutron exceeds the threshold of the neutron-decay into hyperons. See Fig. 9.5(b). Which hyperon appears first depends on the hyperon–nucleon interactions; typically, Σ^- and Λ may appear at 2–3ρ_{nm}, where $\rho_{nm} = 0.16\,\text{fm}^{-3}$ is the normal density of nuclear matter (see Section 9.5.1).

Shown in Fig. 9.6 is an example of the relative populations of various species as a function of the baryon number density, ρ. Interactions among baryons are taken into account in the relativistic mean-field theory (Glendenning, 2000). The deconfinement of the hadronic matter into quark matter is not considered in this

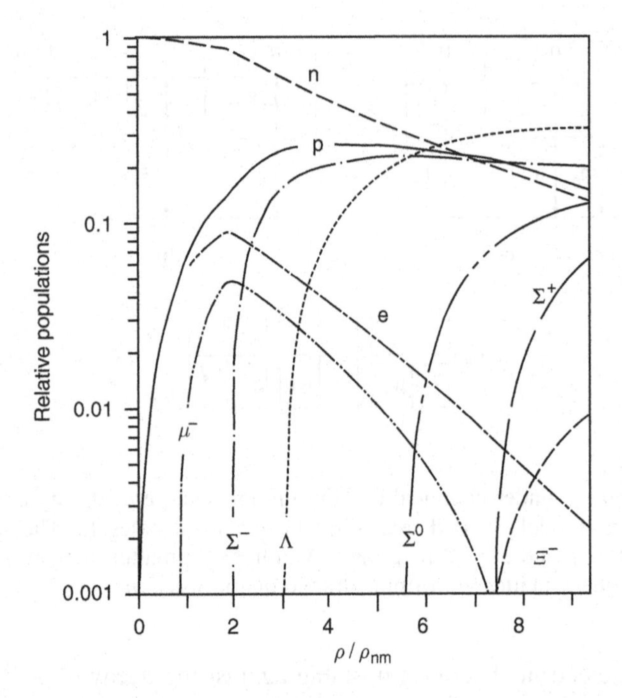

Fig. 9.6. Relative populations of various species as a function of the baryon number density, ρ, scaled by the normal nuclear matter density, $\rho_{\rm nm}(= 0.16\,{\rm fm}^{-3} = 1.6\times 10^{38}\,{\rm cm}^{-3})$ obtained in a relativistic mean-field model of hyperonic matter. Since the hyperon–nucleon and hyperon–hyperon interactions have large uncertainties, the populations of hyperons in this figure are at most qualitative at present. Adapted from Glendenning (2000).

figure. As far as the hadronic matter is concerned, the system starts from the pure neutron matter at low density and approaches the flavor SU(3) symmetric hyperon matter at asymptotic high density.

9.4.2 u, d *quark matter*

Let us now study the deconfined quark matter at high density. We first consider only (u, d) quarks and electrons. The conditions for chemical equilibrium, charge neutrality and baryon number conservation are given by

$$\mu_{\rm d} = \mu_{\rm u} + \mu_{\rm e} \quad ({\rm d} \leftrightarrow {\rm u} + {\rm e}^-), \tag{9.49}$$

$$0 = \frac{2}{3}\rho_{\rm u} - \frac{1}{3}\rho_{\rm d} - \rho_{\rm e}, \tag{9.50}$$

$$\rho = \frac{1}{3}(\rho_{\rm u} + \rho_{\rm d}). \tag{9.51}$$

By using the formula for non-interacting massless quarks (see Exercise 9.1 with $k_{\rm F} = \mu$), we immediately obtain

$$\mu_{\rm u} \simeq 0.80\, \mu_{\rm d}. \tag{9.52}$$

Thus, the Fermi energy of the d quark is slightly higher than that of the u quark. This is different from the situation of neutron matter, where n and p have quite different Fermi energies, as shown in Fig. 9.5(a). The difference originates from the relativistic nature of quark matter.

9.4.3 u, d, s *quark matter*

Next we consider quark matter with (u, d, s) quarks and electrons. Elementary processes which induce chemical equilibration are as follows:

$$\rm d \leftrightarrow u + e^-, \tag{9.53}$$

$$\rm s \leftrightarrow u + e^-, \tag{9.54}$$

$$\rm d + u \leftrightarrow u + s, \tag{9.55}$$

where we have not written neutrinos explicitly since they escape from the system and do not develop the chemical potential in cold quark matter. The equilibrium conditions are given by

$$\mu_{\rm d} = \mu_{\rm u} + \mu_{\rm e}, \tag{9.56}$$

$$\mu_{\rm s} = \mu_{\rm d}, \tag{9.57}$$

$$0 = -\frac{1}{3}(\rho_{\rm d} + \rho_{\rm s}) + \frac{2}{3}\rho_{\rm u} - \rho_{\rm e}, \tag{9.58}$$

$$\rho = \frac{1}{3}(\rho_{\rm u} + \rho_{\rm d} + \rho_{\rm s}). \tag{9.59}$$

For massless u, d and s quarks, Eqs. (9.56) and (9.57) become

$$\mu_{\rm u} = \mu_{\rm d} = \mu_{\rm s}, \quad \mu_{\rm e} = 0. \tag{9.60}$$

Namely, the massless quark matter is charge neutral by itself without leptons. If we have finite $m_{\rm s}$, $\rho_{\rm s}$ becomes smaller than $\rho_{\rm u,d}$. Then the electrons become necessary to make the system charge neutral, as shown in Fig. 9.5(c).

9.5 Quark–hadron phase transition

In Section 3.4, we estimated $T_{\rm c}$ for the deconfinement transition using the percolation model. The same idea can be applied to the system at finite baryon density

(Baym, 1979). If we use an "effective" radius, $R_{\rm N}$, of the nucleon, the close-packing density becomes

$$\rho_{\rm cp} = \left(\frac{4\pi}{3} R_{\rm N}^3\right)^{-1} \simeq \begin{cases} 2.4\ \rho_{\rm nm} & \text{for } R_{\rm N}{=}0.86\,\text{fm} \\ 12\ \rho_{\rm nm} & \text{for } R_{\rm N}{=}0.5\,\text{fm}\ . \end{cases} \tag{9.61}$$

Here, $R_{\rm N} = 0.86$ (0.5) fm corresponds to the charge radius of the proton (hard core radius of the nuclear force).

Since the central baryon density of neutron stars may reach several to ten times the value of $\rho_{\rm nm}$, the transition to quark matter as well as the hadron–quark mixed phase may be possible inside the star. Unfortunately, finite density QCD on the lattice at zero T is not available yet because of the complex phase problem discussed in Section 5.10. Therefore, the studies at finite baryon density have so far been limited to a hybrid model with the "realistic" equation of state (EOS) for hadronic matter at low density and the "phenomenological" bag EOS for quark matter at high density.

To study the phase transition in such a hybrid model at $T = 0$, let us first summarize the basic relations among the pressure, $P(\mu)(= -\Omega(\mu, V)/V)$, the energy density, $\varepsilon(\rho)$, the chemical potential, μ, and the baryon number density, ρ. By simply taking $T = 0$ in Eqs. (3.13)–(3.15) and replacing n by ρ, we obtain

$$P = \mu\rho - \varepsilon = \rho^2 \frac{\partial(\varepsilon/\rho)}{\partial\rho} = -\frac{\partial(\varepsilon/\rho)}{\partial\rho^{-1}}, \tag{9.62}$$

$$\rho = \frac{\partial P}{\partial\mu}, \quad \mu = \frac{\partial\varepsilon}{\partial\rho}. \tag{9.63}$$

The incompressibility modulus, $\mathcal{K}$, is defined as follows:

$$\mathcal{K} \equiv 9\frac{\partial P}{\partial\rho} = 9\frac{\partial}{\partial\rho}\left(\rho^2 \frac{\partial(\varepsilon/\rho)}{\partial\rho}\right). \tag{9.64}$$

It is often useful to define the binding energy per nucleon:

$$-\frac{\mathcal{B}}{A} = \frac{E}{A} - m_{\rm N} = \frac{\varepsilon}{\rho} - m_{\rm N}, \tag{9.65}$$

where E is the total energy of the matter without gravity, A is the total baryon number and $m_{\rm N}$ is the nucleon mass.

9.5.1 Equation of state for nuclear and neutron matter

Baryons are strongly interacting through the nuclear force in high density matter. Therefore it is a difficult task to extract information about high density matter by extrapolating from the known nuclear properties at $\rho_{\rm nm}$. Nevertheless, a number of many-body techniques have been developed to attack the problem, which include non-relativistic potential approaches and relativistic field theoretical approaches (Heiselberg and Hjorth-Jensen, 2000; Heiselberg and Pandharipande, 2000).

Instead of going into detail regarding this developing field, we adopt a phenomenological parametrization of the EOS for matter composed of protons and neutrons (Heiselberg and Hjorth-Jensen, 2000). It contains several parameters which can be fixed by the known nuclear properties and by the results of sophisticated many-body calculations. An example of the parametrized binding energy per nucleon is as follows:

$$-\frac{\mathcal{B}}{A} = \frac{\varepsilon_{\text{H}}}{\rho} - m_{\text{N}} = a_{\text{vol}} \cdot x \cdot \frac{-x+2+\delta}{1+x\delta} + a_{\text{sym}} \cdot x^{\gamma} \cdot (x_{\text{n}} - x_{\text{p}})^2, \tag{9.66}$$

$$\left(x = \frac{\rho}{\rho_{\text{nm}}} = x_{\text{n}} + x_{\text{p}}, \quad x_{\text{n}} = \frac{\rho_{\text{n}}}{\rho_{\text{nm}}}, \quad x_{\text{p}} = \frac{\rho_{\text{p}}}{\rho_{\text{nm}}} \right).$$

The corresponding pressure is given by $P_{\text{H}} = \rho^2 \partial(\varepsilon_{\text{H}}/\rho)/\partial\rho$. Actual values of the parameters are given by

$$a_{\text{vol}} = -15.8 \text{ MeV}, \quad a_{\text{sym}} = 32 \text{ MeV}, \tag{9.67}$$

$$\rho_{\text{nm}} = 0.16 \text{ fm}^{-3}, \quad \delta = 0.2, \quad \gamma = 0.6. \tag{9.68}$$

In Eq. (9.67), $-a_{\text{vol}}$ is the binding energy of the nuclear matter ($x_{\text{n}} = x_{\text{p}}$) at the saturation density, ρ_{nm}, where $\mathcal{B}/A$ takes its maximum; a_{sym} is the symmetry energy which provides a difference between the neutron matter ($x_{\text{p}} = 0$) and the nuclear matter. Since $a_{\text{sym}} > 0$, the binding energy, $\mathcal{B}/A$, has a maximum only at $\rho = 0$ for $x_{\text{p}} = 0$. Namely, the pure neutron matter cannot be bound by the nuclear force alone (Exercise 9.4).

The parameters δ and γ are fixed by fitting the EOS obtained in the non-relativistic calculation with modern two-body nuclear forces, relativistic corrections and the correction from the three-body nuclear force. The incompressibility modulus for nuclear matter at $\rho = \rho_{\text{nm}}$ is given by

$$\mathcal{K}_{\text{nm}} = \frac{18 a_{\text{vol}}}{1+\delta} \simeq 200 \text{ MeV}, \tag{9.69}$$

which is consistent with the empirical value. In Figs. 9.7(a) and (b), the EOS of the pure neutron matter, $x_{\text{p}} = 0$, is shown as a function of the baryon density.

For later purposes, we recapitulate the numerical values of the saturation density, ρ_{nm}, and the corresponding energy density, ε_{nm}, for nuclear matter:

$$\rho_{\text{nm}} = 0.16 \text{ fm}^{-3} = 0.16 \times 10^{39} \text{ cm}^{-3}, \tag{9.70}$$

$$\varepsilon_{\text{nm}} = 0.15 \text{ GeV fm}^{-3} = 2.7 \times 10^{14} \text{ g cm}^{-3}. \tag{9.71}$$

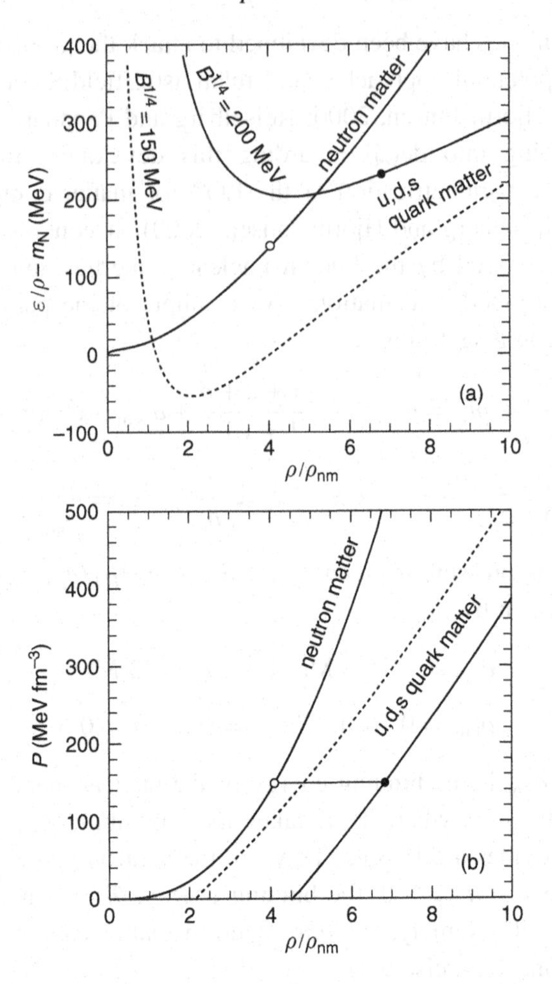

Fig. 9.7. (a) Energy per nucleon relative to the nucleon mass (= −binding energy) as a function of the baryon density for the pure neutron matter and the massless u,d,s quark matter. For the quark matter, the solid (dashed) line corresponds to the bag constant $B^{1/4} = 200$ (150) MeV. The white (black) dot denotes the initial (final) baryon density for the first order transition obtained by the phase equilibrium condition, Eq. (9.88). (b) The pressure for the pure neutron matter and the massless u,d,s quark matter. Lines have the same meaning as in (a). The pressure is constant during the mixed phase, which is represented by the thin solid line connecting the white and black dots.

9.5.2 *Equation of state for quark matter*

In the deconfined quark matter at high density, the EOS may be evaluated in perturbation theory (Freedman and McLerran, 1977; Baluni, 1978) because $\alpha_s(\kappa)$ becomes small due to asymptotic freedom. (Note, however, that the perturbation series may still experience bad convergence, as discussed in Section 4.8.)

The pressure of the non-interacting quarks with mass, m, at finite chemical potential, μ, and at zero temperature is easily evaluated from Eqs. (3.27) and (3.28):

$$P(\mu) = -\left[T\int \frac{d^3k}{(2\pi)^3}\ \ln\left(1+e^{-(E(k)-\mu)/T}\right)\right]_{T=0} \tag{9.72}$$

$$= \int \frac{d^3k}{(2\pi)^3}\frac{1}{3}\boldsymbol{v}\cdot\boldsymbol{k}\ \theta(\mu - E(k)) \tag{9.73}$$

$$= \frac{\mu^4}{4\pi^2}\left[\frac{k_{\rm F}}{\mu}\left(1-\frac{5m^2}{2\mu^2}\right)+\frac{3}{2}\frac{m^4}{\mu^4}\ln\left(\frac{1+k_{\rm F}/\mu}{m/\mu}\right)\right] \tag{9.74}$$

$$\to \frac{1}{4\pi^2}\times\begin{cases}\mu^4\left(1-\frac{3m^2}{\mu^2}+\cdots\right) & (m\ll\mu)\\ k_{\rm F}^4\left(\frac{4k_{\rm F}}{5m}-\frac{2k_{\rm F}^3}{7m^3}+\cdots\right) & (k_{\rm F}\ll\mu).\end{cases} \tag{9.75}$$

Here, $E(k)=(k^2+m^2)^{1/2}$, and $k_{\rm F}=(\mu^2-m^2)^{1/2}$ is the Fermi momentum of the quark.

The total pressure of the quark matter in the leading order $P_{\rm Q}^{(0)}$ is the sum of the pressure from each flavor and the bag constant:

$$P_{\rm Q}^{(0)} = -B + \sum_{q=u,d,s,\ldots} P_q, \tag{9.76}$$

where P_q is defined by Eq. (9.74) with μ, m and $k_{\rm F}$ replaced by μ_q, m_q and k_q, respectively.

The corresponding energy density and baryon number density are given by

$$\varepsilon_{\rm Q}^{(0)} = B - \sum_q (P_q + \mu_q \rho_q), \tag{9.77}$$

$$\rho^{(0)} = \frac{1}{3}\sum_q \rho_{\rm q} = \frac{1}{3}\sum_q \frac{\partial P_{\rm q}}{\partial \mu_q}, \tag{9.78}$$

respectively, where we have assumed that the bag constant, B, is independent of the chemical potential.

The pressure and energy density ($P_{\rm Q}$ and $\varepsilon_{\rm Q}$) with leading $\alpha_{\rm s}$ corrections are written in a particularly simple form if all the quarks are massless ($m_q = 0$):

$$P_{\rm Q} = -B + \sum_q \frac{\mu_q^4}{4\pi^2}\left(1-2\bar{g}^2\right) = \frac{1}{3}(\varepsilon_{\rm Q} - 4B), \tag{9.79}$$

where $\bar{g}^2 = g^2/4\pi^2 = \alpha_s/\pi$. In terms of the quark density ρ_q, the pressure, the energy density and the chemical potential may be rewritten as follows:

$$P_Q = -B + \sum_q \frac{1}{4\pi^2}(\pi^2\rho_q)^{4/3}\left(1+\frac{2}{3}\bar{g}^2\right), \tag{9.80}$$

$$\varepsilon_Q = B + \sum_q \frac{3}{4\pi^2}(\pi^2\rho_q)^{4/3}\left(1+\frac{2}{3}\bar{g}^2\right), \tag{9.81}$$

$$\mu_q = (\pi^2\rho_q)^{1/3}\left(1+\frac{2}{3}\bar{g}^2\right). \tag{9.82}$$

These equations show that the leading α_s correction for massless quarks is repulsive in the sense that the quark–quark interactions increase the energy density and pressure.

Although different flavors decouple up to $O(\alpha_s)$, and the pressure can be written by a sum of each species as shown above, this is no longer true for higher orders. For example, the pressure with common chemical potential, $\mu_q = \mu$, with $m_q = 0$, up to $O(\alpha_s^2)$ in the $\overline{\text{MS}}$ scheme is given by (Fraga *et al.*, 2001)

$$P_Q = -B + \frac{N_f\mu^4}{4\pi^2} \times \left[1 - 2\bar{g}^2 - \left(G + N_f \ln \bar{g}^2 + \left(11 - \frac{2N_f}{3}\right)\ln\frac{\kappa}{\mu}\right)\bar{g}^4\right], \tag{9.83}$$

where $G \simeq 10.4 - 0.536N_f + N_f \ln N_f$ and κ is the renormalization point.

In the following, we neglect $m_{u,d,s}$ altogether to make the discussion as simple as possible. Also, the α_s correction is within the large uncertainty of the bag constant, B; we therefore neglect it for simplicity.

The energy per baryon for the quark matter, $\varepsilon_Q/\rho - m_N$, and the pressure, P_Q, are shown in Figs. 9.7(a) and (b) for $m_q = 0$ and $\alpha_s = 0$. For the canonical value $B^{1/4} \simeq 200$ MeV introduced in Chapters 2 and 3, the quark matter becomes more bound than the neutron matter for $\rho >$ several $\times\, \rho_{nm}$, as can be seen from the solid lines in Fig. 9.7(a). This leads us to consider the possibility of a hybrid star where the quark matter core is surrounded by the neutron matter mantle (Collins and Perry, 1975; Baym and Chin, 1976).

On the other hand, if B is sufficiently small, the quark matter, with an approximately equal number of u, d and s quarks, may even become a stable and self-bound system (the strange matter). Then, a strange quark star (strange star) in which the whole interior of the system is made of strange matter may occur.

9.5.3 Stable strange matter

There may exist either a stable (Bodmer, 1971; Farhi and Jaffe, 1984; Witten, 1984) or a meta-stable (Chin and Kerman, 1979; Terazawa, 1979) state of matter with large strangeness. Within the bag EOS in Eqs. (9.80)–(9.82), it is easy to derive a range of the bag constant which allows the stable strange matter (Exercise 9.5):

$$147 \text{ MeV} < B < 163 \text{ MeV}. \tag{9.84}$$

The upper limit is obtained by requiring that the binding energy, $\mathcal{B}/A$, of the strange matter at its saturation point ($P_{\mathrm{Q}} = 0$) is larger than 8 MeV (which is the binding energy of the most stable nucleus, Fe). The lower limit is obtained from the condition that the $\mathcal{B}/A$ of the u,d quark matter without strangeness is less than 8 MeV. The latter is necessary for the stability of the ordinary nuclei against the decay into the u,d quark matter by the strong interaction. Under normal circumstances, the rate of decay from ordinary nuclei to strange matter is negligible. This is because many u and d quarks must be simultaneously converted to s quarks for the decay, which involves higher orders in the weak interaction.

If the condition in Eq. (9.84) is met, stable strange matter results with the following properties:

$$\varepsilon_{\mathrm{Q}} = 4B \sim 2.4\, \varepsilon_{\mathrm{nm}}, \tag{9.85}$$

$$\rho = \frac{1}{\pi^2}\left(\frac{4\pi^2 B}{3}\right)^{3/4} \sim 2.5\, \rho_{\mathrm{nm}}, \tag{9.86}$$

$$\mu_q = \left(\frac{4\pi^2 B}{3}\right)^{1/4} \sim 310 \text{ MeV}, \tag{9.87}$$

where we have used the upper limit of B in Eq. (9.84) to evaluate the numbers.

9.6 Phase transition to quark matter

Let us study the phase transition from the hadronic matter to the quark matter with finite baryon density at zero temperature. Although it is not entirely realistic, we assume that the hadronic matter is composed only of neutrons for simplicity. Also, let us assume that the quark matter is composed of massless non-interacting u, d and s quarks.

In this situation, the charge neutrality is automatically satisfied without leptons both in the hadronic and quark phases, and the equilibrium conditions, Eq. (3.19), at $T = 0$ reduce to

$$P_{\mathrm{H}} = P_{\mathrm{Q}} (\equiv P_{\mathrm{c}}), \quad \mu_{\mathrm{H}} = \mu_{\mathrm{Q}} (\equiv \mu_{\mathrm{c}}). \tag{9.88}$$

Here, $\mu_{_{\rm H(Q)}} = \partial\varepsilon_{_{\rm H(Q)}}/\partial\rho$ are the baryon chemical potentials evaluated in the hadronic matter (the quark matter). In practice, we have

$$\mu_{_{\rm H}} = \mu_{\rm n}, \quad \mu_{_{\rm Q}} = \mu_{\rm u} + \mu_{\rm d} + \mu_{\rm s} = 3\mu_q. \tag{9.89}$$

The characteristic feature of the phase transition dictated by Eq. (9.88) is analogous to that at finite T, discussed in Section 3.5. In fact, we have one-to-one correspondence if we replace T by μ and s by ρ in Fig. 3.2 and plot $P_{_{\rm H}}(\mu_{\rm n} = \mu)$ and $P_{_{\rm Q}}(\mu_{\rm q} = \mu/3)$, etc. as a function of μ. As in the case at finite T, the phase transition is of first order. The baryon density, $\rho = \partial P/\partial\mu$, has a discontinuity from $\rho_{\rm c1}$ to $\rho_{\rm c2}$ at $\mu_{\rm c}$.

Alternatively, the determination of $\rho_{\rm c1(c2)}$ may be carried out by the double tangent method, which is a variant of the standard Maxwell construction in the liquid–gas phase transition (Reif, 1965). The basic idea is to start with the thermodynamic relation at the critical point,

$$\varepsilon = \mu_{\rm c}\rho - P_{\rm c}. \tag{9.90}$$

This is a straight line in the (ε, ρ)-plane and is simultaneously tangential to the two curves $\varepsilon_{_{\rm H}}(\rho)$ and $\varepsilon_{_{\rm Q}}(\rho)$. Then the points of contact give the critical densities, $\rho_{\rm c1(c2)}$. One may construct a similar double tangent in the (ε/ρ, ρ^{-1})-plane simply by dividing Eq. (9.90) by ρ. The white and black dots shown in Fig. 9.7 correspond to the critical densities thus obtained. (see Exercise 9.6.) The pressure, $P(\rho)$, stays constant at $P_{\rm c}$ in the interval $\rho_{\rm c1} < \rho < \rho_{\rm c2}$.

In a realistic situation with a finite s quark mass and with beta-equilibrium, electrons are not negligible, either in the hadronic phase or in the quark phase. Then, we need to consider not only the baryon chemical potential, but also the charge chemical potential in the phase equilibrium conditions in Eq. (9.88). This may lead to a smooth crossover from the hadronic matter to the quark matter instead of a sharp first order transition (Glendenning, 1992). To describe this situation, however, it is necessary to obtain detailed information of the mixed phase, in particular the surface tension of the quark–hadron interface (Glendening, 2000) .

9.7 Structure of neutron stars and quark stars

9.7.1 Mass–radius relation of neutron stars

Ordinary neutron star matter, which is composed of neutrons with a small fraction of protons and electrons, is not a self-bound system. Therefore, gravity plays an essential role in the formation of neutron stars. Once we have the equation of state, $P = P(\varepsilon)$, obtained using theoretical and/or empirical information, the OV equations, Eqs. (9.28) and (9.29), are easily solved numerically. The pressure, P,

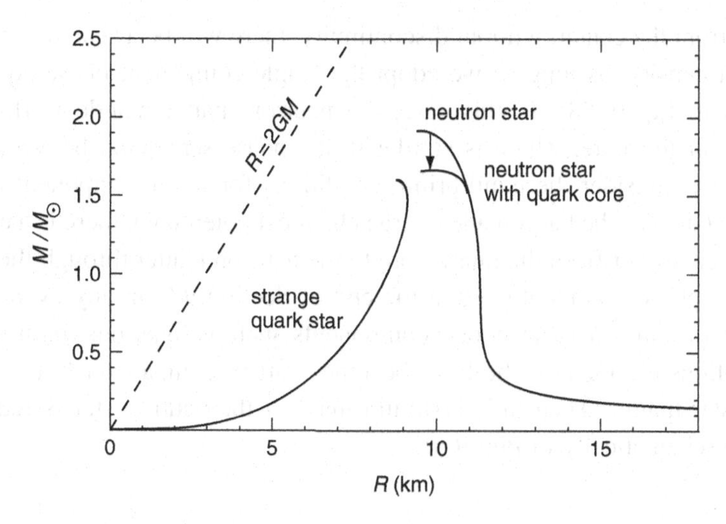

Fig. 9.8. A schematic illustration for the mass (M)–radius (R) relation of the neutron star and the strange quark star. The dashed line shows the Schwarzschild radius, $R = r_g = 2GM$, defined in Eq. (D.44).

as a function of the radial coordinate, r, is a monotonically decreasing function from the center toward the surface. The radius of the star, R, is defined as a point where $P(R)$ reaches zero. Shown in Fig. 9.8 is a schematic relation between the mass of the neutron star, M, and its radius, R. The M–R curve for a neutron star is easily understood by looking at R as a function of M. For small M, there is not much gravitational attraction and the matter is not self-bound. Therefore, R is large. As matter is added and M increases, the size of the star decreases, but the contraction slows down around $R \sim 10\,\text{km}$ because of the repulsive nuclear force among the neutrons at short distances. A further increase in M eventually leads to the general relativistic instability (see Eqs. (9.35) and (9.39)), and the star collapses to form a black hole.

The maximum mass of the neutron star, M_{max}, depends on the EOS inside the star. If we assume a non-interacting gas of degenerate neutrons, protons and electrons in chemical equilibrium, the OV equation yields $M_{\text{max}} \sim 0.7 M_{\odot}$, which is smaller than the observed mass of neutron stars given in Fig. 9.2. Once we take into account the effect of the nuclear forces in the EOS, in particular the repulsive core between the nucleons at short distances, the neutron star with higher mass is stabilized against the gravitational attraction, and M_{max} can be as large as a few times $M_{\odot}$. The soft (stiff) EOS, which is defined as relatively small (large) P at given ε, gives smaller (larger) M given R, and thus smaller (larger) M_{max} in general.

If the central density of the neutron star is high enough, a quark core may be formed at the core of the star. Since the pressure is a monotonically decreasing function from the center, without discontinuity, there will be a sharp discontinuity in baryon density, as long as we adopt the single-component phase equilibrium condition in Eq. (9.88). In this case, the neutron matter mantle is floating on the quark matter core, which is similar to the phase separation between oil and water. If we consider the equilibrium conditions for multi-component chemical potentials (such as the baryon and charge chemical potentials), there is a chance of a smooth crossover from the quark core to the neutron matter through the mixture of the two phases, as mentioned at the end of Section 9.6. In any event, as long as the quark core (or other exotic components such as hyperons, pion and kaon condensations etc.) exists, the EOS becomes soft in comparison to the ordinary neutron star matter. Then the maximum mass of the neutron star is reduced, as indicated schematically in Fig. 9.8.

9.7.2 Strange quark stars

Let us consider a situation in which the strange quark matter is the absolute ground state of matter and is self-bound even without gravity. In this case, the M, R relation becomes qualitatively different from that for neutron stars, as shown schematically in Fig. 9.8. The mass of the star essentially scales like $M \propto R^3$ for small R since the gravitational effect is relatively weak in such a region. The general relativistic effect becomes significant as M increases, and eventually the star becomes unstable at $M = M_{\rm max}$.

For the bag EOS, $P = (\varepsilon - 4B)/3$ (which is valid even with the interactions up to $O(\alpha_s)$ as long as the quarks are massless), the profile of the star is completely dictated by a single parameter, B. Since the pressure is zero at the surface, the energy density jumps abruptly from $\varepsilon_{\rm surf} = 4B$ to the vacuum. Namely, the strange quark star has a very sharp edge in density at the surface. It is also possible that it has a crust composed of ordinary nuclei separated by a vacuum from the bare strange matter surface (Madsen, 1999).

For the bag EOS, the OV equations are made dimensionless by introducing the following barred dimensionless variables (Witten, 1984):

$$P = \bar{P}, \quad \varepsilon = \bar{\varepsilon}, \quad r = \frac{\bar{r}}{\sqrt{GB}}, \quad GM = \frac{\bar{M}}{\sqrt{GB}}. \tag{9.91}$$

By numerical integration of the dimensionless OV equations, we obtain $\bar{M}_{\rm max} = 0.0258$ and the corresponding values $\bar{R} = 0.095$ and $\bar{\varepsilon}_{\rm cent} = 19.2$ (Exercise 9.7). This leads to

$$M_{\rm max} \simeq 1.78 M_\odot \left(\frac{155\ \text{MeV}}{B^{1/4}} \right)^2. \tag{9.92}$$

The corresponding radius and the central energy density are given by

$$R \simeq 9.5 \left(\frac{155\ \mathrm{MeV}}{B^{1/4}} \right)^2 \ \mathrm{km}, \quad \varepsilon_{\mathrm{cent}} \simeq 10 \left(\frac{155\ \mathrm{MeV}}{B^{1/4}} \right)^4 \varepsilon_{\mathrm{nm}}. \tag{9.93}$$

Here we have taken $B^{1/4} = 155\,\mathrm{MeV}$ as a typical value for the bag constant of the absolutely stable strange matter (see Eq. (9.84)).

9.8 Various phases in high-density matter

If there is some attractive channel between the fermions near the Fermi surface, the system may undergo a phase transition to superconductivity or superfluidity in three spatial dimensions (Shankar, 1994). This is indeed the case in the neutron star matter, where the attraction between the neutrons in the $(S, L, J) =$ (spin, orbital angular momentum, total angular momentum) $= (1, 1, 2)$ channel leads to the condensation of ${}^{2S+1}L_J = {}^3\mathrm{P}_2$ Cooper pairs (the neutron superfluidity). (Hoffberg *et al.*, 1970; Tamagaki, 1970) Also, the attraction between the protons in the $(S, L, J) = (0, 0, 0)$ channel leads to the condensation of ${}^1\mathrm{S}_0$ Cooper pairs (the proton superconductivity). For more details on nucleon superfluidity and its implication to the physics of compact stars, see Kunihiro *et al.* (1993).

In the quark matter, the long-range gluonic attraction between the quarks in the color anti-triplet ($\bar{3}$) channel leads to the color superconductivity (CSC) with the condensation of ${}^1\mathrm{S}_0$ Cooper pairs (Bailin and Love, 1984; Iwasaki and Iwado, 1995; Alford *et al.*, 1998; Rapp *et al.*, 1998). An interesting symmetry breaking pattern in the $N_f = 3$ massless QCD at high density is given by

$$\mathrm{SU_c}(3) \times \mathrm{SU_L}(3) \times \mathrm{SU_R}(3) \times \mathrm{U_B}(1) \rightarrow \mathrm{SU_{c+L+R}} \times Z(2), \tag{9.94}$$

where the $\mathrm{SU_{c+L+R}}$ implies a simultaneous rotation of color and flavor. This is called the color–flavor locking (CFL). For more details on the color superconductivity and its implication to the physics of compact stars, consult Alford (2001) and Rajagopal and Wilczek (2001).

Other than the superfluidity and superconductivity, the condensations of pions (π^0 and π^-) (Migdal, 1972; Sawyer and Scalapino, 1973) and of kaons (K^-) (Kaplan and Nelson, 1986) have been studied extensively. Unlike the case of the Bose–Einstein condensation of dilute atomic gases (Pethick and Smith, 2001), strong interactions of the pion and kaon fields with the nucleons are essential in the meson condensation processes. For more details, see Kunihiro *et al.* (1993) and Lee (1996).

The exotic phases discussed above have various implications, not only on the bulk structure, but also on the rotation and cooling of the compact stars. Readers interested in such phenomenological consequences should consult Heiselberg and Hjorth-Jensen (2000) and Weber (2005).

Exercises

9.1 Degenerate Fermi gas. Consider a non-interacting gas of spin 1/2 fermions (the Fermi gas). Show that the fermion number density, ρ, and the Fermi momentum, k_{F}, are related as follows:

$$\rho = 2\int_{k\le k_{\mathrm{F}}} \frac{d^3k}{(2\pi)^3} = \frac{k_{\mathrm{F}}^3}{3\pi^2}.$$

9.2 Gravitational red shift.
(1) Calculate the Schwarzschild radii ($r_{\mathrm{g}} = 2GM$) of the Sun ($M_{\odot} \sim 2 \times 10^{30}$ kg) and of the Earth ($M_{\mathrm{Earth}} \sim 6 \times 10^{24}$ kg). (*Answer*: about 3 km and 9 mm, respectively.)
(2) Derive Eq. (9.34) using the Schwarzschild metric Eq. (D.42).

9.3 Proton fraction in neutron star matter. Derive the exact expression $\rho_{\mathrm{p}}/\rho_{\mathrm{n}}$ without assuming that $m_{\mathrm{n}} - m_{\mathrm{p}}$ and m_{e} are small. (See Sect. 11.4 in Weinberg (1972).)

9.4 Nuclear equation of state. Plot Eq. (9.66) with the parameter set Eqs. (9.67) and (9.68) as a function of ρ/ρ_{nm} for several different values of the proton-to-neutron ratio, $\rho_{\mathrm{p}}/\rho_{\mathrm{n}}$, and confirm that the nuclear (neutron) matter is self-bound (not self-bound).

9.5 Bag constant and dense matter. Derive the lower and upper limits of the bag constant given in Eq. (9.84) for $m_q = 0$ and $\alpha_{\mathrm{s}} = 0$.

9.6 Quark–hadron phase equilibrium. Take the EOS for the pure neutron matter in Eq. (9.66) and that for the u,d,s quark matter in Eq. (9.81) with $\alpha_{\mathrm{s}} = 0$. Apply the double tangent method, Eq. (9.90), and confirm the critical densities given in Fig. 9.7.

9.7 Quark star and scaling solution. Rewrite the OV equations and the bag EOS in terms of the barred quantities in Eq. (9.91) and solve the coupled equations numerically to obtain $\bar{M}_{\mathrm{max}} = 0.0258$, $\bar{R} = 0.095$ and $\bar{\varepsilon}_{\mathrm{cent}} = 19.2$.

III

Quark–Gluon Plasma
in
Relativistic Heavy Ion Collisions

10

Introduction to relativistic heavy ion collisions

From the discussion in Chapter 1, especially Figs. 1.5 and 1.8, we can expect that a quark–gluon plasma is formed by either compressing or heating the hadronic matter. Relativistic heavy ion collisions are thought to provide us with each of these situations. At a few to a few tens of giga-electronvolts per nucleon, the colliding ions stop each other and high baryon densities could theoretically be achieved. At much larger energies, the heavy ions pass through each other, and the region in the middle is high in temperature but low in baryon density. In this chapter, we provide an overview of the basic and simplified pictures of relativistic and ultra-relativistic heavy ion collisions as an introduction to Part III.

10.1 Nuclear stopping power and nuclear transparency

In an attempt to explain production multiplicity, we present sketches of nucleus–nucleus collisions illustrating emitted hadron distributions in (a) AGS, (b) SPS and (c) RHIC experiments in Fig. 10.1. The particle multiplicities in nucleon–nucleon collisions are known empirically to increase as $\sim \ln \sqrt{s}$ (Fig. 16.7), where s is a Mandelstam variable (see Appendix E) and $\sqrt{s}$ corresponds to the total energy in the center-of-mass frame. In head-on nucleus–nucleus collisions (often called central collisions), almost all the nucleons are involved in the reaction and contribute to the particle production equally. Then the particle multiplicities will scale as A, and those in central Pb+Pb collisions will be approximately 200 times larger than proton–proton collisions at the same energy per nucleon.

In high-energy heavy ion collisions, Lorentz-contracted nuclei collide with each other and many nucleon–nucleon collisions with secondary particle productions take place. Incoming nucleons lose their kinetic energies in these collisions, converting the energy into other degrees of freedom. The amount of energy lost during the collisions depends on the thickness of the nuclei and also the collision energy. The degree of energy loss is called the *nuclear stopping power*.

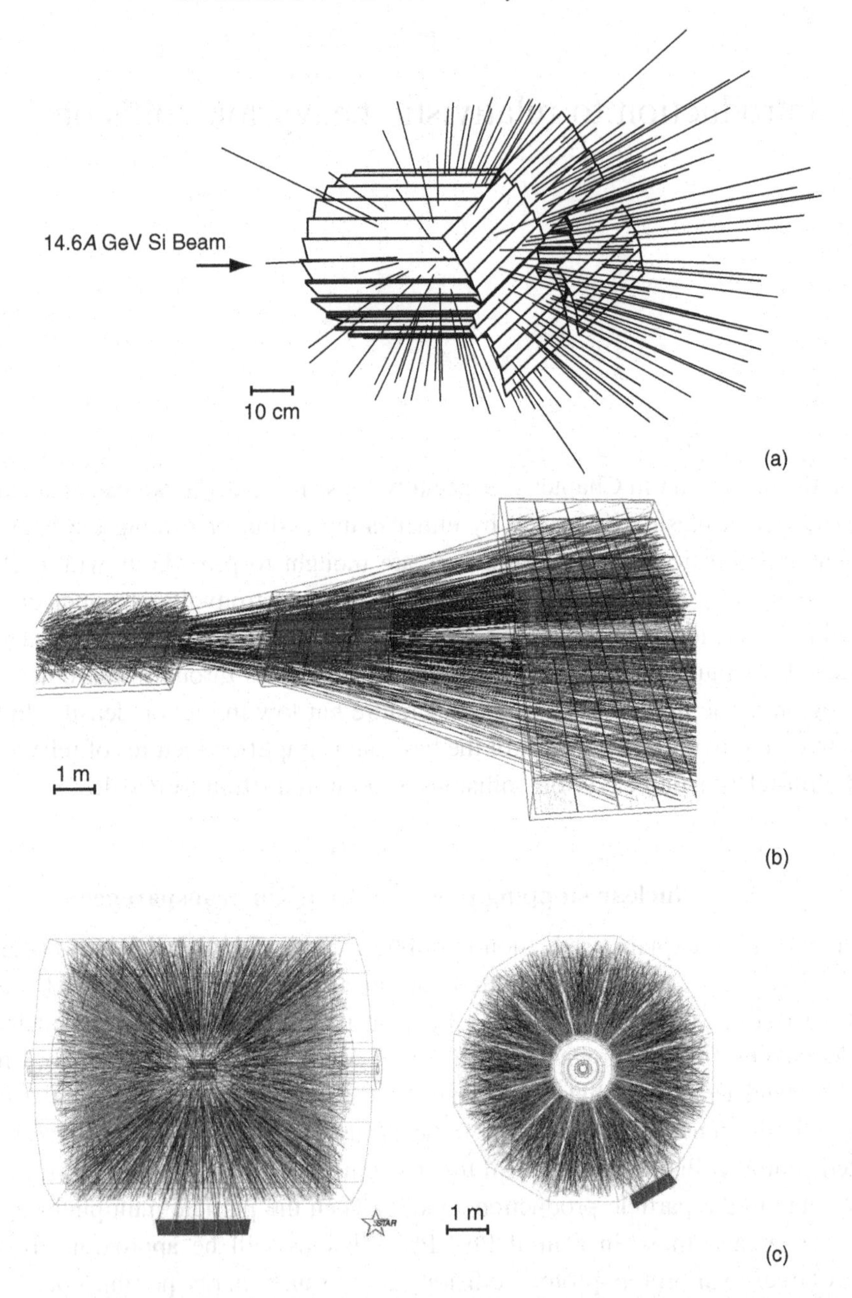

Fig. 10.1. Sketches of charged hadron tracks originating from collisions of (a) 14.6*A* GeV ^{28}Si ions onto a Au target (~220 tracks) (Abbott *et al.*, 1990), (b) 158*A* GeV Pb ions on a Pb target (~1000 tracks), reproduced with permission of the NA49 experiment from Afanasiev *et al.* (1999) and (c) 100*A* GeV ^{197}Au ions on 100*A* GeV ^{197}Au ions (~3000 tracks), reproduced with permission of the STAR experiment from Ackermann, *et al.*, 2003.

In nucleus–nucleus collisions, stopping can be seen as a shift of the rapidity distribution of the incident nucleons towards mid-rapidity, i.e. the center-of-mass of the collision. Thus, the shape of rapidity distributions provides key information in terms of the nuclear stopping power. See Appendix E for the definition of the rapidity, y, which is a relativistic analogue of velocity.

The net proton ($N_{\text{proton}} - N_{\text{anti-proton}}$) rapidity distributions are compared in central Au + Au and Pb + Pb collisions at energies of $\sqrt{s_{NN}} = 5$, 17 and 200 GeV in Fig. 10.2 (Bearden *et al.*, 2004). The distributions yield a strong beam energy dependence. At AGS, the net protons peak at mid-rapidity, but at SPS there is a dip at mid-rapidity. At RHIC, the distribution is almost flat in the middle with small peaks near the beam rapidity. It is clear that the nuclear collision changes from *stopping* to *transparent* in these energies. In other words, there is a saturation of nuclear stopping power; the incident nucleons do not lose all their kinetic energy, but punch through the opponent nucleus. In analogy to optics, one may say that the nucleus becomes transparent in high energy collisions.

The fact that ultra-relativistic nuclei having kinetic energies of ~100 GeV/A or more start to show transparent behavior may, at first sight, be surprising. In fact, the nucleon–nucleon cross-section has a value of about $\sigma_{NN} \approx 40 \sim 50$ mb at such high energies. This implies that the mean free path, ℓ_N, of a nucleon passing through nuclei is smaller than the size of the nucleus:

$$\ell_N = \frac{1}{\sigma_{NN}\rho_{\text{nm}}} = (4.5\ \text{fm}^2 \times 0.16\ \text{fm}^{-3})^{-1} = 1.4\ \text{fm}, \tag{10.1}$$

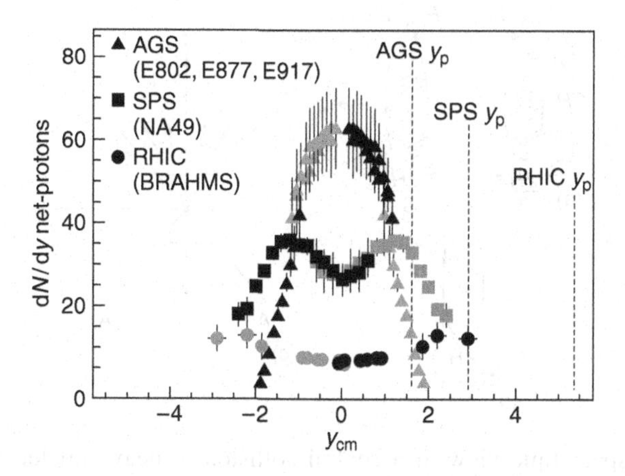

Fig. 10.2. Comparison of the net proton rapidity distributions at AGS ($\sqrt{s_{NN}} =$ 5 GeV Au + Au), SPS ($\sqrt{s_{NN}} = 17$ GeV Pb + Pb) and RHIC ($\sqrt{s_{NN}} = 200$ GeV Au + Au). Taken from Bearden *et al.* (2004). Use Fig. 10.12 to obtain the beam rapidity, y_p, from the center-of-mass energy, $\sqrt{s_{NN}}$, of the three accelerators.

where $\rho_{\rm nm} = 0.16\ {\rm fm}^{-3}$ is the normal density of nuclear matter. To solve this puzzle, one must consider the space-time picture of the collision.

10.2 Space-time picture of collisions

Consider a central collision of two nuclei of mass number A in the center-of-mass frame with center-of-mass energy of $E_{\rm cm}$ per nucleon, i.e. $\sqrt{s_{_{\rm NN}}} = E_{\rm cm}$ (Fig. 10.3). (See Appendix E for an account of the relativistic kinematics.) In this frame, two Lorentz-contracted pancakes of thickness $2R/\gamma_{\rm cm}$ along the longitudinal direction collide with each other; $\gamma_{\rm cm} (= E_{\rm cm}/2m_{_{\rm N}})$ is the Lorentz factor, where $m_{_{\rm N}}$ is the nucleon mass.

In the Landau picture (Landau, (1953)) of high-energy hadron collisions described in more detail in Chapter 11, the colliding nucleons are significantly slowed down and then produce particles, mostly within the thickness of the nuclear matter. Subsequently, the hot and baryon-rich system of particles undergoes a hydrodynamic expansion, primarily along the incident beam axis (the z-axis). Figure 10.3(c) is a schematic figure of this situation in the light-cone representation.

As the incident energy of the nuclei increases significantly, the Landau picture should be replaced by the new reaction picture proposed by Bjorken (1976), which will be discussed in Chapter 11. The Bjorken picture is based on the parton model of hadrons and is different from the Landau picture in two respects: the existence of wee partons and the time dilation of particle productions.

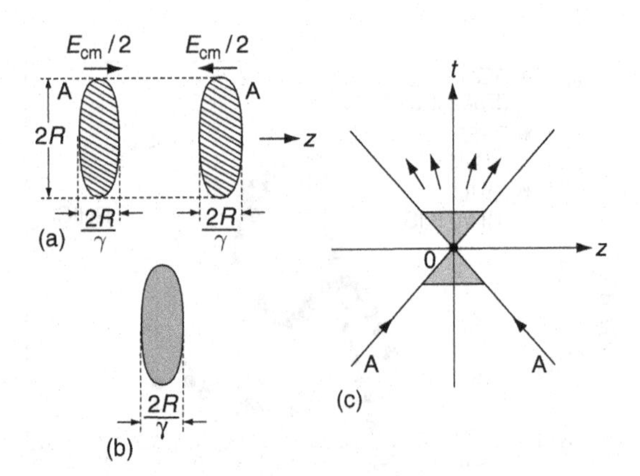

Fig. 10.3. Space-time view of a central collision of heavy nuclei $A+A$ in the Landau picture. (a) Two nuclei approaching each other with relativistic velocities and zero-impact parameters in the center-of-mass frame. (b) They slow down, stick together at the center and produce particles. (c) A light-cone diagram of the collision in the Landau picture, where the particle production takes place in the shaded area.

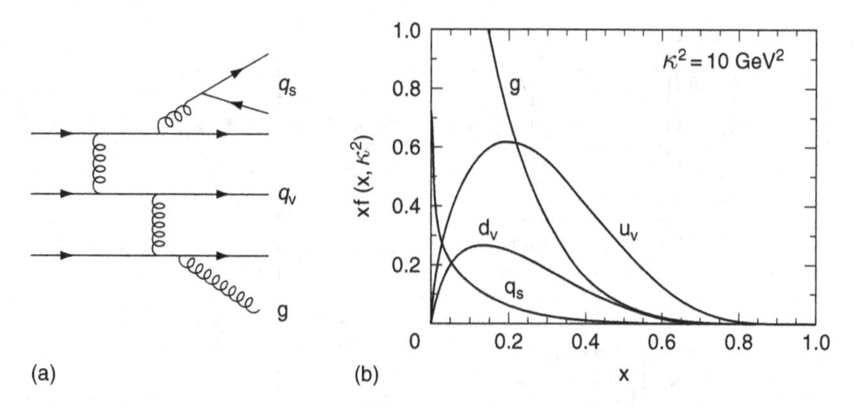

Fig. 10.4. (a) Production of low-momentum gluons (g) and sea-quarks (q_s) from the valence quarks (q_v). (b) Parton distribution function, f(x), in the proton multiplied by x (the longitudinal momentum fraction of the parton) at the renormalization scale $\kappa^2 = 10\,\text{GeV}^2$. Note that u_v and d_v denote the valence u quark and the valence d quark, respectively.

As has been well established by the deep-inelastic lepton–hadron scattering experiments, the nucleon is composed of valence quarks and the wee partons (gluons and sea-quarks). Wee partons have a much smaller momentum fraction (x) of the nucleon compared with the valence quarks, and the number of wee partons increases as x approaches zero. This is illustrated in Fig. 10.4.

The wee partons may be considered as vacuum fluctuations which couple to the fast-moving valence quarks passing through the QCD vacuum (Bjorken, 1976). Alternatively, the wee partons may be regarded as part of a coherent classical field created by the source of fast partons, which is called the color glass condensate (Iancu and Venugopalan, 2004). Because of its non-perturbative nature, the typical momentum, p, of the wee partons is of order $\Lambda_{\rm QCD}$ ($\sim 200\,\text{MeV}$), which characterizes the strong interaction scale of QCD. Since nucleons and nuclei are always associated with these low-momentum wee partons, the longitudinal size of hadrons or nuclei, Δz, can never be smaller than $1/p \sim 1$ fm owing to the uncertainty principle at ultra-high energies, as shown in Fig. 10.5(a):

$$\Delta z \geq \frac{1}{p} \approx 1 \text{ fm}. \tag{10.2}$$

As a consequence, the two incoming nuclei in the center-of-mass frame before the collision wear the "fur coat of wee partons" (Bjorken, 1976) of typical size 1 fm, while the longitudinal size of the wave function of a valence quark is $\sim 2R/\gamma_{\rm cm}$. Therefore, the wee partons are expected to play a vital role at ultra-high energies which fulfil the condition

$$\gamma_{\rm cm} > \frac{2R}{1 \text{ fm}}. \tag{10.3}$$

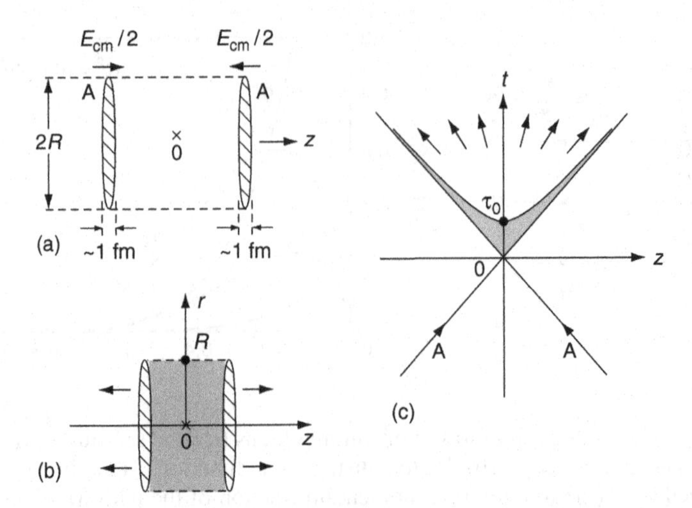

Fig. 10.5. Space-time view of the central AA collisions in the Bjorken picture. (a) Two nuclei approach each other with ultra-relativistic velocities and zero-impact parameters in the center-of-mass frame. (b) They pass through each other, leaving highly excited matter with a small net baryon number (shaded area) between the nuclei. (c) A light-cone diagram of the collision in the Bjorken picture; the highly excited matter is formed in the shaded area.

After the head-on collision of two beams of partons, many virtual quanta and/or a coherent field configuration of the gluons will be excited, as shown schematically in Fig. 10.5(b). It takes a certain proper time, τ_{de} (de-excitation or decoherence time), for these quanta to be de-excited to real quarks and gluons. The de-excitation time, τ_{de}, would typically be a fraction of 1 fm ($\sim 1/\Lambda_{QCD}$), or it could be much less than 1 fm. The state of matter for $0 < \tau < \tau_{de}$ is said to be in the pre-equilibrium stage.

Since τ_{de} is defined in the rest frame of each quantum, it experiences Lorentz dilation and becomes $\tau = \tau_{de}\gamma$ in the center-of-mass frame, where γ is the Lorentz factor of each quantum. This implies that slow particles emerge first near the collision point, while the fast particles emerge last, far from the collision point (Fig. 10.5(c)). This phenomenon, which is not taken into account in the Landau picture, is called the *inside–outside cascade*.

The real partons produced during the de-excitation process interact with each other and constitute an equilibrated plasma (quark–gluon plasma). We define τ_0 ($>\tau_{de}$) as a proper time within which the system is equilibrated, as schematically shown in Fig. 10.5(c); τ_0 depends not only on the basic parton–parton cross-section but also on the density of partons produced in the pre-equilibrium stage. One would expect that τ_0 is of order 1 fm or some fraction of 1 fm. The highly excited matter thus produced cools down and then hadronizes into mesons and baryons, which are eventually observed in the detectors (see Chapter 17).

10.3 Central plateau and fragmentation region

In the central ultra-relativistic AA collision illustrated in Fig. 10.5, the rapidity distribution of hadrons for asymptotically high energies is expected to behave as shown in Fig. 10.6. In between the initial rapidity of the target, y_{T}, and of the projectile, y_{P}, a baryon-free central rapidity region emerges, in which the average multiplicity per unit rapidity interval, dN/dy, is approximately constant and thus forms a central plateau. The production of virtual quanta due to the interactions of wee partons are dominant in this region, and the plateau structure can be understood as a consequence of the de-excitation of each quantum in its proper time, τ_{de}. The baryon-free quark–gluon plasma is expected to be produced in this central plateau region. The height and width of the central plateau depend on the center-of-mass energy of the colliding system and the nuclear mass number, A.

On the other hand, near the initial rapidities of the target and projectile, there are highly excited baryon-rich fragmentation regions. These regions can be best studied by going to the rest frame of either one of the nuclei. In this frame, the spherical nucleus with a radius proportional to $A^{1/3}$ is swept through the Lorentz-contracted nuclear pancake with a surface area $\sim A^{2/3}$ in the head-on collision.

The width of the fragmentation region, Δy_{f}, may be roughly estimated as follows. Consider a particle with Lorentz factor γ produced in the fragmentation region in the rest frame of the target nucleus. Only when $\gamma\tau_{\mathrm{de}} \lesssim R$ (where $R \sim 1.2A^{1/2}$ fm is the nuclear radius) may it re-interact within the nucleus and produce

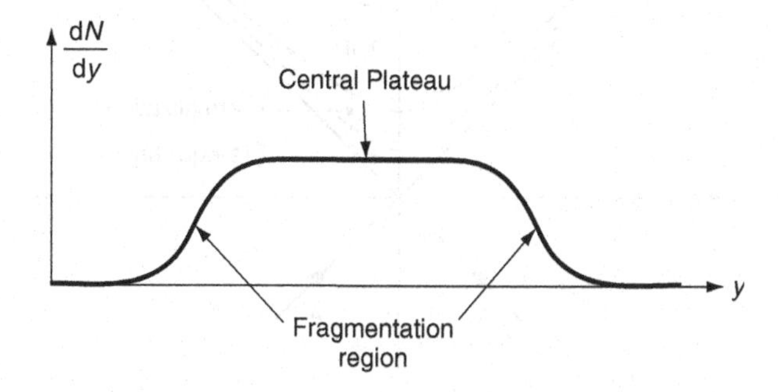

Fig. 10.6. Schematic illustration of hadron rapidity distributions in an ultra-relativistic AA collision at asymptotically high energies. The baryon-free quark–gluon plasma is expected to be produced in the central plateau region.

a cascade to form the fragmentation region. Then we may make the following estimation: $\Delta y_{\rm f} = \frac{1}{2}\ln((1+v_z)/(1-v_z)) \sim \ln 2\gamma$ (Appendix E.2).

10.4 Time history of ultra-relativistic AA collisions

So far we have discussed a gross feature of the heavy ion collisions at relativistic and ultra-relativistic energies. Let us now look at the time history of the central AA collisions in the latter case in more detail. Figure 10.7 is a schematic figure indicating the time history in the space-time diagram.

Pre-equilibrium stage and thermalization: $0 < \tau < \tau_0$

The central collision of ultra-relativistic heavy nuclei is a process of huge entropy production. What is the microscopic mechanism of the entropy production and subsequent thermalization? Theoretically, this is one of the most difficult questions to answer since highly non-equilibrium processes of non-Abelian gauge theory are involved. So far, two classes of models are proposed: the incoherent models and the coherent models.

In the incoherent models, the incoherent sum of collisions of incoming partons produces minijets (semi-hard partons) which subsequently interact with each other to form an equilibrated parton plasma (see Section 13.1.3). The minijet

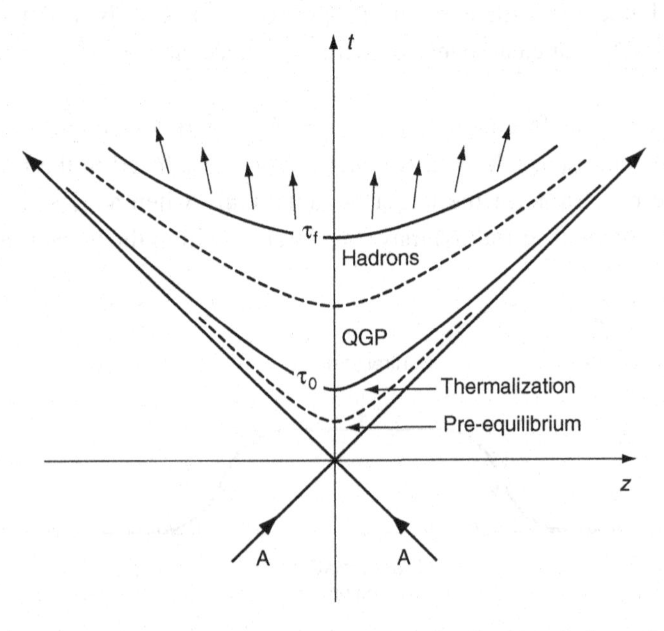

Fig. 10.7. Light-cone diagram showing the longitudinal evolution of an ultra-relativistic AA collision. Contours of constant proper time τ appear as hyperbolas, $\tau = (t^2 - z^2)^{1/2}$.

production is calculated in pertubative QCD (pQCD) with an infrared cutoff of order 1–2 GeV, while the equilibration process is calculated using the relativistic Boltzmann equation with pQCD parton–parton cross-sections (see Chapter 12 and Appendix F).

An example in the coherent models is the formation of color strings and ropes (the coherent color-electric fields) after the impact, which subsequently decay into real partons through the Schwinger mechanism (see Section 13.1.1). Once the real partons are produced, they move toward thermal distribution, for example by obeying the relativistic Boltzmann equation with parton collisions and the background color-electric field. Another example in the coherent models is the one based on the color glass condensate (CGC), which is a coherent classical configuration of low-x gluons associated with incoming nuclei (see Section 13.1.2).

The approaches mentioned above have their own limitations of applicability and are not yet fully developed enough to describe the entropy production and thermalization processes in a quantitative manner. The difficulty originates primarily from the treatment of the time-dependent and non-perturbative processes in non-Abelian gauge theory.

We will come back to the above models in Chapter 13. Here we simply assume that the entropy production and subsequent local thermalization take place before the characteristic proper time, τ_0. This produces the initial condition of hydrodynamical evolution of the system for $\tau > \tau_0$. Hydrodynamical models assuming a perfect fluid indicate that τ_0 of less than 1 fm gives a reasonable account of the RHIC data (Chapter 16).

Hydrodynamical evolution and freeze-out: $\tau_0 < \tau < \tau_f$

Once the local thermal equilibrium is reached at τ_0, we may use the notion of relativistic hydrodynamics to describe the expansion of the system (Chapter 11). Here the basic equations are the conservation of the energy-momentum tensor and the baryon number:

$$\partial_\mu \langle T^{\mu\nu} \rangle = 0, \qquad \partial_\mu \langle j_{\mathrm{B}}^\mu \rangle = 0, \tag{10.4}$$

where the expectation value is taken with respect to the time-dependent state in local thermal equilibrium.

If the system can be approximated by a perfect fluid, the expectation values are parametrized solely by the local energy density, ε, and the local pressure, P. If the system is not a perfect fluid, we require extra information, such as the viscosity, heat conductivity and so on. In the former case, Eq. (10.4) supplemented with the equation of state, $\varepsilon = \varepsilon(P)$, calculated in lattice QCD simulations (Chapter 5) and with appropriate initial conditions at $\tau = \tau_0$ can predict the time development of the system until it undergoes a freeze-out at $\tau = \tau_f$. The most interesting part of the physics is in this period, $\tau_0 < \tau < \tau_f$, where the evolution of the thermalized quark–gluon plasma and its phase transition to the hadronic plasma take place.

Freeze-out and post-equilibrium: $\tau_f < \tau$

Eventually, freeze-out of the hadronic plasma happens at the proper time, τ_f. The freeze-out is defined by a space-time hyper-surface, where the mean free time of the plasma particles becomes larger than the time scale of the plasma expansion so the local thermal equilibrium is no longer maintained. One can think of two kinds of freeze-out: the chemical freeze-out (after which the number of each species is frozen while the equilibration in the phase-space is maintained) and the thermal freeze-out (after which the kinetic equilibrium is no longer maintained). The chemical freeze-out temperature must be higher than that of the thermal freeze-out. Also, these temperatures depend on the hadronic species in principle.

Even after freeze-out, hadrons can still interact in a non-equilibrium way which may be described by the Boltzmann equation at the hadronic level.

10.5 Geometry of heavy ion collisions

The geometric aspects of high-energy heavy ion collisions play an important role in collision dynamics. The de Broglie wavelength of the nucleons in high-energy nucleus–nucleus collisions is much shorter than the size of the nucleus. Then the description of the collision using the impact parameter, b, becomes a good starting point. Figure 10.8 illustrates a distant collision, a peripheral collision and a central collision. In the distant collision ($b > 2R$), due to the electromagnetic interactions between the colliding nuclei, the projectile or the target nucleus may break up. As the impact parameter, b, decreases, the strong interaction produces a sudden rise in inelastic reactions once two colliding nuclei have overlapped geometrically.

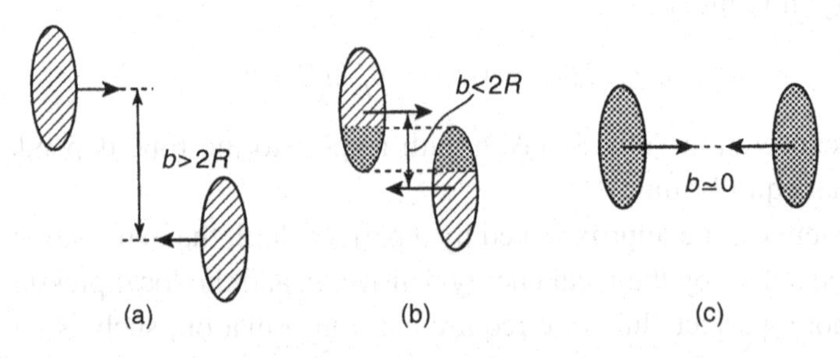

Fig. 10.8. Illustration of high-energy nucleus–nucleus collisions: (a) a distant collision, (b) a peripheral collision and (c) a central collision. Dotted areas denote participant regions and cross-hatched areas denote spectator regions.

The nucleons in high-energy heavy ion collisions are classified into two groups: the *participant*, which is the overlapped region shown by the dotted area in Fig. 10.9, and the *spectator*, shown by the cross-hatched areas in Fig. 10.9. This geometrical treatment of the high-energy nucleus–nucleus collisions is known as a participant–spectator model, and it is successful in describing observed features in experiments. As shown in Fig. 10.9, the size of the participant/spectator is determined by the impact parameter, b, and there is anti-correlation between the size of the participant and that of the spectator. Since the spectator keeps its longitudinal velocity and emerges at nearly zero degrees in the collision, it is relatively easy to separate the spectator and the participant experimentally. In most heavy ion experiments, information about the impact parameter, b, is obtained by measuring the sizes of the spectator and/or the participant. The number of participant nucleons can be evaluated by the semi-classical Glauber model described below.

The Glauber model (Glauber, 1959; Frauenfelder and Henley, 1991) has been successfully applied in the description of high-energy nuclear reactions, and is used extensively in evaluating total reaction cross-sections. It is also employed to evaluate the number of interactions, namely the number of participant nucleons and the number of nucleon–nucleon collisions (binary collisions). It is a semi-classical model, treating the nucleus–nucleus collisions as multiple nucleon–nucleon interactions: a nucleon of incident nucleus interacts with target nucleons with a given density distribution. Nucleons are assumed to travel in straight lines, and are not deflected after the collisions, which holds as a good approximation at very high energies. Also, the nucleon–nucleon inelastic cross-section, $\sigma_{\rm NN}^{\rm in}$, is assumed to be the same as that in the vacuum. In other words, secondary particle production and possible excitation of nucleons are not considered in this model.

Now let us introduce the nuclear overlap function, $T_{AB}(\boldsymbol{b})$, with impact parameter, $\boldsymbol{b}$ (Fig. 10.10) as follows:

$$T_{AB}(\boldsymbol{b}) = \int d^2\boldsymbol{s}\ T_A(\boldsymbol{s})\ T_B(\boldsymbol{s}-\boldsymbol{b}), \tag{10.5}$$

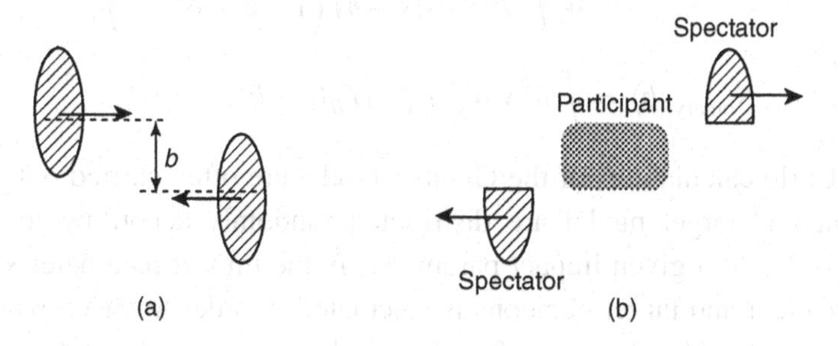

Fig. 10.9. Participant–spectator picture of the high-energy heavy ion collision. (a) Before the collision; impact parameter b. (b) After the collision.

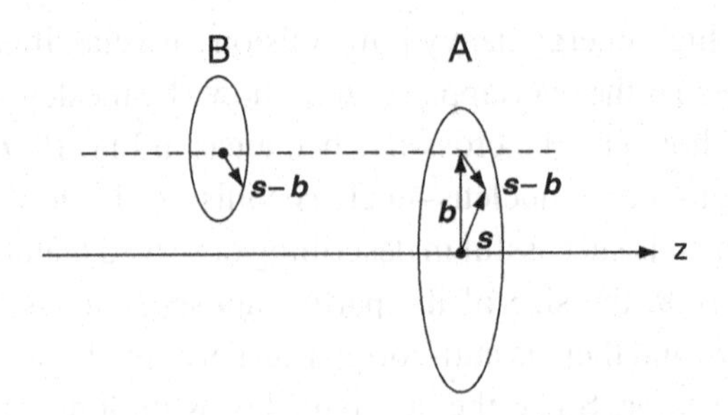

Fig. 10.10. Geometry of a collision between nuclei A and B. Vectors $\boldsymbol{b}$ and $\boldsymbol{s}$ are in the transverse plane with respect to the collision axis, z.

where the thickness function is defined as

$$T_A(\boldsymbol{s}) = \int dz\, \rho_{_A}(z, \boldsymbol{s}). \tag{10.6}$$

Here, $\rho_{_A}$ is the nuclear mass number density normalized to mass number A:

$$\int d^2\boldsymbol{s}\; T_A(\boldsymbol{s}) = A, \quad \int d^2\boldsymbol{b}\; T_{AB}(\boldsymbol{b}) = AB. \tag{10.7}$$

For a spherical nucleus with radius $R_A = r_0 A^{1/3}$, we have $T_{AA}(0) = 9A^2/8\pi R_A^2$, which gives, for example, $T_{\rm AuAu}(0) = 28.4\ \rm mb^{-1}$ for $R_{\rm Au} = 7.0$ fm. In a realistic situation for a heavy nucleus such as Pb or Au, the Woods–Saxon parametrization is a good approximation:

$$\rho_{_A}(r) = \frac{\rho_{\rm nm}}{1+\exp((r-R_A)/a)}. \tag{10.8}$$

The number of participant nucleons, $N_{\rm part}$, and the number of nucleon–nucleon collisions, $N_{\rm binary}$, in the Glauber model are defined, respectively, as follows:

$$N_{\rm part}(b) = \int d^2\boldsymbol{s}\; T_A(\boldsymbol{s}) \left(1 - e^{-\sigma_{\rm NN}^{\rm in} T_B(\boldsymbol{s})}\right) + \int d^2\boldsymbol{s}\; T_B(\boldsymbol{s}-\boldsymbol{b}) \left(1 - e^{-\sigma_{\rm NN}^{\rm in} T_A(\boldsymbol{s})}\right), \tag{10.9}$$

$$N_{\rm binary}(b) = \int d^2\boldsymbol{s}\; \sigma_{\rm NN}^{\rm in}\, T_A(\boldsymbol{s}) T_B(\boldsymbol{s}-\boldsymbol{b}). \tag{10.10}$$

Monte Carlo calculations of the Glauber model are often carried out. Nucleons of incident and target nuclei are distributed randomly according to a nuclear density profile. At a given impact parameter, $\boldsymbol{b}$, the impact parameter $\boldsymbol{s}$ of all the pairs of incident and target nucleons is calculated in order to check whether they interact. An interaction occurs when the nucleon–nucleon impact parameter is less than $\sqrt{\sigma_{\rm NN}^{\rm in}/\pi}$. Figure 10.11 shows the number of nucleon–nucleon collisions

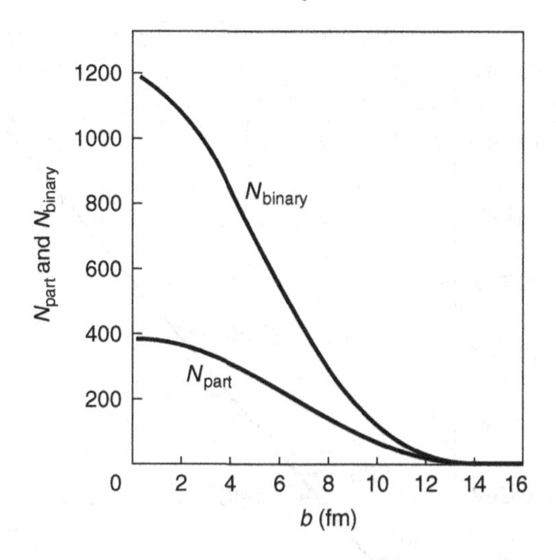

Fig. 10.11. Number of binary collisions and number of participant nucleons as a function of the impact parameter in a Au + Au collision. The Woods–Saxon distribution with parameters $a = 0.53$ fm, $R_{Au} = 6.38$ fm and $\sigma_{NN}^{in} = 42$ mb are used.

(binary collisions) and the number of participants in Au + Au collisions as a function of the impact parameter.

10.6 Past, current and future accelerators

We show in Fig. 10.12 the relation between beam energies per nucleon versus beam rapidities in various heavy ion accelerators. The rapidity, y, transforms additively by a Lorentz boost in the longitudinal direction (Appendix E). When we choose the beam direction as the longitudinal direction, Eq. (E.19) becomes

$$E_{lab} = m_N \cosh y_{lab}, \qquad E_{cm}/2 = m_N \cosh y_{cm}, \tag{10.11}$$

where the beam rapidity in the center-of-mass frame and that of the laboratory frame are related by

$$y_{cm} = (1/2)\, y_{lab}. \tag{10.12}$$

These relations are shown by the solid line in Fig. 10.12. To the upper left (lower right) side of the curve we mark the laboratory (center-of-mass) beam energies per nucleon.

Let us give some examples. The AGS at BNL (SPS at CERN) accelerates Si (O) ions up to 15 GeV/A (200 GeV/A) and Au (Pb) ions up to 11 GeV/A (160 GeV/A), which corresponds to $y_{lab} = 3.5$ (6.1) and $y_{lab} = 3.2$ (5.8), respectively. Corresponding rapidities in the center-of-mass frame are given by $y_{cm} =$ 1.7 (3.1) and $y_{cm} = 1.6$ (2.9), respectively. Then, the curve in Fig. 10.12 gives

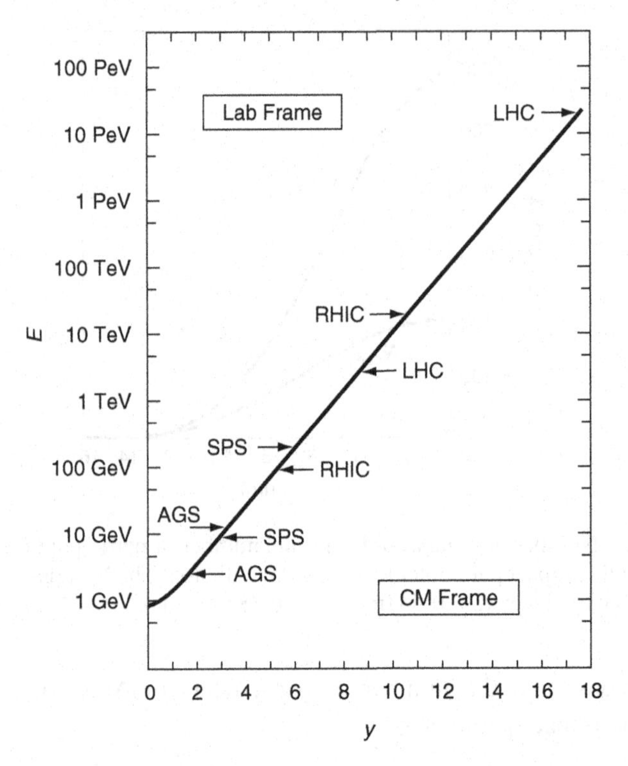

Fig. 10.12. Relation between the beam energy per nucleon, E, and beam rapidity, y: $E = m_N \cosh y$. The rapidity in the center-of-mass frame and in the laboratory frame are simply related by $y_{cm} = y_{lab}/2$. Relativistic and ultra-relativistic heavy ion accelerators are indicated by the arrows.

the center-of-mass energies, $E_{cm}/2$, of 2.8 GeV (10 GeV) and 2.4 GeV (8.6 GeV), respectively.

The RHIC at BNL (LHC at CERN) accelerates heavy ions such as gold (lead) up to $E_{cm}/2 = 100$ GeV (2.8 TeV), i.e. $\sqrt{s_{NN}} = 200$ GeV (5.6 TeV), which gives $y_{cm} = 5.4$ (8.7). This corresponds to $y_{lab} = 10.7$ (17.4), and yields the equivalent laboratory energy of 21 TeV (17 PeV). Before RHIC, nuclear collisions at such high energies had only been observed in cosmic ray events such as the JACEE data (Burnett *et al.*, 1983).

11

Relativistic hydrodynamics for heavy ion collisions

Relativistic hydrodynamics provides a simple picture of the space-time evolution of the hot/dense matter produced in the central rapidity region of relativistic nucleus–nucleus collisions. It allows us to describe, without going into detail about any microscopic model, all stages of the expansion of matter from, possibly, the quark–gluon plasma, through a hadronization transition and freeze-out. With reference to relativistic hydrodynamics, we compare two reaction pictures developed to describe relativistic and ultra-relativistic nucleus–nucleus collisions: the Landau picture and the Bjorken picture. We will emphasize the importance of particle multiplicity and entropy in accessing the conditions of the early stages of matter evolution.

11.1 Fermi and Landau pictures of multi-particle production

In 1950, Fermi published a paper entitled "High energy nuclear events" (Fermi, 1950), which was the first to suggest an ingenious method of applying thermodynamics to multiple meson production in high-energy collisions. The method proposed was based on the assumption that in a collision of high-energy nucleons all of the energy appears at the instant of the collision in a Lorentz-contracted small volume due to the strong interactions (see Fig. 10.3). Fermi proposed that one could then use a statistical method to calculate the multiplicities and spectra of the produced particles. Landau (Landau, (1953); Belensky and Landau, 1955) reexamined Fermi's original idea, arguing that one does not expect the number of finally emitted particles (mesons) to be determined only by the equilibrium condition at the instant of the collision. That is, the system is strongly interacting even after the collision, and the number of particles becomes definite only when the interaction among them becomes small. It was Landau who first introduced relativistic hydrodynamics to describe the expansion stage of the strongly interacting matter. The original papers by Fermi and Landau, as well as other pioneering papers, are collected in Kapusta *et al.* (2003). See also Blaizot and Ollitrault (1990), Csernai (1994) and Rischke (1999) for extensive discussions of the idea applied to relativistic heavy ion collisions.

Let us consider a central (zero-impact parameter) relativistic collision of two equal nuclei of mass number A in the Landau picture (Fig. 10.3). We define the total energy of the system in the center-of-mass frame, $W_{\rm cm}$, as follows:

$$W_{\rm cm} \equiv AE_{\rm cm} = 2Am_{\rm N}\gamma_{\rm cm}. \tag{11.1}$$

Then the initial energy density, ε, is calculated by dividing $W_{\rm cm}$ by the Lorentz-contracted volume of the nucleus, V:

$$\varepsilon = \frac{W_{\rm cm}}{V} = \frac{2Am_{\rm N}\gamma_{\rm cm}}{V_{\rm rest}/\gamma_{\rm cm}} = 2\varepsilon_{\rm nm}\gamma_{\rm cm}^2, \tag{11.2}$$

where $\varepsilon_{\rm nm}$ is approximately the energy density of the nuclear matter,

$$\varepsilon_{\rm nm} \equiv \frac{Am_{\rm N}}{V_{\rm rest}} = 0.15 \text{ GeV fm}^{-3}, \tag{11.3}$$

and $V_{\rm rest}$ is the volume of the nucleus at rest,

$$V_{\rm rest} = \frac{4\pi}{3}R^3 = \frac{4\pi}{3}r_0^3 A \quad (r_0 \approx 1.2 \text{ fm}). \tag{11.4}$$

Similarly, the initial baryon number density is given by

$$\rho = \frac{2A}{V} = \frac{2A}{V_{\rm rest}/\gamma_{\rm cm}} = 2\rho_{\rm nm}\gamma_{\rm cm}, \tag{11.5}$$

where $\rho_{\rm nm}$ is the baryon number density of the normal nuclear matter:

$$\rho_{\rm nm} \equiv \frac{A}{V_{\rm rest}} = 0.16 \text{ fm}^{-3}. \tag{11.6}$$

Landau described the collision process as follows (the quote is from Belensky and Landau, (1955)).

(i) *When two nucleons collide, a compound system is formed, and energy is released in a small volume V subject to a Lorentz contraction in the transverse direction.*[1] *At the instant of collision, a large number of "particles" are formed; the "mean free path" in the resulting system is small compared with its dimensions, and statistical equilibrium is set up.*

(ii) *The second stage of the collision consists in the expansion of the system. Here the hydrodynamic approach must be used, and the expansion may be regarded as the motion of an ideal fluid (zero viscosity and zero thermal conductivity). During*

[1] The word "transverse" here implies "longitudinal" in the notation of the present book.

the process of expansion the "mean free path" remains small in comparison with the dimensions of the system, and this justifies the use of the hydrodynamics. Since the velocities in the system are comparable with that of light, we must use not ordinary but relativistic hydrodynamics. Particles are formed and absorbed in the system throughout the first and second stages of the collision. The high density of energy in the system is of importance here. In this case, the number of particles is not an integral of the system, on account of the strong interaction between the individual particles.

(iii) *As the system expands, the interaction becomes weaker and the mean free path becomes longer. The number of particles appears as a physical characteristic when the interaction is sufficiently weak. When the mean free path becomes comparable with the linear dimensions of the system, the latter breaks up into individual particles. This may be called the "break-up" stage. It occurs with a temperature of the system of the order* $T \approx \mu c^2$, *where* μ *is the mass of the pion. (All temperatures are in energy units.)*

By replacing "nucleons" by "nuclei" in item (i) and "break-up" by "freeze-out" in item (iii), we can also apply the above descriptions to the relativistic nucleus–nucleus collisions. In this case, item (i) corresponds to Figs. 10.3(a) and (b) and items (ii) and (iii) correspond to Fig. 10.3(c). Treating the system as an ideal fluid is an assumption and needs to be examined carefully by microscopic calculations of the transport coefficients such as the viscosities and heat conductivity. In the following, we refer to a "perfect fluid" instead of an "ideal fluid."

If we accept the assumption of the perfect fluid, only an equation of state of matter is necessary for a hydrodynamic description of the system. Let us assume that the energy density, ε, and the pressure, P, of the fluid (which is mainly composed of pions) produced in the relativistic heavy ion collision obey the following familiar relation in black body radiation:

$$P = \frac{1}{3}\varepsilon. \tag{11.7}$$

Further assuming that the baryon chemical potential, $\mu_{\rm B}$, is negligible compared with the temperature, T, the thermodynamic relations, Eqs. (3.13) and (3.15) in Chapter 3, yield

$$Ts = \varepsilon + P, \quad T\, ds = d\varepsilon, \tag{11.8}$$

where s is the entropy density. On combining Eq. (11.7) and Eq. (11.8), we immediately obtain $T \propto s^{1/3}$ and $\varepsilon \propto s^{4/3}$, which yields

$$s \propto \varepsilon^{3/4}, \quad T \propto \varepsilon^{1/4}. \tag{11.9}$$

The results, as expected, coincide with the relations of Stefan–Boltzmann's law discussed in Section 3.5.

Since the initial energy density is written as $\varepsilon \propto \gamma_{\rm cm}^2 \propto E_{\rm cm}^2$ according to Eqs. (11.1) and (11.2), the initial entropy density shortly after the impact is given by

$$s \propto \varepsilon^{3/4} \propto E_{\rm cm}^{3/2}. \tag{11.10}$$

By definition the perfect fluid has no viscosity and does not produce entropy. Therefore, the total entropy of the system, S, stays constant during the course of the hydrodynamical expansion. Also, the number density of the produced particles (pions), n_π, is proportional to s according to the black body formula, and hence the total number of pions produced is $N_\pi \propto S$. Therefore we can relate N_π at the time of the freeze-out to the initial entropy density as follows:

$$N_\pi \propto sV \propto E_{\rm cm}^{3/2} V_{\rm rest}/\gamma_{\rm cm} \propto A E_{\rm cm}^{1/2} \propto A E_{\rm lab}^{1/4}, \tag{11.11}$$

where we have used the relation $E_{\rm cm} \propto E_{\rm lab}^{1/2}$ (from Appendix E). This expression implies a slow growth in production multiplicity with the incident energy. It also shows that the heavy nuclei are much more effective in particle formation than protons.

Detailed evolution of matter in both the longitudinal and transverse directions can be studied by solving the relativistic hydrodynamic equation with the equation of state Eq. (11.7). This predicts a non-isotropic phase-space distribution of produced particles as a result of the hydrodynamical expansion, which takes place predominantly in the longitudinal direction, reflecting the initial conditions of the longitudinally compressed hadronic matter.

A necessary condition for the applicability of the Landau picture to central relativistic nucleus–nucleus collisions is that the nucleons in the front part of each of the colliding nuclei must lose all their kinetic energy in the center-of-mass frame while traversing the other nucleus. This demands that the average energy loss of these nucleons per unit length be greater than the critical value given by

$$\left(\frac{dE}{dz}\right)_{\rm cr} = \frac{E_{\rm cm}/2}{(2R/\gamma_{\rm cm})} \simeq 2\left(\frac{E_{\rm cm}}{10\ {\rm GeV}}\right)^2 \ {\rm GeV\ fm^{-1}}, \tag{11.12}$$

where we have taken $R = 7$ fm (the radius of the Au nucleus). Although Eq. (11.12) yields a number comparable to the typical size of the energy loss in QCD, such as the string tension, $K = 0.9\,{\rm GeV\,fm^{-1}}$, for $E_{\rm cm} < 10\,{\rm GeV}$, it becomes too large to be attainable for ultra-relativistic energies such as $E_{\rm cm} \sim 200\,{\rm GeV}$ at RHIC. We conclude that the Landau picture must break down since the required stopping power becomes too large. Furthermore, in contrast to the Fermi and Landau approaches, the thickness of the colliding nuclei cannot become infinitely small in the ultra-relativistic regime due to the presence of wee partons, as discussed in Section 10.2. Before discussing Bjorken's scaling, which can cure these problems, let us first study the basics of relativistic hydrodynamics.

11.2 Relativistic hydrodynamics

11.2.1 Perfect fluid

Suppose that the local thermal equilibrium is established at a certain stage of the collision. If the mean free path of constituent particles, ℓ, is sufficiently shorter than the length scale characterizing the system, L (i.e. the Knudsen number is much smaller than unity, $K_n = \ell/L \ll 1$), then the later evolution of the system until the freeze-out may be described by relativistic hydrodynamics.

In particular, the perfect fluid is the ideal situation for perfect isotropy to be maintained according to an observer moving with the same local velocity of the fluid because the mean free paths and the times are so short. In this case, the energy-momentum tensor of the fluid is diagonal in the local rest frame of the fluid:

$$\overset{\circ}{T}{}^{\mu\nu}(x) \equiv \begin{bmatrix} \varepsilon(x) & 0 & 0 & 0 \\ 0 & P(x) & 0 & 0 \\ 0 & 0 & P(x) & 0 \\ 0 & 0 & 0 & P(x) \end{bmatrix}, \tag{11.13}$$

where $\varepsilon(x)$ is the local energy density and $P(x)$ is the local pressure. Equation (11.13) is simply Pascal's law, which implies that the pressure exerted by a given portion of fluid is the same in all directions and is perpendicular to the area on which it acts. Recall that $T^{ij}\, df_j = P\, df_i$ is the ith component of the force acting on a surface element, $d\boldsymbol{f}$, whence $T^{ij} = P\delta^{ij}$.

We then proceed to consider a fluid moving with a four-velocity, $u^\mu(x)$, defined by

$$u^\mu(x) = \gamma(x)\,(1, \boldsymbol{v}(x)), \tag{11.14}$$

where $\gamma(x) = 1/\sqrt{1 - \boldsymbol{v}^2(x)}$ and with the normalization

$$u^\mu(x)\, u_\mu(x) = 1. \tag{11.15}$$

Note that u^μ is related to the four-velocity in the local rest frame, $\overset{\circ}{u}{}^\nu = (1, 0, 0, 0)$, by the Lorentz transformation as follows:

$$u^\mu = \Lambda^\mu{}_\nu\, \overset{\circ}{u}{}^\nu. \tag{11.16}$$

Then we have

$$\Lambda^\mu{}_0 = u^\mu. \tag{11.17}$$

Also, the normalization of the four-velocity leads to $g^{\mu\nu}\Lambda^{\rho}{}_{\mu}\Lambda^{\sigma}{}_{\nu} = g^{\rho\sigma}$, which yields

$$\Lambda^{\rho}{}_{i}\Lambda^{\sigma}{}_{i} = \Lambda^{\rho}{}_{0}\Lambda^{\sigma}{}_{0} - g^{\rho\sigma} = u^{\rho}u^{\sigma} - g^{\rho\sigma}. \tag{11.18}$$

In consequence, the energy-momentum tensor of a fluid in motion with velocity u^{μ} is given by

$$T^{\mu\nu} = \Lambda^{\mu}{}_{\rho}\Lambda^{\nu}{}_{\sigma}\overset{\circ}{T}{}^{\rho\sigma} = \Lambda^{\mu}{}_{0}\Lambda^{\nu}{}_{0}\varepsilon + \Lambda^{\mu}{}_{i}\Lambda^{\nu}{}_{i}P \tag{11.19}$$

$$= (\varepsilon + P)\, u^{\mu}u^{\nu} - g^{\mu\nu}P. \tag{11.20}$$

Alternatively, we could derive Eq. (11.20) by simply covariantizing Eq. (11.13).

Now we are in a position to determine the motion of the fluid by using the conservation of the energy-momentum and the baryon number:

$$\partial_{\mu}T^{\mu\nu} = 0, \tag{11.21}$$

$$\partial_{\mu}j^{\mu}_{\mathrm{B}} = 0. \tag{11.22}$$

Here, the baryon number current, $j^{\mu}_{\mathrm{B}}(x)$, is given by

$$j_{\mathrm{B}}{}^{\mu}(x) = n_{\mathrm{B}}(x)\, u^{\mu}(x), \tag{11.23}$$

where the baryon number density, $n_{\mathrm{B}}(x)$, is defined in the local rest frame of the fluid.

The conservation laws, Eqs. (11.21) and (11.22), contain five independent equations, while the equation of state relating ε and P provides an extra equation. By solving these six equations, we determine, for a given initial condition, the space-time evolution of six thermodynamical variables: $\varepsilon(x)$, $P(x)$, $n_{\mathrm{B}}(x)$ and three components of the flow vector, $v_x(x)$, $v_y(x)$ and $v_z(x)$.

To extract a scalar equation, we first contract Eq. (11.21) with the four-velocity,

$$u_{\nu}\partial_{\mu}\left[(\varepsilon + P)\, u^{\mu}u^{\nu} - g^{\mu\nu}P\right] = 0. \tag{11.24}$$

Since the normalization, Eq. (11.15), leads to

$$u_{\nu}\partial_{\mu}u^{\nu} = 0, \tag{11.25}$$

Eq. (11.24) is reduced to

$$u^{\mu}\partial_{\mu}\varepsilon + (\varepsilon + P)\,\partial_{\mu}u^{\mu} = 0. \tag{11.26}$$

We then use the thermodynamic relations $d\varepsilon = Tds + \mu_{\mathrm{B}}dn$ and $\varepsilon + P = Ts + \mu_{\mathrm{B}}n$ given in Eqs. (3.13) and (3.15) together with the baryon number conservation, Eq. (11.22). Then we have

$$\partial_{\mu}(su^{\mu}) = 0, \tag{11.27}$$

which implies that the entropy current is conserved:

$$s^{\mu}(x) = s(x)\, u^{\mu}(x). \tag{11.28}$$

In other words, the fluid motion is adiabatic and reversible. This conclusion is natural because we are considering a perfect fluid with zero viscosity and zero thermal conductivity.

A second set of equations is obtained from the transverse projection of Eq. (11.21):

$$(g_{\rho\nu} - u_{\rho}u_{\nu})\partial_{\mu}T^{\mu\nu} = 0, \tag{11.29}$$

which is reduced to

$$-\partial_{\rho}P + u_{\rho}u^{\mu}\partial_{\mu}P + (\epsilon + P)u^{\mu}\partial_{\mu}u_{\rho} = 0. \tag{11.30}$$

An alternative way of obtaining the same equations is to combine the spatial and temporal components of Eq. (11.21) and use $u^i = u^0 v^i$, which results in (Exercise 11.1)

$$\frac{\partial \boldsymbol{v}}{\partial t} + (\boldsymbol{v}\cdot\nabla)\boldsymbol{v} = -\frac{1-v^2}{\varepsilon + P}\left[\nabla P + \boldsymbol{v}\frac{\partial P}{\partial t}\right]. \tag{11.31}$$

This is a relativistic generalization of the Euler equation. In the non-relativistic case, the right-hand side is $-(\nabla P)/\rho$, where ρ is the mass density.

The basic equations of relativistic hydrodynamics for a perfect fluid are the "continuity equation" for the baryon number density, Eq. (11.22), the "energy equation," Eq. (11.26), or equivalently the "entropy equation," Eq. (11.27), and the "Euler equation," Eq. (11.31). These equations are supplemented by the equation of state (EOS), which relates ε with P and describes the fluid motion starting from a given initial condition.

11.2.2 Dissipative fluid

If the fluid is dissipative, the energy-momentum tensor, Eq. (11.20), the baryon number current, Eq. (11.23), and the entropy current, Eq. (11.28), must be modified to include additional terms, which contain the derivatives of the flow velocity and the thermodynamical variables:

$$T^{\mu\nu} = (\varepsilon + P)u^{\mu}u^{\nu} - Pg^{\mu\nu} + \tau^{\mu\nu}, \tag{11.32}$$

$$j^{\mu}_{\mathrm{B}} = n_{\mathrm{B}}u^{\mu} + \nu^{\mu}_{\mathrm{B}}, \tag{11.33}$$

$$s^{\mu} = su^{\mu} + \sigma^{\mu}. \tag{11.34}$$

Note that $\tau^{\mu\nu}$, $\nu^{\mu}_{\rm B}$ and σ^{μ} are the dissipative parts. The dissipative hydrodynamical equations are derived from $\partial_\mu T^{\mu\nu} = 0$ and $\partial_\mu j^{\mu}_{\rm B} = 0$.

When we have the dissipative corrections, the definition of the flow velocity, u^μ, becomes more arbitrary. Among various possible definitions, the Landau–Lifshitz (LL) definition (Landau and Lifshitz, 1987) and the Eckart definition are useful. In the former (latter), u^μ is identified as an energy (particle) flow defined from $T^{\mu\nu}$ ($j^{\mu}_{\rm B}$). The LL definition is more convenient for application to the central rapidity region in the ultra-relativistic heavy ion collisions, since the baryon number density is small in this region. In the local rest frame of the LL definition ($u^i = 0$), we have $T^{0i} = 0$, and also we define $T^{00} \equiv \varepsilon$ and $j^{0}_{\rm B} \equiv n$. These lead to the following constraints in the general Lorentz frame:

$$u_\mu \tau^{\mu\nu} = 0, \quad u_\mu \nu^{\mu}_{\rm B} = 0. \tag{11.35}$$

It is known that retaining only the linear order of the dissipative quantities in the entropy current, σ^μ, violates causality, and acausal propagation of information takes place. Therefore, to construct a consistent relativistic and dissipative hydrodynamics, we need to go beyond the linear order, as developed in the theories by Müller (1967) and Israel and Stewart (1979). Readers who are interested in such theories should consult Rischke (1999) and Muronga (2002), and references cited therein.

With the above reservations, we will discuss further the approach in linear order to find a physical meaning of the dissipative terms in Eqs. (11.32)–(11.34). For this purpose, we define a second-rank tensor perpendicular to the flow velocity as follows:

$$\Delta^{\mu\nu} = g^{\mu\nu} - u^\mu u^\nu, \quad u_\mu \Delta^{\mu\nu} = 0, \quad \Delta^{\mu}_{\mu} = 0. \tag{11.36}$$

Also, it is convenient to introduce a longitudinal derivative ($\partial_{\|}$) and a transverse derivative ($\partial_{\perp}$) as follows:

$$\partial^\mu = \partial^\mu_{\|} + \partial^\mu_{\perp}, \tag{11.37}$$

$$\partial^\mu_{\|} = u^\mu (u \cdot \partial), \quad \partial^\mu_{\perp} = \partial^\mu - u^\mu (u \cdot \partial). \tag{11.38}$$

Starting from $u_\mu \partial_\nu T^{\mu\nu} = 0$, adopting a similar procedure to that used to derive Eq. (11.27) in the perfect fluid, and using Eq. (11.35), we arrive at (Exercise 11.2(1))

$$\partial_\mu s^\mu = -\nu_{\rm B} \cdot \partial_{\perp} \left(\frac{\mu_{\rm B}}{T} \right) + \frac{1}{T} \tau_{\mu\nu} \partial^{\mu}_{\perp} u^\nu, \tag{11.39}$$

$$\sigma^\mu = -\frac{\mu_{\rm B}}{T} \nu^{\mu}_{\rm B}. \tag{11.40}$$

The explicit forms of $\tau^{\mu\nu}$ and $\nu_{\rm B}$ can be uniquely obtained by assuming (i) that they satisfy the constraints Eq. (11.35), (ii) that they are the functions of the first order derivative $\partial_\mu u_\nu$, and (iii) that the right-hand side of Eq. (11.39) must be positive semi-definite because the second law of thermodynamics indicates that there is an increase in entropy due to dissipative effects, $\partial_\mu s^\mu \geq 0$.

Leaving the derivation as an exercise (Exercise 11.2(2)(3)), we recapitulate the results here:

$$\tau^{\mu\nu} = \eta\left[\partial_\perp^\mu u^\nu + \partial_\perp^\nu u^\mu - \frac{2}{3}\Delta^{\mu\nu}(\partial_\perp \cdot u)\right] + \zeta\Delta^{\mu\nu}(\partial_\perp \cdot u), \tag{11.41}$$

$$\nu_{\rm B}^\mu = \kappa\left(\frac{n_{\rm B} T}{\varepsilon + P}\right)^2 \partial_\perp^\mu \left(\frac{\mu_{\rm B}}{T}\right). \tag{11.42}$$

The shear viscosity, $\eta(\geq 0)$, is a coefficient of the traceless part of the viscosity tensor, $\tau^{\mu\nu}$, while the bulk viscosity, ζ (≥ 0), is a volume effect corresponding to the trace part of the viscosity tensor. Note that κ (≥ 0) denotes the heat conduction which corresponds to the energy flow taking place even without the particle flow (Exercise 11.2(3)).

Finally, the relativistic generalization of the Navier–Stokes equation, which dictates the flow with dissipative effect, is obtained from

$$\Delta_{\rho\nu}\partial_\mu T^{\mu\nu} = 0, \tag{11.43}$$

with $\rho = 1, 2, 3$ (Exercise 11.2(4)). Since some part of the initial energy is converted into the extra entropy, the dissipation generally slows down the expansion of the hot fluid.

11.3 Bjorken's scaling solution

As we have discussed in Section 10.2, owing to the inside–outside cascade, the produced quanta in an ultra-relativistic heavy ion collision materialize far downstream of the target. Namely, the reaction volume is strongly expanded in the longitudinal beam direction (z-axis), as shown in Fig. 10.5. In the first approximation, it is therefore reasonable to drop the transverse spatial dimension (x, y) and to describe the reaction in $(1+1)$ dimensions, z and t. It is then useful to transform the coordinate system from the Cartesian coordinates (t, z) to the proper time, τ, and the space-time rapidity, Y (Fig. 11.1):

$$\tau = \sqrt{t^2 - z^2}, \quad Y = \frac{1}{2}\ln\frac{t+z}{t-z}, \tag{11.44}$$

$$t = \tau\cosh Y, \quad z = \tau\sinh Y. \tag{11.45}$$

Thermodynamic quantities and flow velocity are supposed to be functions of τ and Y only in the $(1+1)$-dimensional flow. For transverse directions, a useful coordinate system is

$$x^\mu = (t, x, y, z) = (\tau \cosh Y, r\cos\phi, r\sin\phi, \tau\sinh Y), \tag{11.46}$$

$$d^4x = dt\ dx\ dy\ dz = \tau d\tau\ dY\ r dr\ d\phi. \tag{11.47}$$

11.3.1 Perfect fluid

Let us now assume that the local thermalization of the matter with short mean free path is realized at a particular early stage of the collision, and we are allowed to use relativistic hydrodynamics for a perfect fluid. Let us construct an ansatz such that the local velocity, $u^\mu(x)$, of the perfect fluid has the same form as the free stream of particles from the origin (Fig. 11.1):

$$u^\mu = \gamma(1, 0, 0, v_z) \tag{11.48}$$

$$\to (t/\tau, 0, 0, z/\tau) = (\cosh Y, 0, 0, \sinh Y), \tag{11.49}$$

where we have taken $v_z = z/t$ in Eq. (11.49). We call this the scaling flow or Bjorken's flow. If it is substituted into the relativistic Euler equation, Eq. (11.31), the left-hand side vanishes automatically, and we obtain

$$\frac{\partial P(\tau, Y)}{\partial Y} = 0, \tag{11.50}$$

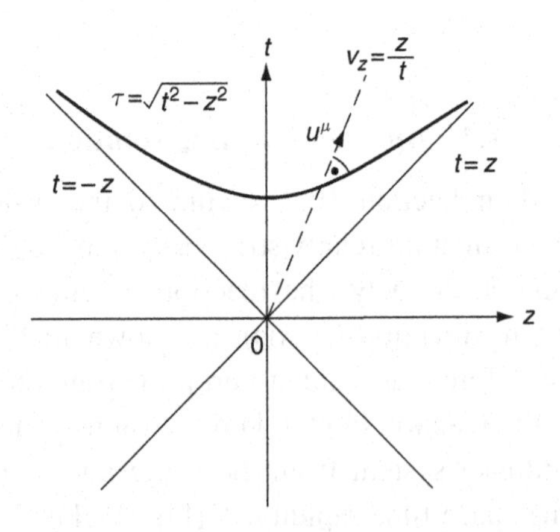

Fig. 11.1. Definition of the $(1+1)$ coordinates. The hyperbola shown by the solid line corresponds to a curve with a constant proper time, $\tau = \sqrt{t^2 - z^2}$. The dashed line represents the direction of the local flow velocity, $u^\mu = (\cosh Y, 0, 0, \sinh Y)$, with the space-time rapidity $Y = \frac{1}{2}\ln[(t+z)/(t-z)]$.

where we have used the transformation law

$$\begin{pmatrix} \frac{\partial}{\partial t} \\ \frac{\partial}{\partial z} \end{pmatrix} = \begin{pmatrix} \cosh Y & -\sinh Y \\ -\sinh Y & \cosh Y \end{pmatrix} \begin{pmatrix} \frac{\partial}{\partial \tau} \\ \frac{1}{\tau}\frac{\partial}{\partial Y} \end{pmatrix}. \tag{11.51}$$

Equation (11.50) implies that the pressure does not depend on the space-time rapidity and hence is constant on the hyperbola in Fig. 11.1. Since Y transforms as $Y \to Y - \tanh^{-1} v_{\rm boost}$ under the Lorentz transformation (Appendix E), it also implies that the pressure, P, is boost-invariant.[2]

By using the relations valid for the scaling flow,

$$u_\mu \partial^\mu = \frac{\partial}{\partial \tau}, \quad \partial^\mu u_\mu = \frac{1}{\tau}, \tag{11.52}$$

the "entropy equation," Eq. (11.27), for the perfect fluid reduces to the simple form:

$$\frac{\partial s(\tau, Y)}{\partial \tau} = -\frac{s(\tau, Y)}{\tau}. \tag{11.53}$$

This is easily solved to give

$$s(\tau, Y) = \frac{s(\tau_0, Y)\tau_0}{\tau}, \tag{11.54}$$

where τ_0 is an initial proper time. Similarly, the "energy" equation, Eq. (11.26), reduces to

$$\frac{\partial \varepsilon}{\partial \tau} = -\frac{\varepsilon + P}{\tau}. \tag{11.55}$$

So far, we have not used the equation of state, and hence the above equations are valid irrespective of the detailed structure of matter. To attain further insights into the solution of the scaling expansion, let us consider a simple form of the equation of state with $\mu_{\rm B} = 0$:

$$P = \lambda \varepsilon, \quad \lambda \equiv c_{\rm s}^2, \tag{11.56}$$

where $c_{\rm s}$ is a numerical constant with the meaning of the speed of sound, $c_{\rm s} = \sqrt{\partial P / \partial \varepsilon}$. For an ideal and relativistic gas, $P = \varepsilon/3$ and $c_{\rm s} = 1/\sqrt{3}$. Combining Eq. (11.56) with the thermodynamics relations, $Ts = \varepsilon + P$ and $dP/dT = s$, we also see that

$$s = a\, T^{1/\lambda}, \quad P = \lambda \varepsilon = \frac{a}{1 + 1/\lambda}\, T^{1+1/\lambda}, \tag{11.57}$$

[2] In local thermal equilibrium, $P(\tau)$ is parametrized by the local temperature, T, and local baryon chemical potential, $\mu_{\rm B}$, as $P(\tau) = P(T(\tau, Y), \mu_{\rm B}(\tau, Y))$. Namely, T and $\mu_{\rm B}$ do not have to be boost-invariant in principle. However, in the central rapidity region of the ultra-relativistic regime, as shown in Fig. 10.6, we have $\mu_{\rm B} \simeq 0$, and hence T becomes boost-invariant.

where a is a constant independent of T (Exercise 11.5). Then the solutions of the hydrodynamic equations become

$$s(\tau) = s_0 \frac{\tau_0}{\tau}, \tag{11.58}$$

$$\varepsilon(\tau) = \varepsilon_0 \left(\frac{\tau_0}{\tau}\right)^{1+\lambda}, \tag{11.59}$$

$$T(\tau) = T_0 \left(\frac{\tau_0}{\tau}\right)^{\lambda}, \tag{11.60}$$

where s_0, ε_0 and T_0 are the values at the initial time, τ_0. The energy density and pressure decrease faster than the entropy under the scaling expansion of the fluid.

Now let us extend our studies on the physical meaning of Eqs. (11.58) and (11.59) (Matsui, 1990). Consider a slice of fluid contained in the region sandwiched by two stream lines:

$$|z| \leq t \frac{\delta v_z}{2}, \tag{11.61}$$

where $\delta v_z \ll 1$. Since the longitudinal thickness of this slice is $\delta z = t\ \delta v_z$, the volume occupied by this fluid element is given by

$$\delta V = \pi R^2 \cdot t \cdot \delta v_z, \tag{11.62}$$

where πR^2 is the transverse area of the nucleus. Then the total entropy contained in this expanding volume is given by

$$\delta S = s \delta V = \pi R^2 \cdot st \cdot \delta v_z = \pi R^2 \cdot s_0 t_0 \cdot \delta v_z, \tag{11.63}$$

where we have used $\tau \simeq t$ for $z \simeq 0$ and Eq. (11.58). Since the right-hand side of Eq. (11.63) is independent of time, the total entropy in this expanding fluid element is a constant of motion.

Next we calculate the internal energy contained in the volume δV by using Eq. (11.59):

$$\delta E = \varepsilon \delta V = \pi R^2 \cdot \varepsilon_0 t_0 \cdot \delta v_z \left(\frac{t_0}{t}\right)^{\lambda}. \tag{11.64}$$

This implies that the internal energy contained in the expanding fluid element is not conserved. The missing part of the internal energy has been converted into the longitudinal flow energy of the adjacent fluid cells by the pressure exerted during the volume expansion. This can be understood better if we rewrite Eq. (11.55) in the following form:

$$\frac{d\,(\tau\varepsilon)}{d\tau} = -P. \tag{11.65}$$

A more realistic analysis of the scaling hydrodynamics with an EOS incorporating the QCD phase transition will be discussed in Chapter 13.

11.3.2 Effect of dissipation

To see the effect of the dissipative corrections to the results in Section 11.3.1, let us consider the first order corrections, $\tau^{\mu\nu}$ and $\nu_{\rm B}$, given in Eqs. (11.41) and (11.42), respectively. In the $(1+1)$-dimensional flow with Eq. (11.49), we obtain

$$u_\mu \partial_\perp^\mu = 0, \qquad \partial_\perp^\mu u_\mu = \frac{1}{\tau}, \tag{11.66}$$

together with Eq. (11.52). Then the non-zero components of the dissipative corrections are given by (Exercise 11.3(1))

$$\tau^{00} = -\left(\frac{4}{3}\eta + \zeta\right)\frac{\sinh^2 Y}{\tau}, \tag{11.67}$$

$$\tau^{33} = -\left(\frac{4}{3}\eta + \zeta\right)\frac{\cosh^2 Y}{\tau}, \tag{11.68}$$

$$\tau^{03} = -\left(\frac{4}{3}\eta + \zeta\right)\frac{\sinh Y \cosh Y}{\tau}, \tag{11.69}$$

$$\nu_{\rm B}^\mu = 0. \tag{11.70}$$

Namely, the shear and bulk viscosities always appear in the combination $\frac{4}{3}\eta+\zeta$, and there is no heat conduction current.

Substituting these in the Navier–Stokes equation, Eq. (11.43), we see that Bjorken's flow is indeed satisfied as long as P and the viscosities are Y-independent and thus boost-invariant (Exercise 11.3(2)). The entropy equation, Eq. (11.39), for the dissipative fluid can then be cast into the following form (Exercise 11.3(3)):

$$\frac{\partial s}{\partial \tau} = -\frac{s}{\tau} + \frac{1}{\tau^2}\frac{\frac{4}{3}\eta+\zeta}{T} \tag{11.71}$$

$$= -\frac{s}{\tau}\left(1 - R_e^{-1}(\tau)\right), \tag{11.72}$$

where R_e is an effective Reynolds number defined by

$$R_e^{-1}(\tau) = \frac{\frac{4}{3}\eta+\zeta}{s}\cdot\frac{1}{T\tau}, \tag{11.73}$$

which is a measure of the relative importance of the dissipative term (the second term on the right-hand side of Eq. (11.71)) over the inertial term (the first term on the right-hand side of Eq. (11.71)). Since the former has an opposite sign to the latter, the dissipative term tends to slow down the dilution of the entropy density.

Note that for large (small) viscosity, R_e^{-1} becomes large (small).[3] Furthermore, R_e is a function of τ: for the relativistic plasma with $c_s^2 = 1/3$, both the viscosity and the entropy density are proportional to T^3. Therefore, $R_e^{-1} \propto (T\tau)^{-1} \sim \tau^{-2/3}$.

We can also derive the energy equation with dissipative corrections from $u_\nu \partial_\mu T^{\mu\nu} = 0$ as follows (Exercise 11.3(3)):

$$\frac{\partial \varepsilon}{\partial \tau} = -\frac{\varepsilon + P}{\tau} + \frac{1}{\tau^2}\left(\frac{4}{3}\eta + \zeta\right). \tag{11.74}$$

The right-hand side of this equation can also be rewritten using the same Reynolds number, $R_e(\tau)$, as given in Eq. (11.73) when $\mu_B = 0$ is satisfied.

The above analyses are, at most, qualitative. To study the realistic effects of dissipation, it is not enough to take simple equations such as Eqs. (11.71) and (11.74). As we have mentioned in Section 11.2.2, we must take into account the second order corrections of the dissipation.

11.4 Relation to the observables

Let us construct the relations between the early-time entropy density, s_0, and the early-time energy density, ε_0, to the observables such as the emitted particle number per unit rapidity, *dN/dy*, and the observed transverse energy per unit rapidity, dE_T/dy, at τ_f in the central AA collisions (Baym *et al.*, 1983; Bjorken, 1983; Gyulassy and Matsui, 1984).

First, we simply assume that the momentum-space rapidity of an observed particle (Appendix E) can be identified with the space-time rapidity, Y, of a fluid element to which the particle belonged at the time of freeze-out, τ_f. Since the volume element on the freeze-out hyper-surface at $\tau = \tau_f$ is $(\pi R^2)\tau_f\, dY$ for $(1+1)$-dimensional expansion, we have

$$\frac{dN}{dy} = \pi R^2 \cdot \tau_f n_f, \tag{11.75}$$

where $n_f = n(\tau_f, Y = y)$ denotes the number density of the constituent particles. Let us further assume the following relation between the entropy density and the number density of particles in the relativistic plasma:

$$s = \xi n, \tag{11.76}$$

[3] At high temperature, the shear viscosity in QCD is evaluated as $\eta \propto T^3/(\alpha_s^2 \ln \alpha_s^{-1})$ (Baym *et al.*, 1990; Arnold *et al.*, 2003). Therefore, η/s increases slowly as T increases, while its behavior near T_c is not precisely known.

where $\xi = 3.6$ (4.2) for an ideal gas of bosons (fermions) at high temperature (Exercise 11.4). Then, by using the fact that $s\tau$ is a constant of motion during the expansion of the perfect fluid, Eq. (11.58), we have

$$s_0 = \frac{\xi}{\pi R^2 \tau_0} \frac{dN}{dy}, \tag{11.77}$$

which relates the initial entropy to the final particles produced. This equation also implies that

$$\frac{dN}{dy} \propto A^{2/3+\delta}, \tag{11.78}$$

if the initial entropy density depends on A as $s_0 \propto A^\delta$.

Similarly, the total energy produced per unit rapidity is given by

$$\frac{dE}{dy} = \pi R^2 \cdot \varepsilon_0 \tau_0 \cdot \left(\frac{\tau_0}{\tau_f}\right)^\lambda, \tag{11.79}$$

where we have used Eq. (11.59). With the relation $dE_T/dy|_{y=0} = dE/dy|_{y=0}$, we find

$$\varepsilon_0 = \frac{1}{\pi R^2 \tau_0} \frac{dE_T}{dy}\bigg|_{y=0} \cdot \left(\frac{\tau_f}{\tau_0}\right)^\lambda. \tag{11.80}$$

On the other hand, Bjorken (1983) first derived an estimate for the case of free streaming:

$$\varepsilon_{0,\mathrm{Bj}} = \frac{1}{\pi R^2 \tau_0} \frac{dE_T}{dy}\bigg|_{y=0}. \tag{11.81}$$

The factor $(\tau_f/\tau_0)^\lambda$ is a measure of the energy transfer due to the work done by the pressure during the hydrodynamical expansion. Therefore, in an attempt to estimate the early energy density, ε_0, using Eq. (11.80), we need information about the freeze-out time, τ_f, in addition to the dE_T/dy data (Exercise 11.5).

Another way to estimate ε_0 is to use Eq. (11.77) for the early entropy density and convert this to the energy density by using the equation of state (Gyulassy and Matsui, 1984). Consider the simple equation of state, $P = \lambda\varepsilon$, given in Eq. (11.56). Because of Eq. (11.57), we have

$$\varepsilon = \frac{1}{(1+\lambda)a^\lambda} s^{1+\lambda}. \tag{11.82}$$

Then, by using Eq. (11.77), we obtain

$$\varepsilon_0 = \frac{1}{(1+\lambda)a^\lambda} \left(\frac{\xi}{\pi R^2 \tau_0} \frac{dN}{dy}\right)^{1+\lambda}. \tag{11.83}$$

This formula tells us that we can estimate the initial energy density, ε_0, by the observed particle number per unit rapidity in the central rapidity region (Exercise 11.5).

Two remarks are in order before closing this chapter. First, hydrodynamics itself cannot tell which value we should take for the initial conditions, i.e. the initial time, τ_0 (the time when the local thermalization is reached), and the spatial profile of the temperature and baryon chemical potential at τ_0. Therefore these parameters are usually adjusted to fit the final observables. A more fundamental approach would be to calculate the initial conditions from microscopic theories without assuming thermal equilibrium. This includes attempts such as the parton cascade model, the color-string model and the color glass condensate model, which will be discussed in Chapter 13. The second remark concerns the use of the realistic equation of state with QCD phase transitions and also the effect of the transverse expansion of the plasma, which will be discussed further in Chapter 13.

Exercises

11.1 Relativistic Euler equation. Derive the relativistic Euler equation, Eq. (11.31), by evaluating $v^i(\partial_\mu T^{\mu 0}) - \partial_\mu T^{\mu i} = 0$.

11.2 Relativistic Navier–Stokes equation.

(1) Derive the equation for the entropy current, Eq. (11.39), for a dissipative fluid.

(2) Derive the dissipative tensor, $\tau^{\mu\nu}$, Eq. (11.41), in the first order inhomogeneities of u^μ. Derive the energy-momentum tensor in the local rest frame of the Landau–Lifshitz definition with the flow velocity ($u^i = 0$), and study the physical meaning of the shear and bulk viscosities in this frame.

(3) Derive the dissipative vector, ν^μ_{B}, given in Eq. (11.42). Taking a frame where particle flux is zero ($n_{\mathrm{B}} u^i + \nu^i_{\mathrm{B}} = 0$), show that the energy flow, T^{0i}, is reduced to

$$-\kappa\left[\nabla T - \frac{T}{\varepsilon + P}\nabla P\right],$$

in the first order inhomogeneities of u^μ.

(4) Derive the Navier–Stokes equation from $\Delta_{\rho\nu}\partial_\mu T^{\mu\nu} = 0$ and $\tau^{\mu\nu}$ in Eq. (11.41).

11.3 Dissipative flow.

(1) Derive the explicit form of the viscosity tensor, Eqs. (11.67)–(11.69) and the heat conduction current, Eq. (11.70), for Bjorken's flow, Eq. (11.49).

(2) Show that Bjorken's flow satisfies the Navier–Stokes equation, Eq. (11.43), as long as P and the viscosities are boost-invariant.

(3) Derive the entropy and energy equations, Eqs. (11.71) and (11.74) for a dissipative fluid, keeping in mind that the viscosities, η and ζ, can still be a function of τ.

11.4 Model equation of state.

(1) Show that $\xi \equiv s/n$ in Eq. (11.76) takes the value 3.6 (4.2) for an ideal gas of massless bosons (fermions) at finite temperature with zero chemical potential. Consult Chapter 3 for help in evaluating the entropy density and number density.

(2) Show that the following relation holds for an ideal gas of massless quarks and gluons at high temperature in equilibrium:

$$\xi = \frac{s}{n} = \frac{2\pi^4}{45\zeta(3)} \frac{d_{\rm g} + \frac{7}{8}d_q}{d_{\rm g} + \frac{3}{4}d_q} = \begin{cases} 3.60 \; (N_f = 0), \\ 3.92 \; (N_f = 2), \\ 3.98 \; (N_f = 3), \\ 4.02 \; (N_f = 4), \\ 4.20 \; (N_f = \infty), \end{cases}$$

where $d_{\rm g}$ and d_q are the degeneracy factors given in Table 3.1.

11.5 Initial energy density in AA collisions. Estimate the early energy density, $\varepsilon_0(\tau_0 = 1\,\text{fm})$, in ultra-relativistic Au + Au central collisions applying the three different formulas in Eqs. (11.80), (11.81) and (11.83). Assume that $dN/dy = 1000$ and $dE_{\rm T}/dy = 500\,\text{MeV} \times dN/dy$ at $y = 0$. Also use $R = 7\,\text{fm}$, $\xi = 4$, $\lambda = c_{\rm s}^2 = 1/3$, $\tau_{\rm f} = 8\,\text{fm}$ and $a = 4d_{\rm QGP}\pi^2/90$ with $d_{\rm QGP} = 37$ in Table 3.1.

12

Transport theory for the pre-equilibrium process

We begin this chapter with a summary of the classical kinetic theory, namely the Boltzmann transport equation and the Boltzmann H-theorem (Huang, 1987). We then present an essential idea of deriving relativistic and quantum mechanical transport theory for a non-Abelian quark–gluon system on the basis of the Wigner function and the semi-classical expansion. The quark–gluon transport theory is one of the possible approaches used to describe the pre-equilibrium stage of the ultra-relativistic heavy ion collisions, as will be discussed further in Chapter 13.

12.1 Classical Boltzmann equation

In order to describe a non-uniform system of many classical particles, let us introduce a one-particle distribution function, $f(\boldsymbol{x}, \boldsymbol{p}, t)$. We do not consider spin, or any other internal degrees of freedom, for simplicity. We assume that, although the particles obey classical mechanics, they are identical and indistinguishable. The function $f(\boldsymbol{x}, \boldsymbol{p}, t)$ describes the number density of particles at time, t, in the phase-space volume, $d^3x\, d^3p$, at the phase-space point, $(\boldsymbol{x}, \boldsymbol{p})$.

The particle density and current can be expressed in terms of $f(\boldsymbol{x}, \boldsymbol{p}, t)$ as follows:

$$n(\boldsymbol{x}, t) = \int d^3p\, f(\boldsymbol{x}, \boldsymbol{p}, t), \tag{12.1}$$

$$\boldsymbol{j}(\boldsymbol{x}, t) = \int d^3p\, \boldsymbol{v}\, f(\boldsymbol{x}, \boldsymbol{p}, \mathrm{t}), \quad \boldsymbol{v} = \frac{\partial \varepsilon(\boldsymbol{p})}{\partial \boldsymbol{p}} = \frac{\boldsymbol{p}}{\varepsilon}, \tag{12.2}$$

where $\boldsymbol{v}$ is the single-particle velocity, with $\varepsilon = \sqrt{\boldsymbol{p}^2 + m^2}$.

The change in distribution with time takes place through two different processes, the drift and the collision:

$$\left(\frac{\partial f}{\partial t}\right) = \left(\frac{\partial f}{\partial t}\right)_{\text{drift}} + \left(\frac{\partial f}{\partial t}\right)_{\text{collision}}. \tag{12.3}$$

The contributions on the right-hand side can be described as follows.

(i) The drift term describes the change in particle distribution through the single-particle motion of the particles flowing into and out of the phase-space volume at $(\boldsymbol{x}, \boldsymbol{p})$:

$$\left(\frac{\partial f}{\partial t}\right)_{\text{drift}} = -\left(\boldsymbol{v}\cdot\nabla_x + \boldsymbol{F}\cdot\nabla_p\right) f, \tag{12.4}$$

where $\boldsymbol{v}$ and $\boldsymbol{F}(\boldsymbol{x})$ are the velocity and the external force, respectively. This is simply a result of the classical equation of motion for independent particles.

(ii) The collision term describes the change in f through kicking-in (gain) and kicking-out (loss) processes due to the particle collisions in the phase-space volume at $(\boldsymbol{x}, \boldsymbol{p})$:

$$\left(\frac{\partial f}{\partial t}\right)_{\text{collision}} = C_{\text{gain}} - C_{\text{loss}}, \tag{12.5}$$

where

$$C_{\text{gain}} = \frac{1}{2}\int d^3p_2\, d^3p_1'\, d^3p_2'\; w(1'2' \to 12) f^{[2]}(\boldsymbol{x}, \boldsymbol{p}_1', \boldsymbol{p}_2', t),$$

$$C_{\text{loss}} = \frac{1}{2}\int d^3p_2\, d^3p_1'\, d^3p_2'\; w(12 \to 1'2') f^{[2]}(\boldsymbol{x}, \boldsymbol{p}_1, \boldsymbol{p}_2, t).$$

Here we have considered only the elastic binary collisions, as shown in Fig. 12.1. Note that $w(12 \to 1'2')\; d^3p_1'\; d^3p_2'$ is the transition rate for two particles with momenta $(\boldsymbol{p}_1, \boldsymbol{p}_2)$ to be scattered into the range $(\boldsymbol{p}_1', \boldsymbol{p}_2') \sim (\boldsymbol{p}_1' + d\boldsymbol{p}_1', \boldsymbol{p}_2' + d\boldsymbol{p}_2')$. The detailed balance relation is then given by

$$w(12 \to 1'2') = w(1'2' \to 12), \tag{12.6}$$

which follows from the time-reversal and rotational invariance of the two-body scattering (Exercise 12.1).

It should be noted that $f^{[2]}(\boldsymbol{x}, \boldsymbol{p}_1, \boldsymbol{p}_2, t)$ is a two-particle distribution function which is proportional to the probability of finding one particle at $(\boldsymbol{x}, \boldsymbol{p}_1)$ and another particle at $(\boldsymbol{x}, \boldsymbol{p}_2)$, at time t. The factor 1/2 in front of the gain and loss terms reflects the fact that the two-particle "state" with momentum $(\boldsymbol{p}_1', \boldsymbol{p}_2')$ is indistinguishable from the "state" with $(\boldsymbol{p}_2', \boldsymbol{p}_1')$ because of the identical nature of the particles.

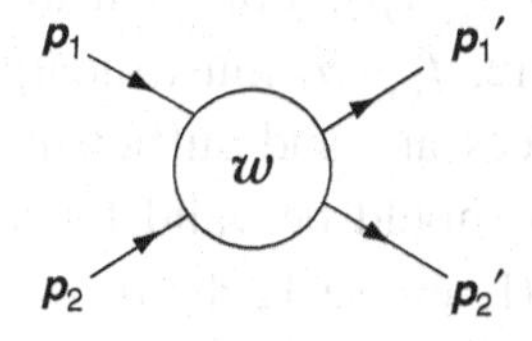

Fig. 12.1. Elastic binary scattering of identical particles with transition rate w.

A crucial step taken by Boltzmann in 1872 was the "Strosszahl Ansatz," in which the correlation between the two particles before the collision is neglected and the two-particle distribution is reduced to a product of the one-particle distributions:

$$f^{[2]}(\boldsymbol{x}, \boldsymbol{p}_1, \boldsymbol{p}_2, t) = f(\boldsymbol{x}, \boldsymbol{p}_1, t) f(\boldsymbol{x}, \boldsymbol{p}_2, t). \tag{12.7}$$

Combining this ansatz with Eq. (12.3), we arrive at the celebrated non-linear integro-differential equation, due to Boltzmann:

$$\left(\frac{\partial}{\partial t} + \boldsymbol{v} \cdot \nabla_x + \boldsymbol{F} \cdot \nabla_p\right) f(\boldsymbol{x}, \boldsymbol{p}, t) = C[f], \tag{12.8a}$$

$$C[f] = \frac{1}{2} \int d^3p_2\, d^3p_1'\, d^3p_2'\, w(12 \to 1'2')$$

$$\times \left[f(\boldsymbol{x}, \boldsymbol{p}_1', t) f(\boldsymbol{x}, \boldsymbol{p}_2', t) - f(\boldsymbol{x}, \boldsymbol{p}_1, t) f(\boldsymbol{x}, \boldsymbol{p}_2, t)\right], \tag{12.8b}$$

where $\boldsymbol{p} \equiv \boldsymbol{p}_1$. The differential cross-section, $d\sigma$, is related to the transition rate as follows:

$$v_{12}\, d\sigma = w(12 \to 1'2')\, d^3p_1'\, d^3p_2', \tag{12.9}$$

where $v_{12} = |\boldsymbol{v}_1 - \boldsymbol{v}_2|$ is the flux factor per collision. With Ω defined as a scattering solid angle between the initial and final relative momenta, $\boldsymbol{p}_1 - \boldsymbol{p}_2$ and $\boldsymbol{p}_1' - \boldsymbol{p}_2'$, we have

$$C[f] = \frac{1}{2} \int d^3p_2 \int d\Omega\, v_{12} \left(\frac{d\sigma}{d\Omega}\right) (f_{1'} f_{2'} - f_1 f_2). \tag{12.10}$$

Here, the shorthand notation, $f_i \equiv f(\boldsymbol{x}, \boldsymbol{p}_i, t)$, is used. Note that most of the integrals, $\int d^3p_1'\, d^3p_2'$, on the right-hand side of Eq. (12.8b) can be carried out due to implicit δ-functions in w representing the conservation of both total energy and total momentum. This is why the integration over the final states reduces to that for the solid angle, $\int d\Omega$, in Eq. (12.10).

The well known Maxwell–Boltzmann distribution for equilibrium systems is derived as a *unique* stationary solution of the transport equation. To see this, consider a uniform and stationary system with no external force, and define the corresponding distribution as $f_{MB}(\boldsymbol{p})$. For such a system the left-hand side of Eq. (12.8a) vanishes. Therefore, $f_{MB}(\boldsymbol{p})$ must satisfy $C[f_{MB}] = 0$.

We can prove that the necessary and sufficient condition for $C[f_{MB}] = 0$ is the following relation, which should be valid for any combination of momenta satisfying $\boldsymbol{p}_1' + \boldsymbol{p}_2' = \boldsymbol{p}_1 + \boldsymbol{p}_2$ (Exercise 12.4(3)):

$$f_{MB}(\boldsymbol{p}_1') f_{MB}(\boldsymbol{p}_2') = f_{MB}(\boldsymbol{p}_1) f_{MB}(\boldsymbol{p}_2). \tag{12.11}$$

Now, by taking the logarithm of Eq. (12.11), we find that $\ln f_{\rm MB}(\boldsymbol{p})$ is an additive and a conserved quantity. Therefore, it is written as a linear combination of the single-particle energy and the momentum:

$$\ln f_{\rm MB}(\boldsymbol{p}) = a + b_0\ \varepsilon(\boldsymbol{p}) + \boldsymbol{b}\cdot\boldsymbol{p}, \tag{12.12}$$

where a, b_0 and $\boldsymbol{b}$ are constants to be determined by the averaged quantities of the system. For example, for non-relativistic particles with averaged momentum $\boldsymbol{p}_0$, the particle density, n, and the pressure parametrized as $P = nT$ lead to the standard Maxwell–Boltzmann distribution (Exercise 12.2):

$$f_{\rm MB}(\boldsymbol{p}) = \frac{n}{(2\pi m T)^{3/2}}\, {\rm e}^{-(\boldsymbol{p}-\boldsymbol{p}_0)^2/2mT}. \tag{12.13}$$

Let us discuss briefly the relaxation time approximation, which is useful for a qualitative study of the transport equation. Suppose that the system is not far from the local equilibrium distribution satisfying $C[f_{\rm eq}(\boldsymbol{x},\boldsymbol{p},t)] = 0$ (note that $f_{\rm eq}$ is not necessarily the Maxwell–Boltzmann distribution, $f_{\rm MB}$). One may then linearize the right-hand side of the transport equation and approximate it as follows:

$$C[f] \simeq -\frac{f - f_{\rm eq}}{\tau}. \tag{12.14}$$

Here, τ characterizes the typical time scale for the system to relax to the local equilibrium distribution. It may be estimated by the mean free time between the collisions,

$$\tau = \frac{1}{n\sigma_{\rm tot} v}, \tag{12.15}$$

where $\sigma_{\rm tot}$ is an averaged total cross-section of the binary collision, v is the averaged relative velocity of particles and n is the averaged particle density in the system. Note that $\ell \equiv (n\sigma_{\rm tot})^{-1}$ is simply the mean free path, which is the average distance traversed by a particle between two successive collisions. (See Exercise 12.3 for more on the connection between τ and the linearized collision term, $C[f]$.)

The reason why τ acts as a relaxation time may be easily understood by considering the example of the spatially uniform distribution without external force. In this case, the solution of the transport equation with the relaxation time approximation is given by

$$f(\boldsymbol{p},t) = f_{\rm MB}(\boldsymbol{p}) + \big(f(\boldsymbol{p},0) - f_{\rm MB}(\boldsymbol{p})\big)\,{\rm e}^{-t/\tau}, \tag{12.16}$$

which shows that the system approaches the equilibrium distribution with a typical time scale τ.

Let us end this section with a remark concerning the derivation of the Boltzmann transport equation from the equations of motion for many interacting particles.

In general, the exact equations are the coupled integro-differential equations for the multi-particle distribution functions:

$$f^{[j]}(\boldsymbol{x}_1, \boldsymbol{x}_2, \ldots, \boldsymbol{x}_j, \boldsymbol{p}_1, \boldsymbol{p}_2, \ldots, \boldsymbol{p}_j, t), \tag{12.17}$$

where j runs from 1 to N (the total number of particles). The coupled equations have a hierarchical structure in which $f^{[j]}$ is related to $f^{[j+1]}$. This is called the Bogoliubov–Born–Green–Kirkwood–Yvon (BBGKY) hierarchy. The Boltzmann transport equation is obtained by truncating this hierarchy by retaining only the slowly varying degrees of freedom and replacing $f^{[2]}$ by a product of $f^{[1]}$, according to the Strosszahl Ansatz.

12.2 Boltzmann's H-theorem

Associated with the time variation of the one-particle distribution function, $f(\boldsymbol{x}, \boldsymbol{p}, t)$, we may introduce the following entropy density and entropy current:

$$s(\boldsymbol{x}, t) = \int d^3p\, f(\boldsymbol{x}, \boldsymbol{p}, t)\,(1 - \ln f(\boldsymbol{x}, \boldsymbol{p}, t)), \tag{12.18}$$

$$\boldsymbol{s}(\boldsymbol{x}, t) = \int d^3p\, \boldsymbol{v}\, f(\boldsymbol{x}, \boldsymbol{p}, t)\,(1 - \ln f(\boldsymbol{x}, \boldsymbol{p}, t)). \tag{12.19}$$

The Boltzmann entropy, $S(t)$, and the associated H-function for a non-equilibrium system, are defined by the spatial integration of $s(\boldsymbol{x}, t)$:

$$S(t) \equiv -H(t) = \int d^3x\, s(\boldsymbol{x}, t). \tag{12.20}$$

The time variation of $s(\boldsymbol{x}, t)$ and the entropy flow are related as follows:

$$\frac{\partial s}{\partial t} + \nabla \cdot \boldsymbol{s} = -\int d^3p\, C[f]\, \ln f, \tag{12.21}$$

where we have used the transport equation, Eq. (12.8a), and we have assumed that the surface integral vanishes at infinity. For a local equilibrium system with $C[f] = 0$, Eq. (12.21) becomes an equation for the conservation of the total entropy, $S(t)$.

Let us rewrite the collision integral on the right-hand side of Eq. (12.21) for a general non-equilibrium situation. Interchanging the variables (p_1, p_2, p_1', p_2'), using the symmetry properties of the scattering probability $w(12 \to 1'2') = w(21 \to 2'1') = w(1'2' \to 12)$, we arrive at the following expression (Exercise 12.4(1)):

$$\frac{dS}{dt} = \frac{1}{8} \int d^3x\, d^3p_1\, d^3p_2\, d^3p_1'\, d^3p_2'\, w(12 \to 1'2') \times (f_{2'} f_{1'} - f_2 f_1)(\ln(f_{2'} f_{1'}) - \ln(f_2 f_1)) \geq 0. \tag{12.22}$$

The inequality is due to the relation $(x-y)(\ln x - \ln y) \geq 0$ together with the positivity of the transition rate, w. Thus, we find that the entropy (the H-function) increases (decreases) in time, and $dS/dt = 0$ is satisfied only when the local equilibrium condition, $f_{2'} f_{1'} = f_2 f_1$ (or equivalently $C[f] = 0$), is met. This constitutes the statement and the proof of Boltzmann's H-theorem.

The theorem implies that an approach to thermal equilibrium is associated with entropy production ($dS/dt \geq 0$), and that its rate is controlled by the particle collision, $C[f]$. The origin of the irreversibility of the Boltzmann equation is the Strosszahl Ansatz, Eq. (12.7), for the binary system *before* the collision in the BBGKY hierarchy. Once the collision has taken place, correlations between the particles are established, which invalidates the ansatz. If the Strosszahl Ansatz is introduced *after* the collision in the BBGKY hierarchy, the sign in front of the collision term is reversed, and the associated entropy decreases in time. (See, for example, Cohen and Berlin (1960).)

12.3 Covariant form of the classical transport equation

The transport equation discussed in Section 12.2 is applicable to both non-relativistic and relativistic particles. For the latter case, however, it is more convenient to rewrite the equation in a covariant form (de Groot *et al.*, 1980).

First, we introduce a distribution function as a function of $x^\mu = (t, \boldsymbol{x})$ and $p^\mu = (p^0 = \varepsilon(\boldsymbol{p}), \boldsymbol{p})$:

$$f(x,p)|_{p^0=\varepsilon(\boldsymbol{p})} \equiv f(\boldsymbol{x}, \boldsymbol{p}, t). \tag{12.23}$$

By using this notation, the particle-number current, the energy-momentum tensor and the entropy flow are written, respectively, as follows:

$$j^\mu(x) = \int \frac{d^3p}{p^0} p^\mu f(x,p), \tag{12.24}$$

$$T^{\mu\nu}(x) = \int \frac{d^3p}{p^0} p^\mu p^\nu f(x,p), \tag{12.25}$$

$$s^\mu(x) = \int \frac{d^3p}{p^0} p^\mu f(x,p)(1 - \ln f(x,p)). \tag{12.26}$$

They are related to the quantities in the previous Section 12.2 as follows:

$$j^\mu(x) = (n(\boldsymbol{x},t), \boldsymbol{j}(\boldsymbol{x},t)), \tag{12.27}$$

$$s^\mu(x) = (s(\boldsymbol{x},t), \boldsymbol{s}(\boldsymbol{x},t)). \tag{12.28}$$

In Eqs. (12.24)–(12.26), $d^3p/p^0 \equiv d^3\bar{p}$ is a Lorentz-invariant volume element satisfying the following identity:

$$\frac{d^3p}{2p^0} = d^4p\ \theta(p^0)\delta(p^2 - m^2). \tag{12.29}$$

We obtain a covariant form of the transport equation by multiplying p^0 and both sides of Eq. (12.8a):

$$\left(p^\mu \partial^x_\mu + \bar{F}^\mu(x,p)\partial^p_\mu\right) f(x,p) = \bar{C}\,[f], \tag{12.30}$$

with

$$\bar{C}\,[f] = \frac{1}{2}\int d^3\bar{p}_2\ d^3\bar{p}'_1\ d^3\bar{p}'_2\ \bar{w}(12 \to 1'2')\ (f_{1'}f_{2'} - f_1f_2). \tag{12.31}$$

Here we have introduced the Lorentz-invariant collision term and the invariant transition rate:

$$\bar{C}[f] = p^0 C[f], \tag{12.32}$$

$$\bar{w}(12 \to 1'2') = \varepsilon_1\varepsilon_2\varepsilon'_1\varepsilon'_2\ w(12 \to 1'2'). \tag{12.33}$$

The external force, $\bar{F}^\mu(x,p)$, in Eq. (12.30) is defined as

$$\bar{F}^\mu(x,p) = (p^0\boldsymbol{v}\cdot\boldsymbol{F},\, p^0\boldsymbol{F}). \tag{12.34}$$

For the Lorentz force in classical electrodynamics, we have $\boldsymbol{F} = q(\boldsymbol{E} + \boldsymbol{v}\times\boldsymbol{B})$, where q is the electric charge of classical particles. In this case, using the relations $F^{0i} = E^i$ and $2F_{ij} = \epsilon_{ijk}B^k$, where $F^{\mu\nu}(x)$ is the field strength tensor of the electromagnetic field, we find

$$\bar{F}^\mu(x,p) = -qF^{\mu\nu}(x)p_\nu, \tag{12.35}$$

and also

$$p_\mu\bar{F}^\mu(x,p) = 0, \quad \partial^p_\mu\bar{F}^\mu(x,p) = 0. \tag{12.36}$$

12.3.1 Conservation laws

Microscopic conservation of energy, momentum and other quantum numbers is related to macroscopic conservation laws through the following theorem. Consider some conserved quantity, $\chi(x,p)$, which satisfies

$$A \equiv (\chi_1 + \chi_2) - (\chi_{1'} + \chi_{2'}) = 0, \tag{12.37}$$

with the definition $\chi_i = \chi(x, p_i)$. Then the following relation holds:

$$\int d^3\bar{p}\ \chi(x, p)\ \bar{C}[f] \tag{12.38}$$

$$= \frac{1}{8}\int d^3\bar{p}_1\ d^3\bar{p}_2\ d^3\bar{p}_{1'}\ d^3\bar{p}_{2'}\ \bar{w}(12 \to 1'2')\ (f_{1'}f_{2'} - f_1 f_2)\ A = 0,$$

where we have applied a similar step as in proof of the H-theorem (see Eq. (12.22) and Exercise 12.4(2)).

Using the Boltzmann transport equation and the relation $\partial^p_\mu \bar{F}^\mu(x, p) = 0$ after the partial integration, we arrive at the macroscopic conservation laws for $\chi = 1$ and $\chi = p^\mu$ as

$$\partial_\mu j^\mu(x) = 0, \quad \partial_\nu T^{\mu\nu}(x) = 0. \tag{12.39}$$

These equations correspond to number conservation and energy-momentum conservation, respectively.

12.3.2 Local H-theorem and local equilibrium

It is obvious from the proof of the H-theorem in Section 12.2 that the theorem is valid even locally:

$$\partial_\mu s^\mu(x) = -\int d^3\bar{p}\ \ \bar{C}[f] \ln f \geq 0. \tag{12.40}$$

This leads to the definition of the local equilibrium, $\partial_\mu s^\mu(x) = 0$, or equivalently

$$f(x, p_1) f(x, p_2) = f(x, p_1') f(x, p_2'). \tag{12.41}$$

As before, this is easily solved by writing

$$\ln f(x, p) = a(x) + b_\mu(x) p^\mu, \tag{12.42}$$

which leads to a local form of the Boltzmann distribution:

$$f_{\mathrm{B}}(x, p) \propto \mathrm{e}^{-(p_\mu u^\mu(x) - \mu(x))/T(x)}. \tag{12.43}$$

Here we have introduced a "local" chemical potential, $\mu(x)$, a "local" temperature, $T(x)$, and a "local" velocity, $u^\mu(x)$, with $u_\mu u^\mu = 1$ so that five parameters a and b^μ $(\mu = 0, 1, 2, 3)$ are given by five independent parameters T, μ and $\boldsymbol{u}$ as $a(x) = \mu(x)/T(x)$ and $b^\mu(x) = u^\mu(x)/T(x)$.

So far, we have not taken into account the Bose or Fermi statistics in the collision term, except for the factor 1/2 in front of the expression originating from the identical nature of the particles. The Bose–Einstein (BE) statistics and the

Fermi–Dirac (FD) statistics may be easily incorporated in the collision term by the following substitution:

$$f_1'f_2' - f_1 f_2 \rightarrow \\ f_{1'}f_{2'}(1\pm f_1)(1\pm f_2) - (1\pm f_{1'})(1\pm f_{2'})f_1 f_2, \tag{12.44}$$

where $+$ $(-)$ is for BE (FD) statistics.

The solution of the local equilibrium condition is again easy to solve, and finally we obtain

$$f_{\mathrm{BE(FD)}}(x,p) = \frac{1}{(2\pi\hbar)^3}\frac{1}{\mathrm{e}^{(p_\mu u^\mu(x)-\mu(x))/T(x)} \mp 1}, \tag{12.45}$$

where we have determined the overall constant by demanding that the integration of $\int d^3p\, f_{\mathrm{BE(FD)}}(x,p)$ in the global equilibrium situation yields the correct definition of the particle number density, n (see Chapter 3).

Once the local equilibrium is established, Eq. (12.45), together with the definition of the particle current and the energy-momentum tensor in Eqs. (12.24) and Eq. (12.25), lead to

$$j^\mu = nu^\mu, \quad T^{\mu\nu} = (\varepsilon + P)u^\mu u^\nu - Pg^{\mu\nu}, \tag{12.46}$$

where n, ε and P are defined as $\int f\, d^3p$, $\int p^0 f\, d^3p$ and $\int \frac{1}{3}(\boldsymbol{v}\cdot\boldsymbol{p}) f\, d^3p$, respectively (Exercise 12.5).

Equation (12.46), together with the conservation laws, Eq. (12.39), thus provide a basis of the hydrodynamics for the perfect fluid discussed in Section 11.2.1. In terms of the Boltzmann equation, this is a leading order of the Chapman–Enskog asymptotic expansion (de Groot *et al.*, 1980) in which the ratio of the mean free path and the spatial non-uniformities of the system act as a small expansion parameter. The dissipative hydrodynamics discussed in Section 11.2.2 is obtained as the next-to-leading and higher orders of this expansion.

12.4 Quantum transport theory

The classical transport theory is a consequence of the classical equations of motion. Similarly, the quantum transport equation may be derived from the quantum Heisenberg equations (Kadanoff and Baym, 1962; de Groot *et al.*, 1980).

In the following, we will outline the derivation of the transport equation of quantum Dirac particles moving in a c-number electromagnetic mean-field as an example (Vasak *et al.*, 1987). The derivation contains the essential ideas of the quantum transport theory with a gauge field and its relation to the classical transport equation. Generalizations to the case with a c-number non-Abelian gauge field are much more tedious, but the basic ideas are similar (Elze and Heinz, 1989; Blaizot and Iancu, 2002).

In the following, we will write $\hbar$ explicitly, unless otherwise stated, so that we can make a direct connection to the classical transport theory.

12.4.1 The density matrix

We will work in the Heisenberg picture, where quantum operators evolve in time while the state vectors remain time-independent. An ensemble average of a Heisenberg operator, $A(t)$, over a mixed state is defined as:

$$\langle A(t)\rangle = \sum_m P_m \langle m|A(t)|m\rangle = \mathrm{Tr}[\rho A(t)], \tag{12.47}$$

where the density matrix, ρ, is given by

$$\rho = \sum_m |m\rangle P_m \langle m|, \tag{12.48}$$

with $\langle m|m\rangle = 1$. Note that the probability is positive semi-definite and is normalized to unity ($P_m \geq 0$ and $\sum_m P_m = 1$).

Note that the ensemble $\{|m\rangle\}$ does not have to be a complete set or an orthonormal set. The density matrix, ρ, satisfies the relations

$$\mathrm{Tr}\ \rho = 1, \quad \mathrm{Tr}\ \rho^2 \leq 1. \tag{12.49}$$

The density matrix for a pure state ($P_m = \delta_{mn}$) is characterized by $\rho^2 = \rho$ or, equivalently, $\mathrm{Tr}\ \rho^2 = 1$.

In the following, an ensemble average over a certain unspecified mixed state is always denoted by the bracket $\langle A(t)\rangle$.

12.4.2 The Dirac equation

A quantum Dirac particle with charge q moving under a c-number Abelian field, $A_\mu(x)$, obeys the Dirac equation:

$$(i\hbar\gamma\cdot D - m)\psi(x) = 0, \ \ \bar{\psi}(x)(i\hbar\gamma\cdot D^\dagger + m) = 0, \tag{12.50}$$

where the covariant derivatives are given by

$$D_\mu = \partial_\mu + i\frac{q}{\hbar}A_\mu(x), \quad D_\mu^\dagger = \partial_\mu^\dagger - i\frac{q}{\hbar}A_\mu(x), \tag{12.51}$$

and the ordinary derivatives are defined as

$$\partial_\mu \equiv \frac{\overrightarrow{\partial}}{\partial x^\mu}, \quad \partial_\mu^\dagger \equiv \frac{\overleftarrow{\partial}}{\partial x^\mu}. \tag{12.52}$$

The Dirac equation is gauge-covariant in the sense that the same equation holds after the local gauge transformation

$$\psi \to \psi' = e^{i\frac{q}{\hbar}\Lambda}\psi, \quad A_\mu \to A'_\mu = A_\mu - \partial_\mu \Lambda. \tag{12.53}$$

There is a back-reaction of the Dirac particles on the c-number Abelian field. After the ensemble average, it is described by

$$\partial_\nu F^{\nu\mu} = q\langle \bar{\psi}\gamma^\mu \psi\rangle. \tag{12.54}$$

12.4.3 The Wigner function

A quantum analogue of the classical distribution function, $f(x, p)$, is the Wigner function, $W(x, p)$. It was originally introduced by Wigner (1932) to study quantum corrections to the thermodynamic quantities in many-particle systems. If it is integrated over p (x), the particle "density" in the coordinate (momentum) space is obtained. Therefore, it is a kind of distribution function in the phase-space. However, $W(x, p)$ cannot really be interpreted as the probability density since it is not always positive.

For a quantum Dirac particle with spin 1/2, one may define the Wigner function as follows:

$$W_{\alpha\beta}(x, p) = \int \frac{d^4y}{(2\pi\hbar)^4}\, e^{-ipy/\hbar}\left\langle \bar{\psi}_\beta\left(x+\frac{y}{2}\right)\psi_\alpha\left(x-\frac{y}{2}\right)\right\rangle, \tag{12.55}$$

where α and β are the spinor indices. In the above equation, p is a momentum conjugate to the relative coordinate, y, of the two operators. Therefore, it represents a local momentum of the particle defined at the center-of-mass located at x.

An apparent problem with Eq. (12.55) is that it is not gauge-covariant because it is defined through the non-local product of operators. Fortunately, we have already discussed the solution in Chapter 5: inserting the Wilson line between the fermion operators cures the problem. Namely,

$$W_{\alpha\beta}(x, p) = \int \frac{d^4y}{(2\pi\hbar)^4}\, e^{-ipy/\hbar}\langle \bar{\psi}_\beta(x_+)U(x_+, x_-)\psi_\alpha(x_-)\rangle, \tag{12.56}$$

$$U(x_+, x_-) = \exp\left[-i\frac{q}{\hbar}\int_{x_-}^{x_+} A_\mu(z)\, dz^\mu\right], \tag{12.57}$$

where $x_\pm \equiv x \pm y/2$. Note that $U(x_+, x_-)$ is defined on a straight line connecting x_- and x_+:

$$\int_{x_-}^{x_+} A_\mu(z)\, dz^\mu = y^\mu \int_0^1 A_\mu\left(x - \frac{y}{2} + sy\right) ds. \tag{12.58}$$

The factor $W_{\alpha\beta}(x, p)$ in Eq. (12.56) has now become gauge-invariant because the phase rotation of the Dirac fields is completely canceled by the gauge transformation of the Wilson line. If $A_\mu(x)$ is not a c-number field but a quantum field, U must be replaced by the path-ordered product, as discussed in Chapter 5.

As a 4×4 matrix in the spinor space, W has the following property:

$$W^\dagger(x, p) = \gamma^0 W(x, p)\gamma^0. \tag{12.59}$$

Suppose we decompose $W(x, p)$ into 16 independent γ-matrices (Appendix B.1):

$$W = W_{\rm S} + \gamma_5 W_{\rm P} + \gamma_\mu W_{\rm V}^\mu + i\gamma_\mu\gamma_5 W_{\rm A}^\mu + \frac{1}{2}\sigma_{\mu\nu} W_{\rm T}^{\mu\nu}. \tag{12.60}$$

Then, Eq. (12.59) implies W_i ($i = {\rm S, P, V, A, T}$) are all real (but not necessarily positive). The induced source current on the right-hand side of Eq. (12.54) is obtained from $W(x, p)$ as

$$\begin{aligned} j^\mu(x) &= \langle \bar\psi \gamma^\mu \psi \rangle \\ &= \int d^4p \ {\rm tr}[\gamma^\mu W(x, p)] = 4 \int d^4p \ W_{\rm V}^\mu(x, p), \end{aligned} \tag{12.61}$$

where tr is over spinor indices.

Finally, to get an intuitive feeling for the Wigner function in the momentum-space, let us derive its explicit form for non-interacting fermions (we take $\hbar = 1$ for simplicity). Using the mode expansion of the free Dirac field given in Appendix B, it is straightforward to show that (Exercise 12.6)

$$\begin{aligned} & W_{\rm free}(x, p) \\ &= \int \frac{d^3k}{(2\pi)^3 2\varepsilon_k} \left[\delta^4(p-k)\Lambda_+(\boldsymbol{k})n_+(\boldsymbol{k}) - \delta^4(p+k)\Lambda_-(\boldsymbol{k})(1-n_-(\boldsymbol{k}))\right] \\ &= (2\pi)^{-3}(\gamma\cdot p + m)\delta(p^2 - m^2) \times \left[\theta(p^0)n_+(\boldsymbol{p}) + \theta(-p^0)(n_-(-\boldsymbol{p}) - 1)\right], \end{aligned} \tag{12.62}$$

where $\Lambda_\pm(\boldsymbol{p}) \equiv \not{p} \pm m$. The distribution function, $n_\pm(\boldsymbol{k})$, is the standard Fermi–Dirac distribution for particles and anti-particles, and it is related to the expectation value of the number operators in the covariant normalization as follows:

$$\langle b_r^\dagger(\boldsymbol{k}) b_s(\boldsymbol{q})\rangle = 2\varepsilon_k (2\pi)^3 \delta_{rs}\delta^3(\boldsymbol{k}-\boldsymbol{q})n_+(\boldsymbol{k}), \tag{12.63}$$

$$\langle d_r(\boldsymbol{k}) d_s^\dagger(\boldsymbol{q})\rangle = 2\varepsilon_k (2\pi)^3 \delta_{rs}\delta^3(\boldsymbol{k}-\boldsymbol{q})(1-n_-(\boldsymbol{k})), \tag{12.64}$$

where r and s are spin indices.

12.4.4 Equation of motion for $W(x,p)$

To derive an equation of motion for $W(x, p)$, let us first consider the response of $\bar{\psi}_\beta U \psi_\alpha$ under the variations of x and y. Allowing ∂_x to act from the left, and making use of the fact that $U(x_+, x_-)$ is defined on the straight line, yields

$$\begin{aligned} &i\hbar\partial_x^\mu[\bar{\psi}_\beta(x_+)U(x_+, x_-)\psi_\alpha(x_-)] \\ &= \bar{\psi}_\beta(x_+)i\hbar D_\mu^\dagger(x_+)U(x_+, x_-)\psi_\alpha(x_-) \\ &+\bar{\psi}_\beta(x_+)U(x_+, x_-)i\hbar D_\mu(x_-)\psi_\alpha(x_-) \\ &+q\bar{\psi}_\beta(x_+)\left[y^\nu \int_0^1 ds\, F_{\mu\nu}\left(x-\frac{y}{2}+sy\right)\right]U(x_+, x_-)\psi_\alpha(x_-). \end{aligned} \quad (12.65)$$

A similar equation holds for $i\hbar\partial_\mu^y[\bar{\psi}_\beta U \psi_\alpha]$. Combining the results, we find that

$$\left[\gamma\cdot\left(p+\frac{i}{2}\hbar\partial_x\right)-m\right]W(x, p) \quad (12.66)$$

$$= i\hbar q\left[\int_0^1 ds\, (1-s)\, e^{i(s-1/2)\hbar\Delta}\right]\gamma^\mu F_{\mu\nu}(x)\partial_p^\nu W(x, p) \quad (12.67)$$

$$= \frac{\hbar}{2}q[ij_0(\hbar\Delta)+j_1(\hbar\Delta)]\gamma^\mu F_{\mu\nu}(x)\partial_p^\nu W(x, p), \quad (12.68)$$

where we have used the relation $F_{\mu\nu}(x+z) = e^{z\cdot\partial_x}F_{\mu\nu}(x)$ and the spherical Bessel functions, $j_n(z)$, with $j_0(z) = z^{-1}\sin z$ and $j_1(z) = z^{-2}\sin z - z^{-1}\cos z$.

Note that Δ in Eq. (12.68) is defined as the following differential operator:

$$\Delta = \frac{1}{2}\partial_x\cdot\partial_p, \quad (12.69)$$

where the x-derivative acts only on $F_{\mu\nu}(x)$, while the p-derivative acts only on $W(x, p)$.

Combining p^μ (∂_x^μ) in Eq. (12.66) with j_1 (j_0) in Eq. (12.68), the above equation of motion for $W(x, p)$ may be rewritten as follows:

$$(\gamma\cdot K - m)W(x, p) = 0, \quad (12.70)$$

where

$$\left.\begin{aligned} K^\mu &\equiv \Pi^\mu + \frac{i}{2}\hbar\nabla^\mu, \\ \Pi^\mu &= p^\mu - \frac{\hbar}{2}qj_1(\hbar\Delta)F^{\mu\nu}(x)\partial_\nu^p, \\ \nabla^\mu &= \partial_x^\mu - qj_0(\hbar\Delta)F^{\mu\nu}(x)\partial_\nu^p, \end{aligned}\right\} \quad (12.71)$$

with $[\nabla_\mu, \Pi^\mu] = 0$.

Equation (12.70) is the basic equation for the Wigner function of quantum Dirac particles moving in a c-number Abelian field. So far, we have made no approximations. Therefore, Eq. (12.70) contains the same information as the original Dirac equation.

For later purposes, it is useful to derive the second order form of Eq. (12.70) by multiplying $\gamma \cdot K + m$ from the left. Using the identity $\gamma_\mu \gamma_\nu = g_{\mu\nu} - i\sigma_{\mu\nu}$, we obtain

$$\left(K^2 - m^2 - \frac{i}{2}\sigma^{\mu\nu}[K_\mu, K_\nu] \right) W(x, p) = 0. \tag{12.72}$$

12.4.5 Semi-classical approximation

What is the relation between the classical transport equation derived in Section 12.3 and the transport equation given in Eq. (12.70)? To construct a close connection between the two, let us expand $j_0(z)$ and $j_1(z)$ in Eqs. (12.71) in terms of z and keep only the lowest order. Then we obtain

$$\Pi_\mu = p_\mu + O(\Delta), \tag{12.73}$$

$$\nabla_\mu = \partial_x^\mu - qF^{\mu\nu}(x)\partial_\nu^p + O(\Delta^2). \tag{12.74}$$

This expansion is justified when $\Delta^{-1} \gg \hbar$, or, more precisely,

$$X_F \cdot P_W \gg \hbar, \tag{12.75}$$

where X_F (P_W) is a typical length (momentum) scale over which $F^{\mu\nu}(x)$ $(W(x, p))$ has a sizable variation in the x-space (p-space). In other words, the field strength and Wigner function should be smooth enough in coordinate space and momentum space, respectively, for this "Δ-expansion" to be valid. We may wonder about the connection between the Δ-expansion and the strict $\hbar$-expansion. Since $\hbar$ and Δ enter into Eqs. (12.71) in different ways, the two expansions are different. In the following, we will adhere to the Δ-expansion, and we call it the "semi-classical expansion."

Let us decompose the transport equation into two types of equations, starting from Eq. (12.72). Up to the lowest order in Δ, we find $[K_\mu, K_\nu] \simeq i\hbar qF_{\mu\nu}$. By adding Eq. (12.72) and its adjoint, and using Eqs. (12.73) and (12.74), we obtain the "constraint equation:"

$$(p^2 - m^2)W - \frac{\hbar^2}{4}(\partial_x^\mu - qF^{\mu\nu}(x)\partial_\nu^p)^2 W = \frac{\hbar}{4} qF^{\mu\nu}(x)\{\sigma_{\mu\nu}, W\}. \tag{12.76}$$

The constraint equation tells us that the on-shell condition, $p^2 = m^2$, which we have assumed in the classical transport equation, is violated by the quantum effects.

By taking the difference of Eq. (12.72) and its adjoint, and using Eqs. (12.73) and (12.74), we obtain an independent equation from Eq. (12.76), namely the"transport equation"

$$i\hbar(p \cdot \partial_x - qp_\mu F^{\mu\nu}(x)\partial_\nu^p)W = \frac{\hbar}{4}qF^{\mu\nu}(x)[\sigma_{\mu\nu}, W]. \tag{12.77}$$

The left-hand side has exactly the same structure as the classical transport equation, Eq. (12.30), while the right-hand side describes the effect of spin on the transport. Since the particles interact only through the c-number Abelian field, the collision term does not appear in Eq. (12.77); namely, Eq. (12.77) corresponds to the Vlasov equation for Dirac particles.

For $p^2 = m^2$ to be a good approximation, not only should the condition Eq. (12.75) be met, but also the gradient and spin-dependent terms in Eq. (12.76) should be small compared with $(\boldsymbol{p}^2 + m^2)W(x, p)$. In physical terms, $F_{\mu\nu}(x)$ and $W(x, p)$ must be slowly varying functions in any of their arguments. Also, $qF_{\mu\nu}(x)$ itself must be small.

Even if all the above conditions are satisfied, Eq. (12.77) still yields coupled equations for W_i $(i = \mathrm{S, P, V, A, T})$ after the decomposition, Eq. (12.60). To simplify the equation further, let us assume that W has the same spinor structure as W_{free} in Eq. (12.62), with the distribution functions $(2\pi)^{-3}n_\pm(\boldsymbol{p})$ replaced by $f_\pm(x, p)$:

$$W(x, p) = (\gamma \cdot p + m)\delta(p^2 - m^2)[\theta(p^0)f_+(x, p) + \theta(-p^0)(f_-(x, -p) - 1)]. \tag{12.78}$$

This obviously satisfies the constraint equation, $(p^2 - m^2)W = 0$. The factor $f_+(x, p)$, for $p^0 > 0$, can be interpreted as a distribution function for positive-energy particles. On the other hand, $f_-(x, -p)$, for $p^0 < 0$, corresponds to the distribution for negative-energy particles, which implies, according to hole theory, that $f_-(x, p)$ for $p^0 > 0$ is interpreted as the distribution for anti-particles.

Substituting Eq. (12.78) into Eq. (12.77) and working out the commutator on the right-hand side, we finally obtain

$$\left(p \cdot \partial_x \mp qp^\mu F_{\mu\nu}(x)\partial_p^\nu\right) f_\pm(x, p) = 0, \tag{12.79}$$

where $p^{0^2} = \boldsymbol{p}^2 + m^2$. The sign in front of the second term reflects the opposite charges carried by the particle and the anti-particle. The induced current obtained from Eq. (12.78) is given by

$$j^{\mu}(x) = 2\int \frac{d^3p}{(2\pi)^3}\left[f_+(x,p) - f_-(x,p)\right], \tag{12.80}$$

where we have neglected a space-time-independent constant. The overall factor of 2 on the right-hand side corresponds to the spin degrees of freedom. Equations (12.79) and (12.80), together with Eq. (12.54), constitute the "derivation" of the Vlasov equation in the leading order of the semi-classical expansion.

For Dirac particles coupled to the quantum gauge field, the transport equation still has a similar form to Eq. (12.70), with $U(x_+, x_-)$ replaced by the path-ordered product. An important difference from the c-number gauge field is that $F_{\mu\nu}(x)$ should be placed inside the ensemble average on the right-hand side of Eq. (12.70). This generates the correlation of photons and Dirac particles, which leads to coupled equations analogous to the BBGKY hierarchy. The collision term is expected to emerge from the truncation of this hierarchy, as in the classical case. For more details on the QED transport equations, see Vasak *et al.* (1987).

12.4.6 Non-Abelian generalization

A transport equation for quarks under a c-number non-Abelian field is derived in a similar way to the Abelian case, although the derivation is much more tedious. The Wigner function with the non-Abelian field is defined, with $\hbar = 1$, as

$$W^{ab}_{\alpha\beta}(x,p) = \int \frac{d^4y}{(2\pi)^4}\, e^{-ipy}\langle [\bar{\psi}_\beta(x_+)U(x_+,x)]_b[U(x,x_-)\psi_\alpha(x_-)]_a\rangle, \tag{12.81}$$

where α and β (a and b) indicate the spinor (color) indices. Note that U is the path-ordered product,

$$U(x_+,x_-) = \mathrm{P}\exp\left(-ig\int_{x_-}^{x_+} A_\mu(z)\,dz^\mu\right), \tag{12.82}$$

where $A_\mu = A^a_\mu t^a$. The path ordering is required because the A_μ do not commute with each other due to their matrix structure in the color-space. Under the local gauge rotation $\psi(x) \to \psi'(x) = V(x)\psi(x)$, the Wigner function transforms covariantly:

$$W(x,p) \to V(x)W(x,p)V^\dagger(x). \tag{12.83}$$

The constraint equation and the transport equation can be derived in a similar way as in the Abelian case. The expansion parameter in the non-Abelian case is written as $\Delta = (1/2)D_x \cdot \partial_p$, where D_μ is the covariant derivative. Taking only the

$O(\Delta^0)$ terms, and neglecting $O(\partial_p \partial_p W)$ and $O(\partial_p D_x W)$, the transport equation is given by

$$i[p \cdot D_x, W(x,p)] - \frac{i}{2} g\{p_\mu F^{\mu\nu}(x), \partial_p^\nu W(x,p)\} = \frac{1}{4} g[\sigma_{\mu\nu} F^{\mu\nu}(x), W(x,p)]. \qquad (12.84)$$

This becomes a non-Abelian generalization of the transport equation in Eq. (12.77). Also, from a constraint equation similar to Eq. (12.76), the on-shell condition, $p^2 = m^2$, is shown to be satisfied only when the spatial and momentum variation of $F^{\mu\nu}(x)$ and $W(x,p)$ are small and $gF^{\mu\nu}(x)$ itself is small. (Here the spatial variation should be defined in terms of the covariant derivative, D_x.)

The Yang–Mills equation with the back-reaction from the induced quark current is given by

$$[D_\nu, F^{\nu\mu}(x)] = g t^a \int d^4 p \, \mathrm{tr}[\gamma^\nu t^a \, W(x,p)] = g j^\mu(x), \qquad (12.85)$$

where tr is taken over color and spin indices.

12.5 Phenomenological transport equation in QCD

When the non-Abelian QCD plasma is locally color-neutral, the induced color current, $j^\mu(x)$, and $F_{\mu\nu}(x)$ in Eq. (12.85) vanish. Then, we may write down phenomenological transport equations for the number distributions of quarks and gluons as follows:

$$p^\mu \partial_\mu f_i(x,p) = \bar{C}_i[f]. \qquad (12.86)$$

Here, $f_i(x,p)$ are the distribution functions for quarks ($i = q$), anti-quarks ($i = \bar{q}$) and gluons ($i = \mathrm{g}$). The right-hand side is the collision term, which represents not only the $2 \to 2$ binary collisions (such as the diagrams shown in Fig. F.1 in Appendix F), but also the decay processes (such as $\mathrm{g} \to \mathrm{gg}$, $\mathrm{g} \to q\bar{q}$ and $q \to q\mathrm{g}$) and the fusion processes (such as $\mathrm{gg} \to \mathrm{g}$, $q\bar{q} \to \mathrm{g}$ and $q\mathrm{g} \to q$).

The quark-number current and the energy-momentum tensor are defined, respectively, as follows:

$$j_{\mathrm{B}}^\mu(x) = \int \frac{d^3p}{p^0} \, p^\mu \, \left[f_q(x,p) - f_{\bar{q}}(x,p) \right], \qquad (12.87)$$

$$T^{\mu\nu}(x) = \int \frac{d^3p}{p^0} \, p^\mu p^\nu \, \left[f_q(x,p) + f_{\bar{q}}(x,p) + f_{\mathrm{g}}(x,p) \right]. \qquad (12.88)$$

In an attempt to solve the phenomenological transport equation, Eq. (12.86), and to describe the time evolution of the parton distributions during the early pre-equilibrium stage of ultra-relativistic nucleus–nucleus AA collisions, we need to introduce further assumptions. First, to justify the use of perturbative QCD in the collision term, a phenomenological cutoff parameter should be introduced to remove the soft processes. Also, the initial parton distribution functions, $f_{i=q,\bar{q},g}(x, p)$, of the colliding nuclei, AA, must be modeled by using the experimental parton distribution function of a nucleon and its nuclear modification. The parton cascade model (Geiger and Müller, 1992; Geiger, 1995; Müller, 2003) is a typical approach which solves the phenomenological transport equation for quarks and gluons by taking into account the points discussed above.

Exercises

12.1 Detailed balance. Using the time-reversal invariance and the rotational invariance, deduce the detailed balance relation for the transition rate, Eq. (12.6).

12.2 Maxwell–Boltzmann distribution.

(1) Calculate the particle density, n, the mean energy, the mean momentum and the pressure, P, using $f_{\text{MB}}(\boldsymbol{p})$ in Eq. (12.12). Then relate a, b_0 and $\boldsymbol{b}$ to the temperature, T, the particle mass, m, the particle density, n, and the mean momentum, $\boldsymbol{p}_0$, to derive the Maxwell–Boltzmann distribution in Eq. (12.13).

(2) Consider the case with a force $\boldsymbol{F}(\boldsymbol{x}) = -\nabla\phi(\boldsymbol{x})$. Show that the equilibrium solution of the Boltzmann equation in this case is given by

$$f_{\text{eq}}(\boldsymbol{x}, \boldsymbol{p}) = \frac{n_0}{(2\pi m T)^{3/2}} \, \mathrm{e}^{-[(\boldsymbol{p}-\boldsymbol{p}_0)^2/2m+\phi(\boldsymbol{x})]/T}.$$

12.3 Relaxation time approximation. Take the collision term $C[f]$ in Eq. (12.10) and linearize it by taking only the first order terms of $\delta f_i = f_i - f_{i,\text{eq}}$. Pick up a loss term proportional to $f_{2,\text{eq}} \times \delta f_1$ and estimate the relaxation time, τ, in Eq. (12.15).

12.4 Boltzmann's H-theorem.

(1) Deduce the second law of thermodynamics, Eq. (12.22), by using the symmetry properties of the transition rate, w, and by making a change of integration variables, $\boldsymbol{p}_1, \boldsymbol{p}_2, \boldsymbol{p}'_1$ and $\boldsymbol{p}'_2$.

(2) Deduce the master equation for the conservation law, Eq. (12.39), by following the same procedure as that in (1).

(3) Show that Eq. (12.11) is a necessary and sufficient condition for the vanishing collision term $C[f_{\text{MB}}] = 0$ by taking the following steps: (i) If Eq. (12.11) is satisfied, $C[f_{\text{MB}}] = 0$ follows; (ii) $C[f_{\text{MB}}] = 0$ is equivalent to $\partial f_{\text{MB}}/\partial t = 0$ by the Boltzmann equation; (iii) $\partial f_{\text{MB}}/\partial t = 0$

leads to $dS/dt = 0$, where S is the Boltzmann entropy; (iv) $dS/dt = 0$ leads to Eq. (12.11) because of the H-theorem Eq. (12.22).

12.5 Relation to hydrodynamics. Confirm Eq. (12.46) by first defining the scalar quantities $j_\mu u^\mu$, $u^\mu u^\nu T_{\mu\nu}$ and $g^{\mu\nu} T_{\mu\nu}$, and then evaluating those in the local rest frame, $\boldsymbol{u} = 0$.

12.6 Wigner function for equilibrium system. By substituting the mode expansion of the Dirac field (Appendix B.2) into the definition of $W(x, p)$ (Eq. (12.55)) and using Eq. (12.29), confirm that the Wigner function for non-interacting fermions, $W_{\text{free}}(x, p)$, has the form of Eq. (12.62).

13

Formation and evolution of QGP

As discussed in Chapter 10, the formation of quark–gluon plasma in ultra-relativistic heavy ion collisions begins with the production of a large number of partons (quarks and gluons) in hard or semi-hard processes and/or in the de-exitation of the soft classical field in the initial stage. The resultant many-parton system passes through the space-time evolution and eventually undergoes a transition to a system composed of color-singlet hadrons. In describing this time evolution and making connections with the final experimental observables, we must utilize the relativistic hydrodynamics, assuming local equilibrium, as discussed in Chapter 11, or the relativistic transport model with collision terms, as discussed in Chapter 12.

At RHIC and LHC, the initial temperature, T_0, of the hot and dense parton system (QGP), provided that it is in a thermal equilibrium, is expected to be higher than the critical temperature of the QCD phase transition ($T_c \sim 170\,\text{MeV}$).

When the QGP cools to T_c, hadronizaton starts, and a drastic decrease in the entropy density takes place because the color degrees of freedom in the hadronic system are confined as latent degrees of freedom. The hadronic gas thus produced continues to expand and cool until a freeze-out transition results, where the mean free path of hadrons exceeds the space-time size of the system. The transverse expansion and flow of the system superimposed on the longitudinal expansion reflect the equation of state of the matter. Observable consequences of the collective transverse flow include a blue shift of the transverse momentum spectrum of emitted hadrons at the freeze-out stage.

In this chapter, we focus on the central collisions of heavy nuclei such as Au + Au ($A = 197$) in the ultra-relativistic regime. Relevant accelerators for such experiments are the Relativistic Heavy Ion Collider (RHIC) at BNL, with an energy of $\sqrt{s} = (100 + 100)\,\text{GeV/nucleon}$ ($\sqrt{s_{\text{NN}}} = 200\,\text{GeV}$), and the Large Hadron Collider (LHC) at CERN, with $\sqrt{s} = (2.8 + 2.8)\,\text{TeV}/A$

($\sqrt{s_{\text{NN}}} = 5.6\,\text{TeV}$). For information about past, present and future heavy ion accelerators, see Table 15.1.

13.1 The initial condition

The parton production at the impact of the heavy ion collisions provides a crucial initial condition for the subsequent development of the parton plasma. However, its mechanism is not well understood, and it is being actively studied both from theoretical and experimental points of view. In the following, we will introduce several models which are the current candidates expected to describe the initial pre-equilibrium stage.

13.1.1 Color-string breaking model

The color-string model for the ultra-relativistic nucleus–nucleus collisions may be summarized as follows (Matsui, 1987); see Fig. 13.1.

(i) Two nuclei collide and pass through each other. Wounded nucleons in nuclei have color excitations and become a source of color strings and color ropes between the two nuclei. The color-string is assumed to be a coherent and classical color electric field.

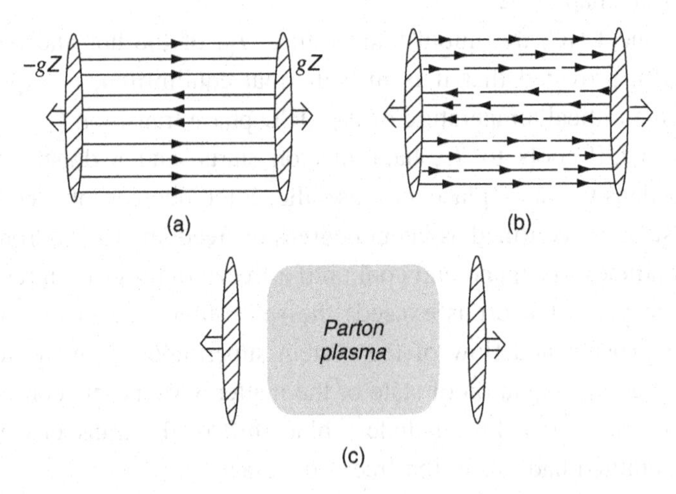

Fig. 13.1. (a) The color strings formed between two nuclei passing through each other. An average color charge, $\pm gZ$, is accumulated in each nucleus due to the exchange of multiple gluons at the time of the collision. (b) Decay of the strings and the production of quark and gluon pairs due to the Schwinger mechanism. (c) The formation of the quark–gluon plasma due to the mutual interaction of the produced partons.

(ii) Due to the Schwinger mechanism (Exercise 13.1) originally introduced in QED (Schwinger, 1951), $q\bar{q}$ and gluon pairs are created under the influence of the strong color electric field between the nuclei. In other words, the coherent, but highly excited, state formed in (a) decays by the spontaneous emission of color quanta (partons). The general pair creation rate per unit space-time volume is given by (Casher *et al.*, 1979; Ambjorn and Hughes, 1983)

$$w(\sigma) = -\frac{\sigma}{4\pi^2} \int_0^\infty dp_{\mathrm{T}}^2 \ln\left[1 \mp \exp\left(-\pi p_{\mathrm{T}}^2/\sigma\right)\right], \tag{13.1}$$

where the upper (lower) sign denotes the creation rate of a pair of massless spin 1/2 fermions (a pair of spin 1 gauge bosons); p_{T} is the transverse momentum of the fermion or the gauge boson perpendicular to the uniform background field; σ characterizes the strength of the external field, and is given by $\sigma = eE$ for QED, where E is the uniform electric field. For QCD, $\sigma \sim gE_{\mathrm{c}}$, where E_{c} is the color electric field in a certain spatial and color orientation; for example, the third direction in space and the third direction in color. By taking into account the number of degrees of freedom in flavor and color, the total emission rates for quarks and gluons are estimated as follows (Gyulassy and Iwazaki, 1985):

$$w_q(\sigma \sim gE_{\mathrm{c}}) \sim N_f \frac{(gE_{\mathrm{c}})^2}{24\pi}, \quad w_{\mathrm{g}}(\sigma \sim gE_{\mathrm{c}}) \sim N_{\mathrm{c}} \frac{(gE_{\mathrm{c}})^2}{48\pi}, \tag{13.2}$$

which implies that they are comparable in magnitude; i.e., $w_{\mathrm{g}}/w_q \sim N_{\mathrm{c}}/2N_f$. For relativistic heavy ion collisions, the magnitude of σ may be also evaluated as $\sigma \sim KZ$, where K is the string tension and gZ is an "effective" color charge, which is accumulated in the nuclei after the collision, as shown in Fig. 13.1.

(iii) Finally, the quark–gluon plasma with local thermal equilibrium is expected to be produced through the mutual interactions of the quarks and gluons just formed. The Boltzmann equation for quarks and gluons coupled with the coherent background field may be a possible starting point for this description (Matsui, 1987, and references thersein).

13.1.2 Color glass condensate

As we have discussed in Chapter 10, partons with small Bjorken x play an important role in the central region of relativistic heavy ion collisions. In particular, the momentum fraction of the gluon in a proton, $xg(x, Q^2)$, which can be measured in deep inelastic lepton–proton scattering experiments, dominates the valence and sea-quark contributions at low x, as shown in Fig. 10.4. Furthermore, $xg(x, Q^2)$ increases very rapidly at low $x < 10^{-2}$ for large Q^2, as shown in Fig. 13.2. Here, $1/Q$ sets the transverse size of the probe (resolution scale perpendicular to the probe's momentum). This rapid increase of the number of gluons cannot continue forever; the gluons with typical transverse size $1/Q$ may eventually overlap each other, interact with each other and saturate

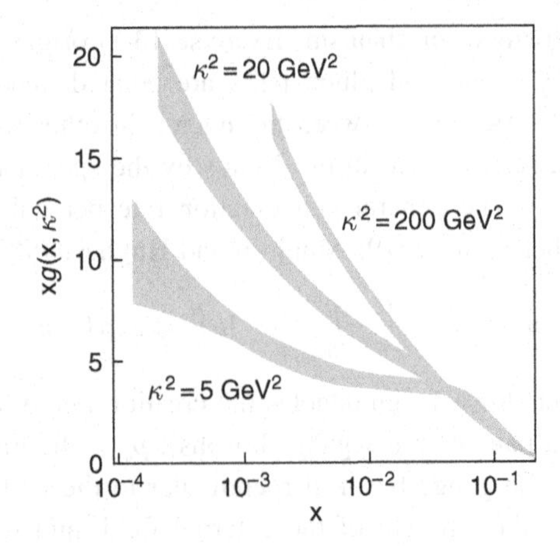

Fig. 13.2. Gluon distribution function, $xg(x, \kappa^2)$, of the proton measured in the deep inelastic electron–proton scattering; κ is taken to be the momentum transfer, Q, to the proton in this case. Adapted from the data of the H1 and Zeus Collaborations (Nagano, 2002).

(Gribov *et al.*, 1983; Mueller and Qiu, 1986). In other words, the gluons form a classical coherent field configuration called the color glass condensate (CGC), as reviewed in Iancu and Venugopalan (2004). The saturation scale, Q_s, is defined in a self-consistent way as follows (Exercise 13.13.5):

$$Q_s^2 \sim \alpha_s(Q_s^2)\frac{xg(x, Q_s^2)}{\pi R^2}, \tag{13.3}$$

where R is the radius of the mother hadron. For a nucleus, $g(x, Q^2)$ should be replaced by the gluon distribution in nuclei, $g_A(x, Q^2) \sim Ag(x, Q^2)$, and R should be replaced by the nuclear radius, $R_A = r_0 A^{1/3}$. Therefore, Q_s^2 increases as $A^{1/3}$, and thus the saturation of the gluons is easier to see in heavy nuclei than in protons.

In the CGC picture (Fig. 13.3), the relativistic nucleus–nucleus collisions are described as a time development of the soft classical field, with the source created by fast moving partons randomly oriented in color-space (McLerran and Venugopalan, 1994a, b). Although the model describes the initial state just before and after the collision, it remains an open question how the decoherence from the classical field to the parton plasma takes place.

13.1.3 Perturbative QCD models

In high-energy AA collisions, hard or semi-hard parton scatterings in the initial stage may result in a large amount of jet production. In particular, the multiple minijets, whose typical transverse momentum is a few GeV, could give rise to

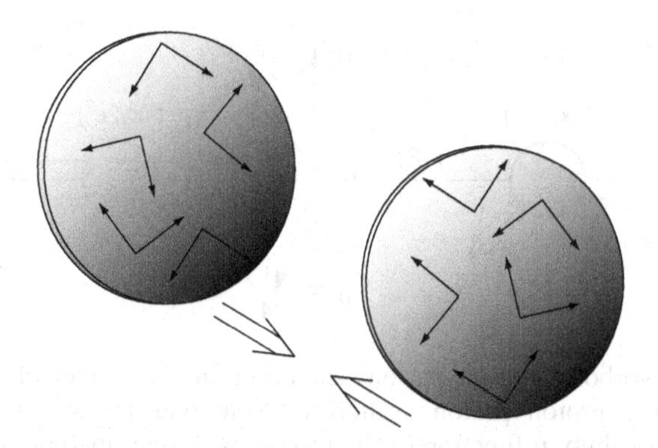

Fig. 13.3. A collision of two Lorentz-contracted plates (nuclei) with color glass condensates. Randomly oriented color electric and magnetic fields reside inside the plates, with the fast moving partons as their source.

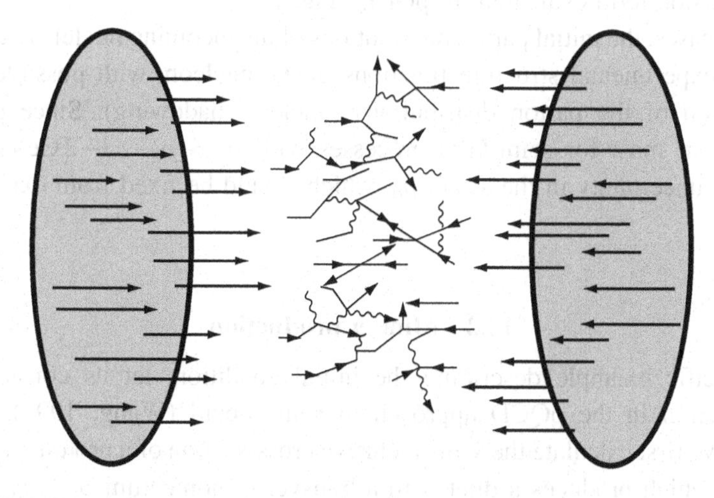

Fig. 13.4. A schematic view of parton cascades in an ultra-relativistic nucleus–nucleus collision.

an important fraction of the transverse energy produced in the heavy ion collisions in the RHIC and LHC (Kajantie *et al.*, 1987; Eskola *et al.*, 1989). Namely, minijets could be good candidates for the initial seeds of the quark–gluon plasma. The minijet production in A+A collisions can be estimated by models based on Monte Carlo event generators. Typical examples of such models include HIJING (heavy ion jet interaction generator) (Wang and Gyulassy, 1994; Wang, 1997) and PCM (parton cascade model) (Geiger, 1995; Müller, 2003). HIJING

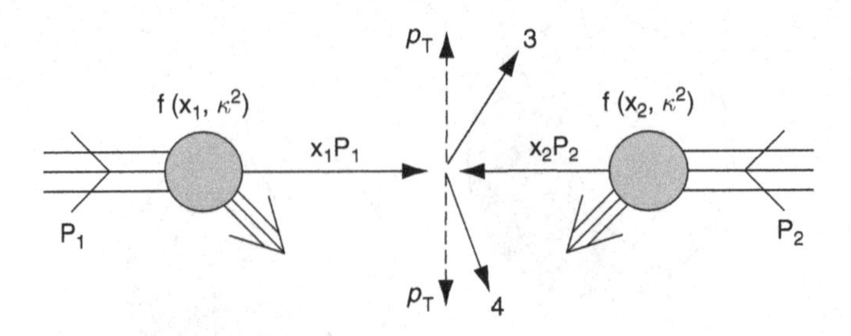

Fig. 13.5. Two-body parton–parton scattering in the center-of-mass system embedded in a proton–proton scattering. Note that $f(x, \kappa^2)$ is the parton-momentum distribution function of the proton, with renormalization scale κ.

is a model which combines a QCD-inspired model of hard and semi-hard jet productions with a phenomenological string model of jet fragmentation, while PCM solves the semi-classical transport equation with parton–parton scatterings in the collision term evaluated in pQCD (Fig. 13.4).

In both cases, the initial parton distributions of the incoming nuclei are evaluated from the experimental structure functions of the nucleon, with possible nuclear modification of the parton distributions (nuclear shadowing). Since pQCD is applicable at most for semi-hard processes with $p_T > p_0 \sim 1-2\,\text{GeV}$, there is always an uncertainty in the scale, p_0, which should be fixed from experimental data.

13.2 Minijet production

As a specific example describing the initial condition, let us consider mini-jet production in the pQCD approach in some detail (Wang, 1997). For this purpose, we first calculate the semi-inclusive cross-section of a proton–proton (pp) collision, which produces a dijet with a transverse momentum $p_T \geq p_0 \simeq$ a few GeV, as shown in Fig. 13.5. The cross-section may be factorized into a long-distance part related to the parton distribution function (PDF), $f(x, \kappa^2)$, of the incoming partons 1 and 2, and a short-distance parton cross-section of the process $1+2 \to 3+4$; κ is the renormalization scale which enters into both the PDFs and the parton-level cross-sections. Although the physical cross-section should not depend on κ, the perturbative calculation introduces κ-dependence in any finite order. Therefore, κ should be chosen appropriately to be a typical scale of the process, as discussed in Chapter 2. In the present case, $\kappa \sim p_T$ is a natural choice.

Then, the differential cross-section is given by

$$\frac{d^2\sigma_{\text{jet}}}{dx_1\, dx_2}(\text{pp} \to 3+4+\text{X}) = \sum_{i,j=q,\bar{q},g} f_i(x_1, p_T^2)\, f_j(x_2, p_T^2)\, \hat{\sigma}^{ij\to kl}, \quad (13.4)$$

where the last factor is the parton-level cross-section of $1+2 \to 3+4$ for various parton species, $(i, j, k, l) = q, \bar{q}, g$.

Bjorken's x is defined in a reference frame where a proton has a very high longitudinal momentum, P. Let the longitudinal momentum of a constituent parton be p. In this frame, we have

$$\mathrm{x} = p/P. \tag{13.5}$$

We need to translate Eq. (13.4) into three observable parameters of the final state: the parton transverse momentum, $p_{_\mathrm{T}}$, and the longitudinal rapidities, y_3, y_4, of the final-state partons. From energy-momentum conservation in simple $2 \to 2$ kinematics, we have (Exercise 13.3(1))

$$\mathrm{x}_1 = \frac{p_{_\mathrm{T}}}{\sqrt{s}}(\mathrm{e}^{y_3} + \mathrm{e}^{y_4}), \quad \mathrm{x}_2 = \frac{p_{_\mathrm{T}}}{\sqrt{s}}(\mathrm{e}^{-y_3} + \mathrm{e}^{-y_4}), \tag{13.6}$$

$$\begin{aligned} \hat{s} &= (p_1 + p_2)^2 = \mathrm{x}_1 \mathrm{x}_2 s = 2p_{_\mathrm{T}}^2[1 + \cosh(y_3 - y_4)], \\ \hat{t} &= (p_1 - p_3)^2 = -p_{_\mathrm{T}}^2(1 + \mathrm{e}^{-y_3 + y_4}), \\ \hat{u} &= (p_2 - p_3)^2 = -(\hat{s} + \hat{t}); \end{aligned} \tag{13.7}$$

$$\text{Jacobian} \equiv \frac{\partial(\mathrm{x}_1, \mathrm{x}_2, \hat{t})}{\partial(y_3, y_4, p_{_\mathrm{T}})} = \frac{2p_{_\mathrm{T}}\hat{s}}{s} = 2p_{_\mathrm{T}}\mathrm{x}_1\mathrm{x}_2, \tag{13.8}$$

where the symbol " ^ " is used to denote the Mandelstam variables for two-body scattering processes at the parton level (Appendix E). The above results are obtained for massless partons.

With the above change of variables, Eq. (13.4) may be rewritten as follows:

$$\frac{d^3\sigma_{\mathrm{jet}}}{dy_3\, dy_4\, dp_{_\mathrm{T}}^2} = \sum_{i,j} \mathrm{x}_1 \mathrm{f}_i(\mathrm{x}_1, p_{_\mathrm{T}}^2)\, \mathrm{x}_2 \mathrm{f}_j(\mathrm{x}_2, p_{_\mathrm{T}}^2)\, \frac{d\hat{\sigma}^{ij \to kl}}{d|\hat{t}|}. \tag{13.9}$$

In the leading order pQCD, there are several $1+2 \to 3+4$ type partonic processes, as summarized in Fig. F.1 and Table F.1 in Appendix F:

$$\mathrm{gg} \to \mathrm{gg},\ q\bar{q}\ ; \quad \mathrm{g}q \to \mathrm{g}q; \quad q\bar{q} \to \mathrm{gg},\ q\bar{q}\ ; \quad qq \to qq. \tag{13.10}$$

Among others, $\mathrm{gg} \to \mathrm{gg}$ has a dominant effect due to a large color factor. From Table F.1, we obtain (Exercise 13.3(2))

$$\frac{d\hat{\sigma}^{\mathrm{gg} \to \mathrm{gg}}}{d|\hat{t}|} = \frac{9\pi\alpha_{\mathrm{s}}^2}{2\hat{s}^2}\left(3 - \frac{\hat{t}\hat{u}}{\hat{s}^2} - \frac{\hat{u}\hat{s}}{\hat{t}^2} - \frac{\hat{s}\hat{t}}{\hat{u}^2}\right) \tag{13.11}$$

$$= \frac{9\pi\alpha_{\mathrm{s}}^2}{2p_{_\mathrm{T}}^4}\left(1 - \frac{p_{_\mathrm{T}}^2}{\hat{s}}\right)^3. \tag{13.12}$$

To take into account the higher order pQCD effects, sometimes a phenomenological K-factor ($K \simeq 2$) multiplies the leading order calculation in Eq. (13.9).

The integrated (semi) hard cross-section for dijet production, $\sigma_{\rm jet}$, in the central rapidity interval, Δy, is thus obtained as follows (Exercise 13.3(3)):

$$\sigma_{\rm jet}(\sqrt{s}; p_0, \Delta y) = \sum_{k,l} \frac{1}{1+\delta_{kl}} \int_{p_0 \le p_{\rm T}} dp_{\rm T}^2 \int_{\Delta y} dy_3 \int_{\Delta y} dy_4 \frac{d^3\sigma_{\rm jet}}{dy_3\, dy_4\, dp_{\rm T}^2}, \tag{13.13}$$

where the Kronecker delta takes care of the symmetry factor of the final state with identical particles. The partons produced in semi-hard parton collisions with transverse momentum transfer $p_{\rm T} \ge 1 \sim 2\,\text{GeV}$ are often referred to as "minijets." We should note that minijets are not jets which are distinctly resolvable in experiments, but rather they produce underlying background yields. It is important, however, to evaluate these background processes because the multiple minijet production plays an important role in creating QGP in ultra-relativistic heavy nucleus collisions.

Let us extend our procedure to a nucleus–nucleus collision by simply introducing the nuclear overlap function, $T_{AB}(b)$, in an AB collision with an impact parameter b (Fig. 10.10). Assuming independent binary parton collisions, the total number of jets, $N_{\rm jet}^{\rm AA}$, for $|y| \le \Delta y/2$ in a central AA collision is evaluated as

$$N_{\rm jet}^{\rm AA}(\sqrt{s}; p_0, \Delta y) \simeq T_{AA}(0)\sigma_{\rm jet}(\sqrt{s}; p_0, \Delta y). \tag{13.14}$$

For a Au+Au collision, $T_{\rm Au+Au}(0) = 9A^2/8\pi R_A{}^2 = 28.4\ \text{mb}^{-1}$, with $R_{\rm Au} = 7.0\,\text{fm}$.

From the kinematical relation Eq. (13.6), partons produced with $p_{\rm T} \sim 2\,\text{GeV}$ in the central rapidity region $y \sim 0$ probe the gluon distributions at

$$\text{x} = \frac{2p_{\rm T}}{\sqrt{s}} \sim \begin{cases} 10^{-3} \text{ for LHC } & (\sqrt{s_{\rm NN}} = 5.6\,\text{TeV}) \\ 10^{-2} \text{ for RHIC } & (\sqrt{s_{\rm NN}} = 200\,\text{GeV}). \end{cases} \tag{13.15}$$

The gluon distribution function for $\text{x} \lesssim 10^{-2}$ happens to have a rapid increase, as shown Fig. 13.2.

There are two nuclear effects which are not included in the above simple formula given in Eq. (13.14): the initial state and final state interactions. The former is related to the nuclear modification of the parton distribution functions (the nuclear shadowing), in particular the depletion of the number of partons inside the nucleus at low x. The latter is related to the energy loss of the produced jets with the surrounding environment (the jet quenching).

Shown in Fig. 13.6 is a HIJING result (Wang and Gyulassy, 1994; Wang, 1997) of charge multiplicity per unit pseudo-rapidity, $dN_{\rm ch}/d\eta$, for Au + Au collisions

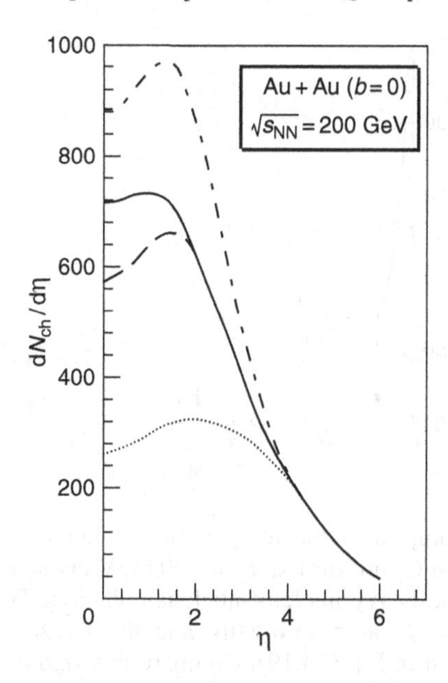

Fig. 13.6. Figure adapted from the HIJING simulation (Wang and Gyulassy, 1994; Wang 1997) on the charge multiplicity per unit pseudo-rapidity, $dN_{ch}/d\eta$, for Au + Au collisions with a zero impact parameter at $\sqrt{s} = 200A$ GeV. The solid line includes all possible effects, namely the soft production, semi-hard minijets, nuclear shadowing and jet quenching. For details, please see the text.

with a zero impact parameter at $\sqrt{s} = 200A$ GeV. (The definition of the pseudo-rapidity, η, and its relation to the rapidity, y, are given in Eqs. (E.20) and (E.21) and Exercise 17.2.) The dotted line denotes the case where only soft interactions are considered without minijet production. The dash-dotted line corresponds to the case for minijets with $p_T > 2$ GeV. The dashed line includes the effect of the nuclear shadowing, with the assumption that the shadowing of the gluons is the same as that of quarks. The solid line includes also the effect of jet quenching, the jet energy loss per unit length being 2 GeV/fm. Approximately 1000 charged particles are produced in the interval $|\eta| < 1$, and at least half of them have minijet origin in this simulation. Also, the initial and final state interactions considerably affect the multiplicity.

13.3 Longitudinal plasma expansion with QCD phase transition

Let us look at the space-time diagram, Fig. 10.7, and also the figure for the time evolution of plasma temperature illustrated in Fig. 13.7. These figures depict the longitudinal space-time evolution of hot matter created at the central rapidity region in an ultra-relativistic nucleus–nucleus collision from the moment of collision

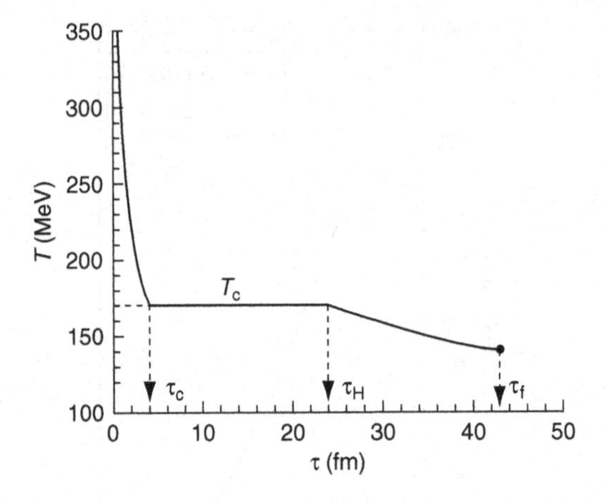

Fig. 13.7. Time evolution of the temperature of hot matter with a first-order QCD phase transition (solid line) at $T_c = 170$ MeV created in the central region of an ultra-relativistic heavy nucleus–nucleus collision. The initial temperature is taken to be $T_0 = 2T_c$ at $\tau_0 = 0.5$ fm, and the freeze-out time is given by $\tau_H/\tau_c = 5.9$, as shown in Eq. (13.19). Compare this figure with that in the early Universe, Fig. 8.11.

to the "freeze-out" stage. We assume that the initial temperature, T_0, at time $\tau = \tau_0 = 0.1 \sim 1$ fm is large enough to produce the quark–gluon plasma, $T_0 > T_c$.

Assuming local thermal equilibrium, the quark–gluon phase expands according to the laws of relativistic hydrodynamics discussed in Chapter 11. It is natural that one-dimensional (longitudinal) expansion dominates initially because of the anisotropic initial condition shown by Fig. 10.5(b) in the Bjorken picture. Superimposed on this motion is a transverse expansion, which will be discussed in Section 13.4. The longitudinal expansion of QGP obeys the simple scaling solution given in Section 11.3.1:

$$s = \frac{s_0 \tau_0}{\tau}, \qquad v_z = \frac{z}{t}. \tag{13.16}$$

This is independent of the specific form of the equation of state, and expresses entropy conservation and a linear growth of the co-moving volume with time. The expansion causes cooling of the plasma according to Eq. (11.60), and T reaches the critical temperature, T_c, of the QCD phase transition at $\tau = \tau_c$. If we assume the Stefan–Boltzmann formula for the QGP entropy, Eq. (3.44), the temperature in this QGP period behaves as

$$T = T_c \left(\frac{\tau_c}{\tau} \right)^{1/3} \qquad (\tau_0 < \tau < \tau_c). \tag{13.17}$$

In the scenario of the first order phase transition, as described by the bag model in Chapter 3, the system becomes a mixture of QGP and hadronic plasma during the phase transition. As we have done for QCD phase transition in the early

Universe (Section 8.6), let us introduce the volume fraction of the hadronic phase, $f(\tau)$, in the mixed phase. Then we have

$$s(\tau) = s_{\rm H}(\tau) f(\tau) + s_{\rm QGP}(1 - f(\tau)) = \frac{s_0 \tau_0}{\tau}, \tag{13.18}$$

where $s_{\rm H}$ ($s_{\rm QGP}$) is the entropy density in the hadronic (QGP) phase. Equation (13.18) determines the τ-dependence of $f(\tau)$ uniquely (Exercise 13.4(1)). If we adopt the Stefan–Boltzmann formulas, Eqs. (3.40) and Eq. (3.44), together with Table 3.1, the lifetime of the mixed phase is given by the ratio

$$\frac{\tau_{\rm H}}{\tau_{\rm c}} = \frac{s_{\rm QGP}}{s_{\rm H}} = \frac{d_{\rm QGP}}{d_\pi} = \begin{cases} 12.3 & (N_f = 2) \\ 5.9 & (N_f = 3). \end{cases} \tag{13.19}$$

Important consequences of the QCD phase transition are (i) to slow down the cooling of the system considerably and (ii) to produce a (longitudinally) large volume of a hot hadronic gas at $T = T_{\rm c}$. The above-mentioned features are qualitatively true as long as the entropy density experiences a rapid change around $T = T_{\rm c}$, and they are not limited to the first order phase transition (Exercise 13.4(2)). The point is that the decrease of the entropy density in a small interval of T has to be compensated for by a time-consuming expansion of the system so as to conserve the total entropy.

After the phase transition is over at $\tau = \tau_{\rm H}$, the interacting hadron plasma undergoes a hydrodynamic expansion. The entropy conservation still governs $T(\tau)$ in this hadronic period. If we assume the simple Stefan–Boltzmann formula, Eq. (3.40), we obtain

$$T = T_{\rm c} \left(\frac{\tau_{\rm H}}{\tau} \right)^{1/3} \qquad (\tau_{\rm H} < \tau < \tau_{\rm f}). \tag{13.20}$$

Eventually, the mean free path of the hadrons exceeds the space-time size of the system, and the local thermal equilibrium is no longer maintained. This is when hadron breakup occurs (the "freeze-out" shown by $\tau_{\rm f}$ in Figs. 10.7 and 13.7).

13.4 Transverse plasma expansion

A transverse hydrodynamic expansion (the r-direction in Fig. 10.5(b)) is caused by a transverse pressure gradient which is significant near the transverse edge of the system ($r \approx R$) (Bjorken, 1983; Blaizot and Ollitrault, 1990; Rischke, 1999). Therefore, the interior parts of the hot matter initially undergo only a longitudinal one-dimensional expansion in the z-direction. In the meantime, a rarefaction wave is created at the transverse edge. The rarefaction wave is a flow where the fluid is continually rarefied as it moves, i.e. the energy density of the fluid element decreases as it moves. A boundary of the rarefaction wave (the wave front)

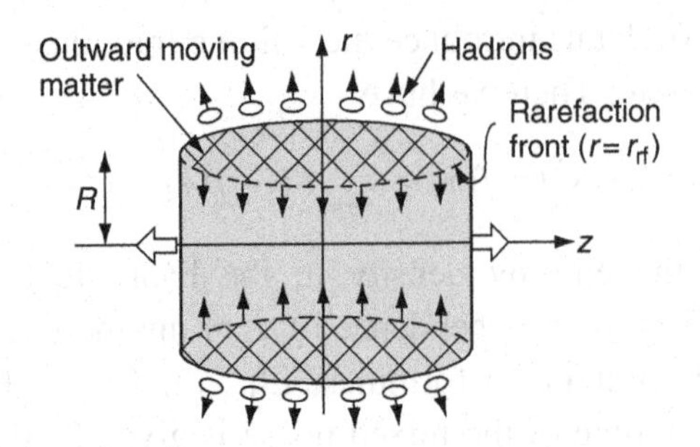

Fig. 13.8. Geometry of the fluid expansion. The rarefaction fronts described by Eq. (13.21) are shown by the dashed lines.

propagates inwards at the velocity of sound, c_s, in the local rest frame of the fluid. Since c_s is determined from $\partial P/\partial\varepsilon$, as expressed by Eqs. (G.4) and (3.59), the transverse expansion is sensitive to the equation of state $P = P(\varepsilon)$. Note that the velocity of sound becomes small near the QCD phase transition. This is because the pressure stays almost constant across the transition while the energy density changes rapidly (see Fig. 3.4).

The rarefaction front that started at the edge travels inward and moves a distance $c_s\tau = c_s\sqrt{t^2 - z^2}$ by time t, for given z. Therefore, the wave front lies on the curve shown by the dashed line in the (z, r)-plane in Fig. 13.8:

$$r_{\rm rf} = R - c_s\sqrt{t^2 - z^2}, \tag{13.21}$$

where c_s is assumed to be a constant. For $r < r_{\rm rf}$, the information that the colliding nuclei are of finite size and that a boundary exists has not been received. The characteristic time scale, $\tau_{\rm T}$, for the transition from one-dimensional expansion to three-dimensional expansion is estimated to be

$$\tau_{\rm T} = \frac{R}{c_s}. \tag{13.22}$$

For the Au + Au collision, we obtain $\tau_{\rm T} = \sqrt{3} \cdot 7 = 12\,{\rm fm}/c$. If we have QCD phase transition, an ordinary rarefaction front finds it hard to propagate in the QGP–hadron mixed region because the velocity of sound, c_s, is small there. This tends to make $\tau_{\rm T}$ even longer.

In an attempt to make the above-mentioned discussion more quantitative, we have to solve the hydrodynamical equations with transverse expansion. The basic equations in the case of cylindrical symmetry are given in Appendix G.2, in which the radial coordinate, r, and the radial velocity, v_r, are essential variables. It is also convenient to introduce a transverse rapidity, α, as follows:

$$\alpha = \tanh^{-1} v_r = \frac{1}{2}\ln\frac{1+v_r}{1-v_r}, \tag{13.23}$$

which is analogous to the longitudinal space-time rapidity, Y, defined in Eq. (11.44).

13.5 Transverse momentum spectrum and transverse flow

To see the observable consequences of the transverse flow of hot matter, let us consider the transverse momentum spectrum of hadrons emitted from the hot matter on the basis of hydrodynamic calculations so far developed. We follow the longitudinal-boost-invariant scenario. The invariant momentum spectrum of hadrons (Appendix E.3) emitted at freeze-out is given by a local thermal distribution, $f(x, p)$, with the freeze-out temperature, $T_{\rm f}$, boosted by a local velocity field, u^{μ}, at the freeze-out hyper-surface, $\Sigma_{\rm f}$ (Cooper and Frye, 1974):

$$E\frac{d^3N}{d^3p} = \frac{d^3N}{m_{\rm T} dm_{\rm T}\, dy\, d\phi_p} = \int_{\Sigma_{\rm f}} f(x,p)\, p^{\mu}\, d\Sigma_{\mu}, \tag{13.24}$$

where p^{μ} is parametrized as Eq. (E.19) in Appendix E.2 (see also Eq. (E.28)) and $d\Sigma_{\mu}$ is a normal vector to the hyper-surface, $\Sigma_{\rm f}$ (see Chap. 1 of Landau and Lifshitz (1988)). The local thermal distribution, $f(x, p)$, is the one given by Eq. (12.45):

$$f(x,p) = \frac{1}{(2\pi)^3} \frac{1}{{\rm e}^{(p_{\mu}u^{\mu}(x) - \mu(x))/T(x)} \mp 1}. \tag{13.25}$$

The rapidity and the transverse mass are denoted by y and $m_{\rm T}$, respectively (Appendix E.2):

$$y = \frac{1}{2}\ln\left(\frac{E+p_z}{E-p_z}\right), \quad m_{\rm T} = \sqrt{p_{\rm T}^2 + m^2}, \tag{13.26}$$

and ϕ_p from Eq. (13.24) is the angle of $\boldsymbol{p}$ in cylindrical coordinates. If there are degeneracies, such as the spin and isospin, the degeneracy factor pre-multiplies the right-hand side of Eq. (13.24).

A cylindrical thermal source expanding in both the longitudinal (z) and transverse (r) directions with boost-invariance in the z-direction leads to the following qualitative formula for the transverse mass spectrum:

$$\frac{dN}{m_{\rm T} dm_{\rm T}} \sim \frac{V_{\rm f}}{2\pi^2} m_{\rm T}\, K_1(\xi_m)\, I_0(\xi_p), \tag{13.27a}$$

where $V_{\rm f}$ is the spatial volume of the cylinder at the freeze-out, K_n (I_n) is the modified Bessel function of the second kind (the first kind) and

$$\xi_m \equiv \frac{m_{\rm T}\cosh\alpha_{\rm f}}{T_{\rm f}}, \quad \xi_p \equiv \frac{p_{\rm T}\sinh\alpha_{\rm f}}{T_{\rm f}}. \tag{13.27b}$$

Here, T_f and α_f are, respectively, the temperature and the transverse rapidity (Eq. (13.23)) at the freeze-out. To obtain the simple qualitative formula given in Eq. (13.27a), we have assumed (i) the Boltzmann distribution with $m_T \gg T$ and $\mu(x) = 0$, (ii) the instant isotherm freeze-out on an r-independent hyper-surface, $\Sigma^0(z) \equiv t_f(z) = \sqrt{\tau_f^2 + z^2}$, and (iii) T and α are r-independent. Assumption (iii) is a rather drastic one. (See Eq. (G.24) and its derivation given in Appendix G.3.)

For $m_T \sim p_T \gg T_f$ with non-zero transverse flow, the arguments of K_1 and I_0 are large, so we can utilize the asymptotic forms of $K_n(\xi_m \to \infty)$ and $I_n(\xi_p \to \infty)$ to simplify Eq. (13.27a) further (Exercise 13.5(1); Schnedermann *et al.*, 1993):

$$\frac{dN}{m_T dm_T} \sim e^{-m_T/T_f^{\text{eff}}}, \quad T_f^{\text{eff}} \simeq T_f \sqrt{\frac{1+v_r}{1-v_r}}. \tag{13.28}$$

The inverse slope for $m_T \sim p_T \gg T_f$ is therefore larger than the original temperature by a *blue shift* factor (see Fig. 13.9), which implies that a rapidly expanding source shifts emitted particles to higher momenta. Such an m-independent asymptotic limit is achieved for m_T larger than 1 GeV (2 GeV) or more for the pion (the proton). In this region, however, non-hydrodynamic power-law contributions to $dN/(m_T\, dm_T)$ (such as the recombination of soft and semi-hard partons and the fragmentation of hard partons) become important.

For moderate values of m_T, the particle mass, m, tends to flatten the slope or equivalently to increase the effective temperature as m increases. This can be seen by defining the effective freeze-out temperature as

$$T_f^{\text{eff}} = -\left[\frac{d}{dm_T} \ln\left(\frac{dN}{m_T\, dm_T}\right)\right]^{-1}. \tag{13.29}$$

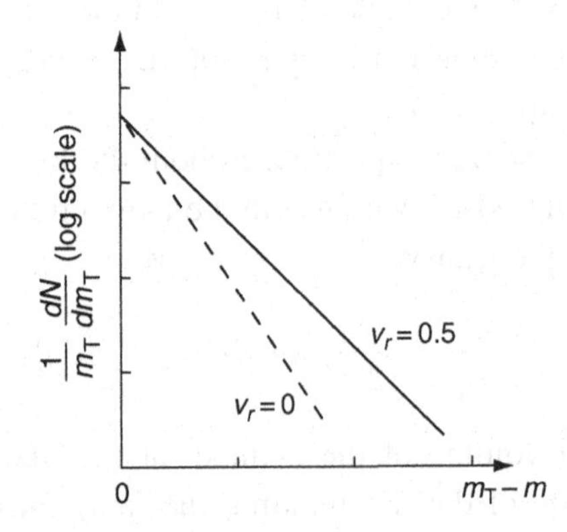

Fig. 13.9. Transverse mass spectrum $(1/m_T)dN/dm_T$ with a transverse flow of $v_r = 0.5$ (solid line) and without a transverse flow (dashed line).

Adopting Eq. (13.27a) and taking the limits $m \gg T_{\rm f}$, $m \gg p_{\rm T}$ and $T_{\rm f} \gg m v_r^2$ (which correspond to $K_n(\xi_m \to \infty)$ and $I_n(\xi_p \to 0)$, we obtain (Exercise 13.5(2)):

$$T_{\rm f}^{\rm eff} \simeq T_{\rm f} + \frac{1}{2} m v_r^2. \tag{13.30}$$

The formula shows that the heavier the particles, the more they gain momenta/energy from the flow velocity, and hence the effective temperature becomes large.

In an actual analysis of the slope of the spectrum of emitted hadrons, one should use the formulas with fewer approximations, as given in Appendix G.3, instead of the crude one given in Eq. (13.27a). Also, we must be careful to take into account additional pions originating from the decay of resonances, for example $\rho^0 \to \pi^+\pi^-$, $\omega \to \pi^+\pi^0\pi^-$, $\Delta \to {\rm N}\pi$, and so on. In contrast to the case of the transverse expansion of the thermal source, these pions contribute to raising the total pion yield at low $m_{\rm T}$.

Single-particle spectra observed in the BNL-AGS, CERN-SPS and BNL-RHIC experiments will be discussed in Chapters 14, 15 and 16 with reference to the approach presented in this chapter.

Exercises

13.1 Schwinger mechanism of particle creation. Following the steps given below, derive the creation rate of e^+e^- pairs under a uniform electric field, E, in the z-direction. For more details, consult Casher *et al.* (1979), Glendenning and Matsui (1983) and Gyulassy and Iwazaki (1985).

(1) Consider the total energy of a positron created at $z = 0$ with longitudinal momentum $p_{\rm L}$ and transverse momentum $p_{\rm T}$; show that $p_{\rm L}^2 = (eEz)^2 - m_{\rm T}^2$, with $m_{\rm T}^2 = p_{\rm T}^2 + m_{\rm e}^2$.

(2) By studying the positron and the electron penetrating from $z = 0$ to $z = \pm z_{\rm c}$ with $p_{\rm L}(z_{\rm c}) = 0$ through the Coulomb potential barrier, show that the Gamow factor is given by $P = \exp(-\pi m_{\rm T}^2/eE)$.

(3) Show that the persistence probability of the vacuum is given by

$$|\langle 0; t = +\infty | 0; t = -\infty \rangle|^2 = \exp\left[\sum_{{\rm spin}, z, t, p_x, p_y} \ln(1 - P) \right].$$

By rewriting the sum in integral form by taking the mesh size $\delta p_x = 2\pi/L$, $\delta p_y = 2\pi/L$ and $\delta t = \pi/m_{\rm T}$, and taking $\Delta z = 2m_{\rm T}/eE$, with L^3 the size of the spatial volume, derive

$$|\langle 0; t = +\infty | 0; t = -\infty \rangle|^2 = \exp(-w \int d^4 x)$$

with

$$w = -\frac{eE}{4\pi^2}\int_0^\infty dp_T^2 \ln\left[1-\exp\left(-\frac{\pi m_T^2}{eE}\right)\right].$$

(4) Generalize the above formula to the $q\bar{q}$ production in QCD, where the quarks have non-Abelian color charge. The interaction between q and $\bar{q}$ just produced tends to reduce the production rate. This effect is not considered in the above formula. How should this effect be taken into account?

13.2 Gluon saturation. The saturation scale given in Eq. (13.3) may be obtained from the condition $\sigma \times n \sim 1$, where σ is a typical parton cross-section, α_s/Q^2, and n is the number of gluons per unit transverse area and unit rapidity, $(dN_{\mathrm{gluon}}/dy)/(\pi R^2)$.

(1) First derive the relation $y = y_{\mathrm{hadron}} - \ln(1/\mathrm{x}) + \ln(m/p_T)$, where y is the rapidity of a massless parton with a transverse momentum p_T, x is the Bjorken variable of the parton and y_{hadron} is the rapidity of the mother hadron with mass m.

(2) Derive Eq. (13.3) by showing that the number of gluons is related to the gluon distribution function as follows:

$$dN_{\mathrm{gluon}} = g\,d\mathrm{x} = \mathrm{x}g\,dy,$$

13.3 Parton kinematics.

(1) Derive the kinematical relations Eqs. (13.6)–(13.8) with the help of Sect. 17.4 of Peskin and Schroeder (1995).

(2) Derive the differential cross-section of the $gg \to gg$ process given in Eq. (13.12) by using the relations $\hat{t}+\hat{u} = -\hat{s} = -\hat{t}\hat{u}/p_T^2$ obtained from Eq. (13.7).

(3) Derive the range of integration for p_T^2, y_3 and y_4 in Eq. (13.13).

13.4 Duration of the mixed phase.

(1) Show that, for the Stefan–Boltzmann gas, the hadron fraction, $f(\tau)$, in Eq. (13.18) is given by

$$f(\tau) = \left(1 - \frac{\tau_c}{\tau}\right)\frac{r}{r-1},$$

where $r \equiv d_{\mathrm{QGP}}/d_\pi$.

(2) Using Eq. (13.16) and the parametrized entropy density for smooth crossover, Eqs. (3.60) and (3.61), plot T as a function of τ with the initial condition $T(\tau = 0.5\,\mathrm{fm}) = 2T_c$.

13.5 The transverse mass spectrum.

(1) Verify Eq. (13.28) by using the asymptotic form of the modified Bessel functions: $K_n(\xi \to \infty) \to \sqrt{\pi/(2\xi)}\ \mathrm{e}^{-\xi}$ and $I_n(\xi \to \infty) \to \sqrt{1/(2\pi\xi)}\ \mathrm{e}^{+\xi}$.

(2) Show that Eq. (13.29) determines the "local" slope of the m_{T} spectrum and indeed has the meaning of the effective temperature. Then, carry out the derivative and rewrite the results using $I_0' = I_1$ and $K_1' = -(1/2)(K_0 + K_2)$. Finally, derive Eq. (13.30) under the conditions given in the text. Note that $I_0(\xi \to 0) = 1 + \xi^2/4 + O(\xi^4)$ and $I_1(\xi \to 0) = \xi/2 + O(\xi^3)$.

14

Fundamentals of QGP diagnostics

We have illustrated the various signatures of QGP proposed in Fig. 1.11 in Chapter 1. In Chapters 15 and 16, the signatures will be discussed in more detail in comparison with experimental results obtained from relativistic heavy ion accelerators. This chapter complements those chapters by summarizing the fundamentals of QGP diagnostics by means of hadrons, jets, leptons and photons.

14.1 QGP diagnostics using hadrons

The QGP formed in nucleus–nucleus collisions with a space-time volume of $O(10^4)$ fm^4 eventually cools and hadronizes into hot hadronic gas. The final-state hadrons are the most abundant and dominant source of information about the early stage of the collisions; however, they suffer from the final-state interactions, which partially mask the early information. In this section, we discuss what we can learn from these remnant hadrons.

14.1.1 Probing the phase transition

As discussed in Chapters 3 and 13, the most significant feature of the QCD phase transition is a drastic change in the numbers of degrees of freedom. For example, for $N_f = 2$, the hot hadron gas has only three light degrees of freedom (π^+, π^0 and π^-), while the QGP has about 37 degrees of freedom (quarks and gluons); see Table 3.1. This is reflected as a rapid change in ε/T^4 or s/T^3 across the critical temperature.

The experimental transverse energy, $dE_{\rm T}/dy$, the hadron multiplicity, dN/dy, and the average transverse momentum, $\langle p_{\rm T} \rangle$, roughly correspond to ε, s and T, respectively (see Sections 11.4 and 13.5). Therefore, a plot of $\langle p_{\rm T} \rangle$ as a function of $dE_{\rm T}/dy$ or dN/dy may show characteristic correlations reflecting the QCD equation of state (van Hove, 1982).

During the phase transition from QGP to the hadron plasma, the total amount of the entropy, S_{total}, is conserved for a perfect fluid (see Chapter 11):

$$S_{\text{total}} = s_{\text{QGP}} V_{\text{QGP}} = s_{\text{H}} V_{\text{H}}, \tag{14.1}$$

where V_{QGP} (s_{QGP}) and V_{H} (s_{H}) are the volume (entropy density) of the QGP and the hadron plasma, respectively. Since entropy density is proportional to the number of degrees of freedom, and s_{QGP} is much larger than s_{H}, we have

$$V_{\text{H}} \gg V_{\text{QGP}}. \tag{14.2}$$

Thus, the QCD phase transition induces an increase in the size of the hadronic gas (Lee, 1998). A possible means of studying this inflation is identical particle interferometry, which will be discussed in Section 14.1.5.

14.1.2 Ratios of particle yields and chemical equilibrium

In high-energy hadron collisions, $q\bar{q}$ pairs are created by hard collisions in which heavy flavors are suppressed due to their masses. According to a systematic study (Wroblewski, 1985), $\mathrm{s\bar{s}}$ yields are only 10–20% of $\mathrm{u\bar{u}}$ or $\mathrm{d\bar{d}}$ yields. This situation may change, however, if secondary interactions take place between the produced particles in heavy ion collisions.

If QGP is created, $\mathrm{s\bar{s}}$ pairs are produced via gluon fusion; the threshold energy of $\mathrm{gg} \to \mathrm{s\bar{s}}$ is about 200 MeV, while the typical thermal energy of a massless gluon is about $3T$. Then the chemical equilibrium among gluons and quarks (u, d, s) could be established, and an enhanced yield in strange and multi-strange particles should be observed. In particular, the probability of forming anti-hyperons by combining $\mathrm{\bar{u}}$, $\mathrm{\bar{d}}$ and $\mathrm{\bar{s}}$ quarks would be greatly enhanced compared with that in nucleon–nucleon collisions; this serves as a signature of QGP (Rafelski and Müller, 1982; Letessier and Rafelski, 2002).

On the other hand, in the hot non-QGP hadronic matter, strangeness production is only slightly enhanced because (i) the threshold energy is large for creating kaons (K) and hyperons (Y) and (ii) the cross-sections of strange-particle production from non-strange particles are small. (For example, the Q-value for $\pi\pi \to \mathrm{K\bar{K}}$ is about ~700 MeV. Also, the cross-section of $\pi\mathrm{N} \to \mathrm{K^+Y}$, where Y denotes a hyperon, is as small as ~1 mb.) Detailed calculations via the rate equations show that the strangeness equilibration time greatly exceeds the reaction time of heavy ion collisions.

Analysis performed on the ratios of the produced hadrons in relativistic heavy ion collisions reveals that they are well described by a simple statistical model

(Cleymans and Satz, 1993; Braun-Munzinger *et al.*, 1999). For a uniform fireball in chemical equilibrium, the particle number density, n_i, is given as an integral over the particle momentum, p:

$$n_i = d_i \int \frac{d^3p}{(2\pi)^3} \frac{1}{\exp[(E_i - \mu_i)/T] \pm 1}, \tag{14.3}$$

where d_i, $\boldsymbol{p}$, E_i, μ_i and T are spin degeneracy, momentum, total energy, chemical potential and temperature, respectively. The $+$ $(-)$ denotes BE (FD) statistics. The chemical equilibrium is given by $\mu_i = \mu_{\rm B} B_i - \mu_{\rm S} S_i - \mu_{\rm I} I_i^3$, where B_i, S_i and I_i^3 are the baryon number, the strangeness and the third component of the isospin quantum numbers of particle species i, respectively. With this model, only two parameters, the temperature, T, and the baryon chemical potential, $\mu_{\rm B}$, are independent, and it has been shown that the ratios of particle production are explained well. To account for the incomplete strangeness equilibration, a factor called the strangeness fugacity, $\gamma_{\rm S}$, has also been introduced in order to describe the experimental data better.

The temperature obtained from the analysis of the number ratios reflects the stage of the chemical freeze-out. After this, no more particles are produced, but there are still collisions among hadrons exchanging energy and momentum. Finally, the mean free path becomes comparable to the size of the system, and then the particles fly away. This is called the kinematical freeze-out or the thermal freeze-out. Observed kinematical distributions of hadrons reflect the stage of this kinematical freeze-out.

For experimental results on chemical equilibrium from SPS and RHIC, see Sections 15.3 and 16.5.

14.1.3 Transverse momentum distributions and hydrodynamical flow

Transverse momentum spectra have been measured for various particle species, not only in nucleus–nucleus collisions but also in pp or pA collisions at various energies. The spectra are usually presented in terms of an invariant cross-section, as given in Eq. (E.28). It is known that the single-particle spectra are well described by an exponential in $m_{\rm T}$ or $m_{\rm T} - m$ rather than in $p_{\rm T}$ in the region $p_{\rm T} < 2\,{\rm GeV}/c$:

$$E \frac{d^3\sigma}{d^3p} = \frac{1}{2\pi m_{\rm T}} \frac{d^2\sigma}{dm_{\rm T}\, dy} \approx \exp(-m_{\rm T}/T). \tag{14.4}$$

This phenomenon is called $m_{\rm T}$ *scaling* (Guettler *et al.*, 1976; Barkte *et al.*, 1977), and the inverse slope parameter, T, in Eq. (14.4) is shown to be common to various particles (such as pions, kaons and protons) in high-energy pp/pA collisions:

$$\text{pp/pA} : T_\pi \sim T_{\rm K} \sim T_{\rm p} \approx 150 \text{ MeV}. \tag{14.5}$$

The inverse slope parameter is sometimes *interpreted* as temperature, which increases with $\sqrt{s}$ and reaches the limiting Hagedorn temperature of 150 MeV (Section 3.6) above $\sqrt{s} \geq 5$ GeV in pp collisions. Such "thermal" behavior is seen only in the transverse direction. In the longitudinal direction, the rapidity distributions of the produced hadrons are flat in high-energy collisions. Note, however, that there is no established mechanism which makes the local thermal equilibrium possible in pp collisions. Therefore, it is not obvious whether Eq. (14.5) is due to the thermalization of the system or due to some phase-space distribution of non-interacting particles.

In contrast to the pp/pA collisions, there is evidence that the produced particles are strongly interacting and create a common outward flow as an explosion takes place in high-energy AA collisions (Siemens and Rasmussen, 1979). This collective radial flow is superposed on the thermal emission of the hadrons, so modifying the transverse mass distributions according to the hadron masses. The effect of the flow can be implemented in the hydrodynamical model, as discussed in Section 13.5 and Appendix G.3, and it leads to a transverse mass spectrum of the form:

$$\frac{dN}{m_{\mathrm{T}}\, dm_{\mathrm{T}}} \sim m_{\mathrm{T}} \int_0^{R_{\mathrm{f}}} r\, dr\, \tau_{\mathrm{f}} K_1 \left(\frac{m_{\mathrm{T}} \cosh \alpha}{T} \right) I_0 \left(\frac{p_{\mathrm{T}} \sinh \alpha}{T} \right). \tag{14.6}$$

A comparison of this model with the measured spectra of various particle species provides us with detailed information on the nature of thermalization in AA collisions as well as on the existence of the freeze-out temperature, T, and the radial velocity, v_r. Additional information on the radial profile of v_r will provide us with a further insight into the origin of the hydrodynamical behavior and the EOS.

For the experimental results on single-particle spectra from SPS and RHIC, see Sections 15.2.1, 15.2.2 and 16.3.

14.1.4 Anisotropic flow and the equation of state

So far, we have mainly focused on central AA collisions. Now let us consider non-central collisions with collision parameter $b \neq 0$ (Fig. 14.1(a)).

In non-central collisions, the overlapped region, which will form a dense/hot matter, has an almond shape, as sketched in Fig. 14.1. Particles are copiously produced and emitted from this region. The emission pattern is very much influenced by the relation of the mean free path, ℓ, and the size of the system, R. When ℓ is not much shorter than R, the particle production is simply a superposition of nucleon–nucleon collisions, so the emission is isotropic in the transverse direction (Fig. 14.1(b)). On the other hand, when $\ell \ll R$ (thus hydrodynamical description can be applicable, as has been discussed in Section 11.1), the emission pattern is affected by the shape of the system (Fig. 14.1(c)).

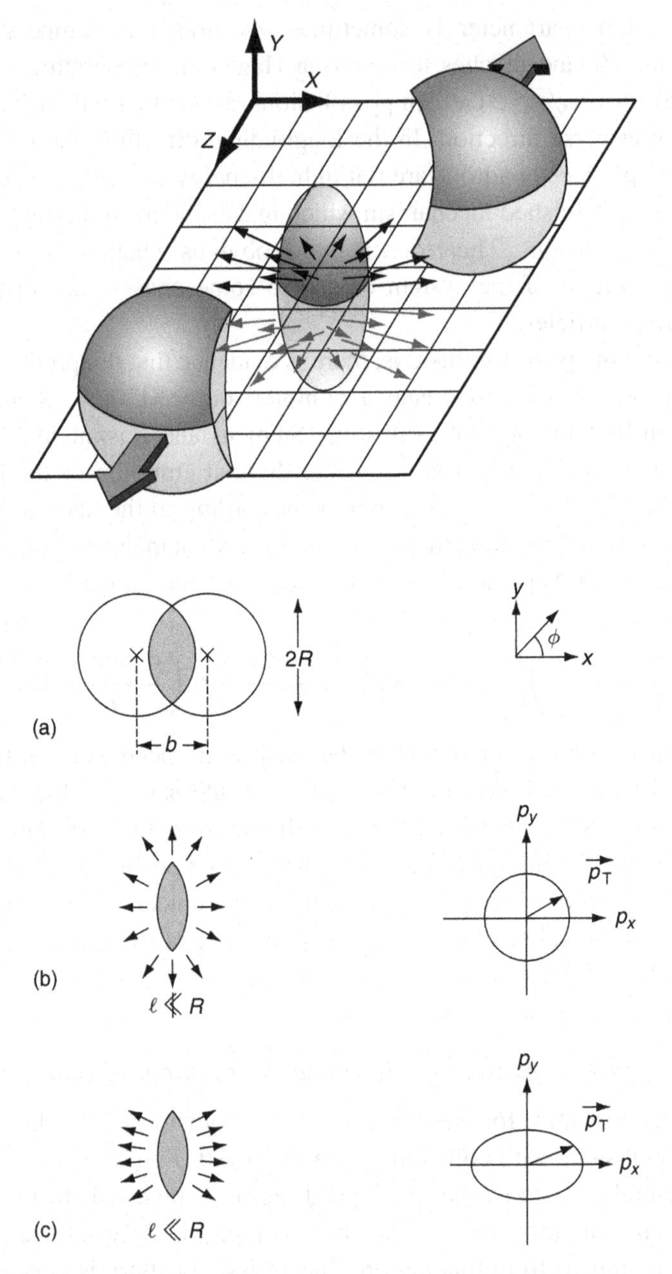

Fig. 14.1. Non-central ($b \neq 0$) relativistic/ultra-relativistic nucleus–nucleus collisions. (a) Transverse view. (b) The almond shape of the participant region (shaded) produces isotropic transverse distributions because $\ell \not\ll R$, whereas (c) it produces non-isotropic (elliptically distributed) flows because $\ell \ll R$. $\vec{p}_{\mathrm{T}} = (p_x, p_y)$ is a transverse momentum vector.

In the hydrodynamical picture, the pressure gradient, ∇P, generates a collective flow: see, for example, Eq. (11.31). In the almond-shaped region, the pressure gradient is expected to be steeper in the direction of the impact parameter, b, and the collective flow will be developed in this direction. Thus, the particle production will have an elliptical azimuthal (ϕ) distribution (Fig. 14.1(c)); see Ollitrault (1992, 1993). Since the pressure gradient is closely related to the EOS, it is important to measure the elliptic flow in order to detect the existence of the QGP pressure in the early stage.

If the phase transition is of first order, the pressure stays constant during the phase transition. This results in a vanishing sound velocity, $c_s = \sqrt{\partial P/\partial \varepsilon}$, which is referred to as a "softening" of the EOS. The collective expansion velocity will be reduced significantly if softening occurs. The study of collective motion in the final state of the produced hadrons is expected to provide key information about the equation of state (Hung and Shuryak, 1995). The effect of the phase transition on the collective flow has also been demonstrated by hydrodynamical calculations (Rischke, 1996).

Experimentally, the azimuthal distribution of particle emission is analyzed with respect to the reaction plane in terms of a Fourier expansion as follows:

$$E\frac{d^3N}{d^3p} = \frac{d^2N}{2\pi p_{\mathrm{T}} dp_{\mathrm{T}}\, dy}\left(1+\sum_{n=1}^{\infty} 2v_n \cos[n(\phi - \Phi_{\mathrm{r}})]\right), \tag{14.7}$$

where ϕ is the azimuthal angle of the particle and Φ_{r} is the azimuthal angle of the reaction plane in the laboratory frame (Fig. 14.1). The first two coefficients in the Fourier decomposition are called the directed and elliptic flows, respectively. Namely, v_1 quantifies the strength of the directed flow, whereas v_2 quantifies the strength of the elliptic flow; it is easily seen that $v_1 = \langle \cos\phi \rangle$ and $v_2 = \langle \cos 2\phi \rangle$ (Exercise 14.1).

Since the orientation of the reaction plane is not known in the experiment, the first task of the analysis is to determine the reaction plane in each collision. If the angular distribution is dominated by the anisotropic flow, as described by Eq. (14.7), then the azimuthal angle of the reaction plane can be determined independently for each harmonic of the flow as follows:

$$\Phi_{\mathrm{r}}^{(n)} = \frac{1}{n}\left(\tan^{-1}\frac{\sum_i w_i \sin n\phi_i}{\sum_i w_i \cos n\phi_i}\right), \tag{14.8}$$

where the sum goes over the ith particles and w_i is the weight. The transverse momentum of the particle, p_{T}, is often used for w_i. Achieving a good angular resolution to determine the reaction plane is one of the key issues in the measurements. If the resolution deteriorates, the observed effects become small, particularly for higher harmonics, and systematic errors become large because larger corrections are needed.

Two-particle azimuthal correlations are also used as a tool to study the anisotropic flow. If they are due to the azimuthal correlation of each particle with respect to the reaction plane, as in Eq. (14.7), the azimuthal angle difference of all the pairs shows the following distribution:

$$\frac{dN_{\text{pair}}}{d\Delta\phi} \propto \left(1 + \sum_{n=1}^{\infty} 2v_n^2 \cos n\Delta\phi\right), \tag{14.9}$$

where $\Delta\phi = \phi_1 - \phi_2$. Note that, with this method, it is not necessary to determine the reaction plane in each event. On the other hand, the observed effect is small since it is proportional to v_n^2.

Caution is needed in this type of flow analysis because there are other sources of azimuthal correlations. First, because the total transverse momentum of all the particles is zero, there is a back-to-back correlation among these momenta. The HBT effect, which will be discussed in Section 14.1.5, creates an azimuthal correlation at an angle corresponding to $\delta p/p \sim R/p$. Resonance decays, such as $\Delta \to \mathrm{p}\pi$ and $\rho \to \pi\pi$, also generate an azimuthal correlation among those products according to the decay kinematics. Final-state Coulomb and strong interactions among particles also produce small modifications. These effects should be evaluated quantitatively, which is often a very complicated process.

For experimental results on anisotropic flows from SPS and RHIC, see Section 16.6.

14.1.5 Interferometry and space-time evolution

The interferometry of identical particles was originally introduced as optical intensity interferometry by Hanbury-Brown and Twiss (1956) and independently as a Bose–Einstein correlation in particle collisions by Goldhaber *et al.* (1960). The method (now called the HBT interferometry) has become a useful tool in the study of the space-time structure of extended sources in astronomy, particle physics, nuclear physics and condensed matter physics (Baym, 1998; Weiner, 2000; Alexander, 2003).

Let us consider two identical Bose particles, for example pions, which are emitted at points a and b and detected at points A and B, as shown in Fig. 14.2(a). The space coordinates of a, b, A and B are denoted by $\boldsymbol{x}_1$, $\boldsymbol{x}_2$, $\boldsymbol{X}_1$ and $\boldsymbol{X}_2$, respectively. The observed particle momentum at A (B), within the uncertainty relation, is denoted by $\boldsymbol{k}_1$ ($\boldsymbol{k}_2$). There are two possibilities for this process in quantum mechanics: either, the particles fly from a to A and b to B (the solid lines), or the particles fly from a to B and b to A (the dashed lines). Since the two particles are identical, and the two possibilities cannot be distinguished, there is an interference term in the transition probability originating from Bose–Einstein statistics.

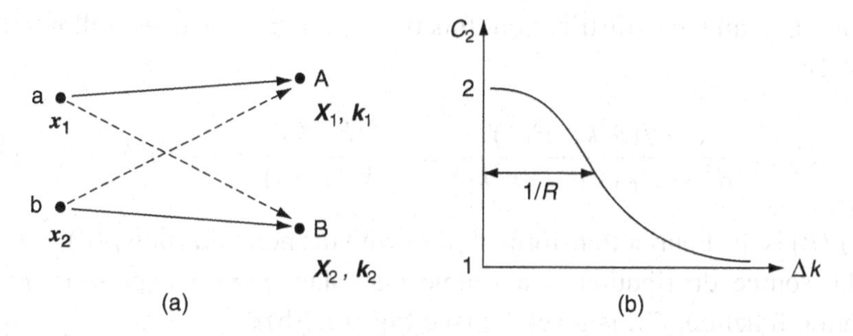

Fig. 14.2. (a) Two identical Bose particles (for example pions) are emitted at points a and b and are detected at points A and B. (b) Typical shape of the correlation function, C_2. Space-time size can be measured as the width of the correlation function.

To relate this idea to experimental observables in high-energy collisions, let us consider the quantity $\Psi_{12} = \langle \boldsymbol{k}_1, \boldsymbol{k}_2 | \boldsymbol{x}_1, \boldsymbol{x}_2 \rangle$, which is a transition matrix element between the two-particle state located at $\boldsymbol{x}_1$ and $\boldsymbol{x}_2$ at the time of emission and the two-particle state with momenta $\boldsymbol{k}_1$ and $\boldsymbol{k}_2$ at the time of detection. Then, we find immediately that

$$|\Psi_{12}|^2 = \frac{1}{2V^2} \left| e^{-i\boldsymbol{k}_1 \cdot \boldsymbol{x}_1} e^{-i\boldsymbol{k}_2 \cdot \boldsymbol{x}_2} + e^{-i\boldsymbol{k}_1 \cdot \boldsymbol{x}_2} e^{-i\boldsymbol{k}_2 \cdot \boldsymbol{x}_1} \right|^2 \tag{14.10}$$

$$= \frac{1}{V^2} (1 + \cos(\Delta\boldsymbol{k} \cdot \Delta\boldsymbol{x})), \tag{14.11}$$

where $\Delta\boldsymbol{k} = \boldsymbol{k}_1 - \boldsymbol{k}_2$, $\Delta\boldsymbol{x} = \boldsymbol{x}_1 - \boldsymbol{x}_2$ and V is the spatial volume of the system. This shows that the probability of having particles with momenta close to each other ($\Delta\boldsymbol{k} \sim 0$) is enhanced due to the Bose–Einstein statistics.

Assuming that the particle sources are distributed in space according to a distribution function, $\rho(\boldsymbol{x})$, and that the particle emissions from different points are incoherent, the probability of observing $\boldsymbol{k}_1$ and $\boldsymbol{k}_2$ is given by

$$P(\boldsymbol{k}_1, \boldsymbol{k}_2) = \frac{1}{2} \int d^3x_1 \, d^3x_2 \, \rho(\boldsymbol{x}_1)\rho(\boldsymbol{x}_2) |\Psi_{12}|^2, \tag{14.12}$$

where the factor 1/2 on the right-hand side is due to the identical nature of the particles, which reduces the integration over $\boldsymbol{x}_1$ and $\boldsymbol{x}_2$ by half. Similarly, the probability of observing a particle $\boldsymbol{k}_i$ $(i = 1, 2)$ is given by

$$P(\boldsymbol{k}_i) = \int d^3x_i \, \rho(\boldsymbol{x}_i) |\Psi_i|^2, \tag{14.13}$$

where $\Psi_i = \langle \boldsymbol{k}_i | \boldsymbol{x}_i \rangle = (1/\sqrt{V}) \, e^{-i\boldsymbol{k}_i \cdot \boldsymbol{x}_i}$.

Suppose $d^6N/(d^3k_1 \, d^3k_2)$ and d^3N/d^3k_i are the two-particle and one-particle phase-space densities, respectively, observed in experiments. Then the correlation function, C_2, and the distribution function, ρ, are related as follows (Exercise 14.2):

$$C_2 \equiv \frac{d^6N/(d^3k_1\, d^3k_2)}{(d^3N/d^3k_1)(d^3N/d^3k_2)} = \frac{2P(\boldsymbol{k}_1, \boldsymbol{k}_2)}{P(\boldsymbol{k}_1)P(\boldsymbol{k}_2)} = 1 + |\tilde{\rho}(\Delta\boldsymbol{k})|^2, \quad (14.14)$$

where $\tilde{\rho}(\boldsymbol{k})$ is the Fourier transform of $\rho(\boldsymbol{x})$ with the normalization $\tilde{\rho}(0) = 1$.

If the source distribution is a simple Gaussian, $\rho(\boldsymbol{x}) \propto \exp(-\frac{1}{2}\boldsymbol{x}^2/R^2)$, the correlation function, C_2, is given by (see Fig. 14.2(b))

$$C_2 = 1 + \lambda\, \mathrm{e}^{-(\Delta\boldsymbol{k})^2 R^2}. \quad (14.15)$$

The parameter λ is introduced for a better fitting of the data. If the particle source is completely incoherent (chaotic), λ is expected to be unity. Therefore, λ is often called the "chaoticity."

Note that $\Delta\boldsymbol{k}$ and $\Delta\boldsymbol{x}$ in Eq. (14.11) can be handled separately in three-dimensional directions. The hot matter produced in the high-energy heavy ion collisions is cylinder-shaped along the beam axis, and the three spatial directions of $\Delta\boldsymbol{k}$ are often taken as outward, sideward and longitudinal, as shown in Fig. 14.3. The inverse widths of the C_2 correlation in the outward, sideward and longitudinal directions provide the source sizes along these directions, namely R_{out}, R_{side} and R_{long}, respectively (the Bertch–Pratt parametrization).

As shown in Eq. (14.2), either an inflation of hadronic gas or an extended emission of hadrons is expected when the QGP is formed, particularly for the first order phase transition. Using the three-dimensional interferometry technique, it is

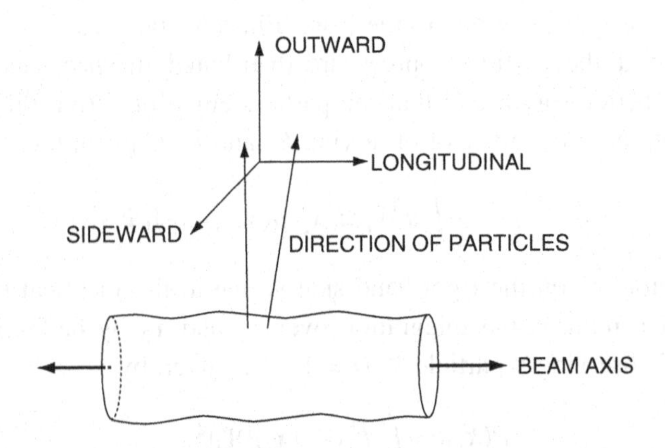

Fig. 14.3. Three-dimensional analysis of the HBT correlation with the Bertch–Pratt parametrization. "Longitudinal" is the direction of the beam axis; "outward" is perpendicular to the beam axis and is toward the direction of the particles; "sideward" is the direction perpendicular to both the longitudinal and outward directions.

proposed that detection of the QGP formation by means of the extended lifetime can be carried out using the ratio of R_{out} and R_{side} (Rischke and Gyulassy, 1996; Pratt, 1984, 1986; Bertch 1989):

$$R_{\text{out}}/R_{\text{side}} > 1. \tag{14.16}$$

The correlation function only depends on the relative momentum, $\boldsymbol{k}_1 - \boldsymbol{k}_2$, and is independent of the sum of the momentum, $\boldsymbol{k}_1 + \boldsymbol{k}_2$, for the static source given in Eq. (14.15). However, if there is a collective expansion, as discussed in Section 13.4, there is a strong correlation in phase-space between the momentum and space coordinates. (See, for example, Fig. 14.1.) In such cases, the correlation function depends on the pair momentum, $K_{\text{T}} = k_{1\text{T}} + k_{2\text{T}}$. With a source expansion of $v_r = v_{\text{surf}}(r/R)$ (Chapman *et al.*, 1995; Heintz *et al.*, 1996), the transverse size, R_{T}, is given by

$$R_{\text{T}} = \frac{R_{\text{G}}}{\sqrt{1 + m_{\text{T}} v_{\text{surf}}^2 / T}}, \tag{14.17}$$

where $m_{\text{T}} = \sqrt{K_{\text{T}}^2 + m^2}$ and m is the mass of the particle, T is the temperature, R_{G} is the geometrical root-mean-squared radius of a Gaussian source distribution and v_{surf} is the expansion velocity at the surface.

For experimental results on HBT interferometry from SPS and RHIC, see Sections 15.2.3 and 16.4.

14.1.6 Event-by-event fluctuations

It is not only the observables obtained after averaging over whole collision events that reflect the equilibrium and non-equilibrium properties of the matter created in relativistic heavy ion collisions but also the event-by-event fluctuations.

One example of a fluctuation observable in a thermal equilibrium situation is the charge fluctuation, $\langle \delta Q^2 \rangle = \langle Q^2 \rangle - \langle Q \rangle^2$. If we use the classical Maxwell–Boltzmann (MB) distribution to evaluate the thermal average, we easily obtain $\langle \delta Q^2 \rangle_{\text{MB}} = q^2 \langle N_{\text{ch}} \rangle$, where q is the particle's electric charge and N_{ch} is the number of charged particles. Since the particles in the hadronic (QGP) phase have charge ± 1 ($\pm\frac{1}{3}$ or $\pm\frac{2}{3}$), the fluctuation in QGP is relatively suppressed (Asakawa *et al.*, 2000; Jeon and Koch, 2000). Indeed, within the MB distribution and with $N_f = 2$, we obtain (Exercise 14.3)

$$\frac{\langle \delta Q_\pi^2 \rangle_{\text{MB}}}{S_\pi} = \frac{1}{6}, \quad \frac{\langle \delta Q_{\text{QGP}}^2 \rangle_{\text{MB}}}{S_{\text{QGP}}} = \frac{1}{24}, \tag{14.18}$$

where S_π (S_{QGP}) is the entropy of the pion (quark–gluon) gas. Since the matter created in the heavy ion collisions comprises a finite system, there is no fluctuation in the conserved total charge, Q. However, if we divide the system into sub-systems, for example in terms of the rapidity, y, the local fluctuation inside the sub-volume may follow the rule in Eq. (14.18), as long as the size of the sub-system is large enough to neglect the quantum fluctuations but small enough to consider the exterior of the sub-system as a heat bath. If these conditions are met, the thermal average, $\langle\cdot\rangle$, in Eq. (14.18) may be replaced by an average over the collision events, and $\langle \delta Q^2 \rangle$ then corresponds to the event-by-event fluctuation. For further details on the theories and experiments of this fluctuation phenomena applied to relativistic heavy ion collisions, see Jeon and Koch (2004) and references therein.

An example of an event-by-event fluctuation in a non-equilibrium situation is the disoriented chiral condensate (DCC). Let us consider the four-component order parameter for $N_f = 2$ QCD, as discussed in Chapter 6: $\vec{\phi} \equiv (\sigma, \boldsymbol{\pi})$. The Landau function (Ginzburg–Landau potential), $V(\vec{\phi}, T)$, is then given by Fig. 14.4(a) in isospin space. The potential is tilted toward the σ-direction because of the quark mass (Section 6.13.4).

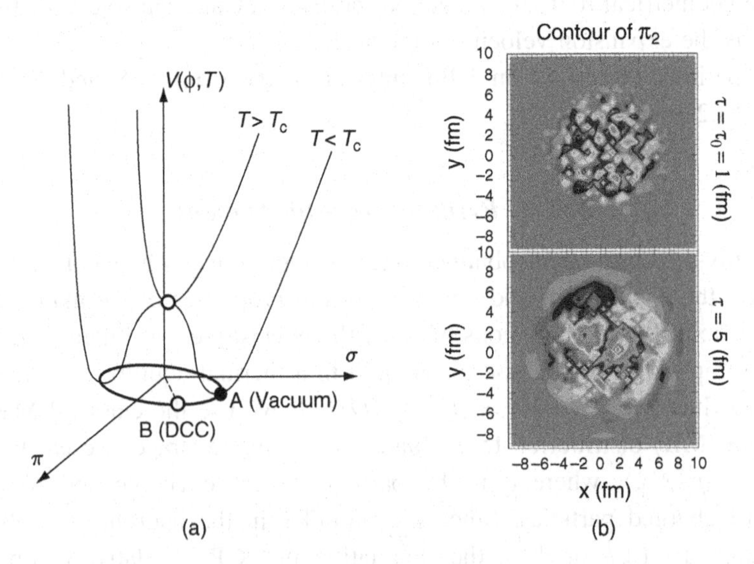

Fig. 14.4. (a) The Ginzburg–Landau potential, $V(\vec{\phi}, T)$, in the isospin space $\vec{\phi} = (\sigma, \boldsymbol{\pi})$. The potential has a wine-glass (wine-bottle) shape for $T > T_c$ ($T < T_c$). A sudden drop of T across T_c (quenching) leads to the system falling down to a point on the chiral circle (B) with an almost equal probability. Then the system evolves from B to the true ground state specified by the point A. (b) Numerical simulations of the DCC formation with longitudinal boost invariance and a quenching initial condition. The contour of the π_2 field is plotted; x and y are the transverse coordinates and τ is the longitudinal proper time. Taken from Asakawa *et al.* (1995).

At high T the chiral symmetry is restored, while at zero T the true ground state is at point A in Fig. 14.4(a). When we cool the system rapidly (quenching) from high to low T, the order parameter may choose any direction on the "chiral circle" and will disorientate $\langle\vec{\phi}\rangle$ (point B) with respect to the QCD vacuum (point A). This DCC eventually decays into the vacuum by emitting a particular combination of pions, depending on the location of B. Assuming that all the directions of the disorientation are equally probable, one finds that the probability, P, of there being an event with the ratio $f = N_{\pi^0}/(N_{\pi^+} + N_{\pi^0} + N_{\pi^-})$ (Exercise 14.4) becomes

$$P(f) = \frac{1}{2\sqrt{f}}. \tag{14.19}$$

This is quite different from the Gaussian distribution centered around $f = 1/3$ obtained from the incoherent pion emission. In particular, the probability of f being small for the DCC case is quite large: $\int_0^{0.01} P(f)\, df = 0.1$. Shown in Fig. 14.4(b) is a result of numerical simulations for a classical chiral field, $\vec{\phi}(t, \boldsymbol{x})$, obeying the equation of motion of the O(4) model with a quenching initial condition (Asakawa *et al.*, 1995). Boost invariance in the longitudinal expansion is assumed to incorporate the Bjorken picture. The growth of large DCC domains with time can be seen.

The disoriented chiral condensate may be closely connected to the so-called "Centauro event" found in cosmic showers at Mt Chacaltaya Observatory in Bolivia (Lattes *et al.*, 1980), where many charged pions but (almost) no neutral pions were observed. Spinodal decomposition in material science (Cahn *et al.*, 1991) and a roll-down transition of an inflaton field in the inflationary cosmology (Linde, 1990) also closely resemble the DCC. For more detailed discussions on the DCC, see the reviews by Rajagopal (1995), Bjorken (1997), and Mohanty and Serreau (2005).

14.1.7 Hadron production by quark recombination

Transverse distributions of produced hadrons observed in high-energy e^+e^- or $p\bar{p}$ collisions show characteristic distributions: these comprise an exponential distribution at low p_T and a flatter one at high p_T, corresponding to soft and hard collisions, respectively.

A well established mechanism for the production of high-p_T hadrons is the fragmentation of parent partons (Sterman *et al.*, 1995; Ellis *et al.*, 1996). Consider a parton, i, with momentum p_i which hadronizes into a hadron, h, with momentum $p = zp_i$. Here, $z(< 1)$ is the momentum fraction of h relative to the parent parton. The fragmentation function, $D_{i\to h}$, is defined such that $D_{i\to h}(z)\, dz$ is the number of hadrons of type h with momentum $zp_i < p < (z + dz)p_i$ in a parton of type i with momentum p_i. The factorization theorem in QCD for the inclusive single-particle production process is given by (Sterman *et al.*, 1995)

$$E\frac{d^3\sigma_h}{d^3p} = \sum_i \int_0^1 \frac{dz}{z^2} E_i \frac{d^3\sigma_i}{d^3p_i} D_{i\to h}(z), \tag{14.20}$$

where $E_i\, d^3\sigma_i/d^3p_i$ is the production cross-section for a parton of type i with momentum $p_i = p/z$. The fragmentation function, $D_{i\to h}(z)$, for various combinations of i and h has been studied in detail in e^+e^-, $p\bar{p}$ and pp collisions (see, for example, Kniehl *et al.* (2001)). Reflecting the fact that $E_i(d^3\sigma_i/d^3p_i)$ has a power-law fall off for high p_i, the high-p_T hadron spectrum exhibits a power-law behavior.

A mechanism which can be important for hadron production at intermediate p_T is quark recombination/coalescence (see, for example, Fries *et al.* (2003), Greco *et al.* (2003) and Hwa and Yang (2003), and references therein). In this picture, quarks and anti-quarks of thermal and/or minijet origin combine to form mesons (M) and baryons (B). This process may become particularly important in nucleus–nucleus collisions because a number of minijets and thermal partons are expected to be produced in comparison to the pp collision, as discussed in Chapter 13. In a simplified version of the recombination model, the production rates of the mesons and baryons are evaluated as follows:

$$E\frac{d^3N_M}{d^3p} \propto \int_{\Sigma_f} p^\mu\, d\Sigma_\mu \int_0^1 dx\; w(r\,;\mathrm{x}p_T)\; \bar{w}(r\,;(1-\mathrm{x})p_T)\; |\phi_M(\mathrm{x})|^2, \tag{14.21}$$

$$E\frac{d^3N_B}{d^3p} \propto \int_{\Sigma_f} p^\mu\, d\Sigma_\mu \int_0^1 d\mathrm{x} \int_0^{1-x} d\mathrm{x}'$$
$$\times\, w(r;\mathrm{x}p_T)\, w(r;\mathrm{x}'p_T)\, w(r;(1-\mathrm{x}-\mathrm{x}')p_T)\, |\phi_B(\mathrm{x},\mathrm{x}')|^2. \tag{14.22}$$

Here, $w(r;p)$ and $\bar{w}(r;p)$ are phenomenological phase-space distributions of constituent quarks and anti-quarks, where $r \in \Sigma_f$; x and x$'$ are the light-cone momentum fractions of the constituent quark, and ϕ_M (ϕ_B) is the light-cone wave function of the meson (baryon); Σ_f denotes the hyper-surface on which the recombination takes place. (Compare the above expressions with the thermal emission of hadrons given in Eq. (13.24).) If we take an equal fraction of the light-cone momentum ($\mathrm{x} = 1/2$ for mesons and $\mathrm{x} = \mathrm{x}' = 1/3$ for baryons), we obtain the following approximate relations:

$$E\frac{d^3N_M}{d^3p} \simeq C_M\, w^2(p_T/2), \quad E\frac{d^3N_B}{d^3p} \simeq C_B\, w^3(p_T/3), \tag{14.23}$$

where $C_{M(B)}$ correspond to the coalescence probabilities for mesons and baryons, respectively.

The momentum of the parent parton (constituent quark) is larger (smaller) than that of the produced hadron in the fragmentation (recombination). For example, consider the production of a meson with $p_T = 5\,\mathrm{GeV}/c$. In the parton fragmentation, the parent parton needs to have a momentum in excess of 5 GeV/c (p_T/z with $z < 1$), while in recombination the constituent quarks are required to have a momentum of only 2.5 GeV/c on average. Therefore, considering the steep fall off of the quark momentum distribution, there may be a region of p_T where the recombination process dominates the fragmentation. Furthermore, for such a p_T region, the yield ratio of baryons to mesons would be enhanced because the p_T of the baryons (mesons) comes from the sum of three (two) quarks.

In the quark recombination picture, the azimuthal anisotropy of the produced hadrons reflects the anisotropy of the constituent quarks and anti-quarks. For simplicity, let us assume that quarks and anti-quarks have a pure elliptic flow:

$$w \propto 1 + 2v_{2,q} \cos 2\phi, \tag{14.24}$$

where $v_{2,q}$ is the strength of the elliptic flow of the quarks and anti-quarks, and ϕ is the azimuthal angle. According to Eq. (14.23), the invariant distribution of mesons is given by

$$\frac{d^2 N_\mathrm{M}}{d\phi\, p_T dp_T} \propto [1 + 2v_{2,q} \cos 2\phi]^2 \simeq 1 + 4v_{2,q} \cos 2\phi, \tag{14.25}$$

for $v_{2,q} \ll 1$. With a similar manipulation for the baryons, we have

$$v_{2,\mathrm{M}}(p_T) \simeq 2v_{2,q}(p_T/2), \qquad v_{2,\mathrm{B}}(p_T) \simeq 3v_{2,q}(p_T/3). \tag{14.26}$$

Thus, the strengths of the elliptic flow for mesons and baryons are scaled according to the number of constituent quarks (*quark number scaling*) (Molnár and Voloshin, 2003). An experimental observation of this type of scaling property should provide a test for the recombination picture.

For the experimental results from RHIC related to the recombination picture of hadron production, see Section 16.9.

14.2 QGP diagnostics using hard probes: jet tomography

A fast charged particle, for example an electron passing through matter, loses its kinetic energy due to collisions with atomic electrons and nuclei via electromagnetic interactions. There are two main sources of energy loss: the collisional source and the radiative source (Jackson, 1999). The former is due to the two-body scattering of the fast particle with the matter constituents, while the latter is due to the Bremsstrahlung (braking radiation) during the collisions with the matter. At high energies, the radiative energy loss becomes dominant, and its rate is dictated by the celebrated Bethe–Heitler formula (Yagi, 1980). If the matter density is high enough, or the energy of the incoming electron is large enough, multiple collisions may take place before the Bremsstrahlung photon in the first collision takes off from the electron. This causes destructive interference of the radiation, and suppression of the Bethe–Heitler energy loss occurs. This is called the Landau–Pomeranchuk–Migdal (LPM) effect (Landau and Pomeranchuk, 1953; Migdal, 1956), and it has been observed in a laboratory experiment at Stanford Linear Accelerator Center (SLAC) (Klein, 1999).

Let us now consider a high-energy parton created by a hard collision in the initial stage of a nucleus–nucleus collision. The fast parton undergoes three major sources of energy loss, as shown schematically in Figs. 14.5(a)–(c).

(a) The fast parton is a colored object and it forms a color flux tube in its wake (Fig. 14.5(a)). Then the parton is decelerated, and the kinetic energy loss is used to form an extra tube. The parton energy loss per unit length in this process is given by $-dE/dx \sim K = 0.9\,\mathrm{GeV\,fm^{-1}}$, where K is the string tension of the flux tube. Eventually, the energy stored in the flux tube is released by particle creation (hadronization) and is observed as jets. This process can even happen in pp or e^+e^- collisions, and has nothing to do with the matter effects, such as the collisional and radiative energy losses.

(b) It was first pointed out in Bjorken (1982) that high-energy jets ($E > 10$ GeV) created in the initial stage of nucleus–nucleus collisions can be affected by the

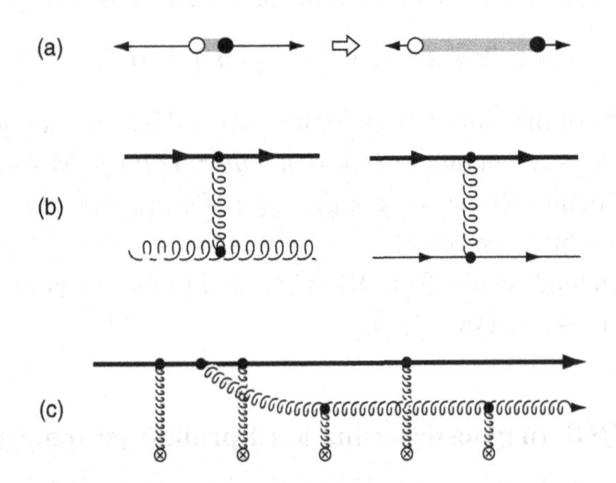

Fig. 14.5. (a) Energy loss in the vacuum. The fast quarks lose their kinetic energy by the formation of a string. Eventually, the string breaks up into other hadrons. (b) Typical diagrams for the collisional energy loss in QGP. Thick solid lines denote fast quarks traversing the QGP. They interact with thermal quarks and gluons inside the plasma. (c) Radiative energy loss in QGP. A fast quark, denoted by the thick solid line, interacts with the randomly distributed color sources ⊗ and emits gluon radiations. Emitted gluons interact further with the color sources.

collisions with soft particles ($E \sim T$) in QGP. This collisional energy loss in QGP can be estimated by the scattering processes shown in Fig. 14.5(b), where the thick lines imply a high-energy quark. The final result for the energy loss per unit length with the hard thermal loop effects discussed in Chapter 4 is given by (Thoma, 1995)

$$-\frac{dE}{dx} \simeq C_2 \pi \alpha_s^2 T^2 \left(1+\frac{N_f}{6}\right) \ln\left(a\frac{ET}{\omega_D^2}\right). \tag{14.27}$$

In Eq. (14.27), a is a constant of $O(1)$; C_2, the quadratic Casimir invariant (see Appendix B.3), is given by $C_2 = C_F = (N_c^2-1)/2N_c = 4/3$ for the quark and $C_2 = C_A = N_c = 3$ for the gluon; $\omega_D = (1+N_f/6)^{1/2} gT$ is as defined in Eq. (4.87). As a rough estimate of the magnitude of the energy loss of a high-energy quark, let us put $T = 0.3$ GeV and $\alpha_s = 0.2$. Then, $-dE/dx \sim 0.3$ GeV/fm for $E = 50$ GeV. This is less than the vacuum energy loss in (a) and does not provide the major part of the in-medium energy loss.

(c) The radiative energy loss with the non-Abelian LPM effect, Fig. 14.5(c), has been considered as an alternative to and more efficient mechanism than the collisional energy loss discussed in (b) (Baier *et al.*, 2000; Gyulassy *et al.*, 2004). In the system created in the relativistic nucleus–nucleus collisions, the plasma is rather thin, in the sense that high-energy partons interact with soft plasma constituents only a few times before escaping from the hot/dense zone. Furthermore, the plasma itself is expanding as a function of time. Taking these effects into account in the $(1+1)$-dimensional Bjorken expansion, we obtain an approximate formula for the radiative energy loss, ΔE, of a fast parton ejected along a direction perpendicular to the beam axis (Accardi *et al.*, 2003):

$$\Delta E(L) \sim \frac{9}{4} C_2 \pi \alpha_s^3 \left(\frac{1}{A_T}\frac{dN_g}{dy}\right) L \ln\left(\frac{2E}{\omega_D^2 L} + \frac{3}{\pi} + \cdots\right). \tag{14.28}$$

Here, $A_T = \pi R^2$ is the transverse size of the nucleus, L ($\sim R$) is the path length of the jet, and dN_g/dy is the number of gluons produced in the central rapidity region. For a Au+Au collision at RHIC with $\sqrt{s_{NN}} = 200$ GeV, $dN_g/dy \sim 2000$. Then, by assuming $T = 0.3$ GeV, $\alpha_s = 0.2$ and $L = 5$ fm, we have $\Delta E/L \sim 1.2$ GeV/fm for a fast quark with $E = 50$ GeV. This exceeds the vacuum energy loss in (a) and is likely to be detected if the QGP is formed in the heavy ion collisions.

The interaction of fast partons and matter is not directly observable but is indirectly reflected in the cluster of hadrons (jets), particularly in the leading high-p_T hadrons. First, the single-particle spectra of high-p_T hadrons in AA collisions may be depleted relative to the superposition of the pp collisions if the in-medium energy loss is at work. Since the high-energy jets in hard processes in a 2→2 collision are produced back-to-back, the absence of the 180° correlations of high-p_T hadrons is also an indication of either the extinction of one of the jets or the deflection of jets inside the matter. For non-central collisions, the initial geometry of the system just after impact has an almond shape in the transverse direction, as discussed in Section 14.1.4. Then the jet energy loss should have an azimuthal angle dependence, since the path length of the fast parton is different for different azimuthal angles. Different hadron species (p, $\bar{\text{p}}$, π, K, ϕ, ...) may also provide detailed information about the properties of energy loss in matter.

The ultimate aim of these studies is to unravel the detailed spatial and temporal structure of QGP, thus it is called *jet tomography* (Gyulassy *et al.*, 2004). It may be also called JET (jet emission tomography) in analogy to positron emission tomography (PET). These studies are applicable not only for the fast parton, but also for heavy quarks and high-energy photons. One promising approach is to study the interplay between the dynamics of the soft plasma particles and the high-energy partons: this is the hydro + jet model (Hirano, 2004), which is a simulation of relativistic hydrodynamics combined with jet production, fragmentation and quenching.

For experimental results on jets from RHIC, see Section 16.7.

14.3 QGP diagnostics using leptons and photons

Leptons and photons can probe the interior of hot plasma because, unlike hadrons, they do not suffer strong final-state interactions. This is why they are called "penetrating probes." There are three sources of leptons and photons in relativistic heavy ion collisions: the initial hard collisions between partons, the thermal production in the hot plasma and the decay of the produced hadrons. The thermal photons and dileptons are particularly important in extracting information from the hot/dense matter (Feinberg, 1976; Shuryak 1978b). Dileptons also can be used to study the production of various neutral vector mesons, such as ρ, ω, ϕ, J/ψ, ψ' and Υ (Tables 7.1 and 7.2).

In the following subsections, we discuss the Drell–Yan production of dileptons, the J/ψ suppression in a hot environment, and thermal photons and dileptons from an expanding plasma.

14.3.1 Drell–Yan production of dileptons

Dileptons via virtual photons (γ^*) can arise from the *initial hard* collisions between partons in the colliding nuclei. This is known as the Drell–Yan mechanism (Drell and Yan, 1970, 1971), and it is well evaluated by means of the perturbative method. We should thus treat Drell–Yan production as a reference to which the emission of dileptons from the other processes described in the following sections can be compared.

First, we consider the a simplest QED reaction, $e^+e^- \to \mu^+\mu^-$. The unpolarized cross-section, σ_0, to lowest order in the extremely relativistic limit ($m_e = m_\mu = 0$) can be easily calculated in a similar way to the $q\bar{q} \to q'\bar{q}'$ process discussed in Appendix F.3:

$$\sigma_0(s) \equiv \sigma(e^+e^- \to \mu^+\mu^-) = \frac{4\pi\alpha^2}{3s}, \quad \alpha = \frac{e^2}{4\pi}, \tag{14.29}$$

where $\sqrt{s}$ is the center-of-mass energy of the collision. If we take into account the effect of finite muon mass, m_μ, a kinematical factor, $F(m_\mu^2/s)$, should be included (Exercise 14.5(c)). The production cross-section of massless quark pairs in the $e^+e^- \to \gamma^* \to q\bar{q}$ process is given in the leading order by

$$\sigma(e^+e^- \to q\bar{q}) = \sigma_0 \cdot 3 \cdot Q_q^2, \tag{14.30}$$

where $Q_q|e|$ is the charge of a quark, q. The factor of 3 arises from the number of allowed colors. We have seen this formula already in Eq. (2.25).

Having made the above preparations, let us now analyze the Drell–Yan process, in which a dilepton, l^+l^-, emerges in a hadron–hadron collision. As long as the invariant mass of the dilepton is large compared with the low-energy QCD scale of $O(1\,\text{GeV})$, the process can be factorized into the hard part ($q\bar{q} \to l^+l^-$) and the soft part (the structure function of the parent hadrons), as shown in Fig. 14.6. In the figure, $f_q(x)(f_{\bar{q}}(x))$ is the quark (anti-quark) distribution function in a proton (Fig. 10.4). To the leading order in QCD, $\sigma(q\bar{q} \to l^+l^-)$ is simply obtained from $\sigma(e^+e^- \to q\bar{q})$ in Eq. (14.30) by averaging instead of summing over color:

$$\sigma(q\bar{q} \to l^+l^-) = \frac{1}{3^2}\sigma(e^+e^- \to q\bar{q}) = \frac{1}{3} \cdot Q_q^2 \cdot \frac{4\pi\alpha^2}{3\hat{s}}, \tag{14.31}$$

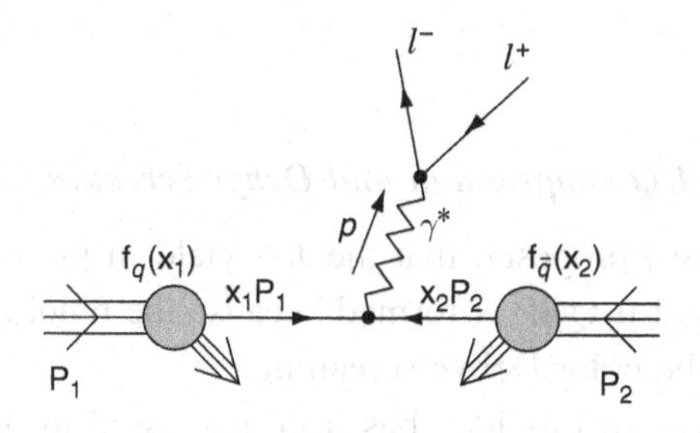

Fig. 14.6. The Drell–Yan process: $pp \to l^+l^- +$ any other particles.

where the symbol ˆ is used for the two-body scattering processes at the parton level (see Section 13.2).

Let us consider the high-energy pp collision in Fig. 14.6 in the center-of-mass system. Neglecting the transverse momentum of the incoming partons in the leading order, the invariant mass, $M = p^2$, and the rapidity, $y = (1/2)\ln((p_0 + p_z)/(p_0 - p_z))$, of the produced dilepton are related to x_1 and x_2 as follows:

$$M^2 = \hat{s} = x_1 x_2 s, \quad y = \frac{1}{2}\ln\left(\frac{x_1}{x_2}\right), \tag{14.32}$$

$$x_1 = \frac{M}{\sqrt{s}}\, e^y, \quad x_2 = \frac{M}{\sqrt{s}}\, e^{-y}, \tag{14.33}$$

where $\hat{s}$ (s) is the invariant mass for the $q\bar{q}$ (pp) system (Exercise 14.6). Thus we have

$$\text{Jacobian} \equiv \frac{\partial(x_1, x_2)}{\partial(M^2, y)} = \frac{x_1 x_2}{M^2}. \tag{14.34}$$

Then, the cross-section for the Drell–Yan process in the pp collision is given by

$$\frac{d^2\sigma_{\mathrm{DY}}^{\mathrm{pp}}}{dM^2\, dy} = \sum_{q,\bar{q}} x_1\, f_q(x_1)\, x_2\, f_{\bar{q}}(x_2) \cdot \frac{1}{3} Q_q^2 \cdot \frac{4\pi\alpha^2}{3M^4}. \tag{14.35}$$

As in the case of Eq. (13.9), the phenomenological K-factor ($K \simeq 2$) multiplies the leading order calculation in Eq. (14.35) to take into account the higher order pQCD contributions.

In an attempt to extend the above-mentioned pp result to the nucleus–nucleus AB collisions, we simply introduce the nuclear overlap function, $T_{AB}(b)$, with impact parameter b, just as in the case of Fig. 10.10 in Section 10.5:

$$\frac{d^2 N_{\mathrm{DY}}^{\mathrm{AB}}}{dM^2\, dy}(b) = \frac{d^2\sigma_{\mathrm{DY}}^{\mathrm{pp}}}{dM^2\, dy} T_{AB}(b). \tag{14.36}$$

14.3.2 J/ψ suppression and Debye screening in QGP

Matsui and Satz (1986) proposed that the J/ψ yield in the relativistic heavy ion collisions is suppressed if QGP is formed because the binding potential becomes short-ranged due to the color Debye screening.

The theoretical basis of this idea has been discussed in detail in Section 7.1. In addition, the following remarks from an observational point of view should be considered.

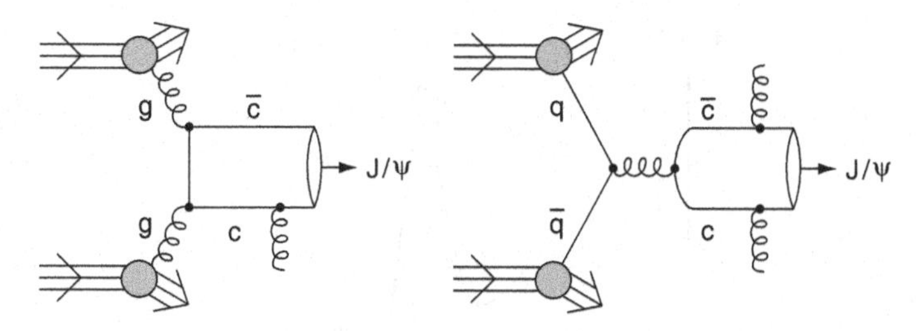

Fig. 14.7. Gluon fusion (left) and quark–anti-quark annihilation (right) processes for J/ψ production.

(i) Due to the large mass of J/ψ (see Table 7.1 and Table H.2), J/ψ is created by hard processes (Fig. 14.7) at the impact of the collision. The thermal production of $c\bar{c}$ pairs at the later stage of the collision is suppressed. For example, $\exp(-2m_c/T) \sim 10^{-5}$ at $T \sim 300$ MeV.

(ii) The fact that J/ψ has a large mass is important for the observation of the J/ψ suppression in the dilepton (e^+e^- and $\mu^+\mu^-$) spectrum. Since $m_{J/\psi} \gg T$, the backgrounds from thermal dileptons are suppressed unless the initial temperature of the QGP is too high. Instead, the Drell–Yan process is expected to dominate over the thermal dileptons around the J/ψ peak.

(iii) Even if QGP is not created in heavy ion collisions, J/ψ (which is color-singlet) or pre-J/ψ (which can be color-octet) produced in hard processes may be destroyed due to final-state interactions with the surrounding hadronic matter. This causes a fake suppression of the J/ψ yield, which should be separated out by systematic measurements in pp, pA and AA collisions (Kharzeev *et al.*, 1997).

Taking the above-mentioned characteristics into account, let us illustrate what we expect to see in the dilepton spectrum, $d\sigma/dM$, in pp and AB collisions (Fig. 14.8). First, the Drell–Yan continuum falls as M^{-3}, which has been explained in the preceding Section 14.3.1. Above this continuum, a J/ψ peak at 3.10 GeV is observed, followed by a ψ' peak at 3.69 GeV. In AB collisions, the Drell–Yan continuum should remain functionally unchanged, while the $c\bar{c}$ binding becomes weak in a deconfining medium, i.e. the J/ψ peak (and also the ψ' peak) should be reduced in size. Matching the functionally identical Drell–Yan continua of the dilepton spectra from pp collisions and from AB collisions, we can observe the behavior illustrated in Fig. 14.8. The experiment thus consists in comparing the corresponding signal-to-continuum ratios, $N_{J/\psi}/N_{DY}$.

The conventional view on the J/ψ suppression, as explained above, may be significantly modified in the following two respects.

- As discussed at the end of Section 7.3, the QGP at temperatures up to about $2T_c$ may still be a strong interacting matter, so it may hold J/ψ as a bound state. In this case, it is not enough to cross the critical point at T_c; we would need to go far beyond T_c to observe the J/ψ suppression.

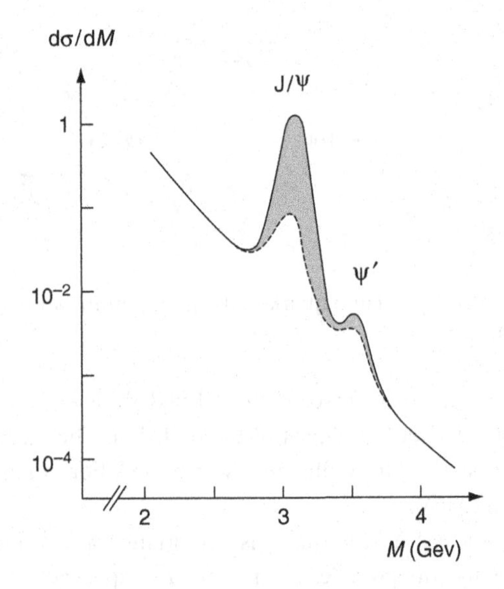

Fig. 14.8. Dilepton spectrum as a function of the invariant dilepton mass, $M(M \geq 2\,\text{GeV})$. The solid (dashed) line is due to pp (nucleus–nucleus) collisions, both have been shifted to match the Drell–Yan continuum of the pp collision.

- At high energies, such as at RHIC or LHC, the J/ψ suppression due to Debye screening may be overcome by the production of J/ψ by the recombination of primordially produced c and $\bar{\text{c}}$. Thermal excitation of charm quarks is rarely allowed, as we mentioned before. However, those produced in primary hard processes may interact with other plasma constituents, and some of them recombine into J/ψ particles at the hadronization stage. This process may lead to a J/ψ enhancement rather than a suppression (Braun-Munzinger and Stachel, 2001; Thews *et al.*, 2001).

For experimental results on J/ψ particles from SPS, see Section 15.4.

14.3.3 Thermal photons and dileptons

General formulas for the photon and dilepton emission rates from hot matter at rest with a fixed temperature, T, have been given in Eqs. (7.44) and (7.50) in terms of the spectrum function, $\rho^{\mu}_{\mu}(\omega, \boldsymbol{p})$. It is convenient for later purposes to choose the variables $p^2(=\omega^2 - \boldsymbol{p}^2)$, ω and T and to express the spectral function as $\rho^{\mu}_{\mu} = \rho^{\mu}_{\mu}(p^2, \omega; T)$. Now we are in a position to extend the results in Section 7.4 to the hot matter in a hydrodynamical expansion with a local four-velocity, u^{μ}. We apply the local Lorentz transformation to take into account the emission from fluid elements with different local velocities. This is equivalent to the replacements $\omega \to p \cdot u$ in the spectral functions and in the thermal factors in Eqs. (7.44) and (7.50). Note that p^2 is Lorentz-invariant and thus is not changed.

Then, for $p \cdot u \gg T$, we have

$$\omega \frac{d^3 N_\gamma}{d^3 p} = \int d^4 x \; S_\gamma(0, p \cdot u \; ; T(x)) \; e^{-p \cdot u / T(x)}, \qquad (14.37)$$

$$\frac{d^4 N_{l^+ l^-}}{d^4 p} = \int d^4 x \; S_{l^+ l^-}(M^2, p \cdot u \; ; T(x)) \; e^{-p \cdot u / T(x)}, \qquad (14.38)$$

where $\omega = |\boldsymbol{p}|$, $S_\gamma(p^2, p \cdot u; T) \equiv -(\alpha/2\pi)\rho^\mu_\mu(p^2, p \cdot u; T)$ and $S_{l^+l^-}(p^2, p \cdot u; T) \equiv -(\alpha^2/3\pi^2 p^2)\rho^\mu_\mu(p^2, p \cdot u; T)$. The lepton mass is neglected. Let us take the one-dimensional Bjorken flow discussed in Chapter 11 with cylindrical symmetry, and assume that S_γ and $S_{l^+l^-}$ are independent of $p \cdot u$ for simplicity. The following changes of variables are also useful (see Eqs.(11.47), (E.28) and (E.29)):

$$\omega^{-1} d^3 p = p_{_\mathrm{T}} dp_{_\mathrm{T}} \; dy \; d\phi_p, \qquad (14.39)$$

$$d^4 p = M_{_\mathrm{T}} dM_{_\mathrm{T}} \; M dM \; dy \; d\phi_p, \qquad (14.40)$$

$$d^4 x = \tau d\tau \; dY \; r dr \; d\phi, \qquad (14.41)$$

where $p_{_\mathrm{T}}$ is the transverse mass of the real photon and M_T (M) is the transverse mass (invariant mass) of the dilepton.

Then, by taking the integration over all variables except for $p_{_\mathrm{T}}$, y and M, we have (Exercise 14.7)

$$\frac{d^2 N_\gamma}{dp_{_\mathrm{T}}^2 \; dy} = (2\pi)^2 A_\mathrm{T} \int_{\tau_0}^{\tau_\mathrm{f}} d\tau \; \tau S_\gamma(0, 0; T) K_0\left(\frac{p_{_\mathrm{T}}}{T}\right), \qquad (14.42)$$

$$\frac{d^2 N_{l^+l^-}}{dM^2 \; dy} = (2\pi)^2 A_\mathrm{T} \int_{\tau_0}^{\tau_\mathrm{f}} d\tau \; \tau M T(\tau) S_{l^+l^-}(M^2, 0; T) K_1\left(\frac{M}{T}\right), \qquad (14.43)$$

where $A_\mathrm{T} = \pi R^2$ is the transverse size of the nucleus, K_n is the modified Bessel function of the second kind, given in Eq. (G.26), and T is a function of the longitudinal proper time, $T = T(\tau)$.

The photon spectrum, Eq. (14.42), as a function of $p_{_\mathrm{T}}$ for the central rapidity region ($y \sim 0$) has a uniform structure dictated only by K_0, whose asymptotic form at large $p_{_\mathrm{T}}$ is essentially the exponential fall off (see Exercise 13.5(1)). This is because the spectral function, S_γ, does not bring any $p_{_\mathrm{T}}$ dependence within the assumption we made. On the other hand, the dilepton spectrum, Eq. (14.43), as a function of M for the central rapidity region ($y \sim 0$) has a non-trivial structure, since $S_{l^+l^-}$ is a function of M and exhibits peak structures associated with the $q\bar{q}$ resonances below and above T_c. In any case, both reflect the different kinematical

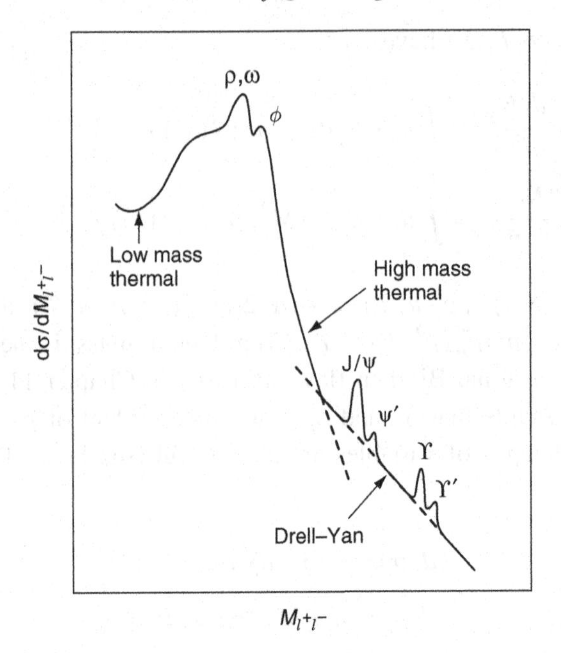

Fig. 14.9. A schematic illustration of the dilepton emission from various sources, Taken from Satz (1992).

region of the same in-medium spectral function, ρ^{μ}_{μ}. Therefore, simultaneous studies of these emission rates will be important.

For experimental results on dileptons and photons from SPS, see Sections 15.5 and 15.6.

Before closing, we show in Fig. 14.9 a schematic figure of the dilepton yield from the relativistic heavy ion collisions. The thermal (Drell–Yan) dileptons dominate in the low (high) invariant mass region. The resonant structures, such as ρ, ω, ϕ, J/ψ, ψ', Υ and Υ', are sitting on top of the background. The medium modifications of these peaks (enhancement, suppression, shift and broadening) can supply useful information on the structure of the hot matter created in the collisions (Alam *et al.*, 1996; Rapp and Wambach, 2000; Gale and Haglin, 2004). Further discussions on these points, together with the data from CERN-SPS, are given in Sections 15.4, 15.5 and 15.6 in Chapter 15.

Exercises

14.1 **Elliptic flow.** Derive the relations $v_1 = \langle \cos\phi \rangle = \langle p_x/p_{\mathrm{T}} \rangle$ and $v_2 = \langle \cos 2\phi \rangle = \langle p_x^2/p_{\mathrm{T}}^2 - p_y^2/p_{\mathrm{T}}^2 \rangle$.

14.2 **HBT correlation for fermions.** Derive the correlation function, C_2, for two identical fermions, such as electrons and protons.

14.3 **Charge fluctuations.** Derive the charge fluctuations in Eq. (14.18) assuming the Maxwell–Boltzmann distribution. What are the numbers if we use Bose–Einstein and Fermi–Dirac distributions?

14.4 **DCC formation.** Derive the probability distribution, Eq. (14.19), for the DCC formation. Identify the physical origin of the difference between this and the Gaussian probability distribution for incoherent pion emission.

14.5 **Cross-section for $e^+e^- \to \mu^+\mu^-$ process.**

(1) Referring to the calculation of the process, $q\bar{q} \to q'\bar{q}'$ (with $m_q = 0$ and $m_{q'} \neq 0$), given in Appendix F.3 (Fig. F.2), prove that the spin-averaged QED amplitude for $e^+e^- \to \mu^+\mu^-$ (with $m_e = 0$ and $m_\mu \neq 0$) in the tree level is given as $|M|^2 = 2e^4 \frac{1}{s^2}\left[(m_\mu^2 - u)^2 + (m_\mu^2 - t)^2 + 2m_\mu^2 s\right]$.

(2) Show that the differential cross-section in the center-of-mass frame results in

$$\left.\frac{d\sigma}{d\Omega}\right|_{\rm cm} = \frac{\alpha^2}{4s}\left(1 - \frac{4m_\mu^2}{s}\right)^{1/2}\left[\left(1 + \frac{4m_\mu^2}{s}\right) + \left(1 - \frac{4m_\mu^2}{s}\right)\cos^2\theta\right].$$

(3) By integrating over θ and ϕ, show that the total scattering cross-section is given by

$$\sigma(e^+e^- \to \mu^+\mu^-) = \frac{4\pi\alpha^2}{3s} F(m_\mu^2/s),$$

with $F(x) \equiv (1+2x)(1-4x)^{1/2}\theta(1-4x)$.

14.6 **Drell–Yan kinematics.** Derive Eqs.(14.32)–(14.34) in the center-of-mass system of the pp collision. Note that the transverse momenta of the incoming quark and anti-quark, as well as that of the outgoing virtual photon, can be neglected in the leading order.

14.7 **Thermal photon and dilepton spectra.** Assuming cylindrical symmetry, derive Eqs.(14.42) and (14.43) by taking into account Eq. (G.18) with $\alpha = 0$ and then performing Y-integration in the interval $(-\infty, +\infty)$ and $M_{\rm T}$-integration in the interval $(M, +\infty)$. Use the indefinite integral $\int d\xi\ \xi K_0(\xi) = -\xi K_1(\xi)$.

15

Results from CERN-SPS experiments

As shown in Chapter 5, phase transitions from hadrons to the QGP can be studied quantitatively in lattice QCD simulations. The critical temperature at which the transition to the QGP phase occurs is estimated to be $\sim$170 MeV. There are a number of proposed signatures of QGP, as discussed in Chapters 1 and 14. However, the nucleus–nucleus collisions involve complex dynamical phenomena which do not allow us to make straightforward comparisons between the predictions and the experimental data. In this respect, we need to collect all the possible experimental data available from the nucleus–nucleus collisions and analyze them carefully before drawing any conclusions about the properties of matter formed during the collision. In other words, it is essential to carry out systematic experimental studies for the progress in QGP physics.

Heavy ion programs have been carried out at BNL-AGS (Alternating Gradient Synchrotron), with heavy ion beams of energy 11*A*–15*A* GeV, and at CERN-SPS (Super Proton Synchrotron), with beam energy 40*A*–200*A* GeV. In these experiments, a collective behavior of nucleus–nucleus collisions was discovered, and the understanding of the collision dynamics in terms of space-time evolution and thermal/chemical equilibrium became possible. Furthermore, several anomalies have been reported, such as the anomalous J/ψ suppression (Abreu *et al.*, 1996, 1999, 2000), the enhancement of low-mass dilepton production (Agakichiev *et al.*, 1998, 1999) and the enhancement of multistrange baryon production (Andersen *et al.*, 1998). Evidently, nucleus–nucleus collisions are qualitatively different from a superposition of nucleon–nucleon collisions.

In this chapter, we present a review of the existing data, obtained mainly from CERN-SPS experiments.

Table 15.1. *Heavy ion accelerators.*

Year[a]	Machine	Beam species	Circumference (km)	$\sqrt{s_{\rm NN}}$ (GeV)
1987	BNL-AGS	Si	0.8	5
1987	CERN-SPS	S	6.9	20
1992	BNL-AGS	Au	0.8	4
1994	CERN-SPS	Pb	6.9	17
2000	BNL-RHIC	Au+Au	3.8	200
2007	CERN-LHC	Pb+Pb	26.7	5600

[a]Year the machine started, or is expected to start, its operation.

15.1 Relativistic heavy ion accelerators

To accelerate heavy ions, it has been necessary to modify the existing proton accelerators as well as to construct new, dedicated, heavy ion machines. We list in Table 15.1 the heavy ion machines currently available and those expected to exist in the future. It can be seen that a new accelerator or heavy ion beam becomes available every few years.

At Brookhaven National Laboratory (BNL), the Tandem Van de Graafs are combined with the AGS synchrotron via a transfer tunnel. At its startup in 1987, heavy ion beams up to Si were accelerated at an energy of 15*A* GeV. Since then, a booster ring has been constructed, which allows acceleration of a gold beam at 12*A* GeV. Furthermore, on the same site, the Relativistic Heavy Ion Collider (RHIC), the first heavy ion collider, has been constructed and has operated successfully since the summer of 2000. In 2001, Au + Au collisions at center-of-mass energies of 200 GeV per nucleon pair ($\sqrt{s_{\rm NN}} = 200$ GeV) were achieved (see Fig. 15.1 (a) and Fig. 16.1). In Europe, CERN-SPS has been modified for heavy ion acceleration; in 1987, oxygen and sulfur beams at 200 GeV/*c* per nucleon were available for the first time. Since 1994, a lead beam at 158*A* GeV has been also available to experimentalists. The Large Hadron Collider (LHC), which is to be constructed in the LEP tunnel at CERN, will provide Pb + Pb collisions at center-of-mass energies of 5.6 TeV per nucleon pair ($\sqrt{s_{\rm NN}} = 5.6$ TeV) in 2007; see Fig. 15.1 (b).

15.2 Basic features of AA collisions

In the first round of pioneering experiments, a significant amount of effort was spent on investigating the gross features and dynamics of collisions. Using a hadronic calorimeter and a charged multiplicity detector (see Chapter 17 for more on detectors), the total transverse energy, $E_{\rm T}$, and charged multiplicity have been studied.

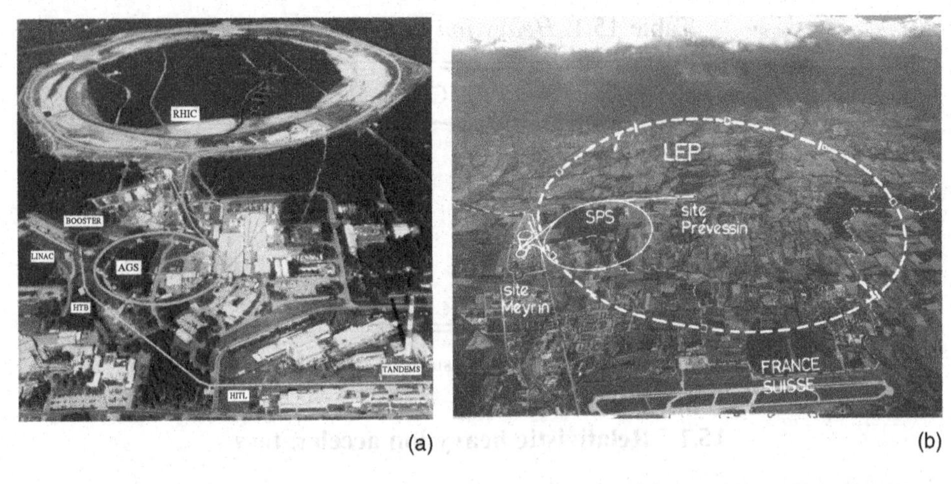

Fig. 15.1. Centers of relativistic heavy ion research. (a) Brookhaven National Laboratory; (b) CERN. See also Fig. 16.1. (Courtesy of BNL and CERN.)

The energy density, ε_0, was evaluated as a basic quantity of the collision, using Bjorken's formula, Eq. (11.81):

$$\varepsilon_0 = \frac{1}{\pi R^2 \tau_0} \frac{dE_\mathrm{T}}{dy}\bigg|_{y\simeq 0}. \tag{15.1}$$

Taking τ_0 as 1 fm/c and R as the nuclear radius, a measured $dE_\mathrm{T}/d\eta$ of 450 GeV in Pb + Pb collisions at CERN-SPS leads to a value for ε_0 as large as 3 GeV fm^{-3}. Although this is a crude estimate, there is a good chance that CERN-SPS has already provided us with a doorway to the new regime, the quark–gluon plasma.

15.2.1 Single-particle spectra

The m_T distributions for identified hadrons around the mid-rapidity region have been intensively studied in nucleus–nucleus collisions at BNL-AGS and CERN-SPS. In Fig. 15.2, the shapes of m_T distributions are compared for pions, kaons and protons from 158A GeV Pb + Pb collisions (Bearden *et al.*, 1997). Invariant cross-sections are plotted as a function of $m_\mathrm{T} - m$, where $m_\mathrm{T} = \sqrt{p_\mathrm{T}^2 + m^2}$ and m are the transverse mass and the rest mass of the particle, respectively. The m_T distributions are found to be exponential, except for a low-m_T region of pions. A steeper component of the pions for $m_\mathrm{T} - m < 0.2$ GeV is known to contain contributions from the decay of the resonances such as Δ. The distributions are limited to $m_\mathrm{T} - m \leq 1$ GeV due to the experimental limitations of particle identification. For neutral pions (π^0), however, the m_T distributions are extended to $m_\mathrm{T} - m \sim 4$ GeV, as shown in of Fig. 15.2(b) (Aggarwal *et al.*, 1999). The π^0s are reconstructed in their two-gamma decay channel using an electromagnetic calorimeter. The local slope of the spectrum, T_local, is shown in the inset of

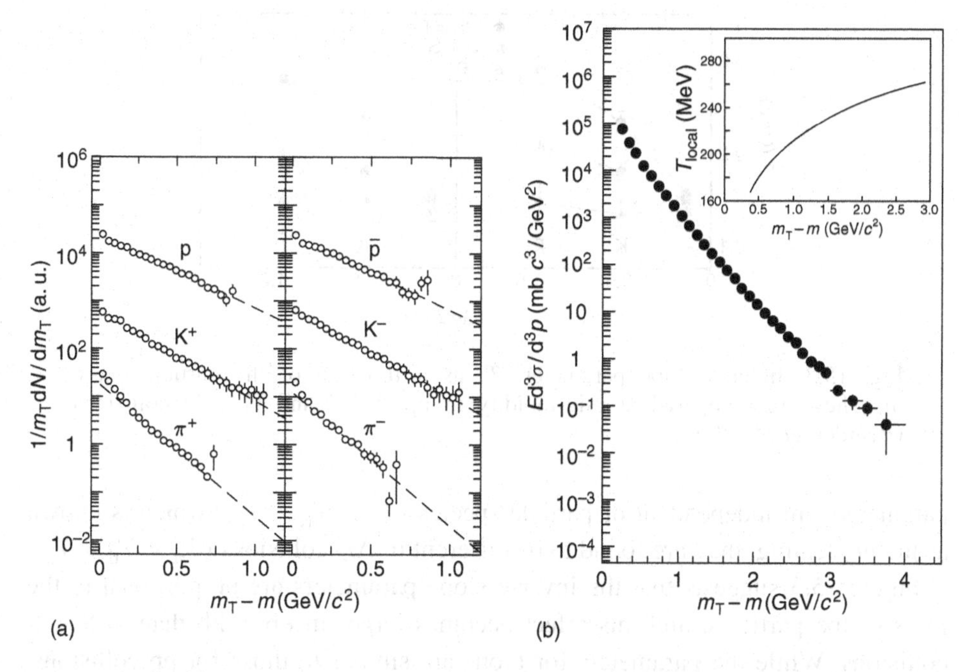

Fig. 15.2. (a) Transverse mass distributions for pions, kaons and protons from 158*A* GeV Pb + Pb collisions. The dashed lines represent the exponential fits (Bearden *et al.*, 1997). (b) Transverse mass distributions of neutral pions. Data are taken from Aggarwal *et al.* (1999).

Fig. 15.2(b). High statistical measurements and wide coverage in m_T have revealed the concave shape of the spectra.[1]

Note that the distributions for the proton and anti-proton are flatter than those for pions and kaons. To compare their slopes, the distributions are fitted to the function

$$E\frac{d^3\sigma}{d^3p} \propto \exp\left(-\frac{m_T}{T}\right), \tag{15.2}$$

where T is the inverse slope parameter. For pions, the range of the fitting is chosen as $m_T - m > 0.25$ GeV in Fig. 15.2(a) to exclude the steeper component. Inverse slope parameters are obtained for pions, kaons and protons of both charges and plotted as a function of particle mass in Fig. 15.3, where they are compared in pp ($\sqrt{s_{NN}} = 23$ GeV), S + S ($\sqrt{s_{NN}} = 19.4$ GeV) and Pb + Pb ($\sqrt{s_{NN}} = 17.2$ GeV) collisions in the mid-rapidity region. While in pp collisions the inverse slope

[1] A flatter component in the high-p_T region is also observed in pp collisions. A deviation from the exponential fall off above 2 GeV/c is understood in terms of hard scatterings with a partonic structure of hadrons. Such hard scattering is also expected in AA collisions in the high-p_T region. Thus, the concave shape of the pion spectra may not only be due to the collective expansion, but also to the existence of hard scatterings, particularly in high-p_T regions.

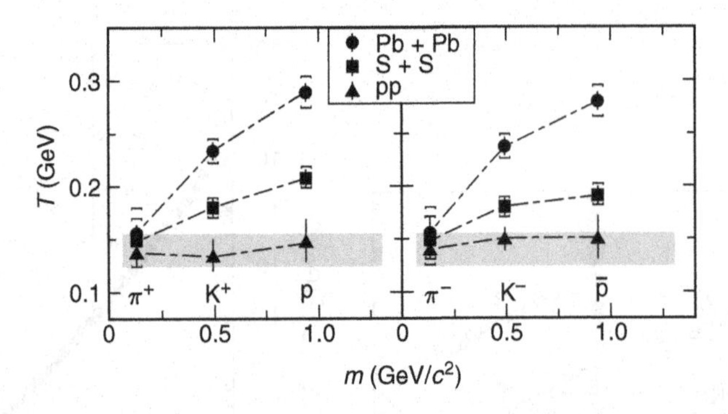

Fig. 15.3. Inverse slope parameter, T, as a function of the particle mass, m. They are compared at mid-rapidity in pp, S+S and Pb+Pb collisions (Bearden *et al.*, 1997).

parameters are independent of particle species ($T_\pi \sim T_K \sim T_p$), which is known as the m_T scaling, they are found to be different in AA collisions ($T_\pi < T_K < T_p$).

Figure 15.3 suggests that the inverse slope parameters are proportional to the mass of the particles, and this effect becomes larger in Pb + Pb than in S + S collisions. While the parameters for pions are similar to those for pp collisions, the slopes of heavier particles become flatter in AA collisions. Such behavior has been further tested with other particle species, as shown in Fig. 15.4, where inverse slope parameters of deuterons, Ω, Ξ and ϕ are plotted in addition to pions, kaons and protons. The parameters of these particles are roughly proportional to the mass of particles, except for Ω and Ξ particles.

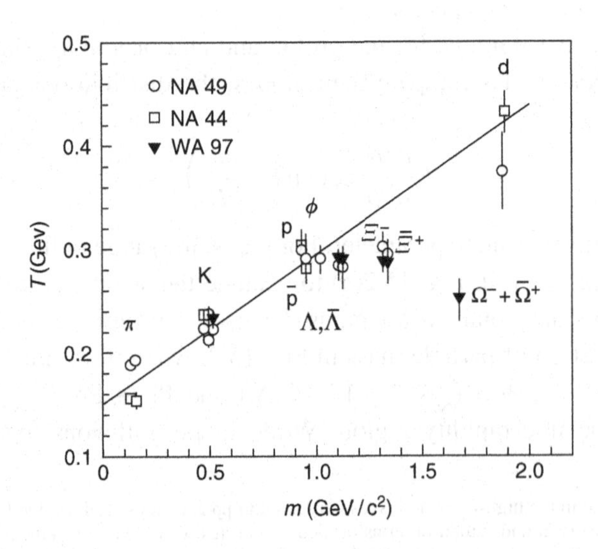

Fig. 15.4. Inverse slope parameter, T, plotted against particle mass, m (Antinori *et al.*, 2000).

15.2.2 Collective expansion

If the matter emitting the particles behaves like a fluid, all the particles will have a common transverse velocity (collective flow) in addition to their thermal/random motion. Kinematically, the inverse slope parameter will then depend on the particle mass as follows:

$$T \approx T_0 + \frac{1}{2} m \langle v_r \rangle^2, \tag{15.3}$$

where T_0 is the parameter of *true* thermal/random motion, $\langle v_r \rangle$ is the averaged collective velocity and m is the mass of particle. This relation can be obtained from the hydrodynamical equation, Eq. (13.30).

The existence of the collective expansion is recognized to be very important in describing the space-time evolution of nucleus–nucleus collisions. Moreover, the deviation of Ξ and Ω from the flow trend may be understood to be due to the fact that they are multi-strange particles (see Table H.2) and are known to have smaller cross-sections with the hadronic medium (van Hecke *et al.*, 1998). Therefore, they decouple from the fireball at the early stage of the collision, and are less influenced by the flow.

Using the phenomenological hydrodynamical model discussed in Section 14.1.3 and Appendix G, it has been shown that single-particle spectra of pions, kaons and protons observed in the mid-rapidity region with a range $m_{\mathrm{T}} - m \leq 1$ GeV are well described by Eq. (14.6), except for pions at low p_{T}. The measured spectra of various particle species have been well described by Eq. (14.6) with a common temperature, T, and a radial velocity, v_r. However, precise determination of T and v_r is found to be rather difficult from the analysis of single-particle spectra alone: for example, large T with small $\langle v_r \rangle$ and small T with large $\langle v_r \rangle$ tend to give similar distributions. Thus it is hard to distinguish between two cases if the experimental data are limited in statistics and/or p_{T} range. We may need additional information, such as two-particle correlation, as discussed below.

15.2.3 HBT two-particle correlation

As discussed in Section 14.1.5, interferometric measurements of identical particles provide a tool with which we can study the space-time evolution of collisions. At BNL-AGS and CERN-SPS, intensive interferometric measurements using pions and kaons have been carried out. From the three-dimensional analysis of two pions at SPS, the transverse radii at mid-rapidity can be seen to increase with the centrality of the collisions: larger radii are observed in larger collision systems. In central Pb+Pb collisions, transverse root-mean-squared (rms) radii of about

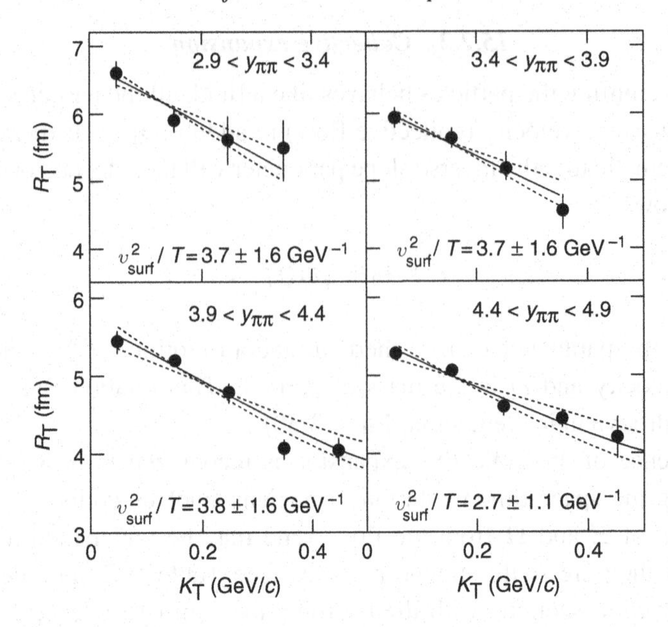

Fig. 15.5. Observed dependence of R_T on K_T for four different intervals of pion–pair rapidity, $y_{\pi\pi}$ (Appelshauser *et al.*, 1998).

5–7 fm are observed, which is twice as large as the geometrical transverse rms size of the colliding nucleus. Thus, the interferometric radii reflect the later stage of the collision, which is preceded by a large expansion.

Interferometric measurements have also been made for kaons: the source radii derived from K^+K^+ and K^-K^- interferometry are the same, but they are consistently smaller than those for pions. This difference might be attributed, at least partially, to the longer mean free path of the kaon than that of the pion. In other words, kaons freeze-out earlier than pions do. In this way, interferometry measurements have revealed the space-time evolution of collisions.

The long duration of hadron emission is one of the QGP signatures expected from the rapid increase of the entropy density across the phase boundary. This may be observed from the ratio between R_{out} and R_{side}, as shown in Eq. (14.16), but significant differences in R_{out} and R_{side} have not been observed.

While pair-momentum (K_T) dependence is not expected in interferometry from a static source (Section 14.1.5), a clear K_T dependence is observed, as shown in Fig. 15.5. The transverse radii of pions decrease with K_T in Pb + Pb collisions at SPS (Alber *et al.*, 1995; Appelshauser *et al.*, 1998). This K_T dependence might be attributed to the presence of the collective expansion, as given Eq. (14.17).[2]

[2] An alternative explanation according to the intra-nuclear cascade picture is that high-p_T particles emerge from the system at an earlier stage, while low-p_T particles are emitted in the later stage after many rescatterings.

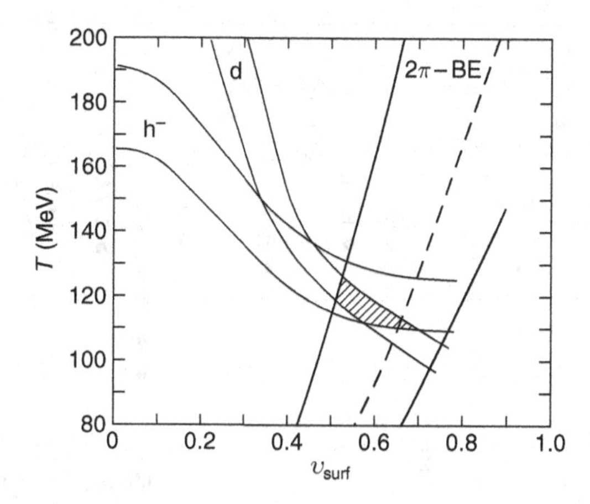

Fig. 15.6. Allowed region of temperature, T, and transverse expansion velocity, v_r, derived from single-particle spectra and HBT two-particle correlations. The dashed curve corresponds to $v_{\text{surf}}^2/T = 3.7\,\text{GeV}^{-1}$ (Appelshauser *et al.*, 1998).

The observed K_{T} dependences have been fitted with Eq. (14.17), and $R_{\text{G}} = (6.5 \pm 0.5)$ fm and $v_{\text{surf}}^2/T = (3.7 \pm 1.6)\,\text{GeV}^{-1}$ have been obtained.

Two distinct features of the data from CERN-SPS are (i) the inverse slope parameters of the transverse mass distribution in AA collisions have a component proportional to the particle mass, and (ii) a clear K_{T} dependence of the HBT effect is seen in AA collisions. Both features are attributed to the expanding fireball (Schnedermann *et al.*, 1993; Chapman *et al.*, 1995; Heintz *et al.*, 1996), in which a local thermal equilibrium and independent transverse and longitudinal motion are assumed. It has been shown that one can fit both features in (i) and (ii) using two parameters: the temperature, T, and the expansion velocity, v_r. In Fig. 15.6, the allowed regions for T and v_r are shown for the single-particle spectra and the K_{T} dependences of HBT, where the profile of expansion velocity, $v_r = v_{\text{surf}}(r/R_{\text{G}})$, is assumed, and where v_{surf} is the expansion velocity at the surface.

The result shows that the fireball is in a state of tremendous explosion, with an expansion velocity of half the speed of light and a temperature of about 120 MeV. Such an explosion would be driven by the strong pressure built up in the early stage of the collision.

15.3 Strangeness production and chemical equilibrium

Hadron abundances have been studied in heavy ion collisions. By integrating particle yields over the complete/wide phase-space, we can evaluate particle ratios. Unlike the momentum distributions, particle ratios are expected to be insensitive to the underlying processes. It is found that the ratios of produced hadrons are

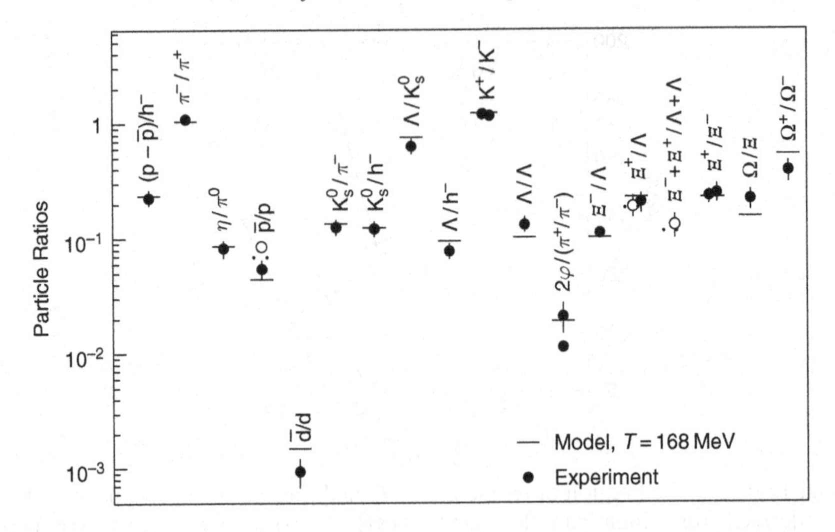

Fig. 15.7. Hadron abundance ratios. A comparison is given between the chemical equilibrium model (horizontal bars) and experimental ratios (filled circles) (Braun-Munzinger *et al.*, 1999).

well described by a simple statistical model (Eq. (14.3)), in which a temperature, T, and a baryon chemical potential, μ_B, play central roles.

The measured particle ratios at CERN-SPS and the model calculations are compared in Fig. 15.7. The effects of resonance decays and excluded volume corrections are included in the model calculation. As seen in the figure, this simple model fits the experimental ratios reasonably well, and a temperature of about 170 MeV and a baryon chemical potential of about 270 MeV are obtained. It is very intriguing that the abundances of the multi-strange particles also show the signs of chemical equilibrium. Since they have small cross-sections with the medium, they would not have enough time to reach chemical equilibrium if they are only produced by hadronic interactions.

Here we note the difference between chemical and thermal freeze-out. During the evolution of the collisions, the temperature of the system decreases due to the expansion. Chemical freeze-out comes first, after which matter composition is frozen. Particles still undergo elastic collisions until the thermal freeze-out, after which no further collisions among particles take place. Thus, the temperature for chemical freeze-out should be higher than that for thermal freeze-out. Indeed, the latter extracted from the single-particle spectra and HBT measurements is 120–140 MeV (Fig. 15.6), while the former, calculated from a statistical fit, is about 170 MeV.

Similar analyses have been carried out for heavy ion collisions at lower beam energies; the obtained parameters are plotted in Fig. 15.8, where the theoretical expectation of the phase boundary between hadronic matter and the QGP

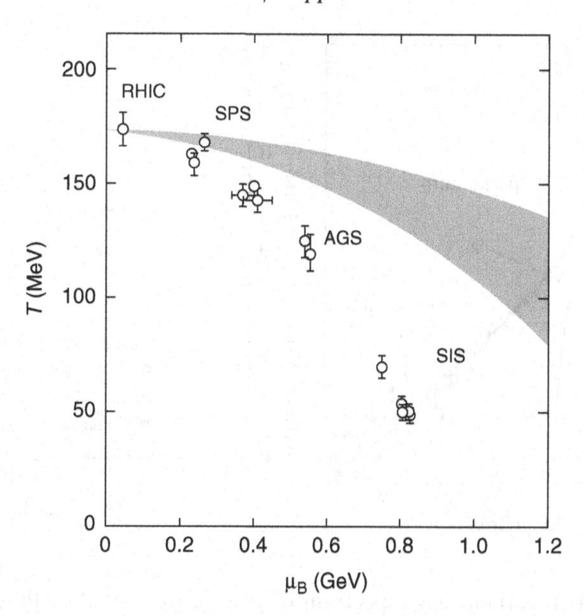

Fig. 15.8. Temperatures and baryon chemical potentials obtained at SIS, AGS, SPS and RHIC from the analysis of hadrochemical equilibrium. The shaded area shows an expected phase boundary between hadronic matter and the quark–gluon plasma (Braun–Munzinger *et al.*, 2004).

phase are also plotted. Note that the data points obtained at CERN-SPS are very close to the boundary. Considering the space-time evolution of the collisions, the system may have crossed the phase boundary shortly before it reaches chemical equilibrium (Braun-Munzinger *et al.*, 1999).

15.4 J/ψ suppression

The formation of QGP would have the effect of screening the color binding potential, preventing the c and $\bar{c}$ quarks forming charmonia states. Production of charmonia would be suppressed, depending on the relation of its spatial size and the screening radius, as discussed in Chapters 7 and 14. Measurement of J/ψ production is therefore important in probing the properties of the plasma.

At CERN-SPS, the NA38/NA50 collaboration constructed a muon spectrometer, and measurements of muon pairs from the continuum and from vector meson decays (ρ, ω, ϕ, J/ψ and ψ') were carried out as shown in Fig. 15.9(a); see Tables 7.1 and 7.2. Measurements were performed systematically in pp, pA and AA collisions (such as pp, p + d, p + Cu, p + Ta, O + U and so on).

To extract the significance of the signature, it is important to be able to differentiate between the background and the primary hard collisions, namely the

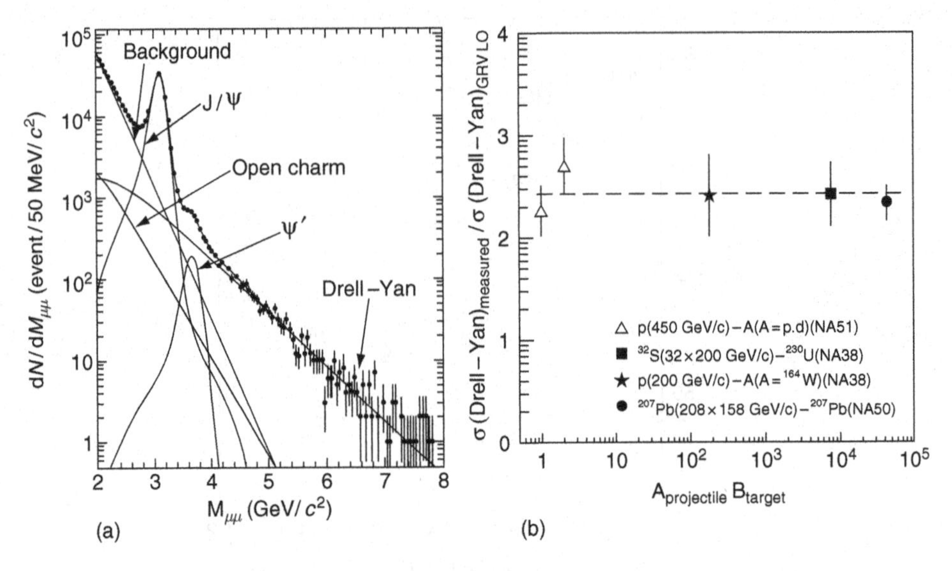

Fig. 15.9. (a) Invariant mass spectrum of μ^+ and μ^- for Pb+Pb collisions at 158*A* GeV/*c* (Abreu *et al.*, 1999). (b) Ratio of the of Drell–Yan cross-section and the normalized theoretical cross-section for pp as a function of the mass number of the projectile (A_{proj}) and the target (B_{targ}) (Abreu *et al.*, 1996).

Drell–Yan process (see Fig. 14.6). Due to the small cross-section, $\sigma_{\mathrm{DY}}^{\mathrm{pp}}$, the total cross-section in $A_{\mathrm{proj}} + B_{\mathrm{targ}}$ is given by

$$\sigma_{\mathrm{DY}}^{\mathrm{AB}} = A_{\mathrm{proj}} B_{\mathrm{targ}} \sigma_{\mathrm{DY}}^{\mathrm{pp}}, \tag{15.4}$$

where A_{proj} and B_{targ} are the mass numbers of the projectile and the target, respectively (Eq. (14.36)). Figure 15.9(b) depicts the ratio of the Drell–Yan cross-section and the normalized theoretical cross-section for pp collisions. As shown, the ratio stays constant in pp, p + d, p + W, S + U and Pb + Pb collisions. Thus, the rate of the Drell–Yan process stays the same in these collisions, and Eq. (15.4) can be used for a systematic comparison among different colliding systems.

Figure 15.10(a) shows $\sigma_{J/\psi}/\sigma_{\mathrm{DY}}$, the ratio of the J/$\psi$ and the Drell–Yan cross-sections. As a measure of centrality, L (the mean path length of the $c\bar{c}$ pair in the projectile/target nucleus) is shown as the abscissa of the figure. The ratio decreases monotonically from pp to most central S + U collisions. A clear departure from this trend, however, is seen in the Pb + Pb collisions. The solid line is an exponential fit of $\exp(-\rho_{\mathrm{nm}}\sigma_{\mathrm{abs}}L)$ to the monotonic decrease part of the data, where $\rho_{\mathrm{nm}} = 0.16\,\mathrm{fm}^{-3}$ is the ordinary nuclear density and σ_{abs} is the absorption cross-section. The exponential fit leads to an absorption cross-section of 6.2 ± 0.7 mb and it is interpreted as ordinary nuclear absorption of the $c\bar{c}$ pairs in nuclear matter (Kharzeev *et al.*, 1995). In the central Pb + Pb collisions, an additional suppression up to $\approx$70% is seen as an anomalous effect.

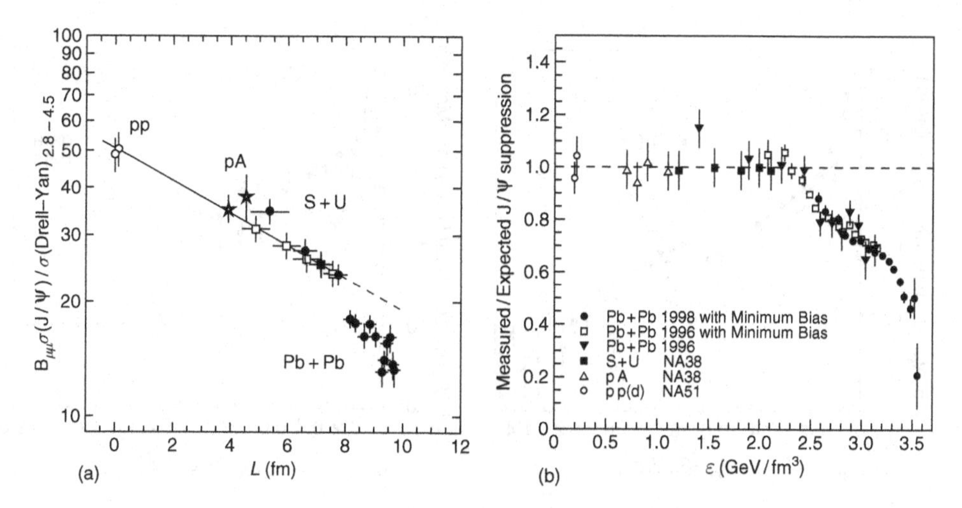

Fig. 15.10. (a) The $\sigma_{J/\psi}/\sigma_{DY}$ ratio as a function of L, the mean path length of the $c\bar{c}$ pairs through the nuclear matter; $B_{\mu\mu}$ is the branching ratio of $J/\psi \to \mu^+\mu^-$ (Abreu *et al.*, 1999). (b) Measured J/ψ production normalized by the ordinary nuclear absorption as a function of the energy density, ε (Abreu *et al.*, 2000).

A detailed pattern of the anomalous suppression is seen in of Fig. 15.10(b), which shows the ratio $\sigma_{J/\psi}/\sigma_{DY}$ divided by the normal nuclear absorption as a function of the energy density, ε, calculated from the Bjorken formula, Eq.(15.1). The first drop in J/ψ production occurs in the Pb + Pb collisions around an energy density of 2.3 GeV fm^{-3}, and even stronger suppression can be seen around 3.1 GeV fm^{-3}. Whether these sudden changes are related to the deconfinement or to something else is not totally clear yet. Experimentally, it is useful to measure the whole charm production and find out whether the branching ratios of the J/ψ particles yield to the whole charm production indeed change above some critical value of ε.

15.5 Enhancement of low-mass dileptons

Electromagnetic probes have a mean free path which is much larger than the reaction volume in heavy ion collisions. Therefore, they carry information about the various stages of the collision, including its early stage. The study of vector mesons (Tables H.2 and 7.2) through their decay into dileptons, in particular the decay of ρ meson, with its lifetime of 1–2 fm/c, is a useful probe, because it decays via dileptons inside the reaction volume and brings with it information about the surrounding matter at the time of the decay.

The CERES (Cerenkov Ring Electron pair Spectrometer) collaboration has constructed a spectrometer optimized to measure low-mass e^+e^- pairs and has

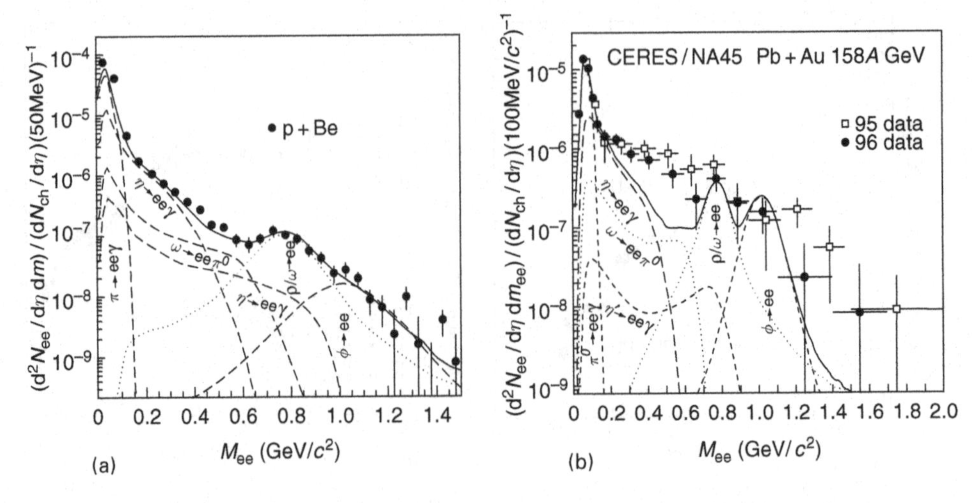

Fig. 15.11. (a) Inclusive invariant e^+e^- mass spectrum observed in p + Be collisions at an energy of 450 GeV/c. The solid line shows the e^+e^- yield from the hadron decays. The contributions of individual decays are also shown as dashed lines. (Agakichiev *et al.*, 1998). (b) Inclusive invariant e^+e^- mass spectrum observed in Pb + Au collisions at an energy of 158*A* GeV. The solid line shows the e^+e^- yield from the hadron decays scaled from p + Be or p + Au collisions (Agakichiev *et al.*, 1999).

carried out measurements in p + Be,[3] pA and AA collisions (Agakichiev *et al.*, 1998, 1999). In p + Be and p + Au collisions, exclusive measurements of π^0 and η have been carried out and then compared with the inclusive e^+e^- pair production. It was demonstrated that the observed spectra are well described in shape as well as absolute yield by the direct decays of the vector mesons and the Dalitz decays of neutral mesons such as π^0 and η (Fig. 15.11(a)). The relative abundances of these mesons in p + Be and p + Au collisions are found to be the same. However, in a Pb + Au collision, the same scenario and the same relative meson abundances do not explain the spectra; a significant excess is seen in the mass region below the ρ meson peak (Fig. 15.11(b)). Also, this excess is shown to increase more rapidly than the increase in charged-hadron multiplicity, and it is concentrated in the low-p_T region.

The annihilation of thermal pions created during the collision process ($\pi + \pi \rightarrow e^+e^-$) provides a contribution to the low-mass region and explains the fraction of the low-mass enhancement. This itself is a strong indication that hot/dense matter is created in the heavy ion collisions. It is also recognized that the $\pi\pi$ annihilation does not explain the whole dilepton excess. The spectral modification

[3] Since the Be nucleus is thin in terms of the mean free path of the projectile nucleon, p + Be collisions can be considered as pp or pn collisions to a good approximation.

in the vector channel discussed in Sections 7.2.5 and 7.3 is a possible candidate for explaining the remaining excess, but further theoretical analyses are necessary (Rapp and Wambach, 2000; Gale and Haglin, 2004).

15.6 Observation of direct photons

The measurement of direct photon production is also considered to provide important information about the collision process. If hot and dense matter, such as QGP, exists, thermal emission of photons takes place. Enhancement of direct photons from thermalized QGP may be seen in the medium-p_T region of a few GeV, while in the high-p_T region the photons from the initial hard collisions are dominant.

Experimentally, the measurement of direct photons is a difficult task: there are large amounts of background photons which arise from decaying hadrons, for example $\pi^0 \to \gamma\gamma$ and $\eta \to \gamma\gamma$. Therefore, the direct photons can be extracted only after a precise comparison of the measured inclusive photon spectra with the background from hadronic decays. The first observation of direct photons in central Pb + Pb collisions at an energy of 158*A* GeV is shown in Fig. 15.12 (Aggarwal *et al.*, 2000). In this measurement, the π^0 s and ηs are reconstructed

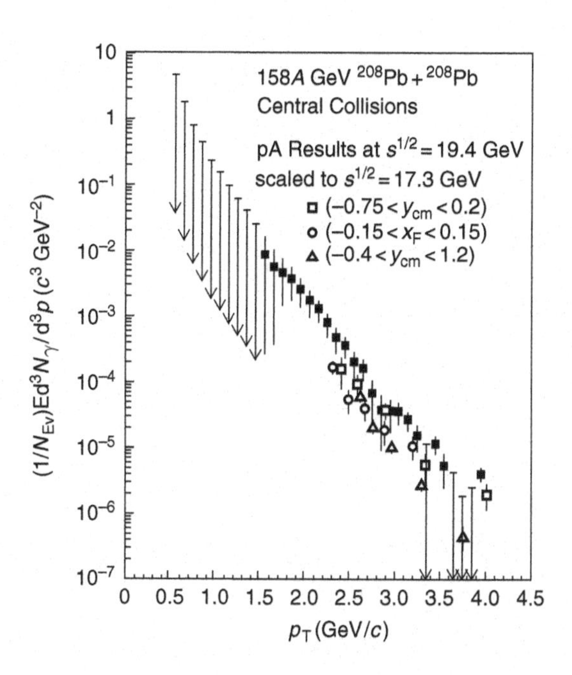

Fig. 15.12. Observed direct photon multiplicity (filled squares) in central 158*A* GeV Pb + Pb collisions compared with that in pA collisions (open symbols) (Aggarwal *et al.*, 2000). Downward arrows show upper limits. Note that N_γ and $N_{\rm Ev}$ are the number of photons and the number of central collisions, respectively.

via their $\gamma\gamma$ decays, which comprise $\sim$97% of the total background. The photon yields from pA collisions scaled by the average number of nucleon–nucleon collisions are also shown in Fig. 15.12. The observed spectrum in the Pb + Pb case has a shape similar to that in the pA case, but the absolute magnitude is enhanced for $p_T > 2\,\text{GeV}/c$. A similar attempt in a smaller system, such as 200*A* GeV ^{32}S + Au collisions, results in no significant excess (Albrecht *et al.*, 1996). This may be another indication that a hot and/or dense matter has been created at CERN-SPS, although it is difficult to tell from this data alone whether it is QGP or a hot/dense hadronic plasma.

16

First results from the Relativistic Heavy Ion Collider (RHIC)

The Relativistic Heavy Ion Collider (RHIC) (Hahn *et al.*, 2003) was the first dedicated heavy ion collider, built at Brookhaven National Laboratory. The machine was completed in 2000 and has provided experimentalists with heavy ion beams of maximum energy since 2001.

In this chapter, we introduce the results from the first three years at RHIC[1], where formation of strongly interacting, hot and dense partonic matter is observed. This field, like the theme of this book, evolves very rapidly, and interested readers are advised to search for latest progress via the internet.[2]

16.1 Heavy ion acceleration and collisions in the RHIC

Heavy ion beams cannot be accelerated to relativistic energies by a single accelerator. This can only be achieved step by step, using a series of accelerators. At the RHIC facility, the Tandem Van de Graaf, the Booster Synchrotron, and the Alternating Gradient Synchrotron (AGS) are used to pre-accelerate heavy ions before injection into the collider, as illustrated in Figs. 16.1 and 15.1(a).

To accelerate a gold beam, for example, negative gold ions are created by a pulsed sputter ion source and are accelerated by the first stage of the Tandem Van de Graaf, where the atomic electrons of the ion are partially stripped off by a foil located inside the high-voltage terminal. The gold ions, now in a positive charge state, are accelerated during the second stage to $\sim 1A$ MeV. These positive ions are transferred through a 540 m transfer line to the Booster Synchrotron, where an RF electric field is applied, the ions are grouped into three bunches and are accelerated to $78A$ MeV. Another foil at the exit of the Booster strips away all of the atomic electrons of the gold ion. The fully stripped positive gold ions

[1] For more details, see the BNL-Report *Hunting the Quark Gluon Plasma*, (BNL-73847-2005). See also Adams *et al.* (2005), Adcox *et al.* (2005), Arsene *et al.* (2005), and Back *et al.* (2005b).

[2] For example, http://www.bnl.gov/rhic/

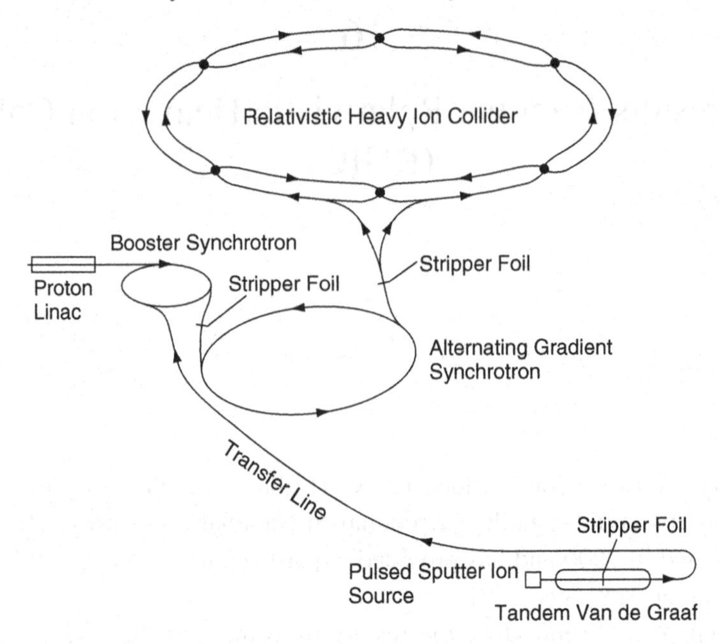

Fig. 16.1. Configuration of the accelerator facility of the RHIC (Hahn *et al.*, 2003). Acceleration of a gold beam is carried out step by step using a chain of accelerators: the Tandem Van de Graaf, the Booster Synchrotron, the Alternating Gradient Synchrotron and the Relativistic Heavy Ion Collider. See also Fig. 15.1(a).

are injected into the AGS, where the three bunches of gold ions are accelerated further to 10.8*A* GeV, which is the required injection energy for the RHIC. The three bunches of gold ions from the AGS are injected into the two 3.834 km long RHIC rings called the blue ring and the yellow ring (Fig. 16.2), where they circulate in opposite directions. By repeating this process, each ring contains 60 bunches of ions; each bunch contains about 10^9 gold ions. The filling of both rings takes a few minutes.

The two RHIC rings are non-circular concentric superconducting magnet rings, located in a common horizontal plane and intersecting one another at six locations along the ring. Each ring consists of three inner and three outer arcs, with six intersection regions in between (Fig. 16.3). There are 396 superconducting dipole magnets and 492 superconducting quadrupole magnets to guide and focus the ion beams into well defined orbits in each beam pipe. Together with other correction magnets, there are 1740 magnets in total. The magnets are cooled to $\lesssim 4.6$ K by circulating supercritical helium supplied by a gigantic 24.8 kW refrigerator system.

The injected bunches of ions from the AGS at an energy of 10.8*A* GeV are accelerated by electric fields in an RF cavity. Two RF cavities are implemented in each ring. On increasing the energy of the ions, the magnetic fields of the rings

Fig. 16.2. RHIC accelerator rings. Courtesy of BNL.

are raised accordingly, up to 3.5 T in the arc dipole, which corresponds to the maximum energy attainable with this machine: 100*A* GeV for gold and 250 GeV for proton beams. The ions are stored in the rings for a period of 6 to 12 hours, and collisions between bunches take place at six locations along the rings. Then, the collision experiments at full beam energy begin.

The collision *rate*, as well as the *energy*, is a very important parameter that describes the performance of an accelerator; the higher the collision rate, the more data and rare events can be obtained. To describe the collision rate of the collider, the *luminosity*, $\mathcal{L}$, is often used. Using $\mathcal{L}$, the collision rate, N, of an event with a cross-section σ is given by

$$N = \mathcal{L}\sigma. \tag{16.1}$$

When counter-circulating bunches come to a crossing point, i.e. a collision point,

$$\mathcal{L} \propto \frac{N_1 N_2}{S}, \tag{16.2}$$

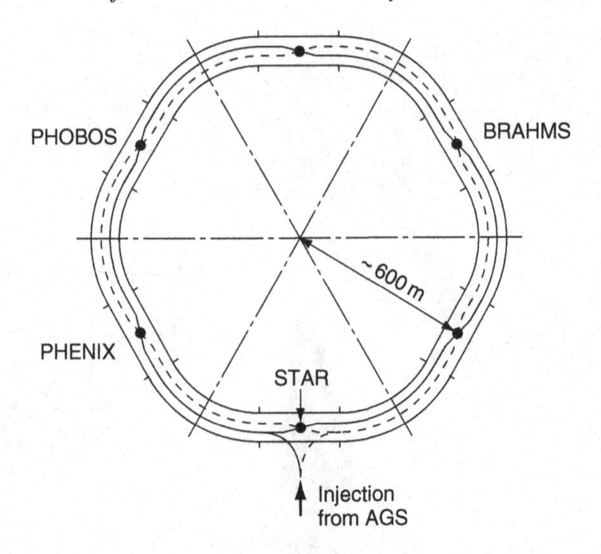

Fig. 16.3. Accelerator rings and experiments at the RHIC (Hahn *et al.*, 2003). The RHIC rings of ~3.8 km circumference consist of six arc and six insertion sections. Four current experiments are at 6 o'clock (STAR (Ackermann *et al.*, 2003), the Solenoidal Tracker At RHIC), 8 o'clock (PHENIX (Adcox *et al.*, 2003), the Pioneering High Energy Nuclear Interaction Experiment), 10 o'clock (PHOBOS (Back *et al.*, 2003a), named after one of the two moons of MARS, the Modular Array for RHIC Spectra) and 2 o'clock (BRAHMS (Adamczyk *et al.*, 2003), the Broad Range Hadron Magnetic Spectrometer).

where $N_{1(2)}$ and S are the number of beam particles in a bunch and the effective size of the bunch, respectively.[3] Thus, more beam particles in the bunch and a well focused beam bunch will result in a higher collision rate. The design luminosity of the RHIC is 2×10^{26} cm^{-2} s^{-1} for Au + Au collisions at full energy.

The collision rate decreases gradually due to the collisions of the beam with residual gas as well as the beam–beam collisions. To maintain the rate, the entire acceleration procedure is repeated every 6–8 hours. It is essential to maintain an ultra-high vacuum (~10^{-10} hPa) in the beam pipe during the lifetime of the beam to reduce collisional losses of the beam.

Having two independent rings allows some flexibility during the operation; for example, collisions between unequal species of ions can be achieved. Bending fields of two rings can be tuned independently to each momentum, but with the same velocity. Collisions between unequal species, as well as a variety of collision energies, provide experimentalists with vital tools for the understanding

[3] For a fixed target experiment, $\mathcal{L} = N_{\text{beam}} N_{\text{target}} D\sigma_{\text{tot}}$, where N_{beam}, N_{target} and D are the number of beam particles per unit time, the number of target nuclei per unit volume and the thickness of the target, respectively.

of complex high-energy collisions and the formation of QGP. (A typical example may be seen in Fig. 16.18.)

16.2 Particle production

Particle production in Au ($A = 197$) + Au collisions has been measured at various energies during the first collision experiments. For example, the PHOBOS experiment has performed a broad and systematic survey of hadronic particle production. Figure 16.4 shows the pseudo-rapidity distributions of charged particles, $dN_{ch}/d\eta$, measured at $\sqrt{s_{NN}} = 19.6$, 130 and 200 GeV (Baker *et al.*, 2003; Back *et al.*, 2003b). As can be read from the figure, the typical values of $dN_{ch}/d\eta$ at mid-rapidity in central Au + Au collisions are 350, 575 and 650 at $\sqrt{s_{NN}} =$ 19.6, 130 and 200 GeV, respectively. Due to the large detector coverage in η, $dN_{ch}/d\eta$ is measured over almost the full range of the collision. As seen in the figure, at higher energies the rapidity densities become larger and the distribution becomes wider.

The tails of the distributions are found to be independent of the collision energy; if the distributions are re-plotted as a function of $y' = y - y_{beam}$, they are identical at around $y' \sim 0$, where y_{beam} is the rapidity of the beam. This feature of particle production away from the mid-rapidity region is known as *limiting fragmentation* (Benecke *et al.*, 1969). At high enough energy, particle production is considered to reach a limiting value, and it becomes independent of the beam energy in a region $y' \sim 0$. It is worthwhile to note that, even at RHIC energies, central plateau and fragmentation regions are not well separated, so a simple application of the Bjorken picture may not be appropriate.

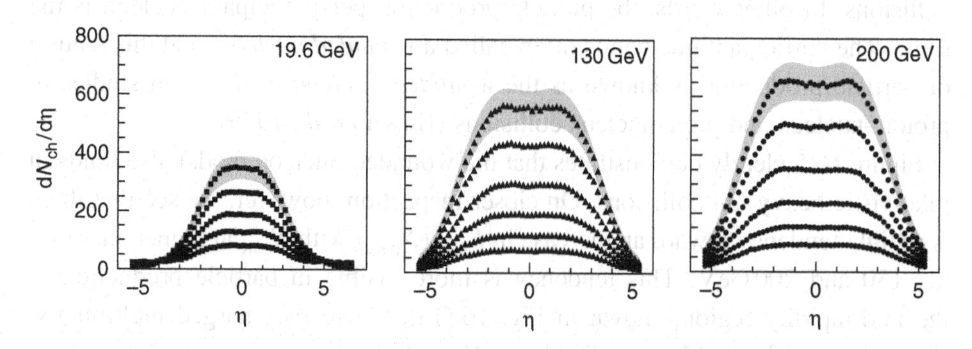

Fig. 16.4. Charged particle pseudo-rapidity distribution, $dN_{ch}/d\eta$, measured for Au + Au collisions at $\sqrt{s_{NN}} = 19.6$, 130 and 200 GeV for the centrality bins of 0–6%, 6–15%, 15–25%, 25–35%, 35–45% and 45–55%. Typical statistical error is shown as a shaded region (Baker *et al.*, 2003; Back *et al.*, 2003b).

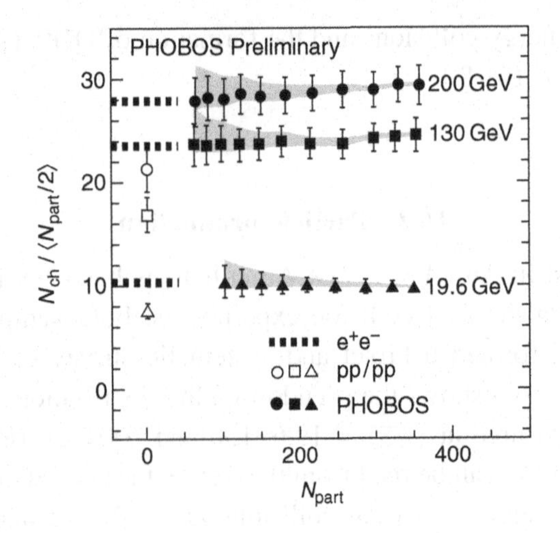

Fig. 16.5. Total charged multiplicity normalized by $\langle N_{part}/2\rangle$ as a function of N_{part} (Back *et al.*, 2003c). Data are shown for $\sqrt{s_{NN}} = 19.6$, 130 and 200 GeV Au + Au collisions.

By integrating the measured $dN_{ch}/d\eta$ distributions, we obtain the total charged multiplicities as a function of the centrality of the collisions. This provides important clues for the mechanisms of particle production. The centrality is characterized by the number of participant nucleons, which is evaluated in each collision as described in Section 17.3. The total charged multiplicities normalized by the participant nucleon pair, $\langle N_{part}/2\rangle$, are shown against N_{part} in Fig. 16.5 (Back *et al.*, 2003c). As is clearly seen in the figure, the normalized total charged multiplicities are almost constant from peripheral (small N_{part}) to central (large N_{part}) collisions. In other words, the particle production per participant nucleon is the same. The participant nucleon is often called a *wounded nucleon*, and this feature of particle production is known as the *wounded nucleon model* from studies in proton–nucleus and pion–nucleus collisions (Busza *et al.*, 1975).

Figure 16.5 clearly demonstrates that the wounded nucleon model also holds in relativistic heavy ion collisions. On closer inspection, however, we see that there is a slight tendency toward an increase in $N_{ch}/\langle N_{part}\rangle$ with N_{part} at higher energies, i.e. 130 and 200 GeV. This tendency is more visible in particle production in the mid-rapidity region, shown in Fig. 16.6(a), where the charged multiplicity, $dN_{ch}/d\eta$, at $|\eta| < 0.35$ normalized by $\langle N_{part}/2\rangle$ is shown as a function of N_{part} (Adcox *et al.*, 2001); pp and p$\bar{\text{p}}$ data are also plotted for comparison. The observed centrality dependence does not simply scale with N_{part}, but shows a steep initial rise from p$\bar{\text{p}}$ to $N_{part} \sim 150$ and then a slow increase with N_{part}.

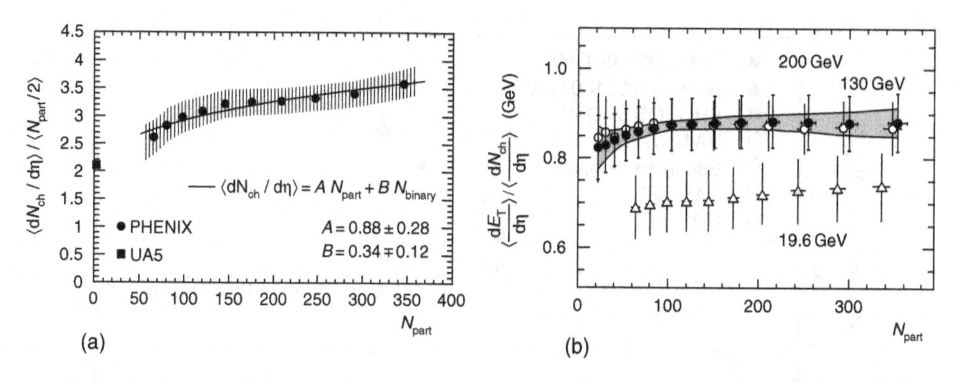

Fig. 16.6. (a) Charged multiplicity density in $\sqrt{s_{\mathrm{NN}}} = 130$ GeV Au+Au collisions at mid-rapidity ($|\eta| < 0.35$) normalized by $\langle N_{\mathrm{part}}/2\rangle$ (Adcox *et al.*, 2001). (b) $\langle dE_{\mathrm{T}}/d\eta\rangle/\langle dN_{\mathrm{ch}}/d\eta\rangle$ plotted as a function of N_{part} (Adcox *et al.*, 2005). Data are shown for $\sqrt{s_{\mathrm{NN}}} = 19.6$, 130 and 200 GeV Au+Au collisions.

A possible origin of the departure from N_{part} scaling is the effect of hard processes on the particle production at high energies, as discussed in Section 13.2. While the particle production in soft processes is expected to be proportional to N_{part}, that from the hard processes should scale with the number of primary nucleon–nucleon collisions (binary collisions), N_{binary}. As described in Section 10.5, N_{part} and N_{binary} can be evaluated using the Glauber model. The solid line shown in Fig. 16.6(a) is a fit of the data with the function

$$\frac{dN_{\mathrm{ch}}}{d\eta} = A \cdot N_{\mathrm{part}} + B \cdot N_{\mathrm{binary}}. \tag{16.3}$$

Obtained values for A and B shown in the figure indicate a large contribution from hard processes in central collisions. This requires further study, however, since this may not be the only interpretation.

Transverse energy, E_{T}, has been also measured, together with the charged particle multiplicities; overall behavior is very similar to that of the charged multiplicity. Figure 16.6(b) shows $\langle dE_{\mathrm{T}}/d\eta\rangle/\langle dN_{\mathrm{ch}}/d\eta\rangle$ as a function of N_{part} (Adcox *et al.*, 2005). The ratios are about 0.8 GeV, which implies that the transverse energy per produced particle does not increase greatly with the beam energy. The average transverse mass per particle is given by

$$\langle m_{\mathrm{T}}\rangle = \frac{\langle dE_{\mathrm{T}}/d\eta\rangle}{\langle dN/d\eta\rangle} \simeq \frac{\langle dE_{\mathrm{T}}/d\eta\rangle}{\frac{3}{2}\langle dN_{\mathrm{ch}}/d\eta\rangle} \sim 0.5\,\mathrm{GeV}. \tag{16.4}$$

Note that N (N_{ch}) is the total number of produced particles (charged particles), and we have assumed $N \simeq (3/2)N_{\mathrm{ch}}$.

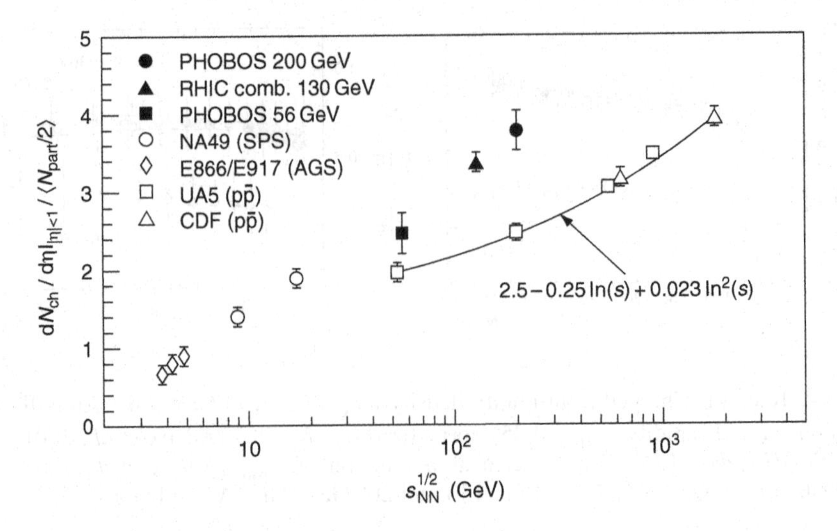

Fig. 16.7. Charged multiplicities at mid-rapidity ($|\eta| < 1$) normalized by $\langle N_{\rm part}/2\rangle$ (Back *et al.*, 2002). Central nucleus–nucleus and $p\bar{p}$ collisions at various beam energies are compared. The solid curve shows a fit to the $p\bar{p}$ data.

Charged particle multiplicities at mid-rapidity ($|\eta| < 1$) are compared for $p\bar{p}$ and AA collisions as a function of the beam energy in Fig. 16.7 (Back *et al.*, 2002), which shows an approximately logarithmic rise with $\sqrt{s_{\rm NN}}$ over the full energy range of collision energies. The RHIC Au + Au collisions at $\sqrt{s_{\rm NN}} = 200\,{\rm GeV}$ create about twice as many multiplicities as that in SPS Pb + Pb collisions at $\sqrt{s_{\rm NN}} = 17\,{\rm GeV}$. It is seen that central Au + Au collisions show significantly larger multiplicities per participant pair than $p\bar{p}$ collisions at comparable values of $\sqrt{s_{\rm NN}}$. The energy evolution of normalized multiplicities in central nucleus–nucleus collisions is found to be different from that for $p\bar{p}$ collisions.

Combining the measurements of the charged particle multiplicities and the average transverse energy of particles, the energy density achieved in the collision, ε_0, may be estimated using the Bjorken formula Eq. (11.81):

$$\varepsilon_0 = \frac{1}{\pi R^2 \tau_0} \frac{dE_{\rm T}}{dy}\bigg|_{y\simeq 0}. \tag{16.5}$$

An estimate with $\tau_0 = 1$ (0.5) fm gives $\varepsilon_0 \sim 4$ (7) GeV fm^{-3} for Au + Au at $\sqrt{s_{\rm NN}} = 200\,{\rm GeV}$. This is high compared with the critical energy density, $\varepsilon_{\rm crit} \sim 1\,{\rm GeV}$ (Eqs. (3.51) and (3.64)), and thus the matter created at RHIC may be well above the threshold for QGP formation.

16.3 Transverse momentum distributions

The transverse energy densities and the particle multiplicities are found to be significantly larger than those of pp collisions at the same energy. A similar behavior has also been seen in heavy ion collisions at lower energies. Measurements of identified hadrons are expected to provide further information on the reaction dynamics, as discussed in Chapter 13. In Fig. 16.8(a), we show the transverse momentum spectra observed at mid-rapidity in Au + Au collisions at $\sqrt{s_{NN}} = 200\,\text{GeV}$. Figure 16.8(a) gives the spectra of $\pi^{\pm}$, $K^{\pm}$, p and $\bar{\text{p}}$ measured with a magnetic spectrometer, and Fig. 16.8(b) shows the spectra of π^0 measured with an electromagnetic calorimeter. Due to experimental limitations, the momentum coverage is different for each species. The spectra of central and peripheral collisions are compared, which is useful in clarifying heavy ion effects, since the peripheral collisions are similar to nucleon–nucleon collisions at the same energy.

The pion spectra exhibit a concave shape in both central and peripheral collisions. The kaon and proton spectra exhibit an exponential shape in peripheral collisions, but protons in central collisions exhibit a so-called “shoulder-arm” shape. A quantitative comparison of the spectra for central and peripheral collisions will be discussed in Sections 16.7 and 16.9.

Two interesting features may be seen immediately in the p_T spectra.

- The anti-proton ($\bar{\text{p}}$) yield is comparable to the proton yield. The yield ratio, $\bar{\text{p}}$/p, is $0.7 \sim 0.8$ and seems to stay constant up to $p_T \sim 4\,\text{GeV}/c$.
- The yields of p and $\bar{\text{p}}$ become comparable to that for pions for $p_T \sim 2\,\text{GeV}/c$. For $p_T > 2\,\text{GeV}/c$, a significant fraction of the total particle yields is from baryons (p and $\bar{\text{p}}$). Particle production at high p_T is, surprisingly, dominated by baryons.

To clarify these features, the yield ratios p/π and $\bar{\text{p}}/\pi$ are shown in Fig. 16.9; p/π and $\bar{\text{p}}/\pi$ are plotted as a function of p_T for central (0–10%), mid-central (20–30%) and peripheral (60–92%) collisions. The ratios increase with p_T, but seem to saturate. The ratio at saturation becomes larger in central than in peripheral collisions. The ratios observed in pp collisions at $\sqrt{s_{NN}} = 53\,\text{GeV}$ and in gluon jets observed in e^+e^- collisions are also shown in the figure. In the high-p_T region ($p_T > 3\,\text{GeV}/c$), these ratios for pp, e^+e^- and the peripheral Au + Au collisions[4] agree with each other. This agreement indicates that in the high-p_T region, the p/π and $\bar{\text{p}}/\pi$ ratios can be well described by the fragmentation. On the other hand, the increase in p/π and $\bar{\text{p}}/\pi$ ratios in the central Au + Au collisions suggests a clear departure from the fragmentation process, even if p_T is as high as $4 \sim 5\,\text{GeV}/c$.

[4] Peripheral Au + Au collisions can be considered as a superposition of a few nucleon–nucleon collisions.

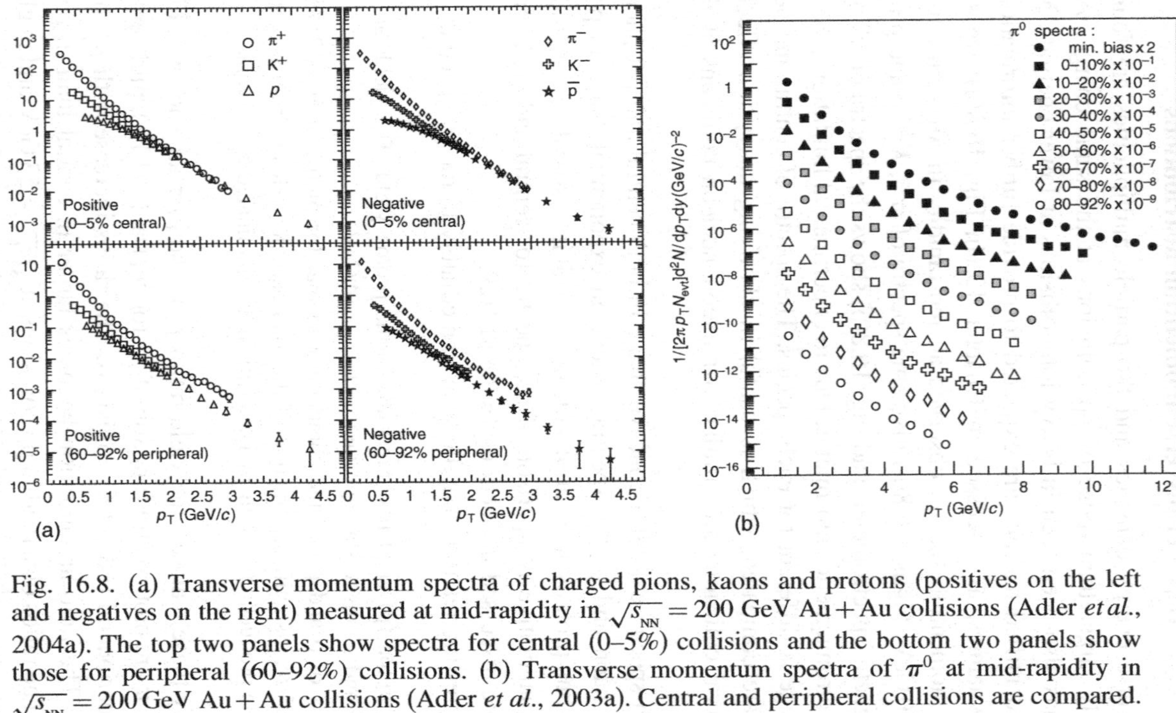

Fig. 16.8. (a) Transverse momentum spectra of charged pions, kaons and protons (positives on the left and negatives on the right) measured at mid-rapidity in $\sqrt{s_{NN}} = 200$ GeV Au + Au collisions (Adler *et al.*, 2004a). The top two panels show spectra for central (0–5%) collisions and the bottom two panels show those for peripheral (60–92%) collisions. (b) Transverse momentum spectra of π^0 at mid-rapidity in $\sqrt{s_{NN}} = 200$ GeV Au + Au collisions (Adler *et al.*, 2003a). Central and peripheral collisions are compared.

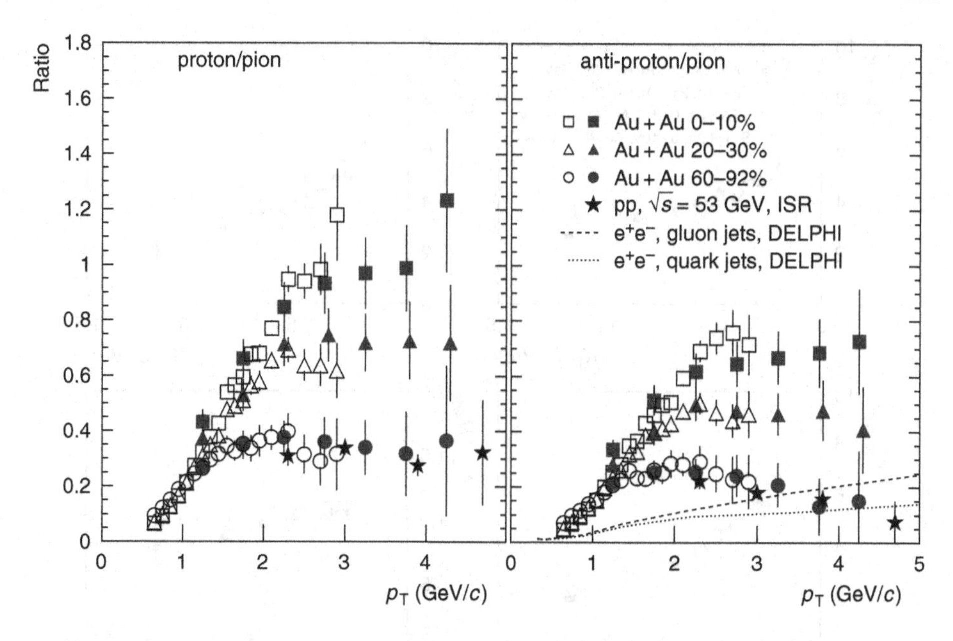

Fig. 16.9. Proton/pion and anti-proton/pion ratios at mid-rapidity in $\sqrt{s_{NN}} =$ 200 GeV Au + Au collisions (Adler *et al.*, 2003c). Ratios in the central (0–10%), mid-central (20–30%) and peripheral (60–92%) collisions are compared. Open (filled) points are for $\pi^{\pm}$ (π^0), respectively. The data from pp collisions at $\sqrt{s_{NN}} = 53$ GeV are shown as stars. Anti-proton/pion ratios measured in gluon and quark jets are shown as dashed and dotted lines, respectively.

16.4 HBT correlations

Motivated by the possibility of detection of a larger source size or a long duration time of particle emission as a signature of QGP formation (Section 14.1.5), the HBT two-particle correlation has been measured at RHIC (Adler *et al.*, 2001b). To extract the three-dimensional information of the source, multi-dimensional Gaussian fits are made to the relative momentum correlations in sideward, outward and longitudinal directions, and $R_{\rm side}$, $R_{\rm out}$ and $R_{\rm long}$ (Section 14.1.5) are obtained. Experimentally, corrections from the Coulomb repulsion between detected particles, and also with other hadrons, are very important, and careful corrections are being made, although they are difficult to handle.

In Fig. 16.10, the measured HBT radii, $R_{\rm side}$, $R_{\rm out}$ and $R_{\rm long}$, are plotted as a function of the sum of the momentum, $K_{\rm T} = k_{1\rm T} + k_{2\rm T}$ (Adler *et al.*, 2004b; Adams, *et al.*, 2004d). As has been seen at SPS (Fig. 15.5), there is a clear dependence of HBT radii on $K_{\rm T}$, which indicates the existence of a collective expansion corresponding to Eq. (14.17). In Fig. 16.10, the ratio $R_{\rm out}/R_{\rm side}$ is also shown. For a prolonged source lifetime, the ratio is predicted to be larger

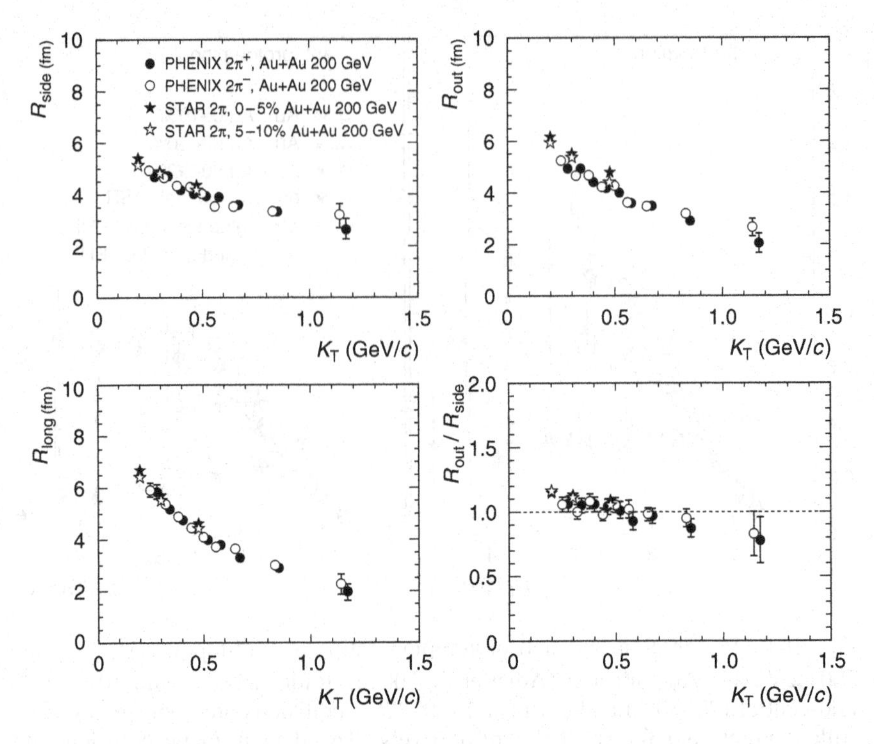

Fig. 16.10. The HBT radii, R_{side}, R_{out} and R_{long}, and the ratio R_{out}/R_{side} as a function of K_T. Data are from Adler *et al.* (2004b) and Adams *et al.* (2003d)

than unity, while the observed ratio is $R_{out}/R_{side} \sim 1$, with a slight decrease in K_T. This has not yet been successfully explained, and is known as the *HBT puzzle*.

16.5 Thermalization

As have been achieved at SPS and described in Section 15.3, the particle ratios provide us with useful information on the thermochemical equilibrium. The STAR experiment has measured yields of various particles using its strong capability to observe decayed particles. Figure 16.11 depicts the ratios of measured yields of various particles and compares them with the values given by the thermochemical model, in which chemical equilibration is assumed to exist at temperature T and baryon chemical potential μ_B. As shown, good agreement is obtained with $T = 176$ (177) MeV and $\mu_B = 41$ (29) MeV at 130 (200) GeV. The observed temperatures in both the 130 and 200 GeV collisions are very similar and slightly larger than that obtained at SPS energy. It is worth noting here that the model requires chemical equilibrium to be established at an early stage of the collisions.

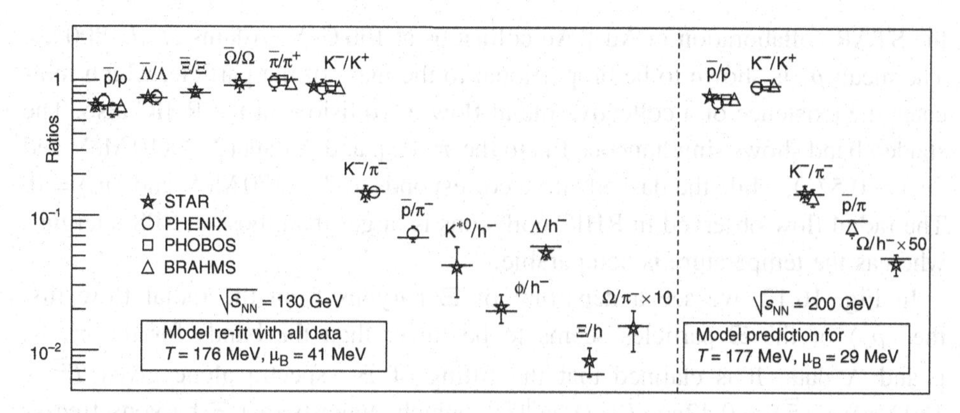

Fig. 16.11. Ratios of measured yields of various particles, compared with the values from the thermochemical model (Braun-Munzinger *et al.*, 2001, 2004).

Then, the question is: what mechanism has brought the system into chemical equilibrium?[5]

The transverse momentum spectra, or transverse mass spectra, have been studied in terms of the thermal equilibrium with a collective radial flow. As is the case at lower energies, the transverse momentum spectra in low-p_T regions are found to be well described by the radial flow model. Figure 16.12 shows the mass dependence of the mean transverse momentum for various particle species measured by

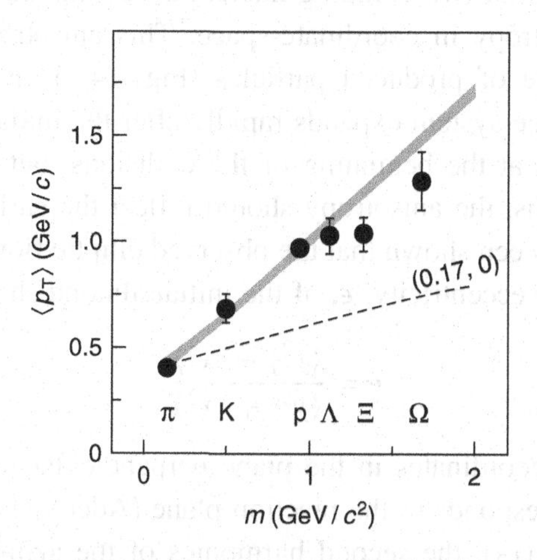

Fig. 16.12. Mean transverse momentum of identified hadrons in Au + Au collisions at 130 GeV as a function of hadron mass. The shaded band shows a hydrodynamical fit to the π, K, p and Λ data ($T \sim 110$ MeV and $\langle v_r \rangle \sim 0.57c$), and the dashed curve corresponds to $T = 170$ MeV and $\langle v_r \rangle = 0$ (Adams *et al.*, 2004a).

[5] The freeze-out temperature is saturated, while the μ_B and the flow velocity increase from SPS to RHIC collisions. The saturation of temperature may imply the existence of the phase boundary of QGP at around $\sim$170 MeV; see Fig. 15.8.

the STAR collaboration in Au + Au collisions at 130 GeV (Adams *et al.*, 2004a). The mean p_T is shown to be proportional to the mass of the particle, which indicates the existence of a collective radial flow in collisions at the RHIC also. The shaded band shows simultaneous fits to the π, K, p and Λ data ($T \sim 110$ MeV and $\langle v_r \rangle \sim 0.57c$), while the dashed curve corresponds to $T = 170$ MeV and $\langle v_r \rangle = 0$. The radial flow observed in RHIC collisions is larger than those at SPS energies, whereas the temperature is comparable.

In Fig. 16.12, we see a departure of Ξ baryons from the radial flow fits; the $\langle p_T \rangle$ of the Ξ particles seems to be lower than the bands for the π, K, p and Λ data. It is claimed that the fitting of $\Xi^{\pm}$ spectra alone gives $T^{\Xi} \sim$ 180 MeV, $\langle v_r^{\Xi} \rangle \sim 0.42c \sim (2/3)\langle v_r^{\text{others}} \rangle$, which suggests that Ξ baryons freeze-out and decouple from the rapidly expanding fireball during an earlier and hotter stage of the collision (Adams *et al.*, 2004a); Ξ baryons might even decouple during chemical equilibrium because T^{Ξ} is close to the chemical freeze-out temperature.

16.6 Azimuthal anisotropy

As discussed in Section 14.1.4, azimuthal anisotropies in the final momentum-space are sensitive to the early evolution of the system. In non-central collisions, the overlapped region of two colliding nuclei has an almond shape, which corresponds to the anisotropy in coordinate-space. This anisotropy is transferred to the momentum-space of produced particles (Fig. 14.1) via the rescattering of constituents. Since the system expands rapidly after the initial impact, the spatial anisotropy is largest at the beginning of the collisions, but it disappears as the system expands. Thus, the anisotropy should reflect the early stage of the collisions. Indeed, it has been shown that the observed elliptic flow, v_2, scales roughly with the geometrical eccentricity, ϵ, of the initial almond shape:

$$\epsilon = \frac{\langle y^2 \rangle - \langle x^2 \rangle}{\langle y^2 \rangle + \langle x^2 \rangle}, \tag{16.6}$$

where x and y are coordinates in the plane perpendicular to the beam axis and the (x, z)-plane corresponds to the reaction plane (Adcox *et al.*, 2002).

The elliptic flow (v_2), the second harmonics of the azimuthal particle distribution in Eq. (14.7), has been measured at AGS, SPS and RHIC. Figure 16.13 shows the pseudo-rapidity (η) distributions of v_2 for Au + Au collisions at energies of $\sqrt{s_{NN}} = 19.6$, 62.4, 130 and 200 GeV (Back *et al.*, 2005a). Data are for mid-central (0–40%) Au + Au collisions, in which $\langle N_{\text{part}} \rangle \sim 200$. Charged particles with p_T values over the entire range are integrated for this measurement since no magnetic field is applied. The η distributions take a roughly triangular

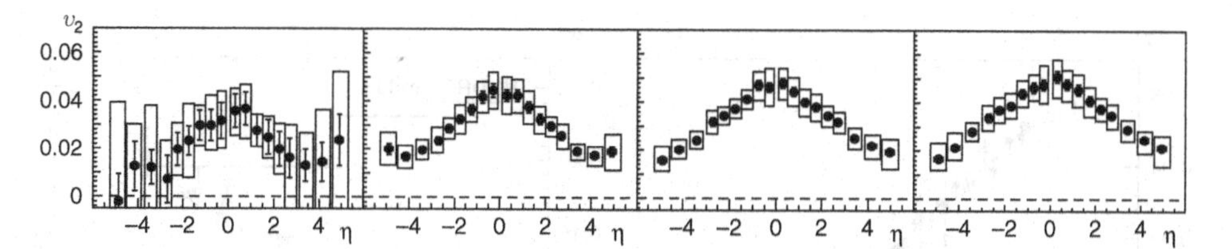

Fig. 16.13. Elliptic flow, v_2, as a function of pseudo-rapidity, η, in mid-central (0–40%) Au + Au collisions at energies of $\sqrt{s_{NN}} = 19.6$, 62.4, 130 and 200 GeV from left to right. Numbers of participant nucleons are 201, 201, 209 and 211, respectively. Error bars show the statistical errors (Back *et al.*, 2005a).

shape, which peaks at mid-rapidity. The peak value of v_2 increases slowly with the beam energy. The observed values of v_2 are consistent with the previous measurements; the maximum v_2 values are ~0.02 at AGS (Barrette *et al.*, 1997), ~0.035 at SPS (Poskanzer *et al.*, 1999) and ~0.06 at RHIC (Ackermann *et al.*, 2001). The larger value of v_2 at RHIC is considered as a striking feature, since it implies a larger amount of thermalization. Indeed, a hadron cascade model, such as relativistic quantum molecular dynamics (RQMD), underestimates the v_2 values, which suggests that other mechanisms for thermalization than those with hadronic rescattering should exist.

The data in Fig. 16.13 are replotted in Fig. 16.14(a) as a function of $\eta' = |\eta| - y_{\text{beam}}$. All the data of four different energies are found to fall on the same curve.

In Fig. 16.14(b), v_2 divided by ϵ (Eq. (16.6)) is plotted as a function of the charged particle density, dN_{ch}/dy, per unit transverse area (Alt *et al.*, 2003), where S is the transverse area of the initial almond shape. The ϵ is evaluated with the wounded nucleon model. Data taken at different energies are shown to fall on a universal curve; it increases with $(1/S)dN_{\text{ch}}/dy$ and seems to attain the hydrodynamical limit at RHIC, which corresponds to the situation that the mean free path in the initial system is short in comparison with the system size.

The v_2 has been measured with particle identification as a function of p_{T}. In Fig. 16.15, the p_{T} dependence of v_2 for $\pi^{\pm}$, $\text{K}^{\pm}$, p and $\bar{\text{p}}$ are shown together with a hydrodynamical calculation (Huovinen *et al.*, 2001). In the low-p_{T} region, ($p_{\text{T}} = 1 \sim 2\,\text{GeV}/c$, depending on particle species), v_2 increases with p_{T}. A clear particle-mass dependence is observed. A particle with a lighter mass has a larger v_2 (Adler *et al.*, 2001a; 2003e; Adams *et al.*, 2004b):

$$v_2^{\pi^{\pm}} > v_2^{\text{K}^{\pm}} > v_2^{\text{p},\bar{\text{p}}}, \tag{16.7}$$

in the low-p_{T} region, where the p_{T} dependence is well reproduced by the hydrodynamical calculation.

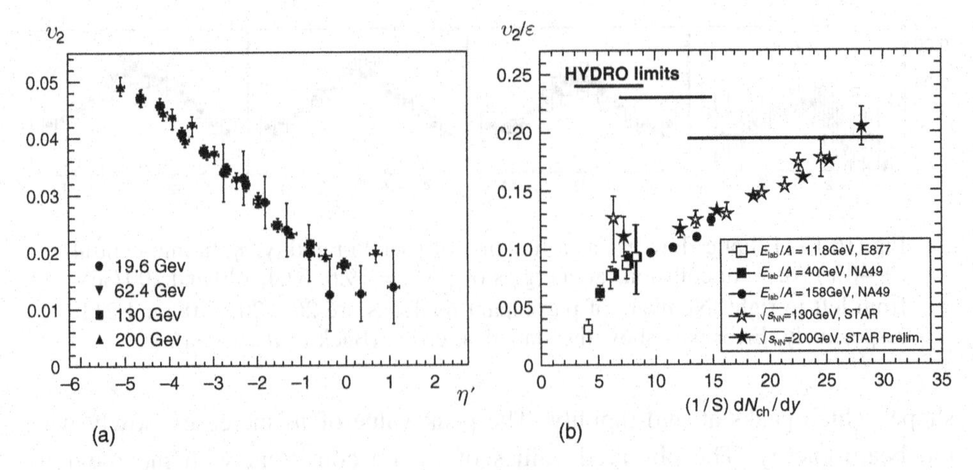

Fig. 16.14. (a) Elliptic flow, v_2, as a function of pseudo-rapidity, η, in mid-central (0–40%) Au + Au collisions at energies of $\sqrt{s_{NN}}$ = 19.6, 62.4, 130 and 200 GeV (Back *et al.*, 2005a). Error bars show the statistical errors. (b) Elliptic flow, v_2, divided by ϵ as a function of the charged particle density per unit transverse area, S (Alt *et al.*, 2003).

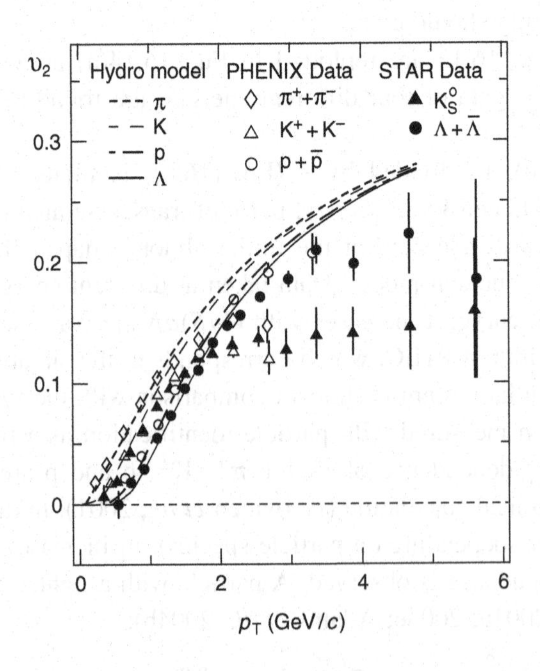

Fig. 16.15. Azimuthal anisotropy, v_2, for pions, kaons, protons and Λs as a function of transverse momentum at mid-rapidity in $\sqrt{s_{NN}}$ = 200 GeV Au + Au collisions. Data are from (Adler *et al.*, 2003e) and (Adams *et al.*, 2004b). Solid curves show results of a hydrodynamical calculation (Huovinen *et al.*, 2001).

It is one of the striking features at RHIC that hydrodynamical calculations, made with the assumption of a perfect fluid (no viscosity), successfully explain the p_T dependence as well as the particle-mass dependence of v_2 in the low-p_T region, where most of the particle yield is included.[6] Hydrodynamical calculations require a set of initial conditions at an initial time τ_0 (see Sections 10.4 and 13.3) when the local thermal equilibrium is established. Reproducing the data with τ_0 as small as 0.6 fm/c implies that the system thermalizes very quickly in the mid-rapidity region of the central collisions at RHIC.

In the higher p_T region, the v_2 of charged particles shows saturation, which suggests a hard scattering regime (Adams *et al.*, 2003c). As we have seen in Fig. 16.15, the v_2 measurement with particle identification, a clear departure from the hydrodynamical behavior, is observed: (i) v_2 for $\pi^\pm$ and $K^\pm$ deviate at lower p_T than v_2 for p and $\bar{p}$, and (ii) $v_2^{\pi^\pm,K^\pm} < v_2^{p,\bar{p}}$ for $p_T > 2$ GeV.

Other harmonics of the azimuthal distributions have been reported in the STAR experiment. The directed flow, v_1, shows a flat pseudo-rapidity dependence at mid-rapidity, and v_4 is about a factor of 10 smaller than v_2 (Adams *et al.*, 2004c). This information will be useful for further studies of the initial structure of the system after the impact.

16.7 Suppression of high-p_T hadrons

Suppression of high-p_T hadrons has been discovered at RHIC. As described below, this effect was not seen during the SPS heavy ion program. Figure 16.16(a) shows the transverse momentum spectra of π^0 in 200 GeV Au + Au central (0–10 %) and peripheral (80–92%) collisions measured by an electromagnetic calorimeter (Adler *et al.*, 2003a). The solid curves are π^0 spectra in 200 GeV pp collisions (Adler *et al.*, 2003d), which are scaled by the number of primary nucleon–nucleon collisions in heavy ion collisions. The number of primary nucleon–nucleon collisions is evaluated by the Glauber model described in Section 10.5. It is shown that in peripheral collisions, the π^0 distributions are consistent with the scaled pp collisions at $p_T > 2\,\text{GeV}/c$. In contrast, in central collisions, the π^0 distributions are found to be noticeably smaller than the scaled pp collisions. Similar behavior is also seen with charged particles at high p_T.

To clarify this effect, the data are shown in terms of a nuclear modification factor, $R_{AA}(p_T)$:

$$R_{AA}(p_T) = \frac{\sigma_{AA}(p_T)}{\langle N_{\text{binary}}\rangle \sigma_{NN}(p_T)}, \tag{16.8}$$

[6] The hydrodynamical calculation shown in Fig. 16.15 further assumes a first order phase transition with a freeze-out temperature of 120 MeV.

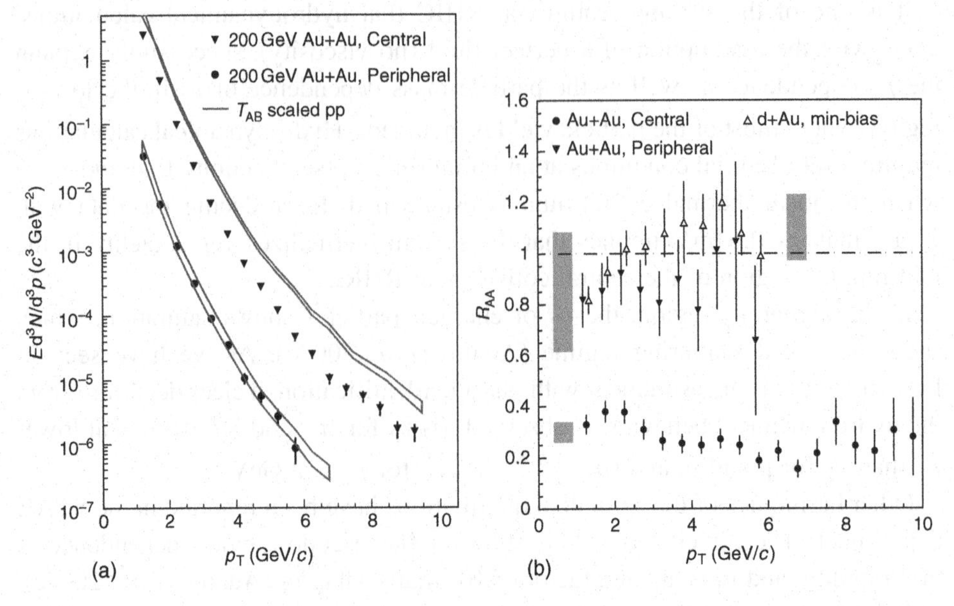

Fig. 16.16. (a) Transverse momentum spectra of π^0 measured in 200 GeV Au + Au central (0–10%) and peripheral (80–92%) collisions (Adler *et al.*, 2003a, b). Solid curves indicate π^0 spectra measured in 200 GeV pp collisions (Adler *et al.*, 2003d), which are scaled by the number of primary nucleon–nucleon collisions in heavy ion collisions. (b) The nuclear modification factor, R_{AA}, in central and peripheral Au + Au collisions compared with the ratio R_{dA} in minimum-bias d + Au collisions (Adler *et al.*, 2003a, b; Adcox *et al.*, 2005). The shaded boxes show the systematic errors. Both collisions are measured at an energy of $\sqrt{s_{NN}} = 200$ GeV.

where $\sigma_{AA}(p_T)$ and $\sigma_{NN}(p_T)$ are the p_T distributions from AA and pp collisions, respectively, and $\langle N_{binary}\rangle$ is the number of primary nucleon–nucleon collisions. In the experiment, $\langle N_{binary}\rangle$ is estimated for central and peripheral collisions from trigger cross-sections using the Glauber model (see Section 10.5).

The ratios observed by the PHENIX experiment in central and peripheral Au + Au collisions are shown in Fig. 16.16(b), compared with the ratios observed in minimum-bias d + Au collisions at $\sqrt{s_{NN}} = 200$ GeV. It is found that the ratios in the central Au + Au collisions are less than unity, but in the d + Au collisions and the peripheral Au + Au collisions the ratio reaches unity in the high-momentum region. Consistent results are reported from the other experiments (Arsene *et al.*, 2003; Back *et al.*, 2003d).

This depletion in central Au + Au collisions is striking. The production of high-p_T hadrons is known to be enhanced in pA collisions compared to that in the scaled pp collisions; due to multiple initial elastic collisions, a random walk of partons in the transverse momentum enhances the production of high-p_T hadrons.

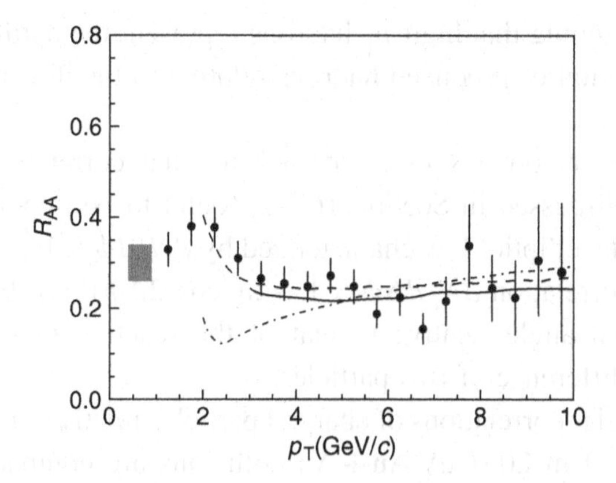

Fig. 16.17. R_{AA} from neutral pions for 200 GeV central Au + Au collisions together with different theoretical calculations with the jet energy loss (dashed, dash-dotted and solid lines). The shaded box denotes a systematic error in the data. Adapted from Adcox *et al.* (2005).

This effect is known as the Cronin effect (Cronin *et al.*, 1975) and is well studied in pA collisions up to 800 GeV.

If the particle production is due to soft processes, it should scale as the number of participant nucleons. In the low-p_T region, where the particles are produced via soft processes, the ratio should scale as the number of participant nucleons, N_{part}, rather than N_{binary}. In this case, R_{AA} is recalculated and 0.2 is obtained. In contrast, in the very high-p_T region, where particle production is due to hard processes, the particle production should scale as N_{binary}, and the value of R_{AA} should be unity if there is no nuclear effect at all. Typical theoretical estimations of R_{AA} are illustrated in Fig. 16.17, together with the observed data in central Au + Au collisions. Assuming that there is an energy loss of partons in the medium created in the collisions (Section 14.2), one may explain the observed R_{AA}. There are different calculations for the energy loss in the medium, which are based on different assumptions. Without the energy loss, the ratio is shown to increase with p_T, and becomes unity in the high-p_T region.

16.8 Modification of the jet structure

As discussed above, RHIC has opened the door to the study of hard scattering in AA collisions. Inspired by the discovery of the high-p_T suppression, detailed analysis of the jet structure has begun.

The method is as follows: partons fragment into jets of hadrons in a cone along the direction of the original parton. Among all the hadrons in the jets, a high-p_T hadron (called a *leading particle*) is likely to be aligned along the original direction of the parton. Assuming that high-p_T hadrons represent the hard-scattered parton, an angular correlation of produced hadrons relative to the direction of the parton can be defined.

However, there are other sources of such angular correlations. In particular, the elliptic flow discussed in Section 16.7 is found to be important, even in the high-p_T region. The elliptic flow characterized by $dN/d\phi \propto 1+2v_2\cos(2\phi)$ leads to an azimuthal correlation $dN/d\Delta\phi \propto 1+2v_2\cos(2\Delta\phi)$ (see Eq. (14.7)), where ϕ is the azimuthal angle relative to that of the reaction plane and $\Delta\phi$ is the azimuthal angle difference of two particles.

Azimuthal angular correlations of charged particles relative to a high-p_T particle ($4 < p_T < 6\,\text{GeV}/c$) in 200 GeV Au + Au collisions are compared in Fig. 16.18 with those seen in central d + Au and pp collisions at the same energy. Constant components and correlations due to the elliptic flow are subtracted in this plot.

Peaks at $\Delta\phi \sim 0$ (nearside peak) are clear in both Au + Au and d + Au collisions, and are very similar to that seen in pp collisions. This is typical of a jet produced by a parton fragmentation process.

On the other hand, the counter-peak at $\Delta\phi \sim 180°$ (back-to-back peak) disappears in central Au + Au collisions, while it shows a typical dijet event in both central d + Au collisions and in pp collisions. Therefore, a clear difference is found between the central Au + Au and the d + Au collisions. If the suppression of the back-to-back peak in the Au + Au collisions is due to initial state effects, independent of the matter created after the collisions, the same effect should also be observed in the

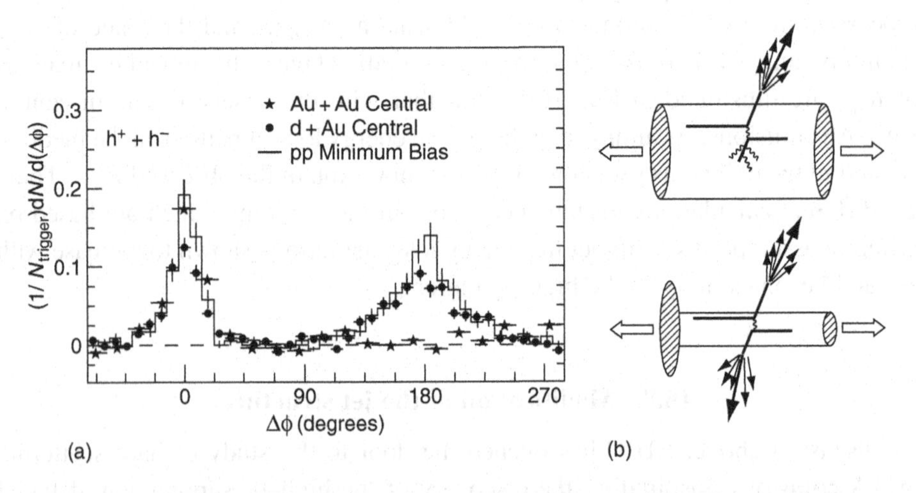

Fig. 16.18. (a) The comparison of two-particle azimuthal angular correlations of charged particles for central Au + Au, central d + Au and pp collisions at $\sqrt{s_{NN}} = 200\,\text{GeV}$ (Adams *et al.*, 2003a; Adler *et al.*, 2003), where N_{trigger} is the number of high-p_T particles. (b) Back-to-back correlations in Au + Au (upper) and d + Au (lower) collisions.

central d + Au collisions. However, this is not the case, and there is little difference between the central d + Au and pp collisions in the mid-rapidity region.

These results indicate that the suppression of the back-to-back peak is due to the final state interactions with the hot and dense medium generated in Au + Au collisions, as illustrated in Fig. 16.18(b). To conclude whether this hot and dense medium is indeed the QGP or not, we need more quantitative studies, both experimental and theoretical.

16.9 Quark-number scaling

Another new feature that has emerged from the RHIC experiments is called *quark-number scaling*. Figure 16.19 shows R_{CP}, the ratio of particle production in central and peripheral Au + Au collisions normalized by the number of binary collisions, N_{binary} (Adams *et al.*, 2004b):

$$R_{\text{CP}} = \frac{(dN^{\text{cent}}/dp_{\text{T}})/N^{\text{cent}}_{\text{binary}}}{(dN^{\text{peri}}/dp_{\text{T}})/N^{\text{peri}}_{\text{binary}}}, \tag{16.9}$$

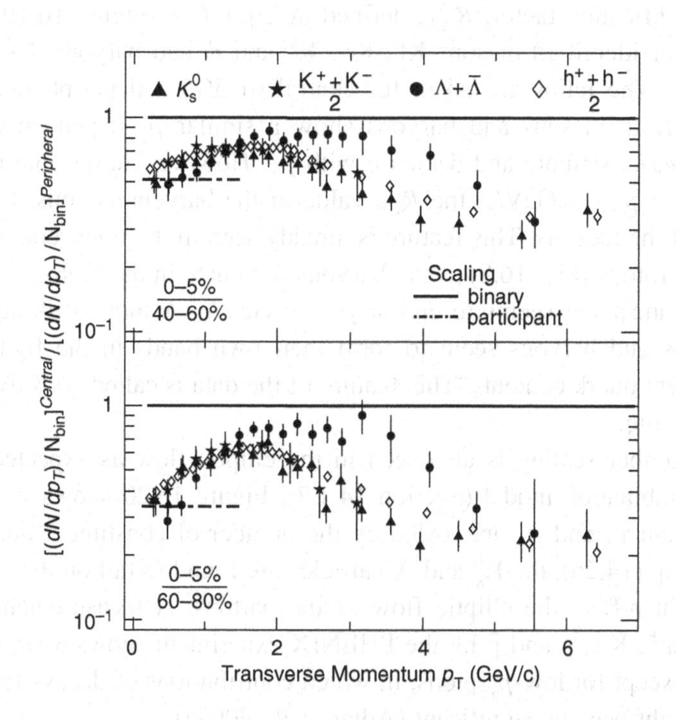

Fig. 16.19. The ratio R_{CP} for mesons (K°_s, $K^\pm$) and baryons (Λ) in the mid-rapidity region in Au + Au collisions at $\sqrt{s_{\text{NN}}} = 200$ GeV (Adams *et al.*, 2004b). Ratios are calculated using centrality of 0%–5% vs. 40%–60% (60%–80%) for top (bottom) panel. Solid (dashed) line shows binary (participant) scaling.

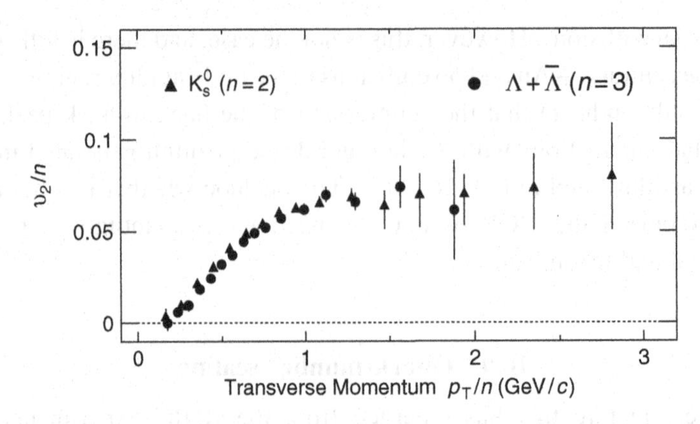

Fig. 16.20. The scaled v_2 for K^0_s and Λ particles; v_2/n is plotted as a function of p_T/n, where n is the number of constituent quarks. (Adams *et al.*, 2004b).

where $N^{\rm cent}_{\rm binary}$ and $N^{\rm peri}_{\rm binary}$ are the number of binary collisions in central and peripheral collisions, respectively. Thus, in the case that the peripheral Au+Au collisions comprise superpositions of the pp collisions, $R_{\rm CP}$ is equivalent to the nuclear modification factor, $R_{\rm AA}$, defined in Eq. (16.8). Figure 16.19 shows the ratio $R_{\rm CP}$ for identified mesons K^0_S, $K^\pm$, K^* and ϕ and baryons $\Lambda+\bar{\Lambda}$, $\Xi+\bar{\Xi}$ and $\Omega+\bar{\Omega}$. The ratios are taken for central (0–5%) and peripheral (40–60%) collisions. Both mesons and baryons show a similar p_T dependence: the $R_{\rm CP}$ values increase, saturate and decrease with p_T. However, in the intermediate p_T region at $2 < p_T < 6\,{\rm GeV}/c$, the $R_{\rm CP}$ value of the baryons is consistently larger than that of the mesons. This feature is already seen in the transverse momentum spectra of protons (Fig. 16.8) as the baryons dominate in the high-p_T region. The baryon enhancement seems to end at $p_T \sim 5\,{\rm GeV}/c$. A more striking feature is that mesons and baryons seem to form their own bands in the figure, despite their different quark contents. This feature of the data is called *constituent quark-number scaling*.

Quark-number scaling is also seen in the elliptic flow as expected from the quark recombination model (Section 14.1.7). Figure 16.20 shows v_2 versus p_T, in which both v_2 and p_T are scaled by the number of constituent quarks, n. As shown in Eq. (14.26), the K^0_s and Λ particles are found to fall on the same curve, which might reflect the elliptic flow of the partons. A measurement of v_2 for identified $\pi^\pm$, $K^\pm$, p and $\bar{p}$ by the PHENIX experiment shows a similar scaling behavior, except for low-p_T pions, in which contributions of decays from various hadrons might become significant (Adler *et al.*, 2003e).

Please return to Section 1.6 in Chapter 1, "Perspectives on relativistic heavy ion experiments," to form your own judgement on the formation of QGP at RHIC.

17

Detectors in relativistic heavy ion experiments

Experimental detector systems used in relativistic heavy ion experiments are similar to those used in high-energy particle physics experiments, where collisions among elementary particles, such as pp or e^+e^- collisions, are studied. Detector systems can be categorized into the hadron spectrometer, the lepton pair spectrometer and the photon spectrometer, and most of the heavy ion experiments consist of combinations of these detector systems. The technologies used in these systems are similar to the high-energy particle physics experiments: a major difference in heavy ion experiments is the large multiplicity of particles and the requirement of the impact parameter information.

In this chapter we discuss the features of relativistic heavy ion collisions and the detectors used in heavy ion experiments.

17.1 Features of relativistic heavy ion collisions

In central heavy ion collisions, particle multiplicities and particle densities at the detector are much larger than those in pp collisions at the same energy. Tracking and particle identification devices often do not work, or the information becomes confused, when there are more than two particles passing through one cell of the detector. Detector systems used in heavy ion experiments have to be designed in such a way that high-multiplicity events can be handled properly.

Another important requirement for heavy ion experiments is impact parameter information. Since a nucleus is an extended object, the geometry of the colliding nucleus plays an important role. In central collisions, two colliding nuclei overlap geometrically, and most of the incoming nucleons are involved in the collisions, while in peripheral collisions, a small fraction of colliding nuclei overlap, and a smaller number of nucleons are involved in the collisions (see Figs. 10.8 and 10.9). The space-time evolution of collisions may be different in central and peripheral collisions. Thus, it is important to sort various observables by the number of participant nucleons or by the impact parameter of the collisions. This information is often called the *centrality* of the collisions.

In pp collisions, which can be considered as an elementary process of high-energy nucleus–nucleus collisions, more particles are produced in collisions with higher beam energies; the produced particles are mostly pions. From earlier measurements of pp collisions, it is known that the number of particles produced does not increase proportionally to the center-of-mass energy of the collision, but scales as

$$\langle N \rangle \propto \ln \sqrt{s}, \tag{17.1}$$

where $\langle N \rangle$ is the average number of multiplicity and s is a Mandelstam variable (see Eq. (E.4) in Appendix E). The particles are uniformly distributed between the rapidities of colliding nucleons, as illustrated in Fig. 10.6. Since the rapidity gap of the colliding nucleus scales very weakly with the collision energy, the rapidity density, dN/dy, also increases logarithmically with s, as shown in Fig. 16.7:

$$\frac{dN}{dy} \propto \ln \sqrt{s}. \tag{17.2}$$

In central heavy ion collisions, most of the incoming nucleons are involved in the collision. For a zeroth order estimation, let us assume that nucleus + nucleus collisions are a simple superposition of pp collisions. In Pb + Pb collisions, for example, 200 nucleons in the Pb nucleus collide. Even with the assumption that an incoming nucleon collides with other nucleons only once, the particle multiplicities in a central Pb + Pb collision are 200 times larger than those in pp collisions. At BNL-AGS, we have 12 GeV per nucleon ($\sqrt{s_{NN}}$ of 4.8 GeV) and a charged particle density of dN_{ch}/dy ~150. At CERN-SPS, we have 150 GeV per nucleon ($\sqrt{s_{NN}} \sim 17$ GeV), $dN_{ch}/dy \sim 270$, and, at BNL-RHIC, $\sqrt{s_{NN}} \sim 200$ GeV, $dN_{ch}/dy \sim 700$ (see Fig. 16.4). Detectors such as hadronic spectrometers trace individual tracks of particles and analyze their momentum and particle species. If more than two tracks pass through an element of the detector, it becomes difficult to reconstruct them. Thus, the minimum element of the detector should be small enough to reconstruct events with high multiplicity. Detectors for nucleus–nucleus collisions should be capable of handling high particle densities according to its expected multiplicities.

The number of charged particles per unit solid angle, $dN_{ch}/d\Omega$, can be evaluated from the rapidity density:

$$\frac{dN_{ch}}{d\Omega} = \frac{1}{2\pi \sin\theta} \frac{dN_{ch}}{d\theta} \simeq \frac{1}{2\pi \sin^2\theta} \frac{dN_{ch}}{dy}, \tag{17.3}$$

where the approximation $dy \simeq d\eta$ and the relations $d\eta = -d\theta/\sin\theta$, $d\Omega = 2\pi \sin\theta \, d\theta$ are used. For the definition of pseudo-rapidity, η, see Eq. (E.20). In the study of QGP formation in heavy ion collisions, it is important to measure the particle production at mid-rapidity ($y \sim 0$), namely at $\theta_{\text{cm}} \sim 90°$.

In collider experiments, $\theta_{\text{cm}} = \theta_{\text{lab}}$ and the $dN_{\text{ch}}/d\Omega$ shows a minimum at $\theta_{\text{cm}} = 90°$ according to Eq. (17.3). Thus, measurements of particle production at around $\theta_{\text{cm}} = 90°$ are relatively easy in collider experiments (Exercise 17.1). On the other hand, for those experiments at BNL-AGS and CERN-SPS, the collisions occur with a target at rest in the laboratory frame, in which $\theta_{\text{cm}} \neq \theta_{\text{lab}}$. The laboratory angle corresponding to $\theta_{\text{cm}} = 90°$ can be evaluated as

$$\theta_{\text{lab}} = 2\tan^{-1}(\mathrm{e}^{-y_{\text{beam}}/2}), \tag{17.4}$$

where y_{beam} is the rapidity of the beam. Mid-rapidity corresponds to $\theta_{\text{lab}} \sim 20°$ and $\theta_{\text{lab}} \sim 6°$ at AGS and SPS energies, respectively. Due to the relativistic effects, the products of the collisions are focused in small forward angles in the laboratory, where the particle density becomes high.

17.2 Transverse energy, E_T

High particle density suggests that measurements of total energy flow are easier than those for individual particles. The transverse energy flow, in particular, provides us with a suitable measure of the amount of center-of-mass energy available in the nuclear reaction for the production of particles in a heavy ion collision experiment, and it is connected with the energy density achieved in the collision (Eq. (11.81)).

Actually, we obtain the transverse energy, E_{T}, by measuring the distribution of the energies of emitted particles, ΔE_i, weighted by $\sin\theta_i$ with respect to the incident beam axis (Appendix E.2) as follows:

$$E_{\text{T}} = \sum_i \Delta E_i \sin\theta_i = \sum_i (\Delta E_{\text{T}})_i. \tag{17.5}$$

Since the pseudo-rapidity, η, of a produced particle, Eq. (E.20), and its derivative with respect to θ are given by

$$\eta = -\ln\left(\tan\frac{\theta}{2}\right), \quad d\eta = -\frac{d\theta}{\sin\theta}, \tag{17.6}$$

$dE_{\text{T}}/d\theta$ as a function of the angle θ is converted into a distribution in pseudo-rapidity η:

$$\frac{dE_{\text{T}}}{d\eta} = -\sin\theta\,\frac{dE_{\text{T}}}{d\theta} = -\sin^2\theta\,\frac{dE}{d\theta}. \tag{17.7}$$

Experimentally the transverse energy distribution, $dE_{\text{T}}/d\eta$, is measured with a segmented calorimeter, an analog of a heat measuring device. Particles entering a segment, i, covering a range around θ_i deposit their energy, ΔE_i.

The transverse energy pseudo-rapidity density, $dE_T/d\eta$, which approximately equals dE_T/dy at extremely relativistic energies, $E_T \gg m$ (Exercise 17.2), plays an important role in analyzing the experimental data from SPS and RHIC.

17.3 Event characterization detectors

It is important to sort data by the centrality of the collisions. For this purpose, most heavy ion experiments are equipped with special detectors. These devices are called event characterization detectors.

The centrality of collisions can be determined either by the size of the spectator or by the size of the participant. According to the participant–spectator model (Section 10.6.1), the beam spectator comes out at $\theta \sim 0°$ with respect to the beam axis. A hadronic calorimeter located at $\theta \sim 0°$, a zero-degree calorimeter, is often used to detect the spectator. The energy measured by the zero-degree calorimeter, E_{ZDC}, is given by

$$E_{\mathrm{ZDC}} = E_{\mathrm{beam}} N_{\mathrm{bs}}, \tag{17.8}$$

where E_{beam} is the beam energy per nucleon and N_{bs} is the number of nucleons in the beam spectator; see Fig. 10.9 (Exercise 17.3). For collisions of nuclei both with mass number A, the total number of participant nucleons, N_{part}, is given by

$$N_{\mathrm{part}} = 2(A - E_{\mathrm{ZDC}}/E_{\mathrm{beam}}). \tag{17.9}$$

To measure the size of the participant, a hadronic calorimeter, an electromagnetic calorimeter or charged particle multiplicity counters at mid-rapidity are implemented. Signals from these detectors are proportional to the number of participant nucleons. In order to reduce the effect of fluctuations, a larger acceptance is required. Both the zero-degree calorimeter and the detectors at mid-rapidity provide the required information on the centrality of collisions.

17.4 Hadron spectrometer

Hadron spectrometers measure the momenta and determine the particle species of charged particles. They provide momentum and rapidity distributions for a large variety of identified particles. Particles with short lives can be observed through detection of their decay products; for example, the momentum and rapidity distributions of ϕ mesons can be studied by detecting their decay products, K^+ and K^-. Also correlation among these observables can be studied with hadron spectrometers if the acceptance of the detector is properly designed. If the acceptance of the detector is large enough, the thermal and chemical analyses of produced particles can be studied on an event-by-event basis.

A typical hadron spectrometer consists of a magnetic field, a tracking detector for the magnetic momentum analysis and particle identification detectors. Key parameters for the performance of the hadron spectrometers are rapidity and momentum coverage, momentum resolution and particle identification capability.

The time projection Chamber (TPC) in a solenoid magnet is one such powerful device for the hadron spectrometer that has been implemented for relativistic heavy ion experiments. As illustrated in Fig. 17.1, the TPC comprises a cylinder filled with a gas (typically a mixture of argon and methane). Uniform electric and magnetic fields are applied parallel to the axis of the cylinder. A beam pipe pokes through the cylinder, as shown in the figure, and the collisions take place at the center. Charged particles created in the collision pass through the chamber and ionize the chamber gas along the trajectories. Electrons produced by the ionization drift along the magnetic field towards the end cap of the TPC due to the electric field. The electron trajectories follow the magnetic field in tiny spirals. On each end cap, the drifting electrons are amplified by a grid of anode wires, and the signals are read out from small pads behind the anode wires. From the location of the pad, the x and y coordinates of the hit are obtained. The arrival time of the signal at the pad gives the z coordinate, since the electric field of the chamber is designed carefully such that the drift velocity of the electron is constant. The trajectory of the charged particle can be determined three-dimensionally from a series of hits along the pads. From the curvature of the trajectory in the magnetic field, the momentum of a particle with unit charge e is given by

$$p_\perp \ [\mathrm{GeV}/c] = 0.3 \left(\frac{B}{1\ \mathrm{T}}\right)\left(\frac{r}{1\ \mathrm{m}}\right), \tag{17.10}$$

where $p_\perp$ is the momentum component perpendicular to the magnetic field, B, and r is the radius of the curved track.

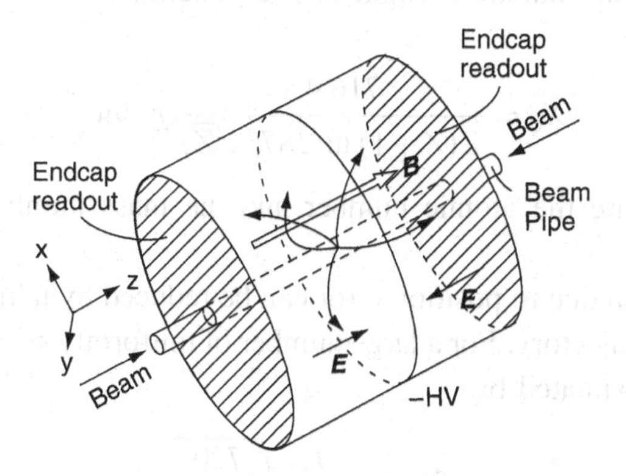

Fig. 17.1. Time projection chamber (TPC).

Table 17.1. *Radiation length, X_0, and critical energy, E_c(Eq. (17.28)), for various materials (Eidelman et al., 2004).* ρ is the density of the material.

Material	Z	ρ (g cm^{-3})	X_0 (g cm^{-2})	X_0/ρ (cm)	E_c (MeV)
H_2(liquid)	1	0.071	61.28	866.0	364.0
He(liquid)	2	0.125	94.32	756.0	250.0
C	6	2.2	42.70	18.8	111.0
Al	13	2.70	24.01	8.9	56.0
Fe	26	7.87	13.84	1.76	29.0
Pb	82	11.35	6.37	0.56	9.6
Water		1.0	36.08	36.08	
Air	(7.2)	0.0012	36.66	30 420.0	95.0

Track reconstruction is usually carried out via a series of position measurements along the trajectory of the track. The error of the curvature can be given by

$$(\delta k)^2 = (\delta k_{\text{mult}})^2 + (\delta k_{\text{pos}})^2, \tag{17.11}$$

where the curvature $k = 1/r$, δk_{mult} is the error due to the multiple scattering of particles in the chamber and δk_{pos} is the error from the measurements of hit positions (Exercise 17.4). The angular deviation due to the multiple scattering can be approximated by

$$\delta k_{\text{mult}} \approx 0.016 \frac{Z}{L\beta} \left(\frac{1\,\text{GeV}/c}{p} \right) \sqrt{\frac{\rho L}{X_0}}, \tag{17.12}$$

where L is the length of material traversed, ρ is the density and X_0 is the radiation length of the material. Here X_0 is given in mass per unit area, the length being given as X_0/ρ. The radiation length of the material is approximately given by (Table 17.1)

$$X_0 = \frac{716.4A}{Z(Z+1)\ln(287/\sqrt{Z})}\ \text{g cm}^{-2}, \tag{17.13}$$

where Z and A are the atomic number and the mass number of the material, respectively.

The contribution due to position error can be reduced by taking many measurements along the trajectory. For a large number of uniformly spaced measurements, the error is approximated by

$$\delta k_{\text{pos}} \propto \frac{\delta_\perp}{L^2} \sqrt{\frac{720}{N+4}}, \tag{17.14}$$

where N is the number of points measured along the trajectory and $\delta_\perp$ is each measurement error of position, perpendicular to the trajectory (Eidelman *et al.*, 2004).

The STAR experiment (Ackermann *et al.*, 2003) at RHIC has a gigantic TPC, 4 m in diameter and 4.2 m long (Fig. 10.1(c)). The TPC is in a 0.5 T solenoidal magnetic field. The magnetic field is surveyed with a precision of 1–2 gauss in order to obtain a good momentum resolution. There are more than 10^5 pads at the end cap, and each pad provides a read-out using a pre-amplifier/shaper followed by a switched capacitor array with an analog to digital converter (ADC) for the arrival timing measurements. Track reconstruction is performed by requiring more than ten pads along the trajectory (Anderson *et al.*, 2003).

17.4.1 Particle identification using dE/dx measurements

Particle identification can be achieved by measuring dE/dx of a charged particle in a gas. The dE/dx of the ionization, known as the Bethe–Bloch formula, is given by

$$-\frac{dE}{dx} \propto \frac{z^2}{\beta^2} \ln \gamma, \tag{17.15}$$

where ze is the charge of the incident particle, $\beta = v/c$ and $\gamma = (1-\beta^2)^{-1/2}$. The dE/dx decreases as $1/\beta^2$ until $\beta \sim 0.95$, where the dE/dx shows the minimum (the minimum ionization):

$$-\frac{dE}{d(\rho x)}(\text{at minimum}) \simeq 1.5\,\text{MeV g}^{-1}\,\text{cm}^2, \quad \text{for } z = 1. \tag{17.16}$$

Then dE/dx increases as $\ln \gamma$ (a relativistic rise). Since the energy loss of a particle is a function of its velocity, we can determine the velocity of the particle from the precise measurements of dE/dx. Using the TPC described above, the size of the signal at the pad corresponds to the ionization energy loss, dE/dx, in the gas.

Momentum analysis in the magnet, together with the velocity information from the dE/dx measurements, determine the mass of the particle. From its mass, the particle can be identified. Observed dE/dx is plotted against the momentum of particles and compared with the predictions of the Bethe–Bloch formula in Fig. 17.2 (Exercise 17.5).

17.4.2 Particle identification using time of flight measurements

Particles may also be identified by studying the precise measurements of their time of flight (TOF). Suppose two counters are separated by a distance L, and

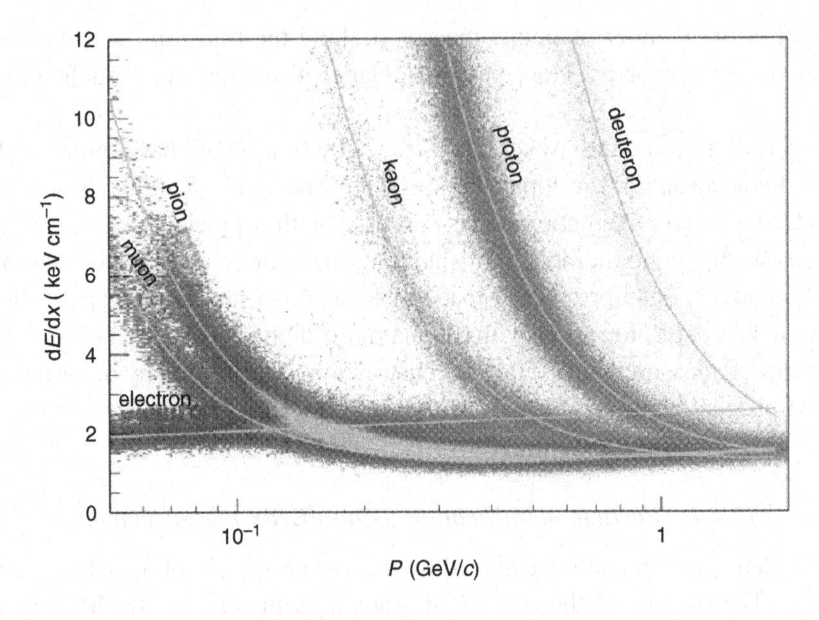

Fig. 17.2. Particle identification using dE/dx measurements with STAR-TPC (Anderson *et al.*, 2003). Curves represent the predictions of the Bethe–Bloch formula. Identification can be achieved for those particles with momentum up to 0.8 GeV/*c*. Reproduced, by permission, from the STAR collaboration (Ackermann *et al.*, 2003).

each counter measures the arrival time of a charged particle. The time of flight, t, the timing difference of the two counters, for a particle with velocity β is given by

$$t = \frac{L}{\beta}. \tag{17.17}$$

For a given momentum, p, the TOF, t, varies according to its mass. In Fig. 17.3, the times of flight of pions, kaons and protons are plotted for a flight path of 5 m. At higher momentum, the timing difference becomes smaller and smaller. In most cases, particle identification is limited by the timing resolution of the TOF measurement. As seen in Fig. 17.3, by ignoring the resolution of the momentum and the path length, one can separate pions and kaons up to 2 GeV/c and kaons and protons up to 4 GeV/c, assuming 3σ separations using a TOF counter with a timing resolution of 100 ps.[1] Measurement of the TOF with good resolution is typically done with plastic scintillators with photomultiplier read-out.

In reality, however, the momentum, p, and the path length, L, are also measured quantities (measured using a magnetic spectrometer), and resolutions of these

[1] Timing distributions measured by a good TOF counter are known to be well described by a clean Gaussian. The timing resolution is often given in σ of the fitted Gaussian.

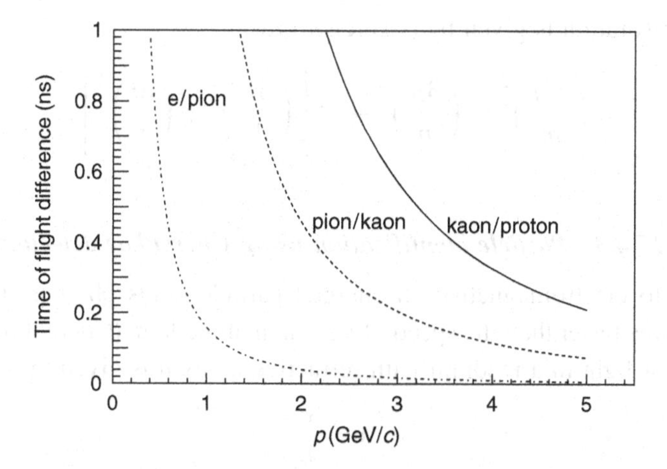

Fig. 17.3. Time of flight of pions, kaons and protons for a flight path of 5 m. Using time of flight counters with a timing resolution of 100 ps, one can separate pions and kaons up to 2 GeV/c and kaons and protons up to 4 GeV/c.

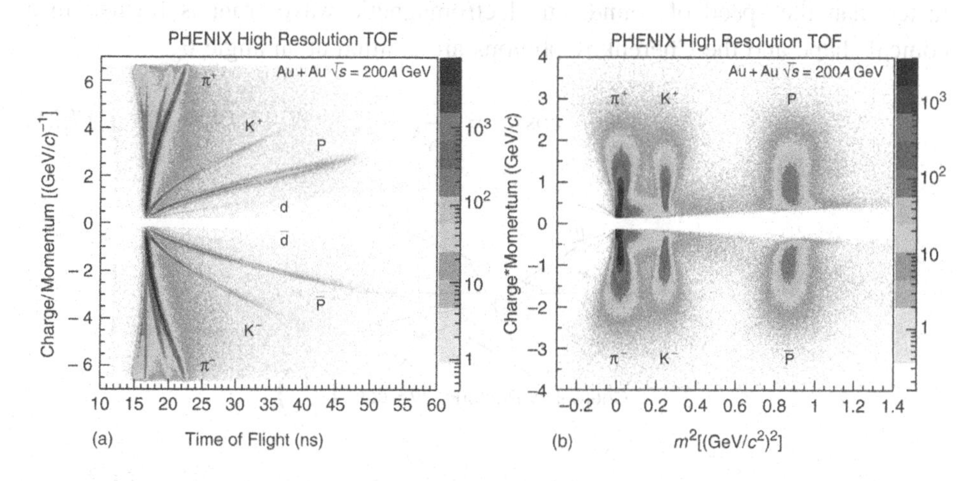

Fig. 17.4. (a) Observed scatter plot between TOF and momentum in the PHENIX experiment (Aizawa *et al.*, 2003). Clear separations of electrons, pions, kaons and (anti)-protons are visible. (b) Scatter plot between m^2 and momentum p by PHENIX (Adler *et al.*, 2004a). Due to the effect of momentum resolution, the distribution of the protons is broader than that of the pions.

need to be considered. As shown in Fig. 17.4(b), particle identification is achieved by determining the mass of the particles as follows:

$$m^2 = p^2\left(\left(\frac{t}{L}\right)^2 - 1\right). \tag{17.18}$$

The mass resolution is given by (Exercise 17.6)

$$\left(\frac{\delta m}{m}\right)^2 = \left(\frac{\delta p}{p}\right)^2 + \gamma^4\left[\left(\frac{\delta L}{L}\right)^2 + \left(\frac{\delta t}{t}\right)^2\right]. \tag{17.19}$$

17.4.3 Particle identification using Cherenkov detectors

According to electromagnetism, a charged particle emits photons in a medium when it moves faster than the speed of light in that medium (Cherenkov radiation). The speed of light in a medium with refractive index n is given by

$$v = \frac{c}{n}, \tag{17.20}$$

where c is the velocity of light in a vacuum. When the velocity of the charged particle exceeds the threshold, $\beta_{\text{thr}} = 1/n$, Cherenkov photons are emitted.

As illustrated in Fig. 17.5, just like a shock wave formed by an aircraft traveling faster than the speed of sound, an electromagnetic wave front is formed in a conical shape and the Cherenkov photons are emitted at an angle θ_c:

$$\cos\theta_c = \frac{1}{\beta n}, \tag{17.21}$$

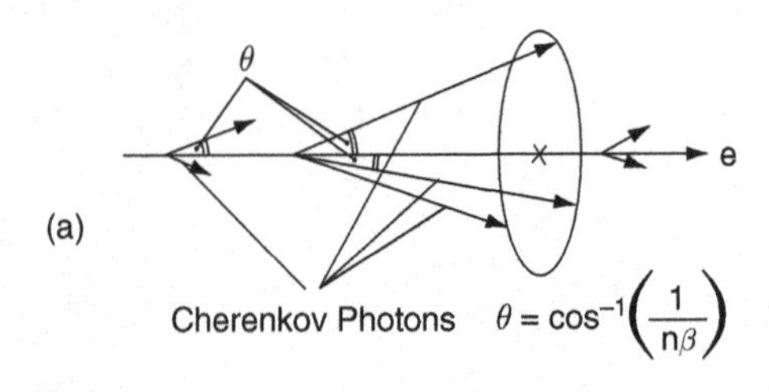

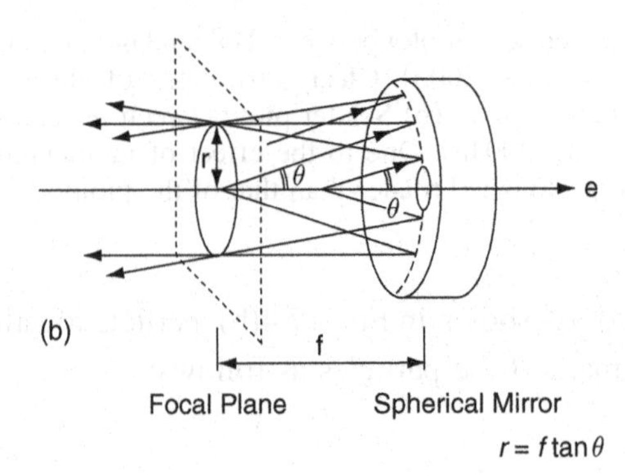

Fig. 17.5. Cherenkov radiation: (a) principle; (b) formation of a ring image.

Table 17.2. *Typical Cherenkov radiators.*

Material	$n-1$	γ_{thr}
He (NTP)[a]	3.5×10^{-5}	120
CO_2 (NTP)	4.1×10^{-4}	35
Aerogel	$0.1\sim0.01$	$2.4\sim7.1$
H_2O	0.33	1.52
Lucite, Plexiglass®	≈1.49	≈1.34

[a]NTP ≡ normal temperature and pressure.

with respect to the particle trajectory. The number of photons with frequencies between ν and $\nu+d\nu$ emitted per unit length by a particle of charge ze is given by (Jackson, (1999) and Exercise 17.7)

$$\frac{Nd\nu}{dx} = z^2\alpha\left[1-\frac{1}{(\beta n)^2}\right]\frac{d\nu}{c}, \quad \alpha=\frac{1}{137}. \tag{17.22}$$

Since the number of photons at a particular frequency is proportional to $d\nu/c = -d\lambda/\lambda^2$, photons with shorter wavelength predominate.

The number of photons obtained from Cherenkov radiation is affected by the optical properties of the radiator, the light collection efficiency, the efficiency of the photon detector, in addition to the index of refraction and the velocity of the charged particle. Using a typical photomultiplier tube with a bi-alkali cathode, the number of photoelectrons, N_{pe}, is given by,

$$N_{pe}/L \sim 90\sin^2\theta_c \ \text{cm}^{-1}, \tag{17.23}$$

where L is the thickness of the radiator.

The simplest method of particle identification with Cherenkov radiation utilizes the existence of a threshold for radiation; the velocity of a particle can be determined as being below or above the threshold, $\beta_{thr}=1/n$. The refractive index needs to be optimized for the particle mass and momentum range of the experiment. Typical materials used for the Cherenkov radiator are listed in Table 17.2.

A more sophisticated use for Cherenkov radiation in particle identification is via ring imaging Cherenkov counters (RICH). With a typical configuration, particles pass through a radiator, and the emitted photons are optically focused onto a position-sensitive photon detector, on which Cherenkov photons are detected on a ring (Fig. 17.5(b)). The velocity of the particle can be determined from the measurement of the ring, θ_c being derived from the radius of the ring.

17.5 Lepton pair spectrometer

Many hadrons have short life times, so they do not reach the detector before its decay. These hadrons can be measured by detecting their decay products. Among many decay channels of these hadrons, their decay into leptons is useful, despite its smaller decay probability; since leptons do not suffer strong interactions, they may escape from the dense hadronic gas, carrying information about their parent particles; see Section 14.3.

The lepton pair spectrometer is designed to detect lepton pairs of e^+e^-, $\mu^+\mu^-$ or $e\mu$ from leptonic decays of hadrons such as ϕ and J/ψ (Chapter 15) and the Drell–Yan process (Chapter 14). The spectrometer utilizes a magnetic field for the momentum measurements together with tracking and particle identification detectors. From the measured momenta and mass of two leptons, we can calculate the invariant mass, m_{12}:

$$m_{12}^2 = (p_1 + p_2)^2, \tag{17.24}$$

where p_1 and p_2 are the four-momenta of each lepton. For every observed lepton pair, m_{12} is calculated and a histogram of m_{12} can be obtained. (See Fig. 15.9(a) as an example of the invariant mass distribution for a $\mu^+\mu^-$ pair.) If the pair is produced via a two-body decay of a hadron with mass m_h, the m_{12} value of the pair will be m_h giving a peak in the histogram. If it is not from a two-body decay, the histogram of m_{12} shows a broad distribution forming a background to the peak at $m_{12} = m_h$. The shape of the background is determined by the inclusive distribution of the lepton. One can distinguish the peak and the background from the difference in the shape if the peak is sharp enough.[2] The width of the peak is determined by the momentum resolution of the measurement and also by the decay time of the mother hadron (natural width).

One of the key issues in designing the lepton pair spectrometer is the particle identification due to the small signal/background ratio. The signals from the lepton pairs of higher mass or the leptonic decay of the short lived hadrons (such as J/ψ) are small. Branching ratios of leptonic decays are much smaller than the hadronic decays. Compared with these small signals, copious pions are produced, and thus background due to the pions becomes a key issue.

As an example of electron detection, the major background processes are $\pi^0 \to \gamma\gamma$ followed by photon conversion ($\gamma \to e^+e^-$) in a detector and $\pi^0 \to \gamma e^+e^-$. In collisions at RHIC, one out of $10^5 \sim 10^6$ electron pairs is from the leptonic decay of hadrons. Thus the background subtraction is a sensitive task. Cherenkov counters or electro-magnetic calorimeters are often used to identify electrons.

[2] There are several ways to evaluate the background from independent measurement/analysis: (i) comparison with pairs of the same charge, for example, $\mu^+\mu^-$ vs $\mu^\pm\mu^\pm$, (ii) background from artificial creation of pairs by mixing events.

Electron identification for the lepton pair spectrometer also requires extra caution, because the rejection of charged pions is important. However, there is a limit to the ability of the detector to reject pions; usually, the detector misidentifies pions as electrons with a probability of 10^3. Delta ray emission of charged pions in a Cherenkov counter produces fake signals. Thus the electron identification is often carried out by various combinations of particle identification detectors.

Since it is necessary to detect both leptons from the decay, the acceptance of the detector is another important issue. A typical opening angle, θ_{12}, of two leptons is given by

$$\theta_{12} = 2\cos^{-1}\sqrt{\frac{p_{\rm T}^2}{m_{\rm h}^2 + p_{\rm T}^2 - 4m^2}}, \tag{17.25}$$

where $p_{\rm T}$ is the transverse momentum of the mother hadron and m is the mass of the lepton. (The momenta of the two leptons are assumed to be the same for simplicity.) In the case of $p_{\rm T} \sim m_{\rm h} \gg m$, we have $\theta_{12} \simeq 90°$. If $m_{\rm h} \gg p_{\rm T}, m$, we have $\theta_{12} \simeq 180°$. Thus, it is important to have a large acceptance detector for the lepton pair measurements.

17.6 Photon spectrometer

The measurement of direct photons in high-energy heavy ion collisions is one of the most challenging experiments. In order to measure direct photons, the subtraction of photons from neutral pions and electron conversions has to be carried out carefully. The direct emission of photons (~100 MeV) from the QGP phase may be visible if the lifetime of the QGP is long enough and/or the temperature of the QGP is high enough, despite the huge background from neutral pions and other particles (Section 15.6).

To measure high-energy photons in the laboratory, an electromagnetic calorimeter is often used. Electrons or positrons with energies above 100 MeV lose energy through Bremsstrahlung, in which a large fraction of the energy is emitted as photons. For photons with energies above 100 MeV, the major interaction with matter is the pair production, in which energetic electron–positron pairs are produced. If the original electron/positron or photon has an energy large enough, these processes, Bremsstrahlung and pair production in turn, result in a cascade, or showers, of electrons, positrons and photons (electromagnetic (EM) shower).

An intuitive picture of the EM shower can be given by simple cascades; suppose we begin with an energetic photon with energy E_0. On average, the photon will convert into an electron–positron pair after traversing matter for one radiation length, X_0, of Eq. (17.13).

The average energy of the produced electron and positron is $E_0/2$. Then, after another one radiation length, each electron and positron will radiate a photon, which carries half the energy of the electron/positron on average. In this manner, after t radiation lengths, the total number, N, of shower members (electrons, positrons and photons) increases as 2^t, and each of the member has energy $E(t)$:

$$E(t) \simeq \frac{E_0}{2^t}. \tag{17.26}$$

The shower develops until the electron/positron energy becomes too low to radiate photons. After that, the electron/positron loses its energy through ionization process (Eq. (17.15)). For simplicity let us assume the development stops immediately, at

$$E(t) \leq E_c, \tag{17.27}$$

where E_c, the critical energy, is defined as

$$\left(\frac{dE}{dx}\right)_{\text{ionization}} = \left(\frac{dE}{dx}\right)_{\text{Bremss}}, \tag{17.28}$$

as given in Table 17.1. Thus, the shower will reach a maximum and then cease abruptly. The maximum will occur at $t = t_{max}$:

$$E(t_{max}) = \frac{E_0}{2^{t_{max}}} = E_c, \tag{17.29}$$

which yields

$$t_{max} \propto \ln(E_0/E_c). \tag{17.30}$$

The assumption of the immediate cessation at t_{max} is certainly overly simplified; in reality, it stops gradually, and a semi-experimental formula of t_{max} for electron-induced and photon-induced showers are given as

$$t_{max}^{(e)} \simeq \ln(E_0/E_c) - 0.5 \quad \text{for electron-induced,} \tag{17.31}$$

$$t_{max}^{(\gamma)} \simeq \ln(E_0/E_c) + 0.5 \quad \text{for photon-induced.} \tag{17.32}$$

The measurement of the number of charged particles in the shower provides information about the incident energy since the number of particles is proportional to the incident energy.

A quantitatively accurate calculation of the EM shower development can be achieved using a Monte Carlo simulation code such as GEANT.[3] Figure 17.6(a) shows the simulation of an EM shower by GEANT. An electron at an initial energy of $E_0 = 30$ GeV is injected to a block of iron (15 cm × 15 cm × 40 cm). One can see many photons (dotted lines) and electrons (solid lines) in the shower and

[3] GEANT is the Detector Description and Simulation Tool, CERN Program Library Long Write up W5013, from the CERN Application Software Group, Geneva, Switzerland.

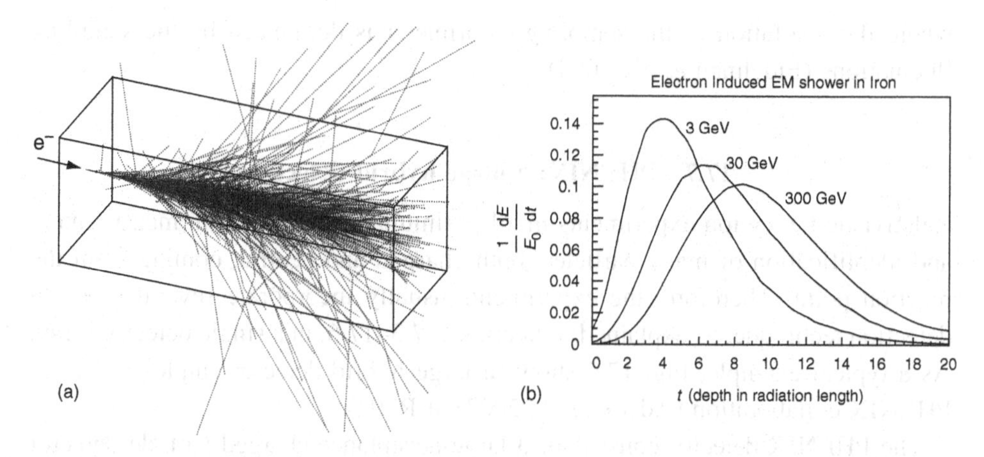

Fig. 17.6. (a) Development of EM shower simulated by GEANT. A 30 GeV electron is injected into a block of iron (15 cm × 15 cm × 40 cm) from the left. Photons and electrons are shown. (b) Longitudinal development of EM shower induced by $E_0 = 3$ GeV, 30 GeV and 300 GeV electrons simulated by GEANT.

how the shower develops in longitudinal and transverse directions in the block. Figure 17.6(b) shows the longitudinal development of the shower in the block. The fractional energy deposition per radiation length, $(1/E_0)(dE/dt)$, is plotted for incident electrons at $E_0 = 3$ GeV, 30 GeV and 300 GeV. An initial exponential rise, a broad maximum at $t_{\max}$, Eq. (17.31), and a gradual decline are seen. Electrons and photons with energies greater than 1.5 MeV are calculated. The development of the EM shower in the transverse direction also occurs, which is predominantly due to multiple scatterings of electrons/positrons. Typically, the transverse size of the shower is given by the *Molière radius*, R_M (Eidelman *et al.*, 2004):

$$R_{\mathrm{M}} \simeq 21 \left(\frac{1\,\mathrm{MeV}}{E_{\mathrm{c}}} \right) X_0 \approx \frac{7A}{Z} \ [\mathrm{g\ cm^{-2}}]. \tag{17.33}$$

Roughly 95% of the shower is contained within $2R_{\mathrm{M}}$.

Electromagnetic shower detectors are constructed from high-Z materials of small X_0, for example a lead-loaded glass (55% PbO and 45% SiO_2) calorimeter or lead-scintillator sampling calorimeter, so that the shower is confined in a small volume. Typical energy resolutions of a lead-loaded glass calorimeter and a lead-scintillator sampling calorimeter are, respectively,

$$\left(\frac{\sigma}{E} \right)_{\mathrm{Pb\ glass}} \simeq 0.05 \sqrt{\frac{1\,\mathrm{GeV}}{E}}, \tag{17.34}$$

$$\left(\frac{\sigma}{E} \right)_{\mathrm{Pb\ scint}} \simeq 0.09 \sqrt{\frac{1\,\mathrm{GeV}}{E}}, \tag{17.35}$$

where the resolution of the sampling calorimeter is dominated by the sampling fluctuations (Eidelman *et al.*, 2004).

17.7 PHENIX: a large hybrid detector

Relativistic heavy ion experiments involve simultaneous detection, measurement and identification of many particles, both charged and neutral, coming from the reaction point. Therefore, the experiments usually incorporate several types of detection technique, as explained in Sections 17.3–17.6, in a single detector array. As a typical example, Fig. 17.7 shows a large hybrid detector employed by the PHENIX collaboration (Adcox *et al.*, 2003) at RHIC.

The PHENIX detector consists of a large-acceptance charged particle detector and of four spectrometer arms: a pair of central spectrometers measuring electrons, photons and charged hadrons at mid-rapidity, and a pair of muon spectrometers at forward rapidities. Each of the four arms has a geometric acceptance of approximately 1 sr. The magnetic field in the volume of the collision region is axial, while the magnets of the muon arms produce radial fields.

The main sources of event characterization information are the beam–beam counter, which consists of two arrays of quartz Cherenkov telescopes surrounding the beam, and the zero-degree hadronic calorimeter, which measures the energy carried by spectator neutrons.

Electromagnetic calorimeters are mounted outermost on each of central arms. PHENIX uses two technologies for EM calorimetry: lead-scintillator with timing capabilities and lead-glass with better energy resolution.

The central arm tracking system uses the information provided by several detectors. Pad chambers yield the three-dimensional space points that are essential for pattern recognition, drift chambers provide precise projective measurements of particle trajectories and time-expansion chambers provide accurate (r, ϕ) information. Using the tracking information, a momentum resolution of

$$\frac{\Delta p}{p} \propto \sqrt{\left(\frac{\sigma_{\rm ms}}{\beta}\right)^2 + (\sigma_{\rm res} p)^2} \tag{17.36}$$

$$\simeq \sqrt{(0.7\%)^2 + \left(1.0\% \frac{p}{1\ {\rm GeV}/c}\right)^2} \tag{17.37}$$

is obtained, where $\sigma_{\rm ms}$ and $\sigma_{\rm res}$ are the terms due to multiple scattering and tracking resolution of the chambers, respectively (Adler *et al.*, 2004a).

Particle identification also depends on several detectors. Arrays of TOF plastic scintillators cover part of the central arm acceptance, and a ~115 ps timing resolution of the TOF system provides identification of kaons and pions up to

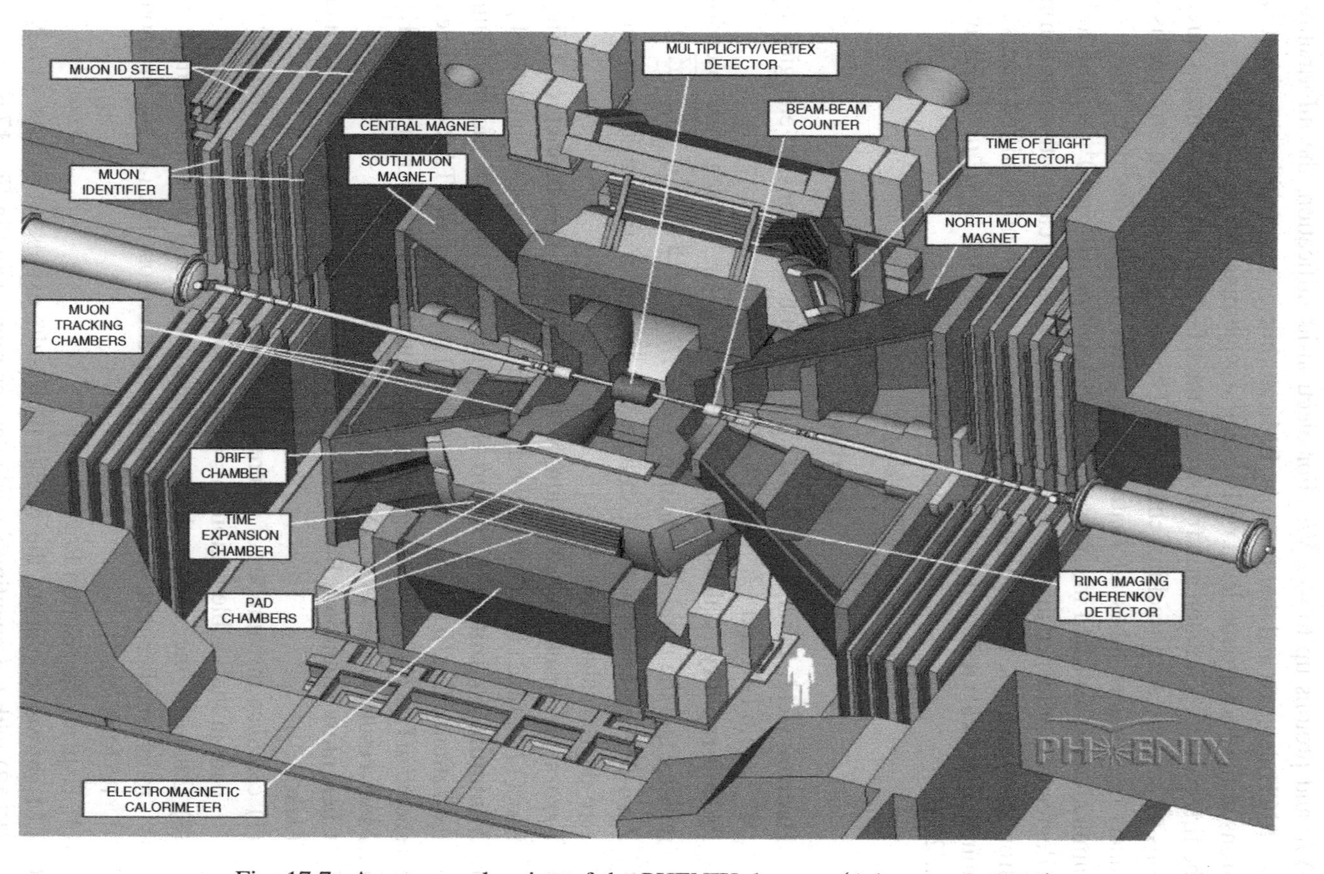

Fig. 17.7. A cutaway drawing of the PHENIX detector (Adcox *et al.*, 2003).

3 GeV/c and protons up to 4 GeV/c. For electron identification, the information from a RICH detector and that from the EM calorimeter are combined to identify electrons at the background level of 10^4 over a wide range in momentum.

The first part of each muon arm following a thick hadron absorber contains three stations of cathode strip tracking chambers. The back part of each arm consists of panels of streamer tubes alternating with plates of steel absorber. The pion contamination of identified muons should be below one part in 10^4, leading to a high degree of confidence in particle identification, as is the case with central arm electron identification. The excellent momentum resolution of identified tracks in the muon arm yields a mass resolution of 100 MeV for $J/\psi \to \mu^+\mu^-$ (Akikawa *et al.*, 2003).

Exercises

17.1 **Hadron yield.** Estimate $dN_c/d\Omega$ for Pb+Pb collisions at energies of $\sqrt{s_{NN}} = 5.6$ TeV and $\sqrt{s_{NN}} = 17$ GeV.

17.2 **Pseudo-rapidity.** Calculate the difference, $\eta - y$, between the pseudo-rapidity and the rapidity. Confirm $\eta > y$. *Hint*: $\frac{\tanh \eta}{\tanh y} = \frac{E}{p} = \sqrt{1 + \frac{m^2}{p^2}} > 1$.

17.3 **Zero-degree calorimeter.** Evaluate the effects of the participants in a zero-degree calorimeter.

17.4 **Errors in track construction with magnetic spectrometer.** Using Eqs. (17.12) and (17.14), describe a typical $\delta p/p$ behavior as a function of p.

17.5 **Minimum ionization.** Confirm an approximate value of the minimum ionization, $dE/d(\rho x)$, given by Eq. (17.16) in the case of STAR experiment shown in Fig. 17.2. The drift gas (NTP) is a mixture of 90% Ar and 10% CH_4.

17.6 **Mass resolution in TOF.** Derive Eq. (17.19).

17.7 **Cherenkov radiation.** For relativistic charged particles of $z = 1$ in water (Aerogel),

(1) calculate the number of Cherenkov photons emitted per centimeter of path in the visible region ($\lambda = 400 \sim 750$ nm).

(2) Calculate the energy loss per centimeter and confirm that it is much smaller than the minimum ionization loss Eq. (17.16) of about 1.5 MeV cm^{-1}.

(3) Explain the relation Eq. (17.23).

17.8 **EM shower.** Compare the qualitative results expressed in Eqs. (17.29)–(17.33) with the quantitative calculations represented in Fig. 17.6.

Appendix A
Constants and natural units

A.1 Natural units

Throughout this book we use the *natural units* $\hbar = c = 1$:

$$\left.\begin{aligned} \hbar \equiv \frac{h}{2\pi} = 6.5821 \times 10^{-25}\ \text{GeV s} = 1, \\ c = 2.9979 \times 10^{8}\ \text{m s}^{-1} = 1. \end{aligned}\right\} \tag{A.1}$$

These units have the following consequences. First, $[c] = [\mathrm{L}][\mathrm{T}]^{-1}$; i.e. $[\mathrm{L}] = [\mathrm{T}]$, where the symbol $[\mathrm{X}]$ implies the dimension of X. Also, $E^2 = p^2c^2 + m^2c^4$ leads to

$$[E] = [m] = [p]. \tag{A.2}$$

Furthermore, $[\hbar] = [E][\mathrm{T}]$ gives

$$[E] = [m] = [\mathrm{L}]^{-1} = [\mathrm{T}]^{-1}. \tag{A.3}$$

Therefore $[m]$, or equivalently $[E]$, can be chosen as a single independent dimension in natural units. It is customary in high-energy physics to speak of mass, momentum and energy in terms of giga-electronvolts (GeV), and to measure length and time in units of $(\mathrm{GeV})^{-1}$.

Next we adopt the rationalized Heaviside–Lorentz system of units for electromagnetic interactions, where the MKSA constants ϵ_0 and μ_0 are set to unity, and the 4π factors appear in the force equations rather than in the Maxwell equations.

The fine structure constant in the quantum electrodynamics, α, is defined as follows:

$$\alpha = \frac{e^2}{4\pi\hbar c} = \frac{e^2}{4\pi} \simeq \frac{1}{137.04}. \tag{A.4}$$

Table A.1. *Transformation among different units in* $\hbar = c = k_B = 1$.

	[J]	[MeV]	[g]	[cm^{-1}]	[K]
1 J	1	6.2415×10^{12}	1.1127×10^{-14}	3.1630×10^{23}	7.2430×10^{22}
1 MeV	1.6022×10^{-13}	1	1.7830×10^{-27}	5.0677×10^{10}	1.1605×10^{10}
1 g	8.9876×10^{13}	5.6096×10^{26}	1	2.8428×10^{37}	6.5096×10^{36}
1 cm^{-1}	3.1615×10^{-24}	1.9733×10^{-11}	3.5177×10^{-38}	1	2.2290×10^{-1}
1 K	1.3807×10^{-23}	8.6173×10^{-11}	1.5362×10^{-37}	4.3670	1

The pion Compton wavelength in natural units is given by

$$\lambda\!\!\!^{-}_{\pi} = \frac{\hbar}{m_\pi c} = \frac{1}{m_\pi} \simeq \frac{1}{140}\ \mathrm{MeV}^{-1} \simeq 1.41\ \mathrm{fm}. \tag{A.5}$$

From (A.1), we obtain a useful numerical relation, namely

$$\hbar c = 197.33\ \mathrm{MeV\ fm} \simeq 200\ \mathrm{MeV\ fm}, \tag{A.6}$$

where 1 fm (femtometer) = 1×10^{-15} m.

A typical hadronic cross-section is of order

$$\sigma \sim \lambda\!\!\!^{-2}_{\pi} = \frac{(\hbar c)^2}{\left(m_\pi c^2\right)^2} \sim \frac{(200)^2\ \mathrm{MeV}^2\ \mathrm{fm}^2}{(140)^2\ \mathrm{MeV}^2} \sim 2\ \mathrm{fm}^2 = 20\ \mathrm{mb}, \tag{A.7}$$

where 1 b (barn) = $10^{-28}\,\mathrm{m}^2 = 100\,\mathrm{fm}^2$ and $1\ \mathrm{mb} = 0.1\,\mathrm{fm}^2$.

The mass of the proton (p), neutron (n) and electron (e) are known to be

$$m_\mathrm{p} = 938.27\,\mathrm{MeV}/c^2 = 1.6726 \times 10^{-24}\ \mathrm{g}, \tag{A.8}$$

$$m_\mathrm{n} = 939.57\,\mathrm{MeV}/c^2 = 1.6749 \times 10^{-24}\ \mathrm{g}, \tag{A.9}$$

$$m_\mathrm{e} = 0.5110\,\mathrm{MeV}/c^2 = 9.1094 \times 10^{-28}\ \mathrm{g}. \tag{A.10}$$

We often set $k_\mathrm{B} = 1$ together with $\hbar = c = 1$ in studying relativistic thermodynamics, where k_B is the Boltzmann constant:

$$k_\mathrm{B} = 8.6173 \times 10^{-14}\ \mathrm{GeV\ K}^{-1} = 1. \tag{A.11}$$

Then we have

$$1\ \mathrm{GeV} = 1.1605 \times 10^{13}\ \mathrm{K}. \tag{A.12}$$

A.2 Frequently used units in astrophysics

The gravitational constant, G, is given by

$$G = 6.673(10) \times 10^{-11}\ \mathrm{m}^3\ \mathrm{kg}^{-1}\ \mathrm{s}^{-2}$$
$$= 6.707(10) \times 10^{-39} \hbar c \left(\frac{\mathrm{GeV}}{c^2}\right)^{-2}, \tag{A.13}$$

where the numbers in the parentheses depict errors in the last two digits.

The Planck mass in natural units is defined as

$$m_{\mathrm{planck}} = \left(\frac{\hbar c}{G}\right)^{\frac{1}{2}} = G^{-1/2}$$
$$\simeq 1.221 \times 10^{19}\ \mathrm{GeV} = [1.616 \times 10^{-20}\ \mathrm{fm}]^{-1}. \tag{A.14}$$

Other units used in astrophysics are defined as follows.

- 1 AU (astronomical unit) is defined as the average distance between the earth and the sun:

$$1\ \mathrm{AU} = 1.4960 \times 10^{11}\ \mathrm{m}. \tag{A.15}$$

- 1 ly (light year) is defined as the distance traveled by light in one year:

$$1\ \mathrm{ly} = 9.461 \times 10^{15}\ \mathrm{m}, \tag{A.16}$$

where $1\ \mathrm{yr} = 3.156 \times 10^7\ \mathrm{s}$.

- 1 pc (parsec) is defined as the distance at which half the major axis of the earth's orbit subtends an angle of one second:

$$1\ \mathrm{pc} = 3.086 \times 10^{16}\ \mathrm{m} = 3.262\ \mathrm{ly}. \tag{A.17}$$

The solar properties (mass, radius and luminosity) are as follows:

$$M_\odot = 1.989 \times 10^{30}\ \mathrm{kg}, \tag{A.18}$$
$$R_\odot = 6.960 \times 10^5\ \mathrm{km}, \tag{A.19}$$
$$L_\odot = 3.85 \times 10^{26}\ \mathrm{W}. \tag{A.20}$$

Appendix B

Dirac matrices, Dirac spinors and SU(N) algebra

B.1 Dirac matrices

In the (1 + 3)-dimensional Minkowski space with the metric $g^{\mu\nu} = \mathrm{diag}(1, -1, -1, -1)$, the Dirac matrices satisfy the relations:

$$\{\gamma^\mu, \gamma^\nu\} = 2g^{\mu\nu}, \quad (\gamma^\mu)^\dagger = \gamma^0 \gamma^\mu \gamma^0. \tag{B.1}$$

It is also convenient to define

$$\gamma^5 = i\gamma^0\gamma^1\gamma^2\gamma^3 = \gamma_5 = (\gamma_5)^\dagger, \tag{B.2}$$

$$\sigma^{\mu\nu} = \frac{i}{2}[\gamma^\mu, \gamma^\nu]. \tag{B.3}$$

In the standard Dirac representation, we have

$$\gamma^0 = \begin{pmatrix} 1 & 0 \\ 0 & -1 \end{pmatrix}, \quad \gamma^j = \begin{pmatrix} 0 & \sigma_j \\ -\sigma_j & 0 \end{pmatrix}, \quad \gamma^5 = \begin{pmatrix} 0 & 1 \\ 1 & 0 \end{pmatrix}, \tag{B.4}$$

where σ_j are the Pauli matrices,

$$\sigma_1 = \begin{pmatrix} 0 & 1 \\ 1 & 0 \end{pmatrix}, \quad \sigma_2 = \begin{pmatrix} 0 & -i \\ i & 0 \end{pmatrix}, \quad \sigma_3 = \begin{pmatrix} 1 & 0 \\ 0 & -1 \end{pmatrix}. \tag{B.5}$$

Some useful relations for the Dirac matrices are as follows:

$$\gamma^\mu \gamma_\mu = 4, \tag{B.6}$$

$$\gamma^\lambda \gamma^\mu \gamma_\lambda = -2\gamma^\mu, \tag{B.7}$$

$$\gamma^\lambda \gamma^\mu \gamma^\nu \gamma_\lambda = 4g^{\mu\nu}, \tag{B.8}$$

$$\not{A}\not{B} = A \cdot B - i\sigma_{\mu\nu} A^\mu B^\nu. \tag{B.9}$$

Some trace identities are given by

$$\mathrm{tr}\,(1) = 4, \quad \mathrm{tr}\,(\gamma^\mu \gamma^\nu) = 4 g^{\mu\nu}, \tag{B.10}$$

$$\mathrm{tr}\left(\text{odd powers of } \gamma^{0,1,2,3}\right) = 0, \tag{B.11}$$

$$\mathrm{tr}\left(\gamma^\mu \gamma^\nu \gamma^\lambda \gamma^\rho\right) = 4(g^{\mu\nu} g^{\lambda\rho} - g^{\mu\lambda} g^{\nu\rho} + g^{\mu\rho} g^{\nu\lambda}), \tag{B.12}$$

$$\mathrm{tr}\left(\gamma^\mu \gamma^\nu \gamma^\lambda \gamma^\rho \gamma^5\right) = 4i\ \epsilon^{\mu\nu\lambda\rho}, \tag{B.13}$$

where the totally anti-symmetric tensor $\epsilon^{\mu\nu\lambda\rho}$ is defined by

$$\epsilon^{\mu\nu\lambda\rho} = -\epsilon_{\mu\nu\lambda\rho}, \quad \epsilon_{0123} = 1. \tag{B.14}$$

In the four-dimensional Euclidean space with the metric $\delta_{\mu\nu} = \mathrm{diag}(1, 1, 1, 1)$, we define the γ matrices as

$$(\gamma_\mu)_{\mathrm{E}} = (\gamma_4 = i\gamma^0, \gamma^i), \tag{B.15}$$

which satisfy the relations

$$\{(\gamma_\mu)_{\mathrm{E}}, (\gamma_\nu)_{\mathrm{E}}\} = -2\delta_{\mu\nu}, \quad (\gamma_\mu)^\dagger_{\mathrm{E}} = -(\gamma_\mu)_{\mathrm{E}}. \tag{B.16}$$

It is also convenient to define Hermitian γ matrices by

$$\Gamma_\mu = -i\,(\gamma_\mu)_{\mathrm{E}}, \quad \Gamma_{-\mu} = -\Gamma_\mu, \tag{B.17}$$

which satisfy

$$\{\Gamma_\mu, \Gamma_\nu\} = 2\delta_{\mu\nu}, \quad \Gamma^\dagger_\mu = \Gamma_\mu. \tag{B.18}$$

B.2 Dirac spinors

The Dirac equation without interaction is given by

$$(i\gamma^\mu \partial_\mu - m)\hat{\Psi}(x) = (i\partial\!\!\!/ - m)\hat{\Psi}(x) = 0, \tag{B.19}$$

where the solution is decomposed as

$$\hat{\Psi}(x) = \sum_{s=1,2} \int \frac{d^3p}{(2\pi)^3 2\varepsilon_p} \left[b_s(\boldsymbol{p}) u_s(\boldsymbol{p})\ \mathrm{e}^{-ipx} + d^\dagger_s(\boldsymbol{p}) v_s(\boldsymbol{p})\ \mathrm{e}^{ipx} \right], \tag{B.20}$$

with $\varepsilon_p = \sqrt{\boldsymbol{p}^2 + m^2}$ and $\bar{\hat{\Psi}} \equiv \hat{\Psi}^\dagger \gamma^0$; $s(= 1, 2)$ denotes the $+1/2$ and $-1/2$ spin projections along a chosen quantization axis.

The anti-commutation relations of the creation and annihilation operators are

$$\{b_s(\boldsymbol{p}), b_{s'}^\dagger(\boldsymbol{p}')\} = \{d_s(\boldsymbol{p}), d_{s'}^\dagger(\boldsymbol{p}')\} = (2\pi)^3 2\varepsilon_p \delta_{ss'} \delta^3(\boldsymbol{p}-\boldsymbol{p}'), \quad \text{(B.21)}$$

$$\{\hat{\Psi}_\alpha(t, \boldsymbol{x}), \hat{\Psi}_\beta^\dagger(t, \boldsymbol{x}')\} = \delta_{\alpha\beta}\delta^3(\boldsymbol{x}-\boldsymbol{x}'), \quad \text{(B.22)}$$

where α and β are spinor indices and all other anti-commutators vanish.

One particle state with a momentum, $\boldsymbol{p}$, and a spin orientation, s, created by the creation operator in Eq. (B.21) has a covariant normalization

$$\langle \boldsymbol{p}, s|\boldsymbol{p}', s'\rangle = (2\pi)^3 2\varepsilon_p \delta_{ss'} \delta^3(\boldsymbol{p}-\boldsymbol{p}'). \quad \text{(B.23)}$$

The normalization and the spin sum of the Dirac spinors u and v are given by

$$u_s^\dagger(\boldsymbol{p}) u_{s'}(\boldsymbol{p}) = 2\varepsilon_p \delta_{ss'}, \quad v_s^\dagger(\boldsymbol{p}) v_{s'}(\boldsymbol{p}) = 2\varepsilon_p \delta_{ss'}, \quad \text{(B.24)}$$

$$\sum_{s=1,2} u_s(\boldsymbol{p})\bar{u}_s(\boldsymbol{p}) = \not{p} + m, \quad \sum_{s=1,2} v_s(\boldsymbol{p})\bar{v}_s(\boldsymbol{p}) = \not{p} - m. \quad \text{(B.25)}$$

For more properties of γ matrices and Dirac spinors, see App. A of Pokorski (2000), in which the same definitions are adopted.

B.3 SU(N) algebra

Let $\mathcal{T}^a$ $(a = 1, \ldots, N^2 - 1)$ be the Hermitian generators of the SU(N) group. They satisfy the Lie algebra

$$[\mathcal{T}^a, \mathcal{T}^b] = i f_{abc} \mathcal{T}^c, \quad \text{(B.26)}$$

where f_{abc} is the structure constant being totally anti-symmetric in its indices. Note that $(\mathcal{T}^b)^2$ commutes with every generator $\mathcal{T}^a$ and is called the quadratic Casimir operator.

For $N = 2$, f_{abc} reduces to the anti-symmetric tensor, ϵ_{ijk}, with $\epsilon_{123} = 1$. For $N = 3$, the non-vanishing components of f_{abc} are given by

$$\begin{aligned} &f_{123} = 1, \\ &f_{147} = -f_{156} = f_{246} = f_{257} = f_{345} = -f_{367} = 1/2, \\ &f_{458} = f_{678} = \sqrt{3}/2. \end{aligned} \quad \text{(B.27)}$$

In the fundamental representation, $\mathcal{T}^a$ is written by the $N \times N$ matrices t^a as

$$t^a = \frac{1}{2}\lambda_a, \quad \text{(B.28)}$$

where λ_a for $N = 2$ reduce to the Pauli matrices σ_i given in Eq. (B.5), while those for $N = 3$ reduce to the Gell-Mann matrices given by

$$\lambda_1 = \begin{pmatrix} 0 & 1 & 0 \\ 1 & 0 & 0 \\ 0 & 0 & 0 \end{pmatrix}, \quad \lambda_2 = \begin{pmatrix} 0 & -i & 0 \\ i & 0 & 0 \\ 0 & 0 & 0 \end{pmatrix}, \quad \lambda_3 = \begin{pmatrix} 1 & 0 & 0 \\ 0 & -1 & 0 \\ 0 & 0 & 0 \end{pmatrix},$$

$$\lambda_4 = \begin{pmatrix} 0 & 0 & 1 \\ 0 & 0 & 0 \\ 1 & 0 & 0 \end{pmatrix}, \quad \lambda_5 = \begin{pmatrix} 0 & 0 & -i \\ i & 0 & 0 \\ 0 & 0 & 0 \end{pmatrix}, \quad \lambda_6 = \begin{pmatrix} 0 & 0 & 0 \\ 0 & 0 & 1 \\ 0 & 1 & 0 \end{pmatrix},$$

$$\lambda_7 = \begin{pmatrix} 0 & 0 & 0 \\ 0 & 0 & -i \\ 0 & i & 0 \end{pmatrix}, \quad \lambda_8 = \frac{1}{\sqrt{3}} \begin{pmatrix} 1 & 0 & 0 \\ 0 & 1 & 0 \\ 0 & 0 & -2 \end{pmatrix}. \tag{B.29}$$

Some useful relations of t^a for general N are as follows:

$$\mathrm{tr}(t^a t^b) = \frac{1}{2}\delta_{ab}, \tag{B.30}$$

$$t^a_{ij} t^b_{kl} = \frac{1}{2}\left(\delta_{il}\delta_{jk} - \frac{1}{N}\delta_{ij}\delta_{kl}\right), \tag{B.31}$$

$$(t^a t^a)_{ij} = C_{\mathrm{F}}\delta_{ij}, \quad \text{with} \quad C_{\mathrm{F}} = \frac{N^2-1}{2N}. \tag{B.32}$$

In the adjoint representation, $\mathcal{T}^a$ is written by $(N^2-1) \times (N^2-1)$ matrices T^a as

$$(T^a)_{bc} = -i f_{abc}, \tag{B.33}$$

which satisfy the relations

$$\mathrm{tr}(T^a T^b) = N\delta_{ab}, \tag{B.34}$$

$$(T^a T^a)_{bc} = C_{\mathrm{A}}\delta_{bc}, \quad \text{with} \quad C_{\mathrm{A}} = N. \tag{B.35}$$

Appendix C
Functional, Gaussian and Grassmann integrals

C.1 Path integral in quantum mechanics

Consider a quantum mechanical particle at position q in one spatial dimension. The transition amplitude of the particle at an initial position $q = q_{\rm I}$ to a final position $q = q_{\rm F}$ is given by the Feynman kernel:

$$K_{\rm FI} \equiv K(q_{\rm F}, t_{\rm F}|q_{\rm I}, t_{\rm I}) = \langle q_{\rm F}|{\rm e}^{-i\hat{H}(t_{\rm F}-t_{\rm I})}|q_{\rm I}\rangle, \tag{C.1}$$

where $\hat{H} = \hat{T} + \hat{V}$ is the Hamiltonian operator and $\hat{T}$ ($\hat{V}$) is the Hermitian kinetic (potential) operator. By dividing the interval $t_{\rm F} - t_{\rm I}$ into n equal steps of length $\epsilon \equiv (t_{\rm F} - t_{\rm I})/n$ and inserting the complete set of states $\int dq|q\rangle\langle q| = 1$, we obtain

$$K_{\rm FI} = \lim_{n\to\infty} \int \prod_{l=1}^{n-1} dq_l \ \langle q_{\rm F}|{\rm e}^{-i\hat{T}\epsilon}\ {\rm e}^{-i\hat{V}\epsilon}|q_{n-1}\rangle \cdots \langle q_1|{\rm e}^{-i\hat{T}\epsilon}\ {\rm e}^{-i\hat{V}\epsilon}|q_{\rm I}\rangle. \tag{C.2}$$

Here we have used the Trotter formula to decompose the exponential operator (Schulman, 1996):

$$\lim_{n\to\infty} \left[({\rm e}^{-\hat{A}\epsilon}\ {\rm e}^{-\hat{B}\epsilon})^n - ({\rm e}^{-\hat{C}\epsilon})^n\right] = 0, \tag{C.3}$$

where $\hat{A} = i\hat{T}$, $\hat{B} = i\hat{V}$ and $\hat{C} = \hat{A} + \hat{B} = i\hat{H}$.

Introducing another complete set in momentum space, $\int dp|p\rangle\langle p| = 1$, and assuming that $\hat{T} = \hat{p}^2/2m$ and $\hat{V} = V(\hat{q})$, we have

$$K_{\rm FI} = \lim_{n\to\infty} \int \prod_{l=1}^{n-1} dq_l \left(\frac{m}{2\pi\epsilon i}\right)^{n/2} {\rm e}^{i\epsilon\sum_{j=1}^{n}\left[\frac{m}{2}\left(\frac{q_j-q_{j-1}}{\epsilon}\right)^2 - V(q_{j-1})\right]}. \tag{C.4}$$

Here, $\langle p|q\rangle = {\rm e}^{-ipq}/\sqrt{2\pi}$ and the Fresnel integral, $\int_{-\infty}^{+\infty} dz\ \exp(-iaz^2/2) = (2\pi/ia)^{1/2}$, are used. Note also that $q_0 \equiv q_{\rm I}$ and $q_n \equiv q_{\rm F}$. Sometimes, Eq. (C.4) is

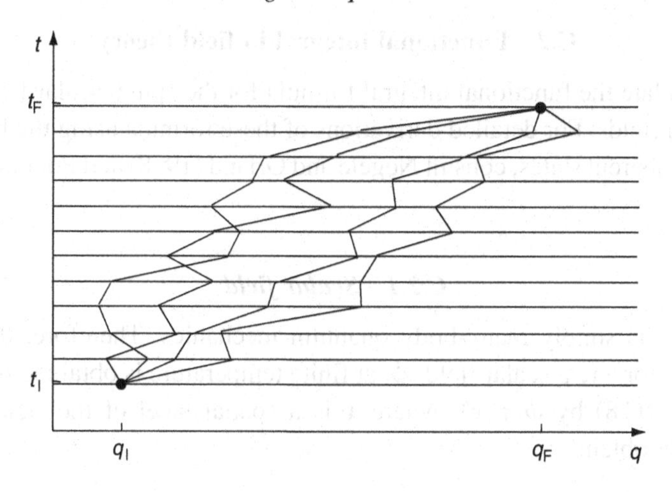

Fig. C.1. Propagation of a particle over many paths from (q_I, t_I) to (q_F, t_F).

written by the less precise notation

$$K_{\rm FI} = \int_{q_{\rm I}}^{q_{\rm F}} [dq]\, {\rm e}^{iS[q(t),\dot{q}(t)]}, \tag{C.5}$$

$$S[q,\dot{q}] = \int_{t_{\rm I}}^{t_{\rm F}} dt \left[\frac{m}{2}\left(\frac{dq}{dt}\right)^2 - V(q)\right], \tag{C.6}$$

where $[dq]$ is the multi-dimensional integral measure. Equation (C.5) may be interpreted as a sum of all trajectories, $q(t)$, with a suitable weight factor, ${\rm e}^{iS[q,\dot{q}]}$, and with the boundary conditions, $q(t_I) = q_I$ and $q(t_F) = q_F$. This is why the formula is called the path integral (Feynman and Hibbs, 1965; Schulman, 1996). Note, however, that the path $q(t)$ does not have to be a smooth function of t (see Fig.C.1).

The partition function, Z, of a single-particle system coupled with a heat bath is also written in path integral form since it is related to the Feynman kernel K as follows:

$$Z = \int dq\, \langle q|{\rm e}^{-\hat{H}/T}|q\rangle = \int dq\, K(q,-i/T|q,0) \tag{C.7}$$

$$= \int [dq]\, {\rm e}^{-S_{\rm E}[q(\tau),\dot{q}(\tau)]}. \tag{C.8}$$

Here, the Euclidean action is defined as

$$S_{\rm E} = \int_0^{1/T} d\tau \left[\frac{m}{2}\left(\frac{dq}{d\tau}\right)^2 + V(q)\right], \tag{C.9}$$

where $t = -i\tau$ is the imaginary time ($0 \le \tau < 1/T$) and $q(\tau)$ is a path satisfying the periodic boundary condition $q(0) = q(1/T)$.

C.2 Functional integral in field theory

We recapitulate the functional integral formula for the spin 0 scalar field and spin 1/2 fermion fields. For detailed derivations of these formula using the bosonic and fermionic coherent states, consult Negele and Orland (1998) and Pokorski (2000).

C.2.1 *Scalar field*

Field theory is simply many-body quantum mechanics. Therefore, the partition function, Z, for a real scalar field, ϕ, at finite temperature is obtained by replacing $q(\tau)$ in Eq. (C.8) by $\phi(\tau, \boldsymbol{x})$, where $\boldsymbol{x}$ is a spatial label of the field. Then we immediately obtain

$$Z = \int [d\phi] \, \mathrm{e}^{-S_{\mathrm{E}}[\phi, \partial\phi]}, \tag{C.10}$$

$$S_{\mathrm{E}} = \int_0^{1/T} d\tau \int d^3x \, \mathcal{L}_{\mathrm{E}}(\phi(\tau, \boldsymbol{x}), \partial\phi(\tau, \boldsymbol{x})), \tag{C.11}$$

with the periodic boundary condition

$$\phi(0, \boldsymbol{x}) = \phi(1/T, \boldsymbol{x}). \tag{C.12}$$

These equations correspond to Eq. (4.3) in Chapter 4.

C.2.2 *Fermion field*

The fermion operator, $\hat{\Psi}$, reflects the Pauli exclusion principle as given by Eqs.(B.21) and (B.22). Therefore, the fermionic functional integral is defined over variables $\psi(\tau, \boldsymbol{x})$, which anti-commute with each other, unlike the classical variables, $\phi(\tau, \boldsymbol{x})$, which commute with each other. Such ψs are called the Grassmann variables. The functional integral over independent Grassmann variables ψ and $\bar{\psi}$ is given by

$$Z = \int [d\bar{\psi} \, d\psi] \, \mathrm{e}^{-S_{\mathrm{E}}[\bar{\psi}, \psi, \partial\bar{\psi}, \partial\psi]}, \tag{C.13}$$

$$S_{\mathrm{E}} = \int_0^{1/T} d\tau \int d^3x \, \mathcal{L}_{\mathrm{E}}(\bar{\psi}(\tau, \boldsymbol{x}), \psi(\tau, \boldsymbol{x}), \partial\bar{\psi}(\tau, \boldsymbol{x}), \partial\psi(\tau, \boldsymbol{x})), \tag{C.14}$$

with the anti-periodic boundary conditions originating from the unusual nature of the trace over Grassmann variables,

$$\psi(0, \boldsymbol{x}) = -\psi(1/T, \boldsymbol{x}), \quad \bar{\psi}(0, \boldsymbol{x}) = -\bar{\psi}(1/T, \boldsymbol{x}). \tag{C.15}$$

C.3 Gaussian and Grassmann integrals

Basic Gaussian and Grassmann integrals are as follows:

$$\int_{-\infty}^{+\infty} \frac{dx}{\sqrt{2\pi}}\, \mathrm{e}^{-ax^2/2} = \frac{1}{\sqrt{a}}, \tag{C.16}$$

$$\int \frac{dz^* dz}{2\pi i}\, \mathrm{e}^{-b|z|^2} = \frac{1}{b}, \tag{C.17}$$

$$\int d\bar{\xi} d\xi \; \mathrm{e}^{-c\bar{\xi}\xi} = c. \tag{C.18}$$

Here, x (z) is a real (complex) number, and $\bar{\xi}$ and ξ are anti-commuting Grassmann numbers ($\{\xi, \bar{\xi}\} = 0$, and $\xi^2 = \bar{\xi}^2 = 0$); a and b are assumed to be real, positive numbers, while c is an arbitrary complex number. Equation (C.17) can be deduced by rewriting the integral in terms of the real and imaginary parts of z or in terms of the polar coordinates of z. Equation (C.18) can be deduced by noting that $\mathrm{e}^{-c\bar{\xi}\xi} = 1 - c\bar{\xi}\xi$ and $\int d\xi = \partial/\partial\xi$ (integral = derivative) for Grassmann variables.

Generalization of the above results to the case of multiple variables is straightforward. For $\boldsymbol{x} = (x_1, \ldots, x_n)$, $\boldsymbol{z} = (z_1, \ldots, z_n)$, $\boldsymbol{\xi} = (\xi_1, \ldots, \xi_n)$ and $\bar{\boldsymbol{\xi}} = (\bar{\xi}_1, \ldots, \bar{\xi}_n)$, with $\{\xi_k, \xi_l\} = \{\bar{\xi}_k, \bar{\xi}_l\} = \{\xi_k, \bar{\xi}_l\} = 0$, we obtain

$$\int \prod_{l=1}^{n} \frac{dx_l}{\sqrt{2\pi}}\, \mathrm{e}^{-\frac{1}{2}{}^t\boldsymbol{x}A\boldsymbol{x}} = \frac{1}{\sqrt{\det A}}, \tag{C.19}$$

$$\int \prod_{l=1}^{n} \frac{dz_l^* dz_l}{2\pi i}\, \mathrm{e}^{-{}^t\boldsymbol{z}^* B\boldsymbol{z}} = \frac{1}{\det B}, \tag{C.20}$$

$$\int \prod_{l=1}^{n} d\bar{\xi}_l \, d\xi_l \; \mathrm{e}^{-{}^t\bar{\boldsymbol{\xi}}C\boldsymbol{\xi}} = \det C. \tag{C.21}$$

Here, A is a non-singular and real-symmetric matrix whose eigenvalues, a_l, satisfy $a_l > 0$ for all l; B is a non-singular complex matrix whose complex eigenvalues, b_l, obtained by the bi-unitary transformation ($UBV^\dagger$) satisfy Re $b_l > 0$ for all l; C is an arbitrary complex matrix with no conditions. Note that B and C do not have to be Hermitian matrices. In field theories, the label "l" summarizes all possible indices including spin, flavor, color, space-time points, etc., and "det" denotes the determinant for all these indices.

Appendix D

Curved space-time and the Einstein equation

D.1 Non-Euclidean space-time

In the general curvilinear coordinates with $x^\mu = (x^0, x^1, x^2, x^3)$, the invariant distance, ds, is expressed in terms of a symmetric metric, $g_{\mu\nu}(x)$, as follows:

$$ds^2 = g_{\mu\nu}(x)\ dx^\mu\ dx^\nu. \tag{D.1}$$

Once the coordinate system is given, the associated metric can be calculated in a unique way. If it is in the Minkowski space-time, the metric reduces to

$$g_{\mu\nu}(x) \to \eta_{\mu\nu} = \text{diag}(+1, -1, -1, -1). \tag{D.2}$$

The contravariant vector, $A^\mu(x)$, transforms in the same way as dx^μ under the general coordinate transformation $x^\mu = x^\mu(x')$:

$$A^\mu(x) = \frac{\partial x^\mu}{\partial x'^\nu} A'^\nu(x'). \tag{D.3}$$

Contravariant tensors such as $A^{\mu\nu\lambda\cdots}(x)$ are defined in a similar way with respect to each space-time index. The covariant vectors and tensors are defined by lowering the space-time indices using the metric $g_{\mu\nu}$, for example $A_\mu(x) \equiv g_{\mu\nu}A^\nu(x)$. In particular, the metric $g_{\mu\nu}(x)$ is a covariant tensor.

Note that $g_{\mu\nu}(x)$ can be diagonalized, at least locally, since it is a real symmetric matrix. When the result of the diagonalization gives one positive eigenvalue and three negative eigenvalues, the corresponding space-time is called the Riemann space, in which the following relation is satisfied: det $g_{\mu\nu}(x) \equiv g(x) < 0$.

Einstein's equivalence principle is simply the statement that one may take the flat metric, $\eta_{\mu\nu}$, locally so that the effect of gravity is canceled out at a given point. However, taking the flat metric globally is not possible in the curved space-time under the gravitational field. The invariant volume element under the general coordinate transformation is $\sqrt{-g(x)}\ d^4x$.

Vectors defined at two different points, x and x', can be compared only by moving the vector at x to the point x' by parallel transport. In other words, the naive derivative of a vector, $\partial_\mu A_\nu(x)$, is not a tensor. The covariant derivative, defined by taking into account the parallel transport, is given by

$$\nabla_\lambda A_\mu = (g^\nu{}_\mu \partial_\lambda - \Gamma^\nu{}_{\lambda\mu}) A_\nu, \quad \nabla_\lambda A^\mu = (g^\mu{}_\nu \partial_\lambda + \Gamma^\mu{}_{\lambda\nu}) A^\nu, \tag{D.4}$$

where the latter relation is obtained by requiring that ∇_μ and ∂_μ are the same operation if they are acting on a scalar such as $A_\mu A^\mu$; $\Gamma^\lambda{}_{\mu\nu}$ is called the Christoffel symbol.

Useful relations with the covariant derivatives obtained from Eq. (D.4) are

$$\nabla_\lambda (A_\mu B_\nu) = (\nabla_\lambda A_\mu) B_\nu + A_\mu (\nabla_\lambda B_\nu), \tag{D.5}$$

$$\nabla_\lambda g_{\mu\nu} = 0. \tag{D.6}$$

From Eq. (D.6), we can derive an explicit form of the Christoffel symbol in terms of the metric tensor:

$$\Gamma^\lambda{}_{\mu\nu} = \frac{1}{2} g^{\lambda\rho} (\partial_\mu g_{\nu\rho} + \partial_\nu g_{\rho\mu} - \partial_\rho g_{\mu\nu}). \tag{D.7}$$

All the components of $\Gamma^\lambda{}_{\mu\nu}$ vanish if the space-time is flat. Note that the Christoffel symbol is not a tensor. Also, $\Gamma^\lambda{}_{\mu\lambda} = \partial_\mu \ln \sqrt{-g}$.

The covariant derivative acting on general tensors is defined in the same way as Eq. (D.4). For example,

$$\nabla_\lambda A_{\mu\nu} = (g^\rho{}_\mu g^\sigma{}_\nu \partial_\lambda - g^\sigma{}_\nu \Gamma^\rho{}_{\lambda\mu} - g^\rho{}_\mu \Gamma^\sigma{}_{\lambda\nu}) A_{\rho\sigma}, \quad \nabla_\lambda A^\mu{}_\nu = g^{\mu\rho} \nabla_\lambda A_{\rho\nu}. \tag{D.8}$$

The covariant derivatives do not commute with each other:

$$[\nabla_\mu, \nabla_\nu] A^\alpha = (\nabla_\mu \nabla_\nu - \nabla_\nu \nabla_\mu) A^\alpha = R^\alpha{}_{\beta\mu\nu} A^\beta, \tag{D.9}$$

where $R^\alpha{}_{\beta\mu\nu}$ is called the Riemann–Christoffel curvature tensor. By explicit calculation,

$$R^\alpha{}_{\beta\mu\nu} = \partial_\mu \Gamma^\alpha{}_{\nu\beta} - \partial_\nu \Gamma^\alpha{}_{\mu\beta} + \Gamma^\alpha{}_{\mu\lambda} \Gamma^\lambda{}_{\nu\beta} - \Gamma^\alpha{}_{\nu\lambda} \Gamma^\lambda{}_{\mu\beta}. \tag{D.10}$$

Note that $R^\alpha{}_{\beta\mu\nu} = 0$ everywhere is a necessary and sufficient condition for the global flatness of space-time.

From the Jacobi identity $[\nabla_\mu, [\nabla_\nu, \nabla_\lambda]] + [\nabla_\nu, [\nabla_\lambda, \nabla_\mu]] + [\nabla_\lambda, [\nabla_\mu, \nabla_\nu]] = 0$, we obtain

$$R_{\alpha\beta\mu\nu} + R_{\alpha\mu\nu\beta} + R_{\alpha\nu\beta\mu} = 0 \tag{D.11}$$

and the Bianchi identity,

$$\nabla_\lambda R^\alpha{}_{\beta\mu\nu} + \nabla_\mu R^\alpha{}_{\beta\nu\lambda} + \nabla_\nu R^\alpha{}_{\beta\lambda\mu} = 0. \tag{D.12}$$

Furthermore, the following symmetry properties follow from the definition of $R_{\alpha\beta\mu\nu}$:

$$R_{\alpha\beta\mu\nu} = -R_{\beta\alpha\mu\nu} = -R_{\alpha\beta\nu\mu} = R_{\mu\nu\alpha\beta}. \tag{D.13}$$

These relations reduce the number of independent components of $R_{\alpha\beta\mu\nu}$ from 256 to 20.

The Ricci tensor, $R_{\mu\nu}$, and the scalar curvature, R, are simply defined as

$$R_{\mu\nu} = R^{\alpha}{}_{\mu\nu\alpha} = R_{\nu\mu}, \quad R = R^{\mu}{}_{\mu}. \tag{D.14}$$

D.2 The Einstein equation

The Einstein tensor, $G_{\mu\nu}$, is defined by

$$G_{\mu\nu} \equiv R_{\mu\nu} - \frac{1}{2} g_{\mu\nu} R. \tag{D.15}$$

From the Bianchi identity and Eq. (D.5), we can prove

$$\nabla_{\lambda} G^{\lambda}{}_{\mu} = 0. \tag{D.16}$$

In terms of $G_{\mu\nu}$, Einstein's equation is given by

$$G_{\mu\nu} = 8\pi G T_{\mu\nu}, \tag{D.17}$$

where G is Newton's constant given by Eq. (A.13) and $T_{\mu\nu}$ is the total energy-momentum tensor of the matter, the radiation and the vacuum.

The coefficient $8\pi G$ is obtained by demanding that the Einstein equation, Eq. (D.17), should reduce to the Poisson equation for the Newtonian potential, $\nabla^2 \phi_{\mathrm{N}} = 4\pi G \varepsilon$, in the limit of a weak gravitational field and in non-relativistic kinematics. In that case, we have $g_{00} \simeq 1 + 2\phi_{\mathrm{N}}$ and $T_{00} \simeq \varepsilon$, where ε is the energy density of matter (Exercise 8.2).

The Einstein equation, Eq. (D.17), relates the curvature of space-time, represented by the gravitational field (left-hand side), to the energy-momentum of all other kinds of particles and fields that exist in the space-time (right-hand side). Note that $T_{\mu\nu}$ is often written as a sum of the matter + radiation contribution and the vacuum contribution by introducing the cosmological constant Λ:

$$T_{\mu\nu} = \bar{T}_{\mu\nu} + \Lambda g_{\mu\nu}. \tag{D.18}$$

Note that the covariantly conserved energy-momentum tensor should appear on the right-hand side of the Einstein equation because of the identity Eq. (D.16); namely,

$$\nabla_{\lambda} T^{\lambda}{}_{\mu} = \nabla_{\lambda} \bar{T}^{\lambda}{}_{\mu} = 0. \tag{D.19}$$

The Einstein equation is non-linear in terms of the metric, $g_{\mu\nu}(x)$, so that the principle of superposition does not hold.

When the metric takes a diagonal form, $g_{\mu\nu}=0$ $(\mu\neq\nu)$, a useful formula for $R_{\mu\nu}$ and R can be derived. For this purpose, let us first parametrize the diagonal components as follows:

$$g_{\mu\mu}=e_\mu\exp(2F_\mu),\quad e_0=+1,\quad e_i=-1\ (i=1,2,3), \tag{D.20}$$

where summation over repeated indices is not taken and will not be taken in the following expressions. With this parametrization, the explicit calculation shows that the non-zero components of the Ricci tensor become (see Sect. 92 in Landau and Lifshitz (1988)):

$$\begin{aligned} R_{\mu\nu} = \sum_{\alpha\neq(\mu,\nu)} &[(\partial_\nu F_\alpha)(\partial_\mu F_\nu)+(\partial_\nu F_\mu)(\partial_\mu F_\alpha) \\ &-(\partial_\mu F_\alpha)(\partial_\nu F_\alpha)-\partial_\mu\partial_\nu F_\alpha],\quad (\mu\neq\nu), \end{aligned} \tag{D.21}$$

$$\begin{aligned} R_{\mu\mu} = &\sum_{\alpha\neq\mu}\left[(\partial_\mu F_\mu)(\partial_\mu F_\alpha)-(\partial_\mu F_\alpha)^2-\partial_\mu\partial_\mu F_\alpha\right] \\ &+\sum_{\alpha\neq\mu} e_\mu e_\alpha e^{2(F_\mu-F_\alpha)}\times\Bigg[(\partial_\alpha F_\alpha)(\partial_\alpha F_\mu)-(\partial_\alpha F_\mu)^2-\partial_\alpha\partial_\alpha F_\mu \\ &\qquad -(\partial_\alpha F_\mu)\sum_{\beta\neq(\mu,\alpha)}\partial_\alpha F_\beta\Bigg]. \end{aligned} \tag{D.22}$$

D.3 The Robertson–Walker metric

The Robertson–Walker metric, Eq. (8.5), reads as follows:

$$ds^2=dt^2-a^2(t)\left[\frac{dr^2}{1-Kr^2}+r^2(d\theta^2+\sin^2\theta\ d\phi^2)\right], \tag{D.23}$$

which leads to

$$g_{tt}=1,\ g_{rr}=\frac{-a^2}{1-Kr^2},\ g_{\theta\theta}=-a^2r^2,\ g_{\phi\phi}=-a^2r^2\sin^2\theta, \tag{D.24}$$

and hence

$$F_t=0,\ F_r=\ln\frac{a}{\sqrt{1-Kr^2}},\ F_\theta=\ln(ar),\ F_\phi=\ln(ar\sin\theta). \tag{D.25}$$

Substituting these into Eqs. (D.21) and (D.22), we arrive at the following relations:

$$R=-6\left[\frac{\ddot a}{a}+\frac{\dot a^2}{a^2}+\frac{K}{a^2}\right], \tag{D.26}$$

$$R^0{}_0 = -3\frac{\ddot{a}}{a}, \tag{D.27}$$

$$R^i{}_i = -\left[\frac{\ddot{a}}{a} + 2\frac{\dot{a}^2}{a^2} + \frac{2K}{a^2}\right] \quad (i = 1, 2, 3), \tag{D.28}$$

$$G^0{}_0 = 3\left[\frac{\dot{a}^2}{a^2} + \frac{K}{a^2}\right], \tag{D.29}$$

$$G^i{}_i = 2\frac{\ddot{a}}{a} + \frac{\dot{a}^2}{a^2} + \frac{K}{a^2} \qquad (i = 1, 2, 3). \tag{D.30}$$

D.4 The Schwarzschild metric

Let us find a static and isotropic solution of the Einstein equation, Eq. (D.17). Using spherical coordinates (r, θ, ϕ), we may express the metric in a standard form (the Schwarzschild metric) :

$$ds^2 = \mathrm{e}^{a(r)}\, dt^2 - \mathrm{e}^{b(r)}\, dr^2 - r^2(d\theta^2 + \sin^2\theta\, d\phi^2), \tag{D.31}$$

$$g_{tt} = \mathrm{e}^{a(r)},\, g_{rr} = -\mathrm{e}^{b(r)},\, g_{\theta\theta} = -r^2,\, g_{\phi\phi} = -r^2\sin^2\theta, \tag{D.32}$$

where the off-diagonal part of the metric can be always eliminated by coordinate transformation; $a(r)$ and $b(r)$ should be determined from the Einstein equation. From Eq. (D.20), we have

$$F_t = a(r)/2, \quad F_r = b(r)/2, \quad F_\theta = \ln r, \quad F_\phi = \ln(r\sin\theta). \tag{D.33}$$

Then, by using Eqs. (D.21) and (D.22), non-vanishing components of the Ricci tensor and the scalar curvature are given by

$$R^0{}_0 = \mathrm{e}^{-b}\left(\frac{a''}{2} + \frac{a'}{r} + \frac{a'^2}{4} - \frac{a'b'}{4}\right), \tag{D.34}$$

$$R^1{}_1 = \mathrm{e}^{-b}\left(\frac{a''}{2} - \frac{b'}{r} + \frac{a'^2}{4} - \frac{a'b'}{4}\right), \tag{D.35}$$

$$R^i{}_i = \mathrm{e}^{-b}\left(\frac{a' - b'}{2r} + \frac{1}{r^2}\right) - \frac{1}{r^2}, \tag{D.36}$$

$$R = \mathrm{e}^{-b}\left(a'' + \frac{2(a' - b')}{r} + \frac{a'^2}{2} - \frac{a'b'}{2} + \frac{2}{r^2}\right) - \frac{2}{r^2}, \tag{D.37}$$

where $i = 2, 3$ and the prime (′) indicates the derivative with respect to r.

In consequence, the non-zero components of the Einstein equation are explicitly given by

$$G^0{}_0 = \mathrm{e}^{-b}\left(\frac{b'}{r} - \frac{1}{r^2}\right) + \frac{1}{r^2} = 8\pi G T^0{}_0, \tag{D.38}$$

$$G^1{}_1 = -\mathrm{e}^{-b}\left(\frac{a'}{r} + \frac{1}{r^2}\right) + \frac{1}{r^2} = 8\pi G T^1{}_1, \tag{D.39}$$

$$G^i{}_i = -\frac{1}{2}\,\mathrm{e}^{-b}\left(a'' + \frac{a'-b'}{r} + \frac{a'^2}{2} - \frac{a'b'}{2}\right) = 8\pi G T^i{}_i. \tag{D.40}$$

Now let us put a spherical object with a total gravitational mass M and a radius R in the vacuum. Since $T^\mu{}_\nu = 0$ for $r > R$, we can integrate Eqs. (D.38) and (D.39) to find

$$\mathrm{e}^{b(r)} = \frac{1}{1 - c_1/r}, \quad \mathrm{e}^{a(r)} = c_2(1 - c_1/r), \tag{D.41}$$

where $c_{1,2}$ are the integration constants. Note that c_2 can be absorbed in the redefinition of time: $\sqrt{c_2}t \to t$. On the other hand, c_1 is related to GM by the Newtonian potential, $\phi_\mathrm{N} = -GM/r$ at $r \to \infty$, and the metric

$$g_{00} = 1 + 2\phi_\mathrm{N} = 1 - \frac{2GM}{r} = 1 - \frac{c_1}{r}. \tag{D.42}$$

Therefore, we obtain $c_1 = 2GM$. As a result, the metric is expressed as

$$ds^2 = \left(1 - \frac{r_\mathrm{g}}{r}\right) dt^2 - \frac{dr^2}{1 - \frac{r_\mathrm{g}}{r}} - r^2(d\theta^2 + \sin^2\theta d\phi^2), \tag{D.43}$$

which is known as the Schwarzschild solution of the Einstein equation outside the star. Here, r_g is called the Schwarzschild radius or the gravitational radius, defined by

$$r_\mathrm{g} \equiv 2GM. \tag{D.44}$$

We may also integrate Eq. (D.38) to obtain

$$M = \int_0^R 4\pi r^2 T^0_0 \, dr = \int_0^R 4\pi r^2 \varepsilon(r)\, dr. \tag{D.45}$$

Note that the integration on the right-hand side is over $4\pi r^2\, dr$ instead of the space-volume element obtained from the metric, $4\pi \mathrm{e}^{b(r)/2} r^2\, dr$. This is because M contains the field energy of gravity as well as the energy of matter.

D.5 The Oppenheimer–Volkoff (OV) equation

In order to calculate the internal structure of a static and spherically symmetric star, we need the energy-momentum tensor, $T^{\mu}_{\ \nu}$, parametrized, for example, by the perfect fluid form (see Eq. (9.27)):

$$T^{\mu}_{\ \nu} = \mathrm{diag}(\varepsilon(r), -P(r), -P(r), -P(r)). \tag{D.46}$$

We assume that the equation of state (EOS), which relates the energy density and the pressure at each r, is known to be

$$P = P(\varepsilon). \tag{D.47}$$

Then the Einstein equations, Eqs. (D.38)–(D.40) become

$$\mathrm{e}^{-b}\left(\frac{b'}{r} - \frac{1}{r^2}\right) + \frac{1}{r^2} = 8\pi G\varepsilon, \tag{D.48}$$

$$\mathrm{e}^{-b}\left(\frac{a'}{r} + \frac{1}{r^2}\right) - \frac{1}{r^2} = 8\pi GP, \tag{D.49}$$

$$\frac{\mathrm{e}^{-b}}{2}\left(a'' + \frac{a' - b'}{r} + \frac{a'^2}{2} - \frac{a'b'}{2}\right) = 8\pi GP. \tag{D.50}$$

We have four equations, so we can solve the four unknown functions $a(r)$, $b(r)$, $P(r)$ and $\varepsilon(r)$. Let us introduce a function $\mathcal{M}(r)$:

$$\mathrm{e}^{-b(r)} \equiv 1 - \frac{2G\mathcal{M}(r)}{r}. \tag{D.51}$$

Outside the surface of the star, the Schwarzschild solution, Eq. (D.43), should be valid such that $\mathcal{M}(r)$ is related to the gravitational mass M:

$$\mathcal{M}(r \geq R) = M. \tag{D.52}$$

By substituting Eq. (D.51) into Eq. (D.48), we obtain

$$\frac{d\mathcal{M}(r)}{dr} = 4\pi r^2 \varepsilon(r), \tag{D.53}$$

or its integral form

$$\mathcal{M}(r) = \int_0^r 4\pi r^2 \varepsilon(r)\, dr. \tag{D.54}$$

In order to obtain the pressure gradient, dP/dr, let us differentiate Eq. (D.49) and eliminate a'' by using Eq. (D.50). In addition, the elimination of b' and a'^2 using Eqs. (D.48) and (D.49) results in

$$-\frac{dP(r)}{dr} = \frac{\varepsilon + P}{2} a'. \tag{D.55}$$

Combining this with Eq. (D.49) and Eq. (D.51), we finally arrive at the following:

$$-\frac{dP(r)}{dr} = \frac{G\varepsilon(r)\mathcal{M}(r)}{r^2} \times \left(1 - \frac{2G\mathcal{M}(r)}{r}\right)^{-1} \left(1 + \frac{P(r)}{\varepsilon(r)}\right) \left(1 + \frac{4\pi r^3 P(r)}{\mathcal{M}(r)}\right). \quad \text{(D.56)}$$

This is called the Oppenheimer–Volkoff (OV) equation (Oppenheimer and Volkoff, 1939; Tolman, 1939).

To obtain the numerical solution of the profile of a star, we solve the first order differential equations, Eqs. (D.53) and (D.56), together with the EOS, Eq. (D.47). The initial conditions for the differential equations are given by

$$\mathcal{M}(0) = 0 \quad \text{and} \quad \varepsilon(0) = \varepsilon_{\text{cent}}. \quad \text{(D.57)}$$

The radius, R, of a star with central energy density $\varepsilon_{\text{cent}}$ is defined by the condition $P(r = R) = 0$. Then the total mass of the star is obtained by Eq. (D.54) with $r = R$.

In the Newtonian limit, we may neglect the gravitational radius and the pressure on the right-hand side of Eq. (D.56). Then, we arrive at

$$-\left(\frac{dP(r)}{dr}\right)_{\text{NR}} = \frac{G\varepsilon(r)\mathcal{M}(r)}{r^2}, \quad \text{(D.58)}$$

which shows a balance between the internal pressure and the gravity acting on a small volume element located at distance r.

Appendix E

Relativistic kinematics and variables

E.1 Laboratory frame and center-of-mass frame

Consider the following reaction with a two-body initial state:

$$a+b \to c+d+e+\cdots, \tag{E.1}$$

and call a the projectile and b the target particle. In the *laboratory frame* (*the target rest frame*), the target is at rest and the projectile strikes it with an energy $E^{\rm lab}$ and a momentum $\boldsymbol{p}^{\rm lab}$. After the collision, the particles in the final state, c, d, $e, \ldots$, are usually moving. In the *center-of-mass frame*, which actually means the *center-of-momentum frame*, the sum of the momentum vectors of all particles in the initial state and that in the final state vanish. Namely, the two frames are defined as follows:

- Laboratory frame

$$\boldsymbol{p}_b^{\rm lab} = \boldsymbol{0}, \quad E_b{}^{\rm lab} = m_b, \tag{E.2}$$

- Center-of-mass frame

$$\boldsymbol{p}_a^{\rm cm} + \boldsymbol{p}_b^{\rm cm} = \boldsymbol{p}_c^{\rm cm} + \boldsymbol{p}_d^{\rm cm} + \boldsymbol{p}_e^{\rm cm} + \cdots = \boldsymbol{0}, \tag{E.3}$$

where m_b is the rest mass of particle b. In the center-of-mass frame, both particles, a and b, in the initial state approach each other with equal but opposite momentum. Only the energy available in the center-of-mass frame can be used to produce new particles or to excite internal degrees of freedom.

In order to obtain the relation between the energies in the laboratory and center-of-mass systems, we utilize Lorentz invariance. We define the following Lorentz scalar quantity, s, which is one of the Mandelstam variables:

$$s \equiv (p_a + p_b)^2 \equiv (p_a + p_b)_\mu (p_a + p_b)^\mu, \tag{E.4}$$

where p_a (p_b) is the four-momentum of the particle a (b). By definition, s is the same in all coordinate systems.

Consider a relativistic collision between two particles with the same rest mass m. Using Eqs. (E.2)–(E.4), we obtain

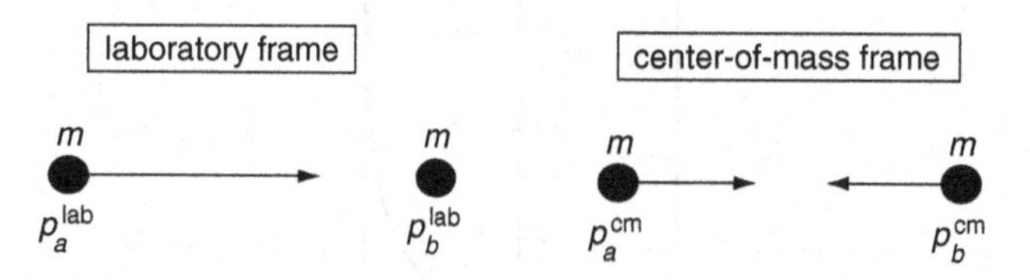

- **Laboratory frame**

$$\begin{aligned} p_a^{\rm lab} &= (E^{\rm lab}, \boldsymbol{p}^{\rm lab}), \\ p_b^{\rm lab} &= (m, \boldsymbol{0}), \\ s &\equiv (p_a + p_b)^2 = (p_a^{\rm lab} + p_b^{\rm lab})^2 \\ &= (E^{\rm lab} + m)^2 - (\boldsymbol{p}^{\rm lab})^2 = (E^{\rm lab} + m)^2 - [(E^{\rm lab})^2 - m^2] \\ &= 2mE^{\rm lab} + 2m^2 \end{aligned} \tag{E.5}$$

- **Center-of-mass frame**

$$\begin{aligned} p_a^{\rm cm} &= (E^{\rm cm}/2, \boldsymbol{p}^{\rm cm}), \\ p_b^{\rm cm} &= (E^{\rm cm}/2, -\boldsymbol{p}^{\rm cm}), \\ s &\equiv (p_a + p_b)^2 = (p_a^{\rm cm} + p_b^{\rm cm})^2 = (E^{\rm cm})^2. \end{aligned} \tag{E.6}$$

By equating Eq. (E.5) and Eq. (E.6), we obtain

$$E^{\rm lab} = \frac{(E^{\rm cm})^2}{2m} - m. \tag{E.7}$$

With $E^{\rm lab} \gg m$, which corresponds to an extremely relativistic or ultra-relativistic case, the energy $E^{\rm cm}$ ($= \sqrt{s}$) becomes

$$\sqrt{s} = E^{\rm cm} \simeq \sqrt{2mE^{\rm lab}}. \tag{E.8}$$

Equation (E.8) shows that the center-of-mass energy which is useful for producing new particles increases only as the square root of the laboratory energy in relativistic energies. This is the reason why we have to construct relativistic collider-type accelerators.

In addition to s in Eq. (E.4), we define two other Mandelstam variables (square of four-momentum transfer) as follows:

$$t \equiv (p_a - p_c)^2, \quad u \equiv (p_a - p_d)^2, \tag{E.9}$$

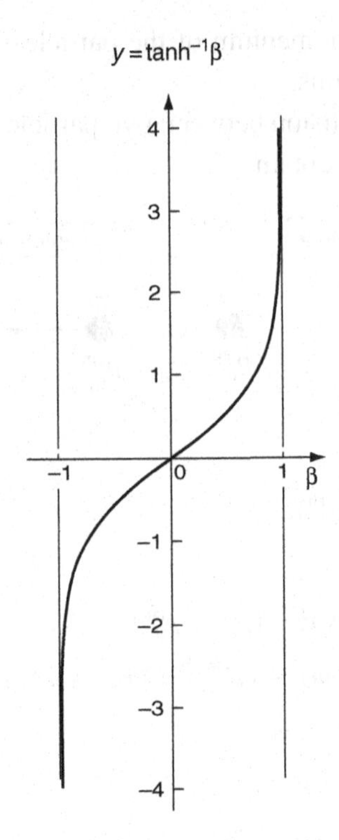

Fig. E.1. Rapidity $y = \tanh^{-1}\beta$ plotted against β.

for the two-body reaction of $a+b \to c+d$. The three variables s, t and u are not independent:

$$s+t+u = m_a^2 + m_b^2 + m_c^2 + m_d^2. \tag{E.10}$$

E.2 Rapidity and pseudo-rapidity

In relativistic mechanics, the addition law of the velocities moving along, for example, the z-axis is non-linear:

$$v = \frac{v_1 + v_2}{1 + \frac{v_1 v_2}{c^2}} \quad \text{or} \quad \beta = \frac{\beta_1 + \beta_2}{1 + \beta_1 \beta_2}, \tag{E.11}$$

where $\beta \equiv v/c = v$. Let us introduce a function $y \equiv y(\beta)$ so as to make the addition law of y linear in y under the condition Eq. (E.11) for β. Recalling the following addition law:

$$\tanh^{-1}\beta_1 \pm \tanh^{-1}\beta_2 = \tanh^{-1}\frac{\beta_1 \pm \beta_2}{1 \pm \beta_1 \beta_2}, \tag{E.12}$$

we obtain

$$y = \tanh^{-1}\beta = \frac{1}{2}\ln\frac{1+\beta}{1-\beta}, \tag{E.13}$$

which we call the *rapidity*, y. For small β, we have $y \simeq \beta$; i.e., the rapidity is a relativistic analogue of the velocity. As shown in Fig. E.1, the rapidity increases without bound as the particle velocity approaches the velocity of light. Since $\beta = p_z/E$, Eq. (E.13) may also be expressed as follows:

$$y = \tanh^{-1}\beta = \frac{1}{2}\ln\frac{E+p_z}{E-p_z}, \tag{E.14}$$

where p_z indicates the momentum component along the z-axis (longitudinal momentum). A Lorentz boost along the longitudinal axis from a frame S to a new frame S′ thus changes the rapidity in a simple additive way:

$$y' = y + \tanh^{-1}\beta, \tag{E.15}$$

where β is the velocity of S′ with respect to S. Consequently, a particle distribution expressed as a function of y has a shape that is invariant under such a boost.

The transverse momentum, p_{T}, and the longitudinal momentum, p_z, of a particle having rest mass m and momentum vector $\boldsymbol{p} = (p_x, p_y, p_z)$ are given by

$$p_{\mathrm{T}} = \sqrt{p_x^2 + p_y^2} = |\boldsymbol{p}|\sin\theta, \quad p_z = |\boldsymbol{p}|\cos\theta, \tag{E.16}$$

where θ is the polar angle of the vector $\boldsymbol{p}$ with respect to the z-axis. By using these variables we define the transverse mass, m_{T}, as

$$m_{\mathrm{T}}^2 = p_{\mathrm{T}}^2 + m^2; \quad E^2 = p_z^2 + m_{\mathrm{T}}^2. \tag{E.17}$$

Then the rapidity, y, in Eq. (E.14) can be rewritten as follows:

$$y = \ln\frac{E+p_z}{m_{\mathrm{T}}}. \tag{E.18}$$

The four-momentum, p^μ, is thus conveniently parametrized as

$$p^\mu = (E, p_x, p_y, p_z) = (m_{\mathrm{T}}\cosh y, p_{\mathrm{T}}\cos\phi_p, p_{\mathrm{T}}\sin\phi_p, m_{\mathrm{T}}\sinh y). \tag{E.19}$$

This parametrization is valid also for an off-shell time-like particle if we define $m^2 \equiv p^2$.

Next we define the pseudo-rapidity, η:

$$\eta \equiv -\ln\left(\tan\frac{\theta}{2}\right). \tag{E.20}$$

From Eq. (E.14), if the particle masses are negligible, i.e. $E^2 = \boldsymbol{p}^2 + m^2 \simeq \boldsymbol{p}^2$, we have

$$y \simeq \frac{1}{2} \ln \frac{p+p_z}{p-p_z} = \frac{1}{2} \ln \frac{1+\cos\theta}{1-\cos\theta} = -\ln\left(\tan\frac{\theta}{2}\right) = \eta. \tag{E.21}$$

Therefore, at extremely high energies ($E \gg m$), the rapidity, y, and the pseudo-rapidity, η, are equivalent. The pseudo-rapidity is useful because it can be determined directly from the particle production angle, θ, measured with respect to the beam axis in experiments.

E.3 Lorentz-invariant differential cross-section

Let us derive the Lorentz-invariant phase-space factor. A transformation of a four-momentum, (E, p_x, p_y, p_z), in the S frame to a four-momentum, (E', p'_x, p'_y, p'_z), in the S$'$ frame moving with velocity β along the z-axis relative to S is

$$\begin{bmatrix} E' \\ p'_x \\ p'_y \\ p'_z \end{bmatrix} = \begin{bmatrix} \gamma & 0 & 0 & -\gamma\beta \\ 0 & 1 & 0 & 0 \\ 0 & 0 & 1 & 0 \\ -\gamma\beta & 0 & 0 & \gamma \end{bmatrix} \begin{bmatrix} E \\ p_x \\ p_y \\ p_z \end{bmatrix}, \tag{E.22}$$

where $\gamma \equiv (1-\beta^2)^{-\frac{1}{2}} = E/m$ is the Lorentz factor. Then, we obtain

$$E' = \gamma\left(E - \beta p_z\right),\ p'_x = p_x,\ p'_y = p_y,\ p'_z = \gamma\left(p_z - \beta E\right). \tag{E.23}$$

Using the fact that $E^2 - p_x^2 - p_y^2 - p_z^2 = m^2$ and that, for p_x, p_y fixed, $p_z dp_z = E\,dE$, we have

$$\frac{dp'_z}{E'} = \frac{\gamma\left(dp_z - \beta\, dE\right)}{\gamma\left(1 - \beta\frac{p_z}{E}\right)E} = \frac{dp_z}{E}, \tag{E.24}$$

so that

$$\frac{d^3p'}{E'} \equiv \frac{dp'_x\, dp'_y\, dp'_z}{E'} = \frac{dp_x\, dp_y\, dp_z}{E} \equiv \frac{d^3p}{E}. \tag{E.25}$$

This can be also obtained directly from the following identity:

$$\frac{d^3p}{2p^0} = d^4p\ \theta(p^0)\delta(p^2 - m^2), \quad p^0 = E. \tag{E.26}$$

Then the Lorentz-invariant differential cross-section for producing a particle of momentum $\boldsymbol{p}$ lying in the phase-space element d^3p and energy E is given by $E(d^3\sigma/d^3p)$. The transverse momentum, p_{T}, the azimuthal angle, ϕ_p, and the rapidity, y, of the produced particle are related to $\boldsymbol{p}$ as follows:

$$dp_x\, dp_y = d\phi_p\, p_{\mathrm{T}} dp_{\mathrm{T}}, \quad dy = \frac{dp_z}{E}, \tag{E.27}$$

where Eq. (E.14) and $dE/dp_z = p_z/E$ are used. Thus we obtain

$$E\frac{d^3\sigma}{d^3p} = E\frac{d^3\sigma}{dp_x\, dp_y\, dp_z} = \frac{d^3\sigma}{p_{\rm T}dp_{\rm T}\; dy\; d\phi_p} = \frac{d^3\sigma}{m_{\rm T}dm_{\rm T}\; dy\; d\phi_p}. \tag{E.28}$$

For the production of a time-like virtual particle with four-momentum p^μ and $m^2 \equiv p^2$, it is also convenient to define the cross-section,

$$\frac{d^4\sigma}{d^4p} = \frac{d^4\sigma}{dp_0\; dp_x\; dp_y\; dp_z} = \frac{d^4\sigma}{m_{\rm T}dm_{\rm T}\; mdm\; dy\; d\phi_p}. \tag{E.29}$$

Appendix F

Scattering amplitude, optical theorem and elementary parton scatterings

F.1 Scattering amplitude and cross-section

The S matrix for the transition from a state $|\alpha\rangle$ to a state $|\beta\rangle$ is given by (Weinberg, 1995)

$$\langle\beta|\hat{S}|\alpha\rangle = \langle\beta|\alpha\rangle + i\langle\beta|\hat{T}|\alpha\rangle, \tag{F.1}$$

$$\langle\beta|\hat{T}|\alpha\rangle \equiv (2\pi)^4\delta^4(P_\alpha - P_\beta)M_{\alpha\to\beta}, \tag{F.2}$$

where $\hat{T}$ is the transition operator ($\hat{S} = 1 + i\hat{T}$) and $M_{\alpha\to\beta}$ is the corresponding invariant amplitude.

For the process $1+2 \to 3+4+\cdots+n$, the invariant differential cross-section, $d\sigma_{\alpha\to\beta}$, is given by

$$d\sigma = d\beta\,\frac{|M_{\alpha\to\beta}|^2(2\pi)^4\delta^4(P_\alpha - P_\beta)}{(2\varepsilon_1)(2\varepsilon_2)\bar{v}_{\text{rel}}} \tag{F.3}$$

$$= d\beta\,\frac{|M_{\alpha\to\beta}|^2(2\pi)^4\delta^4(P_\alpha - P_\beta)}{2\sqrt{(s-(m_1-m_2)^2)(s-(m_1+m_2)^2)}}, \tag{F.4}$$

with

$$d\beta \equiv \prod_{j=3}^{n} \frac{d^3p_j}{2\varepsilon_j(2\pi)^3}. \tag{F.5}$$

Here the Lorentz-scalar flux factor is defined as

$$\bar{v}_{\text{rel}} = \sqrt{(p_1\cdot p_2)^2 - m_1^2m_2^2}\Big/(\varepsilon_1\varepsilon_2), \tag{F.6}$$

which reduces to the relative velocity, $|\boldsymbol{v}_1 - \boldsymbol{v}_2|$, for collinear collision of the initial particles. Note that m_1 and m_2 are the masses of the initial particles, and s is one of the Mandelstam variables, $s = (p_1+p_2)^2$ (see Eqs. (E.4) and (E.9)). The total cross-section is obtained by $\sigma_\alpha^{\text{tot}} = \int d\sigma$, with appropriate symmetry factors for identical particles in the final state.

F.2 Optical theorem

The conservation of probability, or equivalently the unitarity of the S-matrix ($\hat{S}^\dagger \hat{S} = 1$), leads to a constraint, $-i(\hat{T} - \hat{T}^\dagger) = \hat{T}^\dagger \hat{T}$. By taking the forward matrix element of this relation, we have

$$\text{Im } M_{\alpha\to\alpha} = \frac{1}{2}\int d\beta\, |M_{\alpha\to\beta}|^2 (2\pi)^4 \delta^4(P_\alpha - P_\beta) \tag{F.7}$$

$$= 2\varepsilon_1 \varepsilon_2 \bar{v}_{\rm rel} \sigma_\alpha^{\rm tot} \tag{F.8}$$

$$= \sqrt{(s-(m_1-m_2)^2)(s-(m_1+m_2)^2)}\ \sigma_\alpha^{\rm tot}. \tag{F.9}$$

This is called the optical theorem, which relates the imaginary part of the forward scattering amplitude to the total cross-section.

Note that for two-to-two scattering ($1+2 \to 3+4$), both Im $M_{\alpha\to\alpha}$ and $\sigma_\alpha^{\rm tot}$ depend only on s.

F.3 Parton scattering amplitudes

For two-to-two ($1+2 \to 3+4$) scattering, the invariant amplitude M (we drop the suffix α and β for simplicity) is a function of the Mandelstam variables, s and t, defined in Eqs.(E.4) and (E.9). Also, the differential cross-section Eq. (F.4) is rewritten as

$$\frac{d\sigma}{d|t|} = \frac{1}{16\pi\left[s-(m_1+m_2)^2\right]\left[s-(m_1-m_2)^2\right]}|M(s,t)|^2. \tag{F.10}$$

The Feynman diagrams for the two-to-two parton scatterings in the tree level are shown in Fig.F.1. These processes are calculated either by hand or by using automatic computation programs such as GRACE (Yuasa *et al.*, 2000), compHEP (Pukhov *et al.*, 1999) and O'Mega (Moretti *et al.*, 2001).

As a simple illustration, let us consider the cross-section for the process

$$q\{k_1, s_1, i\} + \bar{q}\{k_2, s_2, k\} \to q'\{p_1, s_1', j\} + \bar{q}'\{p_2, s_2', l\}, \tag{F.11}$$

which is shown in Fig. F.1(a) and Fig. F.2. Here, q and q' are quarks with different flavors, (k_1, k_2, p_1, p_2) are the quark momenta, (s_1, s_2, s_1', s_2') are the quark spins and (i, k, j, l) are the quark colors. This is a process quite similar to $e^+e^- \to \mu^+\mu^-$, except for the color factors. The tree-level amplitude, $\mathcal{M}$, in the Feynman gauge is given by

$$\mathcal{M} = \frac{(-i)^3 g^2}{s} t^a_{jl}\, t^a_{ki} \left[\bar{u}\left(p_1, s_1'\right) \gamma_\mu v\left(p_2, s_2'\right)\right]\left[\bar{v}\left(k_2, s_2\right) \gamma^\mu u\left(k_1, s_1\right)\right], \tag{F.12}$$

where $s = (k_1+k_2)^2$ and $t^a = \lambda^a/2$ are the generators of the SU(3) color gauge group (see Appendix B).

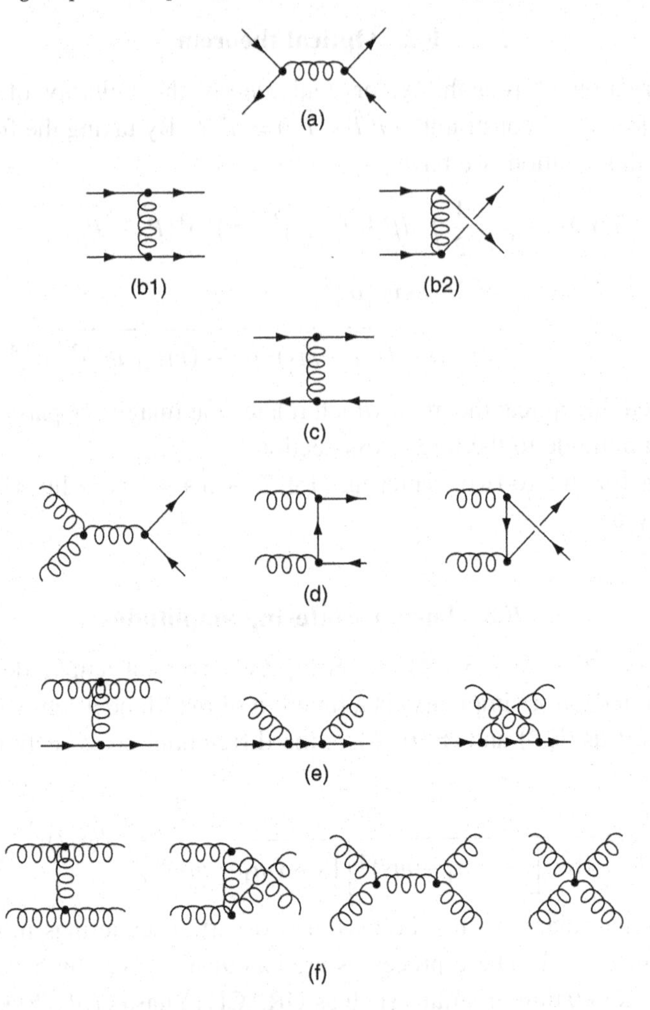

Fig. F.1. Feynman diagrams for two-to-two parton scattering and production in the lowest order, $O(g^2)$.

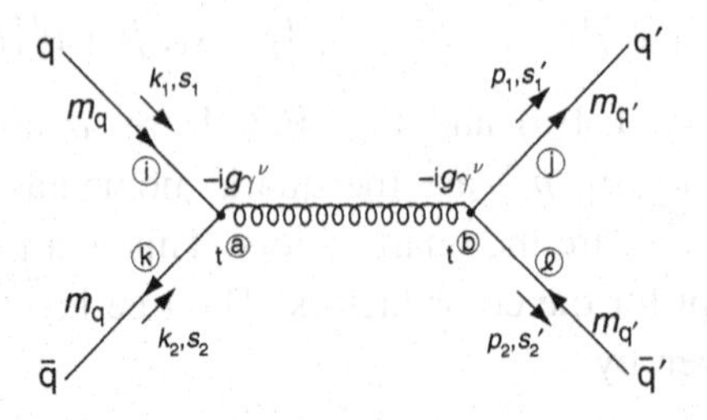

Fig. F.2. The lowest-order Feynman diagram for $q\bar{q} \to q'\bar{q}'$. The factors needed to compute the amplitude are shown. The encircled letters give the color indices of the particles and run from 1 to 3 for quarks and antiquarks, and from 1 to 8 for gluons. Energy-momentum conservation gives $k_1 + k_2 = p_1 + p_2$.

The squared amplitude, $|M|^2$, defined by summing (averaging) over final (initial) states in color and spin is related to $\mathcal{M}$ as follows:

$$|M|^2 \equiv \frac{1}{(2\times 3)\,(2\times 3)} \sum_{\text{spins}} \sum_{\text{colors}} |\mathcal{M}|^2 \tag{F.13}$$

$$= \frac{g^4}{36s^2} \cdot T_4 \cdot \mathrm{tr}\left[\left(\not{p}_1 + m_{q'}\right)\gamma_\mu\left(\not{p}_2 - m_{q'}\right)\gamma_\nu\right] \times \mathrm{tr}\left[\left(\not{k}_2 + m_q\right)\gamma^\mu\left(\not{k}_1 - m_q\right)\gamma^\nu\right]. \tag{F.14}$$

In Eq. (F.13), the summation over spins is taken for s_1, s_2, s_1' and s_2', and the summation over colors is taken for i, j, k and l. In Eq. (F.14), "tr" is taken only for the Lorentz indices. Also, the color part is abbreviated to

$$T_4 \equiv t^a_{jl} t^a_{ki} t^b_{ik} t^b_{lj}, \tag{F.15}$$

where the summation over repeated indices is assumed and we have used the hermiticity $t^a = (t^a)^\dagger$. To carry the spin sum in Eq. (F.13), the completeness relations (see Appendix B) have been used:

$$\sum_{\text{spin}} \psi_\alpha(p,s)\,\bar{\psi}_\beta(p,s) = (\not{p} \pm m)_{\alpha\beta}, \tag{F.16}$$

where $+$ $(-)$ on the right-hand side corresponds to $\psi = u$ $(\psi = v)$.

Note that T_4 is evaluated by using the normalization with respect to the color trace, $\mathrm{tr}(t^a t^b) = \delta^{ab}/2$:

$$T_4 = \mathrm{tr}\left(t^a t^b\right) \mathrm{tr}\left(t^b t^a\right) = \frac{1}{4} \sum_{a=1}^{8} 1 = 2. \tag{F.17}$$

The trace for Lorentz indices in Eq. (F.14) can be carried out straightforwardly using the properties of gamma matrices (see Appendix B):

$$\begin{aligned} &\mathrm{tr}[(\not{k}_2 + m_q)\gamma_\mu(\not{k}_1 - m_q)\gamma_\nu] \\ &= 4\left(k_{1\mu}k_{2\nu} + k_{1\nu}k_{2\mu} - (k_1\cdot k_2 + m_q^2)g_{\mu\nu}\right). \end{aligned} \tag{F.18}$$

Let us consider a situation in which the incoming parton, q, is nearly massless ($m_q = 0$), while the outgoing parton, q', has a finite mass ($m_{q'} \neq 0$). This covers processes such as $u\bar{u}$ (or $d\bar{d}$) $\to s\bar{s}, c\bar{c}$. Then, the energy-momentum conservation $k_1 + k_2 = p_1 + p_2$ leads to

$$\begin{aligned} s &\equiv (k_1+k_2)^2 = (p_1+p_2)^2 = 2k_1\cdot k_2 = 2m_{q'}^2 + 2\,(p_1\cdot p_2), \\ t &\equiv (k_1-p_1)^2 = m_{q'}^2 - 2\,(k_1\cdot p_1) = m_{q'}^2 - 2\,(k_2\cdot p_2), \\ u &\equiv (k_1-p_2) = m_{q'}^2 - 2\,(k_1\cdot p_2) = m_{q'}^2 - 2\,(k_2\cdot p_1). \end{aligned} \tag{F.19}$$

Table F.1. *Two-parton scattering and production sub-processes in the lowest order perturbative QCD.*

The associated matrix elements squared, $|M|^2$, divided by $16\pi^2\alpha_s^2$ and their values at $\theta_{cm} = 90°$ are also shown. The initial (final) colors and spins are averaged (summed). Note that $q(\bar{q})$ and g denote massless quark (anti-quark) and gluon, respectively (Fig.F.1); q and q' denote distinct flavors. All the quark masses are taken to be zero.

Process	Fig.F.1	$\|M\|^2/(16\pi^2\alpha_s^2)$	At $\theta_{cm} = 90°$
$q\bar{q} \to q'\bar{q}'$	(a)	$\frac{4}{9}\frac{t^2+u^2}{s^2}$	0.222
$qq' \to qq'$	(b1)	$\frac{4}{9}\frac{s^2+u^2}{t^2}$	2.22
$q\bar{q}' \to q\bar{q}'$	(c)	$\frac{4}{9}\frac{s^2+u^2}{t^2}$	2.22
$q\bar{q} \to q\bar{q}$	(a)(c)	$\frac{4}{9}(\frac{t^2+u^2}{s^2}+\frac{s^2+u^2}{t^2})-\frac{8}{27}\frac{u^2}{st}$	2.59
$qq \to qq$	(b1)(b2)	$\frac{4}{9}(\frac{s^2+u^2}{t^2}+\frac{s^2+t^2}{u^2})-\frac{8}{27}\frac{s^2}{ut}$	3.26
$\mathrm{gg} \to q\bar{q}$	(d)	$\frac{1}{6}\frac{t^2+u^2}{tu}-\frac{3}{8}\frac{t^2+u^2}{s^2}$	0.146
$q\bar{q} \to \mathrm{gg}$	(d)	$\frac{32}{27}\frac{t^2+u^2}{tu}-\frac{8}{3}\frac{t^2+u^2}{s^2}$	1.04
$q\mathrm{g} \to q\mathrm{g}$	(e)	$-\frac{4}{9}\frac{s^2+u^2}{su}+\frac{s^2+u^2}{t^2}$	6.11
$\mathrm{gg} \to \mathrm{gg}$	(f)	$\frac{9}{2}(3-\frac{tu}{s^2}-\frac{su}{t^2}-\frac{st}{u^2})$	30.4

Using Eqs. (F.17)–(F.19), Eq. (F.14) is rewritten as

$$|M|^2 = \frac{g^4}{36s^2}\cdot 2\cdot 32\cdot[(k_2\cdot p_1)(k_1\cdot p_2)+(k_1\cdot p_1)(k_2\cdot p_2)+m_{q'}^2(k_1\cdot k_2)]$$

$$= 16\pi^2\alpha_s^2\cdot\frac{4}{9}\frac{1}{s^2}\left[\left(m_{q'}^2-u\right)^2+\left(m_{q'}^2-t\right)^2+2m_{q'}^2 s\right], \tag{F.20}$$

where $\alpha_s = g^2/(4\pi)$. If the mass of q' is negligible, we obtain

$$|M|^2 = 16\pi^2\alpha_s^2\cdot\frac{4}{9}\frac{u^2+t^2}{s^2}, \qquad \frac{d\sigma}{d|t|} = \frac{\pi\alpha_s^2}{s^2}\cdot\frac{4}{9}\frac{u^2+t^2}{s^2}. \tag{F.21}$$

The tree level amplitude, $|M|^2/(16\pi^2\alpha_s^2)$, is listed for massless quarks in Table F.1 for all the lowest order processes at $O(g^4)$ shown in Fig. F.1 (Combridge *et al.*, 1977; Ellis *et al.*, 1996). In the center-of-mass system, we have the relations

$$s = 4|\boldsymbol{k}|^2, \quad t = -2|\boldsymbol{k}|^2\,(1-\cos\theta_{cm}), \quad u = -2|\boldsymbol{k}|^2\,(1+\cos\theta_{cm}), \tag{F.22}$$

where $|\boldsymbol{k}| \equiv |\boldsymbol{k}_1| = |\boldsymbol{k}_2|$. Therefore, for large angle scattering, we find

$$-\frac{s}{2} = t = u = -2|\boldsymbol{k}|^2 \quad \text{at} \quad \theta_{cm} = 90°. \tag{F.23}$$

Table F.2. *Virtual-photon production through two-parton processes in the lowest order perturbative QCD.*

The associated matrix elements squared, $|M|^2$, divided by $16\pi^2\alpha_s\alpha Q_q^2$ are shown, where $\alpha = e^2/(4\pi)$ and $Q_q \cdot e$ is the electric charge of a quark with flavor q (i.e. Q_q is either 2/3 or $-1/3$) are given. The initial (final) colors and spins are averaged (summed). All the quark masses are taken to be zero. For real-photon production, $s+t+u=0$.

Process	Fig.F.3	$\|M\|^2/(16\pi^2\alpha_s\alpha Q_q^2)$	At $\theta_{cm}=90°$
$q\bar{q} \to g\gamma^*$	(a)	$\frac{8}{9}\frac{t^2+u^2+2s(s+t+u)}{tu}$	0.889
$qg \to q\gamma^*$	(b)	$-\frac{1}{3}\frac{s^2+u^2+2t(s+t+u)}{su}$	0.833

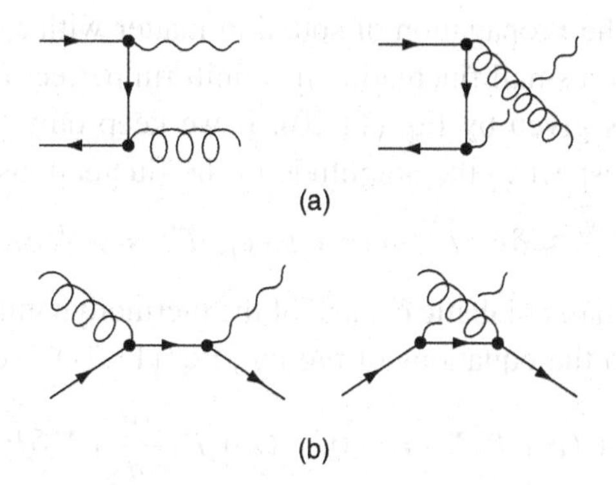

Fig. F.3. The lowest order Feynman diagrams for single-photon production.

Substituting Eq. (F.23) into the third column of Table F.1, allows us to evaluate the relative importance of the processes at $\theta_{cm} = 90°$: this is shown in the right-most column. The cross-section of the gg elastic scattering is prominently large.

The single-photon production by the two-body parton–parton scattering at $O(g^2)$ can also be calculated from the diagrams in Fig. F.3, and the results are shown in Table F.2.

Appendix G

Sound waves and transverse expansion

G.1 Propagation of sound waves

In order to discuss the propagation of sound in matter with a relativistic equation of state, we consider a small fluctuation in a uniform perfect fluid, whose energy-momentum tensor is given by Eq. (11.20). If we keep only the terms which are of first order with respect to the amplitude of the fluctuations, we have

$$T^{00} \approx \delta\varepsilon, \; T^{i0} \approx (\varepsilon + P)\, v_i, \; T^{ij} \approx -g^{ij}\delta P, \tag{G.1}$$

where δ denotes a small adiabatic change of the thermodynamic quantities. Substituting Eq. (G.1) into the equations of motion, Eq. (11.21), we obtain

$$\frac{\partial \delta\varepsilon}{\partial t} + (\varepsilon + P)\, \boldsymbol{\nabla} \cdot \boldsymbol{v} = 0, \quad (\varepsilon + P)\, \frac{\partial \boldsymbol{v}}{\partial t} + \boldsymbol{\nabla}\delta P = \boldsymbol{0}. \tag{G.2}$$

Eliminating $\boldsymbol{v}$, we find

$$\frac{\partial^2 \delta\varepsilon}{\partial t^2} - \boldsymbol{\nabla}^2 \delta P = 0. \tag{G.3}$$

Let us assume a fluid with zero baryon density. Then we have $\delta\varepsilon = (\partial\varepsilon/\partial P)\delta P$, and Eq. (G.3) reduces to the wave equation describing the propagation of a small pressure fluctuation with the velocity of sound:

$$c_{\mathrm{s}}^2 = \frac{\partial P}{\partial \varepsilon} = \frac{\partial \ln T}{\partial \ln s}. \tag{G.4}$$

This follows from the thermodynamical relations, Eqs. (3.14) and (3.15). For a simple equation of state such as $P = \lambda\varepsilon$, we have $c_{\mathrm{s}} = \sqrt{\lambda}$ (see Eq. (11.56)).

G.2 Transverse hydrodynamical equations

Let us assume that two colliding nuclei have a zero-impact parameter ($b = 0$) and that the colliding system is axially symmetric with respect to the collision axis. We employ cylindrical coordinates (r, ϕ, z), where z is the coordinate along the collision axis and $r \equiv \sqrt{x^2+y^2} = z = 0$ is the collision point. In addition, we define the longitudinal proper time, $\tau = \sqrt{t^2 - z^2}$. To obtain the boost-invariant solutions in the longitudinal direction, we generalize the Bjorken flow, Eq. (11.49), to the following form:

$$u^\mu = \left(u_t, u_r, u_\phi, u_z\right) = \frac{1}{\sqrt{1-v_r^2}} \left(\frac{t}{\tau}, v_r(\tau, r), 0, \frac{z}{\tau}\right), \qquad \text{(G.5)}$$

where the overall factor on the right-hand side is obtained from the normalization $u^\mu u_\mu = 1$. We define the transverse rapidity α as follows:

$$\alpha = \tanh^{-1} v_r = \frac{1}{2} \ln \frac{1+v_r}{1-v_r}, \qquad \text{(G.6)}$$

which gives $\cosh\alpha = 1/\sqrt{1-v_r^2}$ and $\sinh\alpha = v_r/\sqrt{1-v_r^2}$. Together with Eq. (11.45), we then find

$$u^\mu = \cosh\alpha \,(\cosh Y, \tanh\alpha, 0, \sinh Y). \qquad \text{(G.7)}$$

In the present framework of the two-dimensional expansion of a fluid, the identities in Eq. (11.52) are modified to

$$u^\mu \partial_\mu = \cosh\alpha \frac{\partial}{\partial\tau} + \sinh\alpha \frac{\partial}{\partial r}, \qquad \text{(G.8)}$$

$$\partial_\mu u^\mu = \frac{1}{\tau r}\left[\frac{\partial}{\partial\tau}(\tau r \cosh\alpha) + \frac{\partial}{\partial r}(\tau r \sinh\alpha)\right]. \qquad \text{(G.9)}$$

Then the entropy equation, Eq. (11.53), is generalized to

$$\frac{\partial}{\partial\tau}(\tau r s \cosh\alpha) + \frac{\partial}{\partial r}(\tau r s \sinh\alpha) = 0. \qquad \text{(G.10)}$$

This is independent of the specific form of the equation of state.

Another equation necessary to determine the transverse acceleration of the fluid can be obtained from Eq. (11.30):

$$-\partial_\rho P + u_\rho u^\mu \partial_\mu P + (\varepsilon + P) u^\mu \partial_\mu u_\rho = 0, \qquad \text{(G.11)}$$

together with Eqs. (G.8), and (G.9). In particular, for an equation of state such as that given in Eqs. (11.56) and (11.57), we may reduce Eq. (G.11) to

$$\frac{\partial}{\partial\tau}(T \sinh\alpha) + \frac{\partial}{\partial r}(T \cosh\alpha) = 0. \qquad \text{(G.12)}$$

Eqs. (G.10) and (G.12) provide us with a basis to determine the boost-invariant cylindrical expansion for the equation of state, $P = \lambda\varepsilon$.

G.3 Transverse mass spectrum

Let us consider the emission of hadrons. Their transverse mass spectrum is dictated by the Cooper–Frye formula, Eq. (13.24) (Cooper and Frye, 1974) with the transverse expansion (Ruuskanen, 1987; Heinz *et al.*, 1990). We assume that the freeze-out takes place on $\Sigma_{\rm f}$, which is a three-dimensional hyper-surface specified by the cylindrical coordinate system:

$$\Sigma^0 \equiv t_{\rm f}(r,z), \quad \Sigma^1 = r\cos\phi, \quad \Sigma^2 = r\sin\phi, \quad \Sigma^3 = z. \tag{G.13}$$

A normal vector, $d\Sigma_\mu$, to the surface is then given by (see Chap. 1 in Landau and Lifshitz (1988))

$$d\Sigma_\mu = \epsilon_{\mu\nu\lambda\rho}\frac{\partial\Sigma^\nu}{\partial r}\frac{\partial\Sigma^\lambda}{\partial\phi}\frac{\partial\Sigma^\rho}{\partial z}\,dr\,d\phi\,dz \tag{G.14}$$

$$= \left(1, -\frac{\partial\Sigma^0}{\partial r}\cos\phi, -\frac{\partial\Sigma^0}{\partial r}\sin\phi, -\frac{\partial\Sigma^0}{\partial z}\right) rdr\,d\phi\,dz. \tag{G.15}$$

The four-momentum of the emitted hadron may be expressed in terms of the momentum-space rapidity, y, and the transverse mass, $m_{\rm T}$, as in Eq. (E.19) in Appendix E.2. Thus, we obtain

$$p^\mu\,d\Sigma_\mu = \left[m_{\rm T}\cosh y - p_{\rm T}\cos(\phi-\phi_p)\frac{\partial\Sigma^0}{\partial r} - m_{\rm T}\sinh y\frac{\partial\Sigma^0}{\partial z}\right] rdr\,d\phi\,dz. \tag{G.16}$$

The expansion flow velocity, u^μ, is obtained as a generalization of Eq. (G.7) to the case of $\phi \neq 0$:

$$u^\mu = (u_t, u_x, u_y, u_z) = \cosh\alpha\,(\cosh Y, \tanh\alpha\cos\phi, \tanh\alpha\sin\phi, \sinh Y). \tag{G.17}$$

Then we have

$$p^\mu u_\mu/T = \xi_m\cosh(Y-y) - \xi_p\cos(\phi-\phi_p), \tag{G.18}$$

with

$$\xi_m \equiv \frac{m_{\rm T}\cosh\alpha}{T}, \quad \xi_p \equiv \frac{p_{\rm T}\sinh\alpha}{T}. \tag{G.19}$$

For a local thermal distribution of Eq. (13.25), the invariant momentum spectrum is thus given by an integration over the freeze-out hyper-surface, $\Sigma_{\rm f}$:

$$E\frac{d^3N}{d^3p} = \frac{d^3N}{m_{\rm T} dm_{\rm T}\, dy\, d\phi_p}$$
$$= \frac{m_{\rm T}}{(2\pi)^2}\sum_{n=1}^{\infty}(\mp)^{n+1}\int_{\Sigma_{\rm f}} dz\; r\; dr\; {\rm e}^{n(\mu/T-\xi_m\cosh(Y-y))}$$
$$\times\left[\left(\cosh y - \sinh y\frac{\partial\Sigma^0}{\partial z}\right) I_0(\xi_p) - \frac{p_{\rm T}}{m_{\rm T}}\frac{\partial\Sigma^0}{\partial r} I_1(\xi_p)\right]. \qquad ({\rm G.20})$$

The integration over ϕ is carried out by making use of the modified Bessel functions of the first kind, $I_\nu(\xi)$, defined in Eq. (G.25). The upper (lower) sign in $(\mp)^{n+1}$ corresponds to the boson (fermion).

Let us make further integrations over ϕ_p and y. The former integration is trivial. As for the latter, we assume that the detector system has full y-acceptance and shift the integration variable y to $y - Y$. Then the result is expressed by the modified Bessel function of the second kind, $K_\nu(\xi)$, defined in Eq. (G.26):

$$\frac{dN}{m_{\rm T} dm_{\rm T}} = \frac{m_{\rm T}}{\pi}\sum_{n=1}^{\infty}(\mp)^{n+1}\int_{\Sigma_{\rm f}} dz\; r\; dr\; {\rm e}^{n(\mu/T)}$$
$$\times\left[\left(\cosh Y - \sinh Y\frac{\partial\Sigma^0}{\partial z}\right) K_1(n\xi_m)I_0(n\xi_p)\right.$$
$$\left. - \frac{p_{\rm T}}{m_{\rm T}}\frac{\partial\Sigma^0}{\partial r} K_0(n\xi_m)I_1(n\xi_p)\right]. \qquad ({\rm G.21})$$

The boost-invariance in the longitudinal direction implies $T = T(r, \tau)$, $\mu = \mu(r, \tau)$ and $\alpha = \alpha(r, \tau)$. Let us further parametrize the freeze-out hyper-surface as $\Sigma^0(r, z) = t_{\rm f}(r, z) = \sqrt{\tau_{\rm f}^2(r) + z^2}$. Then, we have $\partial\Sigma^0/\partial z = z/\sqrt{\tau_{\rm f}^2 + z^2} = \tanh Y$ and $\partial\Sigma^0/\partial r = (d\tau_{\rm f}/dr)/\cosh Y$. Knowing $dz = \tau\cosh Y dY$, we may integrate over z to obtain

$$\frac{dN}{m_{\rm T} dm_{\rm T}} = \frac{m_{\rm T}}{\pi} 2Y_{\rm max}\sum_{n=1}^{\infty}(\mp)^{n+1}\int_0^{R_{\rm f}} rdr\; \tau_{\rm f}(r)\,{\rm e}^{n(\mu/T)}$$
$$\times\left[K_1(n\xi_m)I_0(n\xi_p) - \frac{p_{\rm T}}{m_{\rm T}}\frac{d\tau_{\rm f}}{dr} K_0(n\xi_m)I_1(n\xi_p)\right]. \qquad ({\rm G.22})$$

If we further assume an instant freeze-out independent of r at $\tau_f = \text{const}$, the second term inside the bracket of Eq. (G.22) vanishes. Note that the volume of the cylindrical matter at the freeze-out, V_f, is given by the following integral:

$$V_f \equiv 2Y_{\text{max}} \cdot 2\pi \int_0^{R_f} r\ \tau_f(r)\ dr. \tag{G.23}$$

If we further assume that (i) $m_{\text{T}} \gg T$ (i.e. the dominance of the $n = 1$ term), (ii) an instant isotherm freeze-out takes place on an r-independent hyper-surface, $\Sigma^0(z) \equiv t_f(z) = \sqrt{\tau_f^2 + z^2}$, and (iii) T and α are r-independent, Eq. (G.22) reduces to a crude but a simple formula,

$$\frac{dN}{m_{\text{T}} dm_{\text{T}}} \sim \frac{V_f}{2\pi^2}\ m_{\text{T}}\ e^{\mu_f/T_f} \cdot K_1(\xi_m) I_0(\xi_p), \tag{G.24}$$

where T_f, μ_f and α_f are the temperature, chemical potential and the transverse rapidity at the freeze-out.

Finally, we recapitulate the integral representations of the modified Bessel functions of the first kind, $I_\nu(\xi)$, and of the second kind, $K_\nu(\xi)$:

$$I_\nu(\xi) = \frac{1}{2\pi} \int_0^{2\pi} d\phi\ \cos(\nu\phi)\ e^{\xi \cos\phi}, \tag{G.25}$$

$$K_\nu(\xi) = \frac{1}{2} \int_{-\infty}^{+\infty} dy\ \cosh(\nu y)\ e^{-\xi \cosh y}. \tag{G.26}$$

Appendix H
Tables of particles

Table H.1. *Electric charge, spin, parity and mass of elementary particles.*

Particle	Symbol	Charge	J^P	Mass (GeV)
Fermion, generation 1				
Up quark	u	$\frac{2}{3}$	$\frac{1}{2}^+$	$(1.5 \sim 4.5)\times 10^{-3}$
Down quark	d	$-\frac{1}{3}$	$\frac{1}{2}^+$	$(5 \sim 8.5)\times 10^{-3}$
Electron	e	-1	$\frac{1}{2}^+$	0.511×10^{-3}
Electron neutrino	ν_e	0	$\frac{1}{2}^+$	$\leq 3 \times 10^{-9}$
Fermion, generation 2				
Charm quark	c	$\frac{2}{3}$	$\frac{1}{2}^+$	$1.0 \sim 1.4$
Strange quark	s	$-\frac{1}{3}$	$\frac{1}{2}^+$	$0.08 \sim 0.155$
Muon	μ	-1	$\frac{1}{2}^+$	0.106
Muon neutrino	ν_μ	0	$\frac{1}{2}^+$	$\leq 0.19 \times 10^{-3}$
Fermion, generation 3				
Top quark	t	$\frac{2}{3}$	$\frac{1}{2}^+$	174.3 ± 5.1
Bottom quark	b	$-\frac{1}{3}$	$\frac{1}{2}^+$	$4.0 \sim 4.5$
Tau	τ	-1	$\frac{1}{2}^+$	1.78
Tau neutrino	ν_τ	0	$\frac{1}{2}^+$	$\leq 18.2 \times 10^{-3}$
Gauge bosons				
Photon	γ	0	1^-	0
W boson	$W^\pm$	± 1	1^-	80.4
Z boson	Z^0	0	1^-	91.2
Gluon	g	0	1^-	0

A more extensive and updated listing may be found in the most recent issue of *The Review of Particle Physics* (Eidelman *et al.*, 2004) via the internet: http://pdg.lbl.gov/

Table H.2. *Properties of selected baryons and mesons.*

Particles	Mass (MeV)	J^P	Valence quarks
p	938.3	$\frac{1}{2}^+$	uud
n	939.6	$\frac{1}{2}^+$	udd
Λ	1115.6	$\frac{1}{2}^+$	uds
Σ^+	1189.4	$\frac{1}{2}^+$	uus
Σ^0	1192.6	$\frac{1}{2}^+$	uds
Σ^-	1197.4	$\frac{1}{2}^+$	dds
Ξ^0	1314.8	$\frac{1}{2}^+$	uss
Ξ^-	1321.3	$\frac{1}{2}^+$	dss
Δ^{++}	1232	$\frac{3}{2}^+$	uuu
Δ^+	1232	$\frac{3}{2}^+$	uud
Δ^0	1232	$\frac{3}{2}^+$	udd
Δ^-	1232	$\frac{3}{2}^+$	ddd
Σ^+	1382.8	$\frac{3}{2}^+$	uus
Σ^0	1383.7	$\frac{3}{2}^+$	uds
Σ^-	1387.2	$\frac{3}{2}^+$	dds
Ξ^0	1531.8	$\frac{3}{2}^+$	uss
Ξ^-	1535.0	$\frac{3}{2}^+$	dss
Ω^-	1672.5	$\frac{3}{2}^+$	sss
π^0	135.0	0^-	$(u\bar{u}-d\bar{d})/\sqrt{2}$
π^+, π^-	139.6	0^-	$u\bar{d}$, $d\bar{u}$
K^+, K^-	493.7	0^-	$u\bar{s}$, $s\bar{u}$
$K^0, \overline{K}^0$	497.7	0^-	$d\bar{s}$, $s\bar{d}$
η	547.3	0^-	$(u\bar{u}+d\bar{d})/\sqrt{2}$
η'	957.8	0^-	$s\bar{s}$
ρ^+, ρ^-	767	1^-	$u\bar{d}$, $d\bar{u}$
ρ^0	769	1^-	$(u\bar{u}-d\bar{d})/\sqrt{2}$
K^{*+}, K^{*-}	891.7	1^-	$u\bar{s}$, $s\bar{u}$
$K^{*0}, \overline{K}^{*0}$	896.1	1^-	$d\bar{s}$, $s\bar{d}$
ω	782.6	1^-	$(u\bar{u}+d\bar{d})/\sqrt{2}$
ϕ	1019.5	1^-	$s\bar{s}$
D^+, D^-	1869.3	0^-	$c\bar{d}$, $d\bar{c}$
$D^0, \overline{D}^0$	1864.5	0^-	$c\bar{u}$, $u\bar{c}$
D_s^+, D_s^-	1968.5	0^-	$c\bar{s}$, $s\bar{c}$
B^+, B^-	5279.0	0^-	$u\bar{b}$, $b\bar{u}$
$B^0, \overline{B}^0$	5279.4	0^-	$d\bar{b}$, $b\bar{d}$
η_c	2979.7	0^-	$c\bar{c}$
J/ψ	3096.9	1^-	$c\bar{c}$
ψ'	3686.0	1^-	$c\bar{c}$
Υ	9460.3	1^-	$b\bar{b}$
Υ'	10 023.3	1^-	$b\bar{b}$

References

Codes, for example hep-ph/0310274 in Accardi *et al.* (2003), refer to the article in the e-print archive, http://www.arxiv.org/.

Abbott, T. *et al.* (1990). E802 Collaboration. *Nucl. Instrum. and Meth.* **A290**, 41.
Abreu, M. C. *et al.* (1996). NA50 Collaboration. *Nucl. Phys.* **A610**, 404c.
(1999). NA50 Collaboration. *Phys. Lett.* **B450**, 456.
(2000). NA50 Collaboration. *Phys. Lett.* **B477**, 28.
Abrikosov, A. A., Gor'kov, L. P. and Dzyaloshinskii, I. E. (1959). *Sov. Phys. JETP* **9**, 636.
Accardi, A. *et al.* (2003). hep-ph/0310274.
Ackermann, K. H. *et al.* (2001). STAR Collaboration. *Phys. Lett.* **86**, 402.
(2003). STAR Collaboration. *Nucl. Instrum. and Meth.* **A499**, 624.
Adamczyk, M. *et al.* (2003). BRAHMS collaboration. *Nucl. Instrum. and Meth.* **A499**, 437.
Adams, J. *et al.* (2003a). STAR Collaboration. *Phys. Rev. Lett.* **91**, 72304
(2003b). STAR Collaboration. *Phys. Rev. Lett.* **91**, 172302.
(2003c). STAR Collaboration. *Phys. Rev. Lett.* **90**, 032301.
(2003d). STAR Collaboration. *Phys. Rev. Lett.* **93**, 012301.
(2004a). STAR Collaboration. *Phys. Rev. Lett.* **92**, 182301.
(2004b). STAR Collaboration. *Phys. Rev. Lett.* **92**, 052302.
(2004c). STAR Collaboration. *Phys. Rev. Lett.* **92**, 062301.
(2004d). STAR Collaboration. *Phys. Rev. Lett.* **93**, 012301.
(2005). STAR Collaboration. *Nucl. Phys.* **A757**, 102.
Adcox, K. *et al.* (2001). PHENIX Collaboration. *Phys. Rev. Lett.* **86**, 3500.
(2002). PHENIX Collaboration. *Phys. Rev. Lett.* **89**, 212301.
(2003). PHENIX Collaboration. *Nucl. Instrum. and Meth.* **A499**, 469.
(2005). PHENIX Collaboration. *Nucl. Phys.* **A757**, 184.
Adler, C. *et al.* (2001a). STAR Collaboration. *Phys. Rev. Lett.* **87**, 182301.
(2001b). STAR Collaboration. *Phys. Rev. Lett.* **87**, 082301.
(2003). STAR Collaboration, *Phys. Rev. Lett.* **90**, 082302.
Adler, S. S. *et al.* (2003a). PHENIX Collaboration. *Phys. Rev. Lett.* **91**, 072301.
(2003b). PHENIX Collaboration. *Phys. Rev. Lett.* **91**, 072303.
(2003c). PHENIX Collaboration. *Phys. Rev. Lett.* **91**, 172301.
(2003d). PHENIX Collaboration. *Phys. Rev. Lett.* **91**, 241803.
(2003e). PHENIX Collaboration. *Phys. Rev. Lett.* **91**, 182301.
(2004a). PHENIX Collaboration. *Phys. Rev.* **C69**, 034909.
(2004b). PHENIX Collaboration. *Phys. Rev. Lett.* **93**, 152302.
Afanasiev, S. *et al.* (1999). NA49 Collaboration. *Nucl. Instrum. and Meth.* **A430**, 210.
Agakichiev, G. *et al.* (1998). CERES/NA45/TAPS Collaboration. *Eur. Phys.* J. **C4**, 231.
(1999). CERES Collaboration. *Nucl. Phys.* **A661**, 23c.
Aggarwal, M. M. *et al.* (1999). WA98 Collaboration. *Phys. Rev. Lett.* **83**, 926.
(2000). WA98 Collaboration. *Phys. Rev. Lett.* **85**, 3595.
Aizawa, M. *et al.* (2003). PHENIX Collaboration. *Nucl. Instrum. and Meth.* **A499**, 508.
Akikawa, H. *et al.* (2003). PHENIX Collaboration. *Nucl. Instrum. and Meth.* **A499**, 537.
Alam, J., Sinha, B. and Raha, S. (1996). *Phys. Rept.* **273**, 243.
Alber, T. *et al.* (1995). NA35 and NA49 Collaboration. *Nucl. Phys.* **A590**, 453c.

Albrecht, R. *et al.* (1996). WA80 Collaboration. *Phys. Rev. Lett.* **76**, 3506.
Alexander, G. (2003). *Rept. Prog. Phys.* **66**, 481.
Alford, M. G., Rajagopal, K. and Wilczek, F. (1998). *Phys. Lett.* **B422**, 247.
(2001). *Ann. Rev. Nucl. Part. Sci.* **51**, 131.
Alpher, R. A., Bethe, H. A. and Gamow, G. (1948). *Phys. Rev.* **73**, 803.
Alt, C. *et al.* (2003). NA49 Collaboration. *Phys. Rev.* **C68**, 034903.
Ambjorn, J. and Hughes, R. J. (1983). *Ann. Phys.* **145**, 340.
Amit, D. J. (1984). *Field Theory, the Renormalization Group and Critical Phenomena*. Singapore: World Scientific.
Andersen, E. *et al.* (1998). WA97 Collaboration. *Phys. Lett.* **B433**, 209.
Anderson, M. *et al.* (2003). *Nucl. Instrum. and Meth.* **A499**, 659.
Antinori, F. *et al.* (2000). WA97 Collaboration. *Eur. Phys. J.* **C14**, 633.
Aoki, S., *et al.* (2000). CP-PACS Collaboration. *Phys. Rev. Lett.* **84**, 238.
(2003). CP-PACS Collabration. **D67**, 034503.
Appelshauser, H. *et al.* (1998). NA49 Collaboration. *Eur. Phys. J.* **C2**, 661.
Arnold, P. and Zhai, C. (1995). *Phys. Rev.* **D51**, 1906.
Arnold, P., Moore, G.D. and Yaffe, L.G. (2003). *JHEP* **0305**, 051.
Arsene, I. *et al.* (2003). BRAHMS Collaboration. *Phys. Rev. Lett.* **91** 072305.
(2005). BRAHMS Collaboration. *Nucl. Phys.* **A757**, 1.
Asakawa, A. and Hatsuda, T. (1997). *Phys. Rev.* **D55**, 4488.
Asakawa, M. and Yazaki, K. (1989). *Nucl. Phys.* **A504**, 668.
Asakawa, M., Huang, Z. and Wang, X-N. (1995). *Phys. Rev. Lett.* **74**, 3126.
Asakawa, M., Heinz, U. W. and Muller, B. (2000). *Phys. Rev. Lett.* **85**, 2072.
Asakawa, M., Hatsuda, T. and Nakahara, Y. (2001). *Prog. Part. Nucl. Phys.* **46**, 459.
Back, B. B. *et al.* (2002). PHOBOS Collaboration. *Phys. Rev. Lett.* **88**, 022302.
(2003a). PHOBOS Collaboration. *Nucl. Instrum. and Meth.* **A499**, 603.
(2003b). PHOBOS Collaboration. *Phys. Rev. Lett.* **91**, 052303.
(2003c). PHOBOS Collaboration, nucl-ex/0301017.
(2003d). PHOBOS Collaboration. *Phys. Rev. Lett.* **91**, 072302.
(2005a). PHOBOS Collaboration, *Phys. Rev. Lett.* **94**, 122303.
(2005b). PHOBOS Collaboration. *Nucl. Phys.* **A757**, 28.
Baier, R. and Niégawa, A. (1994). *Phys. Rev.* **D49**, 4107.
Baier, R., Schiff, D. and Zhaharov, B. G. (2000). *Ann. Rev. Nucl. Part. Sci.* **50**, 37.
Bailin, D. and Love, A. (1984). *Phys. Rept.* **107**, 325.
Baker, G. A. Jr. and Graves-Morris, P. (1996). *Padé Approximants*, 2nd edn. Cambridge: Cambridge University Press.
Baker, M. D. *et al.* (2002). PHOBOS Collaboration. *Phys. Rev. Lett.* **88**, 022302.
(2003). PHOBOS Collaboration *Nucl. Phys.* **A715**, 65c.
Bali, G. S. (2001). *Phys. Rept.* **343**, 1.
Baluni, V. (1978). *Phys. Rev.* **D17**, 2092.
Banks, T. and Ukawa, A. (1983). *Nucl. Phys.* **B225**, 145.
Bardeen, J., Cooper, L. N. and Schrieffer, J. R. (1957). *Phys. Rev.* **108**, 1175.
Barrette, J. *et al.* (1997). E877 Collaboration. *Phys. Rev.* **C55**, 1420.
Barkte, J. *et al.* (1977). *Nucl. Phys.* B **120**, 14.
Baym, G. (1979). *Physica* **96A**, 131.
(1998). *Acta Phys. Polon.* **B29**, 1839.
Baym, G. and Chin, S. A. (1976). *Phys. Lett.* **62B**, 241.
Baym, G. and Mermin, J. (1961). *J. Math. Phys.* **2**, 232.
Baym, G., Friman, B. L., Blaizot, J.-P., Soyeur, M. and Czyż, W. (1983). *Nucl. Phys.* **A407**, 541.
Baym, G., Monien, H., Pethick, C. J. and Ravenhall, D. G. (1990). *Phys. Rev. Lett.* **64**, 1867.
Bearden, I.G. *et al.*, (1997). NA44 Collaboration. *Phys. Rev. Lett.* **78**, 2080.
(2004). BRAHMS Collabration. *Phys. Rev. Lett.* **93**, 102301.
Becchi, C., Rouet, A. and Stora, R. (1976). *Ann. Phys.* **98**, 287.
Belensky, S. Z. and Landau, L. D. (1955). *Ups. Fiz. Nauk.* **56**, 309; reprinted (1965) in *Collected Papers of L.D. Landau,* ed. D. T. ter Haar. New York: Gordon & Breach, p. 665.
Benecke, J., Chou, T. T., Yang, C-N. and Yen, E. (1969). *Phys. Rev.* **188**, 2159.
Bennett, C. L. *et al.* (2003). WMAP Collaboration. *Astrophys. J. Suppl.* **148**, 1.
Bertch, G. (1989). *Nucl. Phys.* **A498**, 173c.
Bertlemann, R. (1996). *Anomalies in Quantum Field Theory*. Oxford: Oxford University Press.
Bjorken, J. D. (1976). In *Current Induced Reactions*, Lecture Notes in Physics vol. 56. New York: Springer, p. 93.
(1982). http://library.fnal.gov/archive/test-preprint/fermilab-pub-82-059-t.shtml
(1983). Phys. Rev. **D27**, 140.
(1997). *Acta Phys. Polon.* **B28**, 2773.

Blaizot, J.-P. and Iancu, E. (2002). *Phys. Rept.* **359**, 355.
Blaizot, J.-P. and Ollitrault, J.-Y. (1990). In *Quark-Gluon Plasma*, ed. R.C. Hwa. Singapore: World Scientific, p. 393.
Bodmer, A. R. (1971). *Phys. Rev.* **D4**, 1601.
Bombaci, I. (2001). In *Physics of Neutron Star Interiors*, Lecture Notes in Physics vol. 578. New York: Springer.
Borgs, C. and Seiler, E. (1983a). *Nucl. Phys.* **B215**, 125
(1983b). *Commun. Math. Phys.* **91**, 329.
Box, G. E. P. and Tiao, G. C. (1992). *Bayesian Inference in Statistical Analysis*. New York: John Wiley and Sons.
Braaten, E. and Nieto, A. (1996a). *Phys. Rev.* **D53**, 3421.
(1996b). *Phys. Rev. Lett.* **76**, 1417.
Braaten, E. and Pisarski, R.D. (1990a). *Nucl. Phys.* **B337**, 569.
(1990b). *Phys. Rev.* **D42**, 2156.
(1992a). *Phys. Rev.* **D46**, 1829.
(1992b). *Phys. Rev.* **D45**, 1827.
Braun-Munzinger, P. and Stachel, J. (2001). *Nucl. Phys.* **A690**, 119.
Braun-Munzinger, P., Specht, H. J., Stock, R. and Stocker, H., eds. (1996). Quark Matter '96 *Nucl. Phys.* **A610**, 1c.
Braun-Munzinger, P., Heppe, I. and Stachel, J. (1999). *Phys. Lett.* **B465**, 15.
Braun-Munzinger, P., Magestro, D., Redlich, K. and Stachel, J. (2001). *Phys. Lett.* **B518**, 41.
Braun-Munzinger, P., Redlich, K, and Stachel, J. (2004). In *Quark-Gluon Plasma 3*, eds. R. C. Hwa and X. N. Wang. Singapore: World Scientific, p. 491.
Brown, L. S. (1995). *Quantum Field Theory*. Cambridge: Cambridge University Press.
Brown, G. E. and Rho, M. (1991). *Phys. Rev. Lett.* **66**, 2720.
(1996). *Phys. Rept.* **269**, 333.
Brown, L. S. and Weisberger, W. I. (1979). *Phys. Rev.* **D20**, 3239.
Burnett, T. H. *et al.* (1983). *Phys. Rev. Lett.* **50**, 2062.
Busza, W. *et al.* (1975). *Phys. Rev. Lett.* **34**, 836.
Cahn, R. W., Haasen, P. and Kramer, E. J., eds. (1991). *Phase Transformations in Material Science and Technology*, vol. 5. New York: Weinheim.
Casher, A., Neuberger, H. and Nussinov, S. (1979). *Phys. Rev.* **D20**, 179.
Celik, T., Karsch, F. and Satz, H. (1980). *Phys. Lett.* **B97**, 128.
Chandrasekhar, S. (1931). *Astrophys. J.* **74**, 81.
(1943). *Rev. Mod. Phys.* **15**, 1.
Chapman, S., Nix, J. R. and Heinz, U. (1995). *Phys. Rev.* **C52**, 2694.
Chin, S. A. and Kerman, A. (1979). *Phys. Rev. Lett.* **43**, 1292.
Chodos, A., Jaffe, R. L., Johnson, K., Thorn, C. B. and Weisskopf, V. F. (1974). *Phys. Rev.* **D9**, 3471.
Cleymans, J. and Satz, H. (1993) *Z. Phys.* **C57**, 135.
Cohen, E. G. D. and Berlin, T. H. (1960). *Physica* **26**, 95.
Colangelo, P. and Khodjamirian, A. (2001). In *At the Frontier of Particle Physics/Handbook of QCD*, vol. 3, ed. M. Shifman. Singapore: World Scientific, p. 1495.
Coleman, S. (1973). *Commun. Math. Phys.* **31**, 259.
Coleman, S. and Gross, D. J. (1973). *Phys. Rev. Lett.* **31**, 851.
Collins, J. C. and Perry, M. J. (1975). *Phys. Rev. Lett.* **34**, 1353.
Collins, J. C., Duncan, A. and Joglekar, S. D. (1977). *Phys. Rev.* **D16**, 438.
Combridge, L., Kripfganz, J. and Ranft, J. (1977). *Phys. Lett.* **B70**, 234.
Cooper, F. and Frye, G. (1974). *Phys. Rev.* **D10**, 186.
Corless, R. M. *et al.* (1996). *Adv. Comp. Math.* **5**, 329.
Creutz, M. (1985). *Quarks, Gluons and Lattices*. Cambridge: Cambridge University Press.
Cronin, J. W. *et al.* (1975). *Phys. Rev.* **D11**, 3105.
Csernai, L. P. (1994). *Introduction to Relativistic Heavy Ion Collisions*. New York: John Wiley & Sons.
De Groot, S. R., Leeuwen, W. A. van and Weert, Ch.G. van (1980). *Relativistic Kinetic Theory*. Amsterdam: North-Holland.
De Rujula, A., Georgi, H. and Glashow, S. L. (1975). *Phys. Rev.* **D12**, 147.
Debye, P. and Hückel, E. (1923). *Z. Physik* **24**, 185.
DeGrand, T., Jaffe, R. L., Johnson, K. and Kiskis, J. E. (1975). *Phys. Rev.* **D12**, 2060.
Dey, M., Eletsky, V. L. and Ioffe, B. L. (1990). *Phys. Lett.* **B252**, 620.
Di Pierro, M. (2000). *From Monte Carlo Integration to Lattice Quantum Chromodynamics: An Introduction*. Lectures at the GSA Summer School on Physics on the Frontier and in the Future, Batavia, Illinois; hep-lat/0009001.
Drell, S. D. and Yan, T.-M. (1970). *Phys. Rev. Lett.* **25**, 316.
(1971). *Ann. Phys.* **66**, 578.

Ei, S-I., Fujii, K. and Kunihiro, T. (2000). *Ann. Phys.* **280**, 236.
Eichten, E. and Feinberg, F. (1981). *Phys. Rev.* **D23** 2724.
Eidelman, S. *et al.* (2004). Particle Data Group. *Phys. Lett.* **B592**, 1.
Ellis, R. K., Stirling, W. J. and Webber, B. R. (1996). *QCD and Collider Physics*. Cambridge: Cambridge University Press.
Elze, H. T. and Heinz, U. (1989). *Phys. Rept.* **183**, 81.
Engels, J., Karsch, F., Satz, H. and Montvay, I. (1981). *Phys. Lett.* **B101**, 89.
Eskola, K. J., Kajantie, K. and Lindfors, J. (1989). *Nucl. Phys.* **B323**, 37.
Ezawa, H., Tomozawa, Y. and Umezawa, H. (1957). *Nuovo Cimento* **5**, 810.
Faddeev, L. D. and Popov, V. N. (1967). *Phys. Lett.* **25B**, 29.
Farhi, E. and Jaffe, R. L. (1984). *Phys. Rev.* **D30**, 2379.
Feinberg, E. L. (1976). *Nuovo Cim.* **A34**, 391.
Fermi, E. (1950). *Prog. Theor. Phys.* **5**, 570.
Fetter, A. and Walecka, J. (1971). *Quantum Theory of Many-Particle Systems*. New York: McGraw-Hill.
Feynman, R. and Hibbs, A. R. (1965). *Quantum Mechanics and Path Integrals*. New York: McGraw-Hill.
Fixsen, D. J. *et al.* (1996). *Astrophys. J.* **473**, 576.
Fowler, R. H. (1926). *Mon. Not. Roy. Astro. Soc.* **87**, 114.
Fradkin, E. S. (1959). *Sov. Phys. JETP* **9**, 912.
Fraga, E. S., Pisarski, R. D. and Schaffner-Bielich, J. (2001). *Phys. Rev.* **D63**, 121702.
Frauenfelder, H. and Henley, E. M. (1991). *Subatomic Physics*. New York: Prentice.
Freedman, B. A. and McLerran, L. D. (1977). *Phys. Rev.* **D16**, 1169.
Freedman, W. L. *et al.* (2001). *Astrophys. J.* **553**, 47.
Frenkel, J. and Taylor, J. C. (1992). *Nucl. Phys.* **B374**, 156.
Friedmann, A. (1922). *Z. Phys.* **10**, 377.
Fries, J. B., Müller, B., Nonaka, C. and Bass, S. A. (2003). *Phys. Rev.* **C68**, 044902.
Fujikawa, K. (1980a). *Phys. Rev.* **D21**, 2848.
(1980b). *Phys. Rev.* **D22**, 1499.
Fukugita, M., Okawa, M. and Ukawa, A. (1989). *Phys. Rev. Lett.* **63**, 1768.
(1990). *Nucl. Phys.* **B337**, 181.
Gale, C. and Haglin, K. L. (2004). In *Quark Gluon Plasma 3*, eds. R. C. Hwa and X. N. Wang. Singapore: World Scientific, p. 364.
Gasser, J. and Leutwyler, H. (1984). *Ann. Phys.* **158**, 142.
(1985). *Nucl. Phys.* **B250**, 465.
Geiger, K. (1995). *Phys. Rept.* **258**, 237
Gell-Mann, M. and Brueckner, K. A. (1957). *Phys. Rev.* **106**, 364.
Gerber, P. and Leutwyler, H. (1989). *Nucl. Phys.* **B321**, 387.
Gilmore, R. (1994). *Lie Groups, Lie Algebras and Some of Their Applications*. New York: RE Krieger Publishing.
Ginsparg, P. H. and Wilson, K. G. (1982). *Phys. Rev.* **D25**, 2649.
Ginzburg, V. L. (1961). *Sov. Phys. Solid State* **2**, 1824.
Glauber, R. J. (1959). *Lectures on Theoretical Physics*, vol. 1. New York: Interscience, p. 315.
Glendenning, N. K. (1992). *Phys. Rev.* **D46**, 1274.
(2000). *Compact Stars: Nuclear Physics, Particle Physics and General Relativity*, 2nd edn. New York: Springer-Verlag.
Glendenning, N. K. and Matsui, T. (1983). *Phys. Rev.* **D28**, 2890.
Goity, J. L. and Leutwyler, H. (1989). *Phys. Lett.* **B228**, 517.
Goldenfeld, N. (1992) *Lectures on Phase Transitions and the Renormalization Group*. Frontiers in Physics 85. New York: Addison-Wesley.
Goldhaber, G., Goldhaber, S., Lee, W-Y. and Pais, A. (1960). *Phys. Rev.* **120**, 300.
Greco, V., Ko, C. M. and Levai, P. (2003). *Phys. Rev. Lett.* **90**, 202302.
Gribov, L. V., Levin, E. M. and Ryskin, M. G. (1983). *Phys. Rept.* **100**, 1.
Gross, D. J. (1981). In *Methods in Field Theory*, eds. R. Balian and J. Zinn-Justin. Singapore: World Scientific.
Gross, D. J. and Wilczek, F. (1973). *Phys. Rev. Lett.* **30**, 1343.
Guettler, K. *et al.* (1976). *Phys. Lett.* **64**, 111.
Gupta, R. (1999). In *Probing the Standard Model of Particle Interactions*. vol. 1, eds. R. Gupta, A. Morel, E. de Rafael and F. David. Amsterdam: Elsevier.
Gutbrod, H., Aichelin, J. and Werner, K., eds. (2003). Quark Matter '02, *Nucl. Phys.* **A715**, lc.
Guth, A. H. (1980). *Phys. Rev.* **D21**, 2291.
(1981). *Phys. Rev.* **D23**, 347.
Gyulassy, M. and Iwazaki, A. (1985). *Phys. Lett.* **B165**, 157.
Gyulassy, M. and Matsui, T. (1984). *Phys. Rev.* **D29**, 419.
Gyulassy, M., Vitev, I., Wang, X-N. and Zhang, B-W. (2004). In *Quark-Gluon Plasma 3*, eds. R. C. Hwa and X. N. Wang. Singapore: World Scientific, p. 123.

Hagedorn, R. (1985). *Lecture Notes Phys.* **221**, 53.
Hahn, H. *et al.* (2003). *Nucl. Instrum. and Meth.* **A499**, 245.
Halasz, M. A., Jackson, A. D., Shrock, R. E., Stephanov, M. A. and Verbaarschot, J. J. M. (1998). *Phys. Rev.* **D58**, 096007.
Hallman, T. J., Kharzee, D. E., Mitchell, J. T. and Ullrich, T., eds. (2002). Quark Matter '01. *Nucl. Phys.* **A698**, lc.
Halperin, B. I., Lubensky, T. C. and Ma, S.-K. (1974). *Phys. Rev. Lett.* **32**, 292.
Hanbury-Brown, R. and Twiss, R. Q. (1956). *Nature*, **178**, 1046.
Harada, M. and Yamawaki, K. (2003). *Phys. Rept.* **381**, 1.
Harrison, B. K., Thorne, K. S., Wakano, M. and Wheeler, J. A. (1965). *Gravitation Theory and Gravitational Collapse*. Chicago: University of Chicago Press.
Hasenbusch, M. (2001). *J. Phys.* **A34**, 8221.
Hasenfratz, P. and Karsch, F. (1983). *Phys. Lett.* **B125**, 308.
Hashimoto, T., Hirose, K., Kanki, T. and Miyamura, O. (1986). *Phys. Rev. Lett.* **57**, 2123.
Hashimoto, T., Nakamura, A. and Stamatescu, I. O. (1993). *Nucl. Phys.* **B400**, 267.
Hatsuda, T. (1997). *Phys. Rev.* **D56**, 8111.
Hatsuda, T. and Asakawa, M. (2004). *Phys. Rev. Lett.* **92**, 012001.
Hatsuda, T. and Kunihiro, T. (1985). *Phys. Rev. Lett.* **55** 158.
(1994). *Phys. Rept.* **247**, 221.
(2001). nucl-th/0112027.
Hatsuda, T. and Lee, S. H. (1992). *Phys. Rev.* **C46**, 34.
Hatsuda, T., Koike, Y. and Lee, S. H. (1993). *Nucl. Phys.* **B394**, 221.
Hatsuda, T., Miake, Y., Nagamiya, S. and Yagi, K., eds. (1998). Quark Matter '97, *Nucl. Phys.* **A638**, lc.
Hatta, Y. and Ikeda, T. (2003). *Phys. Rev.* **D67**, 014028.
Hayashi, C. (1950). *Prog. Theor. Phys.* **5**, 224.
Hecke, H. van, Sorge, H. and Xu, N. (1998). *Phys. Rev. Lett.* **81**, 5764.
Heintz, U., Tomášik, B., Wiedemann, U. A. and Wu, Y.-F. (1996). *Phys. Lett.* **B382**, 181.
Heinz, U., Lee, K. S. and Schnedermann, E. (1990). In *Quark-Gluon Plasma*, ed. R. C. Hwa. Singapore: World Scientific, p. 471.
Heiselberg, H. and Hjorth-Jensen, M. (2000). *Phys. Rept.* **328**, 237.
Heiselberg, H. and Pandharipande, V. (2000). *Ann. Rev. Nucl. Part. Sci.* **50**, 481.
Hewish, A., Bell, S. J., Pilkington, J. D. H., Scott, P. F. and Collins, R. A. (1968). *Nature* **217**, 709.
Hirano, T. (2004). *J. Phys.* **G30**, S845.
Hirano, T. and Nara, Y. (2004). *J. Phys.* **G30**, S1139.
Hirata, H. *et al.* (1987). KAMIOKANDE-II Collaboration. *Phys. Rev. Lett.* **58**, 1490.
Hoffberg, M., Glassgold, A. E., Richardson, R. W. and Ruderman, M. (1970). *Phys. Rev. Lett.* **24**, 775.
(1982). *Phys. Lett.* **118B**, 138.
Hove, L. van (1982). *Phys. Lett.* **118B**, 138.
Huang, K. (1987). *Statistical Mechanics*. New York: Wiley.
(1992). *Quarks, Leptons & Gauge Fields*, 2nd edn. Singapore: World Scientific, chap. 12.
Hubble, E. (1929). *Proc. Natl Acad. Sci. (USA)* **15**, 168.
Hughes, R. J. (1981). *Nucl. Phys.* **B186**, 376.
Hulse, R. A. and Taylor, J. H., (1975). *Astrophys. J.* **L51**. 195.
Hung, C. M. and Shuryak, E. V. (1995). *Phys. Rev. Lett.* **75**, 4003.
Huovinen, P., Kolb, P. F., Heinz, U., Ruuskanen, P. V. and Voloshin, S. A. (2001). *Phys. Lett.* **B503**, 58.
Hwa, R. C. and Yang, C. B. (2003). *Phys. Rev.* **C67**, 034902.
Iacobson, H. H. and Amit, A. D. (1981). *Ann. Phys.* **133**, 57.
Iancu, E. and Venugopalan, R. (2004). In *Quark-Gluon Plasma 3*, eds. R. C. Hwa and X. N. Wang. Singapore: World Scientific, p. 249.
Iofa, M. Z. and Tyutin, I. V. (1976). *Theor. Math. Phys.* **27**, 316.
Isichenko, M. B. (1992). *Rev. Mod. Phys.* **64**, 961.
Israel, W. and Stewart, J. M. (1979). *Ann. Phys.* **118**, 341.
Itoh, N. (1970). *Prog. Theor. Phys.* **44**, 291.
Iwasaki, M. and Iwado, T. (1995). *Phys. Lett.* **B350**, 163.
Jackson, J. D. (1999). *Classical Electrodynamics*, 3rd edn. New York: John Wiley & Sons.
Jeon, S. and Koch, V. (2000). *Phys. Rev. Lett.* **85**, 2076.
(2004). In *Quark-Gluon Plasma 3*, eds. R. C. Hwa and X. N. Wang. Singapore: World Scientific, p. 430.
Kadanoff, L. and Baym, G. (1962). *Quantum Statistical Mechanics*. New York: Benjamin.
Kahrzeev, D., Lourenco, C., Nardi, M. and Staz, H. (1997). *Z. Phys.* **C74**, 307.
Kajantie, K., Landshoff, P. V., Lindfors, J. (1987). *Phys. Rev. Lett.* **59**, 2527.
Kajantie, K., Laine, M., Rummukainen, K. and Schröder, Y. (2003). *Phys. Rev.* **D67**, 105 008.
Kaplan, D. (1995). *Lectures Given at 7th Summer School in Nuclear Physics Symmetries*, Seattle, USA; nucl-th/9506035.

Kaplan, D. B. and Nelson, A. E. (1986). *Phys. Lett.* **B175**, 57.
Kapusta, J. I. (1979). *Nucl. Phys.* **B148**, 461.
(1989). *Finite-Temperature Field Theory*. Cambridge: Cambridge University Press.
(2001). In *Phase Transitions in the Early Universe: Theory and Observations*, eds. H. J. de Vega, I. Khalatnikov and N. Sanchez. New York: Kluwer.
Kapusta, J., Müller, B. and Rafelski, J. (2003). *Quark-Gluon Plasma: Theoretical Foundations: An Annotated Reprint Collection*. Amsterdam: Elsevier Science.
Karsch, F. (2002). *Lecture Notes Phys.* **583**, 209.
Karsch, F., Laermann, E. and Peikert, A. (2001). *Nucl. Phys.* **B605**, 579.
Karsten, L. H. (1981). *Phys. Lett.* **104B**, 315.
Kazanas, D. (1980). *Astrophys. J. Lett.* **241**, L59.
Kharzeev, D., Lourenco, C., Nardi, M. and Satz, H. (1997). *Z. Phys.* **C74**, 307.
Kharzeev, D. and Satz, H. (1995). *In Quark-Gluon Plasma Z*, ed. R. C. Hwa. Singapore: World Scientific, p. 395.
Klein, S. (1999). *Rev. Mod. Phys.* **71**, 1501.
Klevansky, S. P. (1992). *Rev. Mod. Phys.* **64**, 649.
Klimov, V. V. (1981). *Sov. J. Nucl. Phys.* **33**, 934.
Kniehl, B. A., Kramer, G. and Pötter, B. (2001). *Nucl. Phys.* **B597**, 337.
Kobayashi, M. and Maskawa, T. (1970). *Prog. Theor. Phys.* **44**, 1422.
Kolb, E. W. and Turner, M. S. (1989). *Early Universe*, Frontiers in Physics, vol. 69. New York: Perseus Books.
Koshiba, M. (1992). *Phys. Rept.* **220**, 229.
Kraemmer, U. and Rebhan, A. (2004). *Rept. Prog. Phys.* **67**, 351.
Kronfeld, A. S. (2002). In *At the Frontiers of Particle Physics: Handbook of QCD*, vol. 4, ed. M. Shifman. Singapore: World Scientific.
Kugo, T. and Ojima, I. (1979). *Prog. Theor. Phys. Suppl.* **66**, 1.
Kunihiro, T., Muto, T., Takatsuka, T., Tamagaki, R. and Tatsumi, T. (1993). *Prog. Theor. Phys. Suppl.* **112**, 1.
Kuti, J., Polonyi, J. and Szlachanyi, K. (1981). *Phys. Lett.* **B98**, 199.
Laermann, E. and Philipsen, O. (2003). *Ann. Rev. Nucl. Part. Sci.* **53**, 163.
Landau, L. D. (1953). *Izv. Akad. Nauk SSSR* **17**, 51; reprinted (1965) in *Collective Papers of L. D.Landau*, ed. D. T. ter Haar New York: Gordon & Breach, p. 569.
Landau, L. D. and Lifshitz, E. M. (1980). *Statistical Physics*, 3rd edn. Oxford: Pergamon.
(1987). *Fluid Mechanics*, 2nd edn. Oxford: Pergamon.
(1988). *Classical Theory of Fields*, 6th edn. Oxford: Pergamon.
Landau, L. D. and Pomeranchuk, I. J. (1953). *Dokl. Akad. Nauk. SSSR* **92**, 92.
Landsman, N. P. and van Weert, G. W. (1987). *Phys. Rept.* **145**, 141.
Lattes, C. M. G., Fujimoto, Y. and Hasegawa, S. (1980). *Phys. Rept.* **65**, 151.
Lawrie, I. and Sarnach, S. (1984). In *Phase Transitions and Critical Phenomena*, vol. 9, eds. C. Domb and J. Lebowitz. New York: Academic Press, p. 1.
Le Bellac, M. (1996). *Thermal Field Theory*. Cambridge: Cambridge University Press.
Le Guillou, J. C. and Zinn-Justin, J. (1990). *Large-order Behaviour of Perturbation Theory. Current Physics-Sources and Comments,* vol. 7. Amsterdam: Elsevier Science.
Lee, C. H. (1996). *Phys. Rept.* **275**, 255.
Lee, T. D. (1975). *Rev. Mod. Phys.* **47**, 267.
(1998). In *T. D. Lee: Selected Papers, 1985–1996*, eds. H.-C. Ren and Y. Pang. New York: T&F STM, p. 583.
Lepage, G. P. (1990). In *From Actions to Answers (TASI '89)*, eds. T. DeGrand and D. Toussiant. Singapore: World Scientific, p. 483.
Letessier, J. and Rafelski, J. (2002). *Hadrons and Quark–Gluon Plasma*. Cambridge: Cambridge University Press.
Leutwyler, H. (2001a). *Nucl. Phys. Proc. Suppl.* **94**, 108.
(2001b). In *At the Frontier of Particle Physics/Handbook of QCD*, vol. 1, ed. M. Shifman. Singapore: World Scientific, p. 271.
Leutwyler, H. and Smilga, A. V. (1990). *Nucl. Phys.* **B342**, 302.
Linde, A. D. (1980). *Phys. Lett.* **B96**, 289.
(1990). *Particle Physics and Inflationary Cosmology*. New York: Harwood.
Lüscher, M. (2002). In *Theory and Experiment Heading for New Physics: Proceedings of the International School of Subnuclear Physics (Subnuclear Series, vol. 38)*, ed. A. Zichichi. Singapore: World Scientific.
McLerran, L. D. and Svetitsky, B. (1981a). *Phys. Lett.* **B98**, 195.
(1981b). *Phys. Rev.* **D24**, 450.
McLerran, L. D. and Venugopalan, R. (1994a). *Phys. Rev.* **D49**, 3352.
(1994b). *Phys. Rev.* **D50**, 2225.
Mather, J. C. *et al.* (1999). *Astrophys. J.* **512**, 511.

Madsen, J. (1999). *Lecture Notes Phys.* **516**, 162.
Matsubara, T. (1955). *Prog. Theor. Phys.* **14**, 351.
Matsui, T. (1987). *Nucl. Phys.* **A461**, 27c.
(1990). In *Proceedings of the Second Symposium on Nuclear Physics: Intermediate Energy Nuclear Physics*, ed. D. P. Min. Korea: Han Lim Won, p. 150.
Matsui, T. and Satz, H. (1986). *Phys. Lett.* **B178**, 416.
Mermin, N. D. and Wagner, H. (1966). *Phys. Rev. Lett.* **17**, 1133.
Metropolis, N., Rosenbluth, A. W., Rosenbluth., M. N., Teller, A. H. and Teller, E. (1953). *J. Chem. Phys.* **21**, 1087.
Migdal, A. B. (1956). *Phys. Rev.* **103**, 1811.
(1972). *Nucl. Phys.* **A210**, 421.
Mills, R. (1969). *Propagators for Many Particle Systems.* New York: Gordon and Breach.
Mohanty, B. and Serreau, J. (2005). hep-ph/054154.
Molnár, D. and Voloshin, S. A. (2003). *Phys. Rev. Lett.* **91**, 092301.
Montvay, I. and Munster, G. (1997). *Quantum Fields on a Lattice.* Cambridge: Cambridge University Press.
Moretti, M., Ohl, T. and Reuter, J. (2001). hep-ph/0102195.
Mueller, A. H. and Qiu, J-W. (1986). *Nucl. Phys.* **B268**, 427.
Müller, B. (2003). *Int. J. Mod. Phys.* E12, 165
Müller, I. (1967). *Z. Phys.* **198**, 329.
Muronga, A. (2002). *Acta Phys. Hung. Heavy Ion Phys.* **15**, 337; nucl-th/0105046.
Muroya, S., Nakamura, A., Nonaka, C. and Takaishi, T. (2003). *Prog. Theor. Phys.* **110**, 615.
Muta, T. (1998). *Foundation of Quantum Chromodynamics: An Introduction to Perturbative Methods in Gauge Theories*, 2nd edn. Singapore: World Scientific.
Nadkarni, S. (1986a). *Phys. Rev.* **D33**, 3738.
(1986b). *Phys. Rev.* **D34**, 3904.
Nagano, K. (2002). *J. Phys.* **G28**, 737.
Nakamura, A. (1984). *Phys. Lett.* **B149**, 391.
Nakamura, A. *et al.*, eds. (2004). *Prog. Theor. Phys. Suppl.* **153**, 1.
Nambu, Y. (1960). *Phys. Rev. Lett.* **4**, 380.
(1966). In *Preludes in Theoretical Physics, in Honor of Weisskopf, V. F.*, eds. A. de-Shalit, H. Feshbach and L. van Hove. Amsterdam: North-Holland, p. 133.
Nambu, Y. and Jona-Lasinio, G. (1961a). *Phys. Rev.* **122**, 345.
(1961b). *Phys. Rev.* **124**, 246.
Negele, J. W. and Orland, H. (1998). *Quantum Many-particle Systems.* New York: Harper Collins.
Neuberger, H. (2001). *Ann. Rev. Nucl. Part. Sci.* **51**, 23.
Nielsen, H. B. and Ninomiya, M. (1981a). *Nucl. Phys.* **B185**, 20; erratum *ibid.*, p. 541.
(1981b). *Nucl. Phys.* **B193**, 173.
Nielsen, N. K. (1977). *Nucl. Phys.* **B120**, 212.
(1981). *Am. J. Phys.* **49**, 1171.
Okamoto, M. *et al.* (1999). CP-PACS Collaboration. *Phys. Rev.* **D60**, 094510.
(2002). CP-PACS Collaboration. *Phys. Rev.* **D65**, 094508.
Ollitrault, J. Y. (1992). CP-PACS Collaboration. *Phys. Rev.* **D46**, 229.
(1993). *Phys. Rev.* **D48**, 1132.
Oppenheimer, J. R. and Volkoff, G. (1939). *Phys. Rev.* **55**, 374.
Osterwalder, K. and Seiler, E. (1978). *Ann. Phys.* **110**, 440.
Patel, A. (1984). *Nucl. Phys.* **B243**, 411.
Paterson, A. J. (1981). *Nucl. Phys.* **B190**, 188.
Peebles, P. J. E. (1993). *Principles of Physical Cosmology.* Princeton: Princeton University Press.
Pelissetto, A. and Vicari, E. (2002). *Phys. Rept.* **368**, 549.
Penzias, A. A. and Wilson, R. W. (1965). *Astrophys. J.* **142**, 419.
Peskin, M. E. and Schroeder, D. V. (1995). *An Introduction to Quantum Field Theory.* Reading, MA: Perseus Books.
Pethick, C. and Smith, H. (2001). *Bose-Einstein Condensation in Dilute Gases.* Cambridge: Cambridge University Press.
Pisarski, R. D. (1982). *Phys. Lett.* **B110**, 155.
Pisarski, R. D. and Wilczek, F. (1984). *Phys. Rev.* **D29**, 338.
Pokorski, S. (2000). Gauge Field Theories, 2nd edn. Cambridge: Cambridge University Press.
Polchinski, J. (1992). In *Recent Directions in Particle Theory: From Superstrings and Black Holes to the Standard Model (TASI'92)*, eds. J. Harvey and J. Polchinski. Singapore: World Scientific, p. 235.
Politzer, H. (1973). *Phys. Rev. Lett.* **30**, 1346.
Polyakov, A. M. (1978). *Phys. Lett.* **B72**, 477.
Poskanzer, A. M. *et al.* (1999). NA49 Collaboration. *Nucl. Phys.* **A661**, 341c.

Pratt, S. (1984). *Phys. Rev. Lett.* **53**, 1219.
(1986). *Phys. Rev.* **D33**, 1314.
Pukhov, A. *et al.* (1999). hep-ph/9908288.
Rafelski, J. and Müller, B. (1982). *Phys. Rev. Lett.* **48**, 1066.
Rajagopal, K. (1995). In *Quark-Gluon Plasma 2*, ed. R. C. Hwa. Singapore: World Scientific, p. 484.
(2001). In *At the Frontier of Particle Physics: Handbook of QCD*, ed. M. Shifman, Singapore: World Scientific.
Rajagopal, K. and Wilczek, F. (2001). In *At the Frontiers of Particle Physics: Handbook of QCD*, vol. 3, ed. M. Shifman. Singapore: World Scientific, p. 2061.
Rapp, R. and Wambach, J. (2000). *Adv. Nucl. Phys.* **25**, 1.
Rapp, R., Schafer, T., Shuryak, E. V. and Velkovsky, M. (1998). *Phys. Rev. Lett.* **81**, 53.
Reif, F. (1965). *Fundamentals of Statistical and Thermal Physics*. London: McGraw-Hill.
Reisz, T. (1989). *Nucl. Phys.* **B318**, 417.
Riccati, L., Masera, M. and Vercellin, E., eds. (1999). Quark Matter '99. *Nucl. Phys.* **A661**, lc.
Ring, P. and Schuck, P. (2000). *The Nuclear Many-body Problem*. Berlin: Springer-Verlag.
Rischke, D. H. (1996). *Nucl. Phys.* **A610**, 88c.
(1999). In *Hadrons in Dense Matter and Hadrosynthesis*, eds. J. Cleymans, H. B. Geyer and F. G. Scholtz. Lecture Notes in Physics vol. 516. Berlin: Springer, p. 21.
Rischke, D. H. and Gyulassy, M. (1996). *Nucl. Phys.* **A608**, 479.
Ritter, H. G. and Wang, X.-N., eds. (2004). Quark Matter '04. *J. Phys.* **G30**, 633.
Ruuskanen, P. V. (1987). *Acta Phys. Pol.* **B18**, 551.
Sato, K. (1981a). *Mon. Not. Roy. Astron. Soc.* **195**, 467.
(1981b). *Phys. Lett.* **B99**, 66.
Satz, H. (1992). *Nucl. Phys.* **A544**, 371.
Sawyer, R. F. and Scalapino, D. J. (1973). *Phys. Rev.* **D7**, 953.
Schenk, A. (1993). *Phys. Rev.* **D47**, 5138.
Schnedermann, E., Sollfrank, J. and Heinz, U. (1993). *Phys. Rev.* **C48**, 2462.
Schramm, D. N. and Turner, M. S. (1998). *Rev. Mod. Phys.* **70**, 303.
Schulman, L. S. (1996). *Techniques and Applications of Path Integration*. New York: Wiley.
Schwinger, J. (1951). *Phys. Rev.* **82**, 664.
Seiler, E. (1978). *Phys. Rev.* **D18**, 482.
Shankar, R. (1994). *Rev. Mod. Phys.* **66**, 129.
Shapiro, S. L. and Teukolsky, S. A. (1983). *Black Holes, White Dwarfs and Neutron Stars: The Physics of Compact Objects*. New York: John Wiley & Sons.
Shifman, M. A., Vainstein, A. I. and Zakharov, V. I. (1979). *Nucl. Phys.* **B147**, 385, 448.
Shuryak, E. V. (1978a). *Sov. Phys. JETP* **47**, 212.
(1978b). *Phys. Lett.* **B78**, 150.
Siemens, P. J. and Rasmussen, J. O. (1979). *Phys. Rev. Lett.* **42**, 880.
Smit, J. (2002). *Introduction to Quantum Fields on a Lattice*. Cambridge: Cambridge University Press.
Stauffer, D. and Aharony, A. (1994). *Introduction to Percolation Theory*, revised 2nd edn. London: Taylor and Francis.
Stephanov, M. A., Rajagopal, K. and Shuryak, E. V. (1998). *Phys. Rev. Lett.* **81**, 4816.
Sterman, G. *et al.* (1995). CTEQ Collaboration. *Rev. Mod. Phys.* **67**, 157.
Susskind, L. (1977). *Phys. Rev.* **D16**, 3031.
(1979). *Phys. Rev.* **D20**, 2610.
Svetitsky, B. (1986). *Phys. Rept.* **132**, 1.
Takahashi, Y. and Umezawa, H. (1996). *Int. J. Mod. Phys.* **B10**, 1755.
Tamagaki, R. (1970). *Prog. Theor. Phys.* **44**, 905.
Terazawa, H. (1979). INS-Rep.-336. Tokyo: University of Tokyo.
Thews, R. L., Schroedter, M. and Rafelski, J. (2001). *Phys. Rev.* **C63**, 054905.
Thoma, M. H. (1995). In *Quark-Gluon Plasma* 2, ed. R. C. Hwa, Singapore: World Scientific, p. 51.
't Hooft, G. (1972). Unpublished.
(1985). *Nucl. Phys.* **B254**, 11.
(1986). *Phys. Rept.* **142**, 357.
Thorsett, S. E. and Chakrabarty, D. (1999). *Astrophys. J.* **512**, 288.
Titchmarsh, E. C. (1932). *The Theory of Functions*. Oxford: Oxford University Press.
Tolman, R. C. (1939). *Phys. Rev.* **55**, 364.
Tomboulis, E. T. and Yaffe, L. G. (1984). *Phys. Rev. Lett.* **52**, 2115.
(1985). *Commun. Math. Phys.* **100**, 313.
Tonks, L. and Langmuir, I. (1929). *Phys. Rev.* **33**, 195.
Toublan, D. (1997). *Phys. Rev.* **D56**, 5629.
Tsuruta, S. (1998). *Phys. Rept.* **292**, 1.
Tsuruta, S., Teter, M. A., Takatsuka, T., Tatsumi, T. and Tamagaki, R. (2002). *Astrophys. J.* **571**, L143.

Turner, M. S. and Tyson, J. A. (1999). *Rev. Mod. Phys.* **71**, S145.
Tytler, D. *et al.* (2000). *Physica Scripta* **T85**, 12.
Ukawa, A. (1995). In *Phenomenology and Lattice QCD: Proceedings of the 1993 Vehling Summer School*, eds. G. Kilcup and S. Sharpe. Singapore: World Scientific.
Vasak, D., Gyulassy, M. and Elze, H. T. (1987). *Ann. Phys.* **173**, 462.
Vogl, U. and Weise, W. (1991). *Prog. Part. Nucl. Phys.* **27**, 195.
Vogt, R. and Jackson, A. (1988). *Phys. Lett.* **B206**, 333.
Wang, X.-N. (1997). *Phys. Rept.* **280**, 287.
Wang, X.-N. and Gyulassy, M. (1994). *Comp. Phys. Comm.* **83**, 307.
Weber, F. (2005). *Prog. Part. Nucl. Phys.* **54**, 193.
Weinberg, S. (1972) *Gravitation and Cosmology: Principles and Applications of the General Theory of Relativity*. New York: John Wiley & Sons.
(1977). *First Three Minutes*. London: Deutsch and Fontana.
(1979). *Physica* **A96**, 327.
(1995). *The Quantum Theory of Fields, Vol 1: Foundations*. Cambridge: Cambridge University Press.
(1996). *The Quantum Theory of Fields, Vol 2: Modern Applications*. Cambridge: Cambridge University Press.
Weiner, R. M. (2000). *Phys. Rept.* **327**, 249.
Weldon, H. A. (1982). *Phys. Rev.* **D26**, 2789.
(1990). *Phys. Rev.* **D42**, 2384.
Wigner, E. (1932). *Phys. Rev.* **40**, 749.
Wilson, K. (1974). *Phys. Rev.* **D10**, 2445.
Wilson, K. G. and Kogut, J. B. (1974). *Phys. Rept.* **12**, 75.
Witten, E. (1984). *Phys. Rev.* **D30**, 272.
Wroblewski, A. (1985). *Acta Phys. Pol.* **B16**, 379.
Wu, N. (1997). *The Maximum Entropy Method*. Berlin: Springer.
Yaffe, L. G. and Svetitsky, B. (1982). *Phys. Rev.* **D26**, 963.
Yagi, K. (1980). *Nuclei and Radiations*. Tokyo: Asakura.
Yang, C. N. and Mills, R. L. (1954). *Phys. Rev.* **96**, 191.
Ynduráin, F. J. (1993). *Theory of Quark and Gluon Interactions*. New York: Springer.
Yuasa, F., *et al.* (2000). *Prog. Theor. Suppl.* **138**, 18.
Zhai, C. and Kastening, B. (1995). *Phys. Rev.* **D52**, 7232.
Zinn-Justin, J. (2001). *Phys. Rept.* **344**, 159.
(2002). *Quantum Field Theory and Critical Phenomena*, 4th edn. London: Oxford University Press.

Index